BEST PRACTICES SERIES

Server Management

THE AUERBACH

BEST PRACTICES SERIES

Broadband Networking, James Trulove, Editor,
ISBN: 0-8493-9821-5

Electronic Messaging, Nancy Cox, Editor,
ISBN: 0-8493-9825-8

Financial Services Information Systems, Jessica Keyes, Editor,
ISBN: 0-8493-9834-7

Healthcare Information Systems, Phillip L. Davidson, Editor,
ISBN: 0-8493-9963-7

Internet Management, Jessica Keyes, Editor,
ISBN: 0-8493-9987-4

Multi-Operating System Networking: Living with UNIX, NetWare, and NT, Raj Rajagopal, Editor,
ISBN: 0-8493-9831-2

Network Manager's Handbook, John Lusa, Editor,
ISBN: 0-8493-9841-X

Project Management, Paul C. Tinnirello, Editor,
ISBN: 0-8493-9998-X

Server Management, Gilbert Held, Editor,
ISBN: 0-8493-9823-1

Enterprise Systems Integration, John Wyzalek, Editor,
ISBN: 0-8493-9837-1

Web-to-Host Connectivity, Lisa Lindgren and
Anura Guruge, Editors
ISBN: 0-8493-0835-6

Network Design, Gilbert Held, Editor,
ISBN: 0-8493-0859-3

AUERBACH PUBLICATIONS

www.auerbach-publications.com

TO ORDER: Call: 1-800-272-7737 • Fax: 1-800-374-3401

E-mail: orders@crcpress.com

BEST PRACTICES SERIES

Server Management

Editor

GILBERT HELD

Boca Raton London New York Washington, D.C.

Library of Congress Cataloging-in-Publication Data

Server management/ Gilbert Held, editor.
p. cm.
Includes bibliographical references and index.
ISBN 0-8493-9823-1 (alk. paper)
1. Client/server computing--Management. I. Held, Gilbert, 1943– .
QA76.9.C55S38 1999
658′.05436--dc21 99-16234
CIP

Direct all inquiries to CRC Press LLC, 2000 Corporate Blvd., N.W., Boca Raton, Florida, 33431.

Auerbach is an imprint of CRC Press LLC

International Standard Book Number 0-8493-9823-1
Printed in the United State of America 1 2 3 4 5 6 7 8 9 0
Printed on acid-free paper

Contributors

GERALD L. BAHR, *Networking Consultant, Roswell, GA*
LEE BENJAMIN, *Managing Consultant, New Technology Partners, Bedford, NH*
ROBERT BRAINARD, *Director of Internet Products, Microtest, Phoenix, AZ*
DANIEL CARRERE, *Open Systems Consulting, Milledgeville, GA*
STEVEN P. CRAIG, *Managing Partner, Venture Resources Management Systems, Lake Forest, CA*
JON DAVID, *President, The Fortress, New York, NY*
PAUL DAVIS, *Director of Strategic Marketing, WRQ, Seattle, WA*
DENNIS DILLMAN, *Senior Consultant and Instructor, USConnect Milwaukee, Milwaukee, WI*
JAMES E. GASKIN, *Independent Consultant, Dallas, TX*
M. DAVID HANSON, *Principal, MDH Communications, Monsey, NY*
GILBERT HELD, *Director, 4-Degree Consulting, Macon, GA*
JONATHAN HELD, *U.S. Navy, Monterey, CA*
TYSON HEYN, *Product Communications Manager, Seagate Technology, Inc., Scotts Valley, CA*
LAWRENCE E. HUGHES, *Independent Consultant, Santa Barbara, CA*
BRIAN JEFFERY, *Managing Director and Co-Founder, International Technical Group, Mountain View, CA*
DENISE M. JONES, *Marketing Communications Manager, Cubix Corporation, Carson City, NV*
SCOTT KOEGLER, *Vice President, Information Technology, Lake Worth, FL*
JEFF LEVENTHAL, *Founder and President, Remote Lojix, New York, NY*
JOHN LILLYWHITE, *Director of Marketing, ChatCom Corporation, Chatsworth, CA*
JACK T. MARCHEWKA, *Professor, Operations Management and Information Systems, Northern Illinois University, Dekalb, IL*
HOWARD MARKS, *Chief Scientist, Networks Are Our Lives, Inc., Sherman, CT*
KRISTIN MARKS, *Chief Scientist, Networks Are Our Lives, Inc., Sherman, CT*
STEVEN MARKS, *Chief Information Officer, White and Case, New York, NY*
STEWART MILLER, *President, Executive Information Services, Carlsbad, CA*
NATHAN J. MULLER, *Consultant, The Oxford Group, Sterling, VA*
JUDITH M. MYERSON, *Independent Consultant, Jenkintown, PA*
NETWORK ASSOCIATES, INC., *Santa Clara, CA*

A. PADGETT PETERSON, *Senior Staff, Corporate Information Security, Lockheed Martin Corporation, Orlando, FL*
ROY REZAC, *Vice President for Development, Seagate Software, Marlboro, MA*
LAWRENCE D. ROGERS, *Vice President and General Manager, Software Systems Division, Emerald System Corp., San Diego, CA*
RON ROSEN, *Enterprise Consultant and Project Manager, ENTEX Information Services, Inc., New York, NY*
RICHARD ROSS, *Principal, CSC Index, New York, NY*
MAX SCHROEDER, *Chief Operating Officer, Optus Software, Inc., Somerset, NJ*
ERIC STRAL, *Partner, EK&C Partners, Portland, OR*
ALLEN WAUGERMAN, *Lexmark International, Lexington, KY*
BILL WONG, *Independent Consultant, Yardley, PA*

Introduction

The role of the server in industry, government, and academia has considerably expanded over the past few years. The growth in the World Wide Web, the increased use of electronic mail, and the servicing of communications access by Internet service provider facilities significantly increased the diversity and role of servers in organizations. While servers have always played an important role in the computational efforts of organizations, today they have reached the point where they are literally indispensable, for no modern organization can do without operating and accessing the facilities of different types of servers. Recognizing the vital role of servers, this new edition of the *Server Management* has been significantly revised for the new millennium.

In this new edition you will find a considerable amount of material covering application-specific server-related topics. Recognizing the growing role of Linux and Windows NT/Windows 2000, several articles are included that cover specific server operating system information as well as introduce the operation and use of key operating system components. Other major additions in this edition include articles on evolving server technologies to include virtual servers and voice servers. Recognizing the important role of security and performance monitoring, several sections containing relevant and timely articles covering these extremely important topics were added to provide specific how-to information.

Because a handbook represents a compilation of articles and papers oriented toward the requirements of a diverse readership, special care was taken in selecting both contributing authors and the topics to be covered. The contributors to this book were selected based upon their knowledge and expertise. Each contributor is an industry expert who has demonstrated his or her skills through hands-on experience working with different operating systems, building Web sites, constructing sophisticated databases, implementing organizational security policies, and performing other server-related tasks. Some of our authors serve as consultants to industry, government, and academia. Other authors are experienced knowledge disseminators, and some are well-known public speakers. All of our authors have firsthand experience in the topics they authored, and we are truly fortunate to be able to share their expertise with you.

Articles were grouped into sections based upon their primary topic or focus. While this grouping is designed to facilitate your ability to locate related information, it should also be noted that a comprehensive index is provided to facilitate locating specific information. Thus, you can use both the table of contents and index to facilitate locating specific information.

Acknowledgments

Regardless of the type of book or subject matter, a printed volume represents the collaborative effort of many individuals. Thus, I would be remiss if I did not acknowledge as well as thank the persons involved in the effort that resulted in the book you are now reading.

First and foremost, it is important to recognize that a handbook is a collection of related papers and articles focused upon a core series of related topics. Without the efforts of authors that enjoy the ability to disseminate information this handbook would not be possible. Thus, I am truly indebted to the contributors whose efforts made this new edition possible.

A second group of key players involved in producing a book is the organization sponsoring the writing project. From the publisher to the acquisitions editor, from the copy editors to the galley page reviewers, they all work as a team to ensure the production of a high quality book that we can use for a substantial period of time. Thus, I am indebted to Rich O'Hanley, Theron Shreve, Claire Miller, and the other folks at Auerbach for their efforts in making this new edition a reality.

As a frequent flyer who travels to locations where electrical wall outlets represent a challenge for every adapter kit I have purchased, I've found it easier to use pen and paper than a battery-dependent notebook computer. Thus, once again I am indebted to Mrs. Linda Hayes for her fine efforts in deciphering my handwriting and preparing a professional manuscript which resulted in this book. Last, but not least, I would also be remiss if I did not thank my family for their understanding during long evenings and weekends when I worked on this handbook.

Contents

Section I
The Roles of the Server

Unlike a typewriter or basic personal computer, a server is not ubiquitous nor does it serve, no pun intended, a ubiquitous role within an organization. As the title of this section implies, there are many roles that servers perform. Recognizing this fact, as well as the fact that servers can range in scope from personal computer and minicomputer-based systems to mainframes, the purpose of the chapters included in this section is to provide you with a detailed overview of the many and varied roles servers perform within an organization. To accomplish this, this introductory section contains four chapters that provide you with an understanding of how servers can be used within an organization, with topics ranging from a basic overview of the potential roles of a server within an organization to specifics that illustrate that, contrary to prior so-called conventional wisdom, the mainframe's death has been highly exaggerated and it retains a critical role in the enterprise.

In the first chapter in this section, entitled "The Client/Server Architecture," David Hanson provides an introduction to client/server architecture concepts. Hanson describes several typical server configurations, reviews a variety of services different types of servers commonly provide to network users — from PC to mainframe — and describes and discusses such key server-related topics as availability, security, Java support, scalability, and load-balancing. In concluding this chapter, Hanson polishes his crystal ball and provides a peek at the future, discussing the evolving requirements of Web site development and trends in networked server technologies.

The second chapter in this section, "The Server at the Heart of Your Business," continues to provide information on the role of servers. In this chapter Ron Rosen examines the criteria for deciding to acquire a server and the issues a business must confront to include economic and technical aspects and compatibility of new products with existing equipment.

In continuing our examination of the multifaceted role of servers within an organization, Kristin Marks' chapter, "What Does a Server Do? A General Overview," takes us on a tour of network-related terms, defining centralized, distributed, and collaborative computing and explaining network elements and services necessary to support server operations.

Concluding this section on the role of servers is the chapter "The Case for Mainframe Servers in the Enterprise," authored by Brian Jeffery. To paraphrase Mark Twain, the death of the mainframe is greatly exaggerated, and Jeffery succinctly points out in his chapter that the mainframe continues to run core information systems without which most businesses could not function. Thus, it is extremely important for you to understand the role of the mainframe as a server in the enterprise, which is the focus of this section-concluding chapter.

Chapter 1
The Client/Server Architecture

M. David Hanson

The client/server model is the most common form of network architecture used in data communications today, and its popularity can be seen in the phenomenal expansion of the World Wide Web. A dictionary definition of a client is a system or a program that requests the activity of one or more other systems or programs, called servers, to accomplish specific tasks. A server is a system or program that receives requests from one or more client systems or programs to perform activities that allow the client to accomplish certain tasks. Exhibit 1.1 shows a typical local area network based on client/server technology. Shown are three PC clients, one server, and a network printer.

In a client/server environment, the PC or workstation is usually the client. The client/server concept functionally divides the execution of a unit of work between activities initiated by the end user (client) and resource responses (services) to the activity request. Client/server is an application of cooperative processing in which the end-user interaction with the computing environment is through a programmable workstation that executes some portion of the application (beyond terminal emulation).

In the 1980s, when PCs were first introduced commercially, the file systems were on the PCs themselves. Local Area Networks (LANs) provided access to file systems on local servers, and with LAN interconnection that access has extended to remote servers on corporate networks and the Web. Exhibit 1.2 shows how clients on a LAN can access remote Web host servers through the Internet. Exhibit 1.3 shows the internal configuration of a client PC and how it mirrors the client/server technology being discussed.

0-8493-9823-1/00/$0.00+$.50

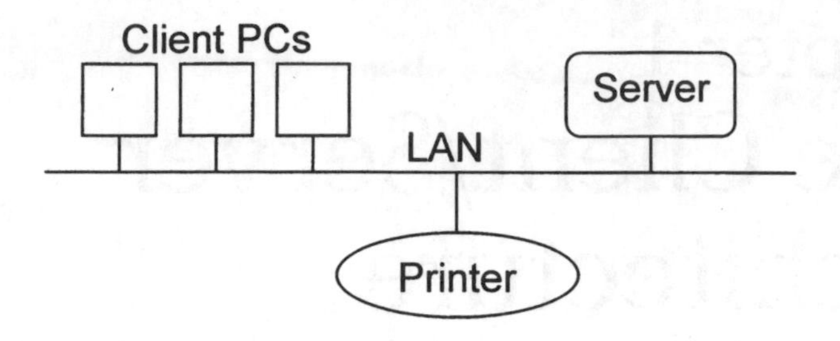

Exhibit 1.1 Typical client/server LAN.

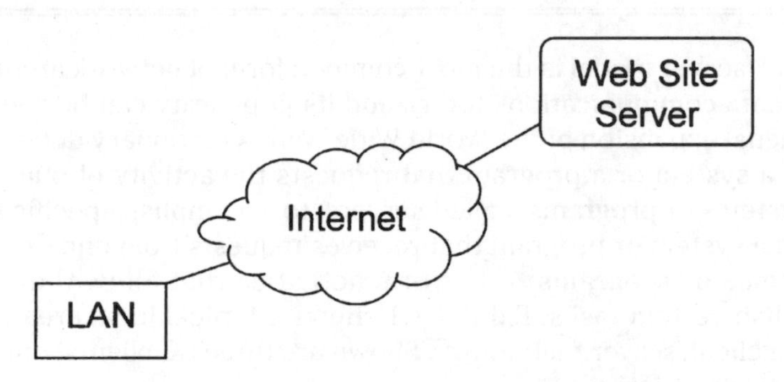

Exhibit 1.2 Remote server configuration.

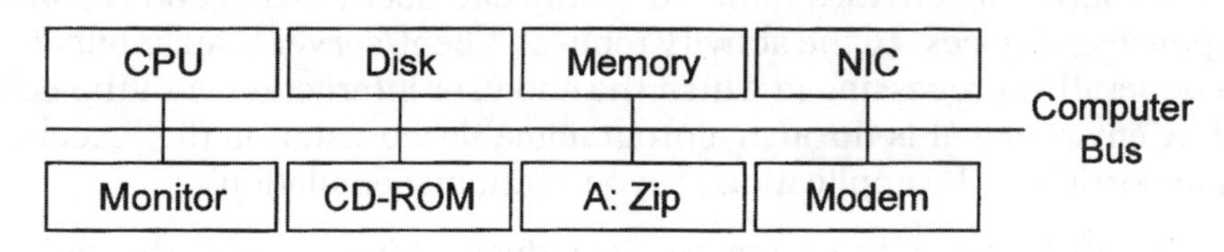

Exhibit 1.3 Typical PC configuration.

TYPICAL SERVER CONFIGURATIONS

How does a server differ from a PC client? The PC usually has one hard drive, one processor, one L2 cache, and one user at the keyboard. The typical server may have the following configuration:

- Four Pentium processors, each with 512K of L2 cache
- At least 512 MB of RAM
- Two 100 MB Ethernet NICs
- Five 2 GB or larger hard drives

One of the disks contains the operating system and the remaining are organized in a RAID level 0 (data striping, no parity) configuration. The server's memory and processor configurations can be changed through software rather than by physically adding or removing hardware, if so desired for testing or other purposes. Typically, the server would be running either Windows NT 4 with service pack 3 or the Netware operating systems.

SERVER TYPES

A server provides specific services to systems on the network:

- Routing servers connect subnetworks of like architecture.
- Gateway servers connect networks of different architectures by performing protocol conversions.
- Terminal, printer, and file servers provide interfaces between peripheral devices and systems on the network.
- Application servers usually host the systems used by the client PCs.
- Local access servers provide connections for terminals to LANs.
- Remote access servers (RAS) provide connections to serial and high-speed asynchronous lines for e-mail or dialup access to Web-based product information.

Building a client/server system is as much an art as a science. There are the nitty-gritty, concrete questions such as which application development environment to build in, which servers and clients to use, which NOS to adopt, and so on. None of these decisions is simple, but all of them can be made with some sort of data such as benchmarks, testing, and the benefit of past experiences. Then there are the more theoretical matters to figure out: If this is a new client/server environment, how will the new server change the business? How can it be made more acceptable to the users? What new uses will be found for it?

Mainframes as Hosts

From the 1950s through 1970s, mainframe computers that were connected to corporate networks were referred to as *hosts.* With the advent of PCs in the 1980s, along with local area networks (LANs) and client/server architecture, hosts have evolved into specialized PC-versions of mainframe hosts called *servers,* each of which served specific functions for their *clients* such as printing, applications, files, and communications. It should be noted that hosts are seldom intentionally configured to be used as servers, and therefore do not perform optimally as such. Some purposes for using the host as a server are file storage, security, and central backup.

It should be noted, however, that while most of today's critical business applications (finance and accounting) reside on advanced UNIX-based client/server systems, by some estimates at least 70 percent of the world's data are still on venerable, sometimes referred to as *legacy,* mainframe computers. Web site or no Web site, the wheezing mainframe from another era is still a factor to be reckoned with. They have their niche along with all the other specialized servers existing on corporate intranets as well as the Internet and World Wide Web. Today that famous, sometimes overused phrase, "the convergence of technology," has never been more appropriate for the blending of client/server, Internet, and mainframe.

Corporate networks are evolving into intranets within the Internet infrastructure, or are maintaining their individual corporate identity as extranets but still using Internet protocols and naming conventions.

Every server has its own unique address that identifies it to the network. This address is called an IP address and consists of a set of four numbers between 0 and 255. Each set of numbers is separated by a period. All data being sent over the backbone carry the destination IP address, which enables the backbone intersections to route the data toward its goal. In addition to an IP address, each server has a unique IP name, also known as a domain name, which is a name that makes it easier to find a Web site. It would be difficult for the average user to remember IP addresses, but domain names are usually easier to remember.

Servers are the computing engines of the Internet. Every month, 53,000 new servers connect to the Internet. That's 1.2 servers per minute going online and this is only the beginning!

On the Web, everything depends on the server. Specifically, in the world of *e-business*, aka e-commerce, the ability of customers to find electronic commerce and business suppliers, the experience they have once they get there, the security of the transactions, and the integrity of the data all depend on the servers. In an e-business world, you're only as good as your servers.

So, just as the proliferation of affordable personal computers transformed the desktop, servers have facilitated the capability of these PCs to interact with the world. In this new transaction-based world, server design, performance, and security are critical. A server that goes down, that has security breaches, and that cannot scale up to increased traffic leads to reduced customer confidence and increased frustration.

Powerful servers now cost as little as basic PCs did just a few short years ago, but the issues involved in making a purchasing decision are radically different. How are maintenance and upgrading costs controlled when servers are spread over multiple departments and cities? Is the best server solution for company A the same as for company B?

Servers come in various architectural and operating system solutions: Windows NT or UNIX operating systems, midrange or enterprise architectures. Each one has strengths and weaknesses, depending where and how used. The point is that there is no single server solution that's right, especially when the environment contains legacy servers and applications. Knowing how to preserve and extend these investments while adding new servers and new capabilities is critical.

The current goal of today's servers is "Five Nines" (99.999 percent availability), with a design point of just five minutes of planned or unplanned downtime a year. Availability, however, isn't just a measure of the time a network is up and running. It also refers to a network's ability to provide consistent data transmission at higher data rates. Software plays a major role in this area. Software solutions facilitate the management of powerful information networks, working across multiple platforms at significant cost effectiveness.

Security issues require that servers be designed to protect the data against corruption from outsiders (often referred to as *cyber-mischief*). Consequences of a vulnerable server are serious, ranging from loss of online customers to clients that won't share information. Servers must be designed to keep intruders out. Servers can be configured with security ranging from passwords to certificates on smart cards, which establish virtual walls that only designated employees, vendors, and customers can scale. Encryption minimizes risk still further but at the expense of response time (encrypting every word takes time).

Servers, like their smaller cousins the ordinary desktop PC, are also at risk for exposure to a computer virus. On the Web especially, viruses are a constant threat. A good antivirus product, such as Norton AntiVirus or McAfee, is a must.

Every server must be designed to run Java applications. Servers are usually clustered in "farms." Moving from these farms to server consolidation can save money, improve availability, and increase service levels.

As with built-in availability and security, servers have to be designed to be scalable for enormous growth. Scalability is about being able to distribute the workloads of multiple servers to thousands of simultaneous users. Scalability also means providing a structure that can grow to support thousands of users at a reasonable cost. Servers have to be scalable. As a company's Web site becomes a primary source of interaction with its customers, scalability becomes a major issue.

While most companies think of servers as being large PCs, large legacy mainframes such as the IBM S390 are sometimes used to host Web sites because of their superior scalability. Some companies use minicomputers such as the IBM AS/400 to connect their retailers nationwide.

What is important is to have a strategy in place for quick expansion so that the servers will allow for both a planned rollout of more capabilities and the ability to handle huge volumes of transactions while ensuring high customer satisfaction. Electronic commerce is about transactions, and every day billions of transactions are conducted on servers worldwide.

Server Load Balancing

Content-smart switching and global server load balancing are important attributes for optimizing Internet performance. Server load balancing usually distributes traffic within a server farm at a single location, but as large users — especially ISPs — adopt caching as a way of moving Web content closer to users, load balancing is being extended across the wide area. Often, the customers of large ISPs are likely to have Web content mirrored on dozens of different sites around the world. However, networks haven't had much ability to send end user traffic to the most appropriate server at a given moment.

A content site like Disney may give each Web server an individual host name such that the Domain Name System (DNS) cycles through the different discrete servers, without knowing their status. The difficulty with this round-robin DNS system is that it doesn't take into account if the server is even up, how the server is performing, or the load at the site. If there is a problem with the server to which a particular user has been directed, the user doesn't know until a "server unavailable" message is received in response to the request.

By contrast, server load-balancing systems make decisions based on more relevant criteria than does a simple round-robin system. A typical transaction would occur as follows:

- A DNS request goes out from the client to a local DNS server, which in turn queries an upstream DNS.
- So-called "content-smart switches," putting more smarts at the edges of the network cloud, receive the DNS request, determine which Web servers are closest to the user, and then select the one among these candidates that's judged to be the best site to receive the traffic.
- The switch then sends the selected Web server's IP address back to the local DNS, which responds to the client.
- The client sets up the session with the selected Web server based on the criteria of availability, server health, geographic proximity, response time, and throughput.

Load-balancing criteria are dynamically updated to reflect changing conditions in the network. This guards against a scenario in which the switch detects a lightly loaded server, starts bombarding that server with traffic until it becomes overloaded, then backs off until it's nearly idle, and then repeats the process in a never-ending cycle.

One more thing about load balancing: content-smart switches can utilize their ability to snoop HTTP headers and URLs to distinguish dynamic requests that need to go to the application servers and static Web-page-serving functions that can go to the static page servers. That information is important because of the way distributed Web sites typically operate; the sites closest to most users probably contain only static, mirrored information, while requests for dynamic sessions such as Web searches must go to a main server.

Without a feature called URL load balancing, a user nearest a mirror site would send all its traffic to that site first, even dynamic requests that end up getting passed on to the main site. With URL load balancing, the smart-content switch detects the dynamic nature of the request and immediately sends it to the main site, bypassing the mirrored site.

Server Consolidation

Thin clients and Web technology are creating new options and economies of scale. In the 1980s mainframe data centers around the United States, and later around the world, went through a wave of consolidation. Networks made it possible to deploy mainframes further away from the users without sacrificing performance. Savings in people and facilities paid for the networks, T1 backbones in particular.

Today it has become technically feasible and economically attractive to concentrate servers within a few national or regional centers. The Internet and Web technologies have made more applications and servers accessible over wide area networks. With the rise of thin clients and multitiered application architecture, the application code can now run on a server rather than on the client, and servers can now be put almost anywhere without users knowing or caring.

As software and data are pulled back from clients and larger servers are concentrated in fewer centers, there has been a tendency to put many applications and databases on the same server. But it should be noted that keeping to one application per server makes it easier to tune the system for optimal performance, while raising reliability and simplifying scheduled downtime.

Optimizing the Server

Optimizing a server's performance is not an exact science, especially given the variety available. In general there are three server types:

- File and print servers
- Application servers
- Web servers

In the traditional file and print server, users run their applications from their local machines, but their data files reside on the server. In addition, the server functions as a centralized conduit to one or more shared printers. While usually least demanding of resources, when a single server has to accommodate a large number of clients, it can hardly be considered an insubstantial machine.

The application server is usually referred to as a true client/server environment because the server contains not only the user's data files (word processing documents, spreadsheets, databases, etc.) but also the applications themselves. The program loads from the server into the client's RAM and runs there. The application server configuration generally requires much more of a hardware investment than does a file and print server, particularly when it comes to processor power.

The Web server is by far the most complicated. It typically functions in one of two ways. It can house static HTML files that it transfers to a client in response to HTTP requests, or it can use applications, such as CGI executables and ISAPI libraries to dynamically generate HTML code.

Web Application Servers: A Peek at the future

The client/server-based applications of today may soon be usurped by browser-based and thin client applications, which will allow businesses to take advantage of the considerable capabilities of Internet technology. The two-tier client/server model will give way to a three-tier Web-based computing paradigm.

The application server (see Exhibit 1.4) is the centerpiece of the new model. It can basically be described as middleware, located between clients and data sources and (or) legacy systems, that manages and processes applications in a three-tier fashion. Application servers help enable the development and deployment of increasingly complex Web applications.

The requirements of Web site development have grown considerably more complex over the past few years. Whereas at one time Web site development meant no more than creating HTML files and simple CGI scripts, today the Web has become the preferred user interface to handle a host of new application types. As a result, Web developers have to deal with many of the same requirements and problems associated with traditional applications development.

Many of those problems have nothing to do with development, but rather with deployment. The challenge of building Web sites with reliability, scalability, and manageability is just now being addressed with the application server. The main function of the application server is to ensure that developed and deployed applications are scalable, reliable, and accessible.

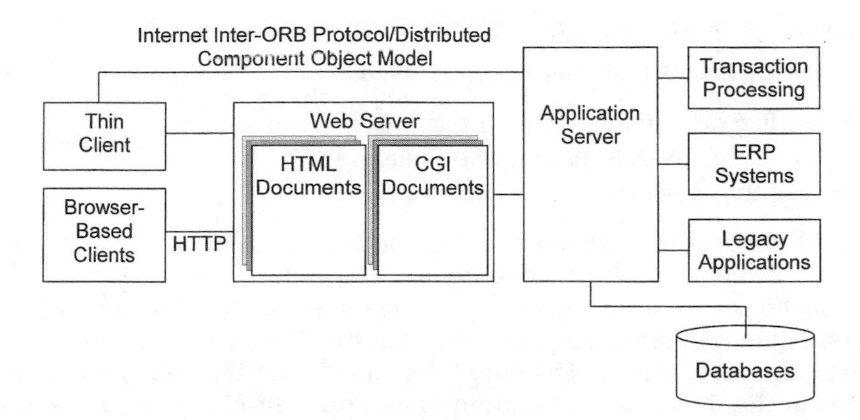

Exhibit 1.4 Application server architecture.

By integrating this server with a Web server, the application server can also be used to manage Web-enabled applications.

One of the main attractions of application servers is that they do not require businesses to replace their existing systems; instead, they help to tie together all of a company's diverse computer systems, both old and new, at one central point. In this way, application servers make it easier for IT departments to share data from many disparate sources with both their internal and external customers.

Successfully taking advantage of this technology raises many complex technical issues, and IT decision makers will have many important choices to make when it comes to deciding on the best application server solutions for their companies. What is the optimum application server platform? What are the most suitable development tools for a given IT environment? What is the best object framework to work with?

Software Servers

Can a server be only in hardware? What does it mean to be a software server? Some vendors package a comprehensive set of client/server products into a suite, usually under the NT operating system, and call the result a *software server*. These suites allow for the building of a high-function, reliable environment for decision support, messaging and groupware, transaction processing, and other client/server functions vital to today's businesses. Software server suites help application developers build sophisticated applications quickly and reliably, and powerful new Java gateways can deliver them to users over the Internet in a flash. The brave new world of network computing has arrived!

Current Trends in Networked Server Technologies

The critical issues that have to be resolved are

- settling on a standard server computer model,
- increasing overall server performance, and
- adopting a 64-bit server architecture.

With respect to a standard server model, the current server-centric nature of client/server network architecture, with the server serving the applications to a lower level client, has resulted in a trend toward thin clients. There is another school, of thought, however, that sees a benefit in having clients that are not arbitrarily excluded from performing significant roles in the design and operation of an application. According to this school, client-routed transactions make complete sense, particularly in the age of the Internet with Virtual Private Network (VPN) technology, leaving application developers free to implement an application or transaction in the client or the server. Applications may move from the client up the middleware architecture hierarchy as appropriate.

Migrating to a 64-bit server will be very important to increasing overall server performance. Databases, in particular, will be a leading-edge driver for implementing the 64-bit server since they consume so much address space due to disk storage buffering and disk caching. With 64-bit RAM, it will be possible to have completely memory-resident databases. With the entire database in memory, moving pointers will replace much slower repeated disk accesses. The result will be nearly real-time responses.

SERVER PERFORMANCE

SMP (symmetric multiprocessing) is the core of today's infrastructure. In the case of clustering, SMP-based systems are key building blocks. Clustering architecture supplements SMP by facilitating the use of SMP building blocks to build high performance configurations of servers.

NUMA (non-uniform memory access) is always going to have a low volume niche. Fiber channel products provide much more scalability, performance, capacity, redundancy, distance, and remote-mirroring capabilities.

Applications place unique demands on server platforms. OLTP, for example, drives a significant amount of CPU demands. A later chapter explores in more detail the impact of server components on server performance, but this chapter will close by considering processor speed in that context.

In the case of an OLTP environment, CPU performance is critical. But it's not just a matter of processor speed, megahertz. Other factors also affect performance, such as the size of the cache (primary and secondary), the speed of those caches, and finally the memory bandwidth between those caches and the actual memory. For most compute-intensive applications, particularly in an SMP environment, the memory bandwidth is the primary factor that determines performance. However, to take full advantage of increased cache, the server does need more processing power. Adding processors, just as with adding RAM, will significantly increase performance. A single CPU simply isn't sufficient for most application servers. In fact, increasing the number of processors has a much greater effect on performance than does adding RAM.

Chapter 2
The Server at the Heart of Your Business

Ron Rosen

WHAT'S A SERVER?

A server is both a function and a device. In most communications, it is referred to as a device, since in corporate society today large networks are a given, where *super-server* is the term du jour. But in smaller companies and in organizations where English is spoken (or the native tongue of any technological society), server is more a function performed by one or more devices.

The term *server* implies networked (connected), shared, and centralized, as opposed to the politically correct and gender-neutral term for waiter or waitress. It is generally used to refer to a file server, or file and print server — one computer with an operating system to which others are connected, which provides a sharable, central repository for files (programs and data). Often, since it's already centralized, a printer or two is attached to capitalize on the shared capabilities of the server. There are many different shapes, sizes, speeds, and functions of servers. These functions include fax, dial-in, dial-out, print, file, database, CD-ROM, Intranet, Web, e-mail, etc.

In a peer-to-peer network — several PCs connected to each other without a central computer — the server function still exists. For example, if part of one PC's disk drive is set up to hold data files that others can access, that becomes a file server function. If another PC has a printer that is set up for sharing so others can print to that printer, that PC is performing a print server function.

BUSINESS NEEDS ANALYSIS

Background

The decision to buy (or create) a server has many implications, especially if it's your first one. It starts with your business. The big trend to downsizing and outsourcing of Information Technology (IT) is based on the fact

0-8493-9823-1/00/$0.00+$.50

that companies have been spending too much energy (besides money) on IT and want to refocus their company resources on their core businesses, rather than on technology. By outsourcing basic technology functions, their internal IT staff can focus more on how to use technology for the company, rather than on improving/maintaining the technology itself.

One caveat here is that technologists often have a tendency to overdo technology because that's their business (and often their hobby), and companies end up with more technology or state-of-the-art equipment than they really need for the required functions. This is appropriate for an IT company — one that's in the computer business — but not for a company that needs computers simply to keep track of its regular business.

A small business with a few computers connected to share a printer and a modem generally does not need the biggest, fastest, or best (read most expensive) equipment for their functional sharing — especially if the reason for the shared devices is to avoid the cost of several individual devices. Most times the one or two shared devices can be attached to one of the existing computers, and special, inexpensive software can make those devices sharable.

Why a Server?

There are two basic scenarios that beg this question:

1. You've got a network with some shared functionality, be it peer-to-peer or server-centric, with a number of local functions (nonshared resources), such as modems at each desk, individual printers, etc., and the cost, or management, or coordination, or support, of one (or more) functions is becoming increasingly expensive, difficult, uncoordinated, confusing, ineffective, etc.
2. You've been using unconnected PCs for some time and are considering a network of some kind.

The difference between the two scenarios is the additional costs and complexity associated with the second (installing a network), and the technical compatibility issues (in addition to costs) associated with the first. In both cases, the reasons for a server are the same: connectivity, centralization, and sharability. Let's address each one separately. We'll address the common issues later.

Connectivity. There are only two basic reasons to want to have computers connected to each other: to save money and to enjoy the convenience. All other reasons fall within these two.

Leaving someone a voice mail involves waiting for the five rings, their message, then the machine's message, then leaving your message. Taking the elevator upstairs or walking down the hall to leave someone a Post-it when the message is not time-critical but gets the thought out of your head and passes the buck to the other person, or copying a file to a floppy and handing it to someone else to print because you don't have a printer on your PC, doesn't really cost a lot of money, but it is very inconvenient. These activities cause a break in concentration and are distracting.

Wouldn't it be easier to pop into e-mail, write a brief note, send it to the person, and pop back to what you were doing? It maintains concentration, lets you get the idea out of your head, and avoids the possibility of other distractions.

If that walk down the hall really involves walking to the end of the hall, waiting for the elevator, going to another floor, walking back to the same end of the hall (of course), looking for a pad to leave the note, and then retracing all your steps to get back to your desk while permitting other distractions to occur (such as meeting the boss in the elevator and having to explain what you were doing), it could become costly in terms of time — and therefore money. Then the reason becomes a matrix of economics as well as one of convenience.

Centralization. If the disk drive in the machine that has all your customer records crashes, but you have a maintenance contract, you can often get the disk drive replaced by your maintenance vendor within a day or two (assuming a quick turnaround agreement). But what about the data? You told your accountant to back up his files regularly, and he promised he would, but he's on vacation and the temp forgot to do it. How much would the reconstruction of those files cost you? What if your office had a small fire, and all your papers and computers were damaged? Wouldn't it be safer (and easier) to have all those files — for everyone — in one place, with one or two people, one of whose primary responsibilities is to back up the files every single night? And rotate versions of the backups off site on a controlled basis?

Suppose you were on vacation in the Cayman Islands and the sales rep leaves an urgent message at the front desk for you that a client finally responded to a proposal you sent out last month, requesting a slightly modified bid in the same proposal. Do you tell her to go to your PC (on another floor), call you back, find the file (which would be the same in either case), go back to her desk to get the new information, go back to your desk to enter the information, then copy the file to a floppy and take it to someone else's desk to print it? Or just have her log into the file server from her desk, access your file, make the changes from her notes, re-save the file, and print it on the shared printer. If all data files were in one place, they'd be much more accessible.

If all data, all modems, all CD-ROMs, etc. are in one primary location, that location can be physically secured in a locked environment, without anyone worrying about the cleaning staff accidentally (or intentionally) damaging or borrowing the equipment.

Sharability. Suppose you had someone working on a large, sophisticated spreadsheet; or a large proposal for a major client opportunity; or working on entering new shipments or invoices into the account database. Suddenly you realized you forgot to change the date on your desk calendar, and today was Friday, not Wednesday, and you promised the end results by 3 P.M. today. Do you tell the person at the computer to downshift (we usually say shift into high gear)? Work faster? What if your business has expanded to the point where you need two people to maintain records. Do you copy the database and have two people working on two computers enter the data to the same database? Split the database into two parts: A–M and N–Z? Then consolidate the numbers on both sets of reports every day? Wouldn't it be easier if both could access the same database at the same time on a central computer (that, by the way, would be backed up properly one time every night)?

Rather than purchasing a fax machine for every three or four people, wouldn't it be more efficient to have a centralized, sharable fax machine on the network? Then all you need to do is add the additional phone line and modem when you want to expand the service, instead of purchasing another stand-alone fax machine (since most outbound faxes originate on PCs these days)?

How about not installing a phone line, modem, and software on the desktop PC for every salesperson in the field to dial into in order to access the version of the product catalog they happen to have copied to their desktop (or laptop) that day, but rather have them all dial into a sharable dial-in server that is centrally controlled to maintain security and consistency, to access the same, sharable catalog database and provide their customers with accurate, timely, and consistent information?

These are just a tiny sampling of the myriad reasons for a server (or a network).

If you've already got a peer-to-peer network, these server functions can often be added to existing equipment with some software (and usually consulting services). If you've already got a server-centric network, they can be added to the existing server or by the installation of an additional server. In both cases, of course, there is usually some work that has to be done, such as installing "client" software on each PC that will be using the added functionality; but that's just part of the installation process.

Why NOT a Server?

Believe it or not, there are a number of good reasons for NOT sharing a service, or having shared, centralized, and connected servers. One outstanding reason, of course, is money. The cost of installing a network significantly exceeds the cost of installing one computer (or one more computer), in most cases. Security is another issue. Although having a centralized file server under lock and key is a benefit, having your own desktop PC under lock and key in your own office is even more secure. Having an easily removable disk drive (they do exist) provides security, mobility, and, to a very controlled extent, sharability. It's actually more secure than a server in some cases, and much less expensive.

Having one or two computers with modems in one location permits visibility as to who's using those facilities, rather than permitting everyone access to a (often) nonvisible modem server (a/k/a dial-in/out server). This "peer pressure," or "boss visibility," can sometimes be more effective in controlling telephone costs than the most sophisticated software package.

COMMON ISSUES

What Business Are You In?

One of the difficult decisions business people often make is defining what business they're in. Although startup businesses are usually in several lines of business, at different times in the growth of a business, decisions have to be made as to whether you want to continue doing some of the activities you've been doing previously.

The owner of a startup business usually keeps the books, does the taxes, writes all the checks, balances the checkbook, etc. At some point, he hires a bookkeeper to offload some of those tasks. At some other point in time, an accountant is usually retained to offload the increasingly difficult and complex aspects of taxes, financial planning, benefits packages, administration, etc.

The same applies to technology. Many companies are in the computer business and are therefore (often) comfortable making all computer-related decisions and performing computer-related tasks. Most aren't. The first activity usually offloaded to a vendor is maintenance — hardware support/repair. Then, after the office staffers get to the point that they're spending 30 percent of their time trying new software or trying to fix software problems, comes offloading (outsourcing) various other computer-related services: support (a/k/a Help Desk), installation, reconfiguration, upgrades, etc. The list continues and grows with the complexity of the environment. However, there are a variety of costs associated with these decisions, and they must be identified, quantified, and evaluated in order to make effective decisions.

These are the decisions that have to be made in order for your company's staff to focus on your company's business. Making these decisions, however, is not always easy.

Whether you've got stand-alone PCs or you already have a network, there are two primary considerations to evaluating the addition of a server: economic and technical. Keep in mind that they all boil down to the primary driver of business: dollars (or yen, rubles, francs, pesos, pounds, etc.). Every issue discussed can ultimately be interpreted in terms of the costs.

Costs

The most easily measured costs are the actual costs of purchase, taxes, shipping, installation, service, support, training, maintenance, replacement, repair, etc. These are the actual dollar outlays, whether rented, leased, or purchased outright.

The more difficult costs to determine/estimate are the costs of not making the purchase: lost business and (or) employees, morale, competitive situations, the additional time associated with the "old" method of performing the service, over the time saved with the new function, etc., and these contrived numbers must be compared to the real costs, factoring in the time cost of money (interest) to make a reasonably accurate business decision. Also, given a particular product line, the cost analysis of the features/benefits of the variations in that product line requires further analysis. Are the added benefits of the additional memory feature in the new server worth the incremental cost?

Each of these issues, and others, comprise volumes of reference material in business libraries (and on the Internet these days), with case studies, instructions, theories, formulas, etc. But with the degree of popularity of this technology today and the ever-declining costs associated with them, these types of analyses can often be accomplished by the owner of a small business or an accountant for a medium-sized business — with a little homework.

What about rate of return on investment? Is that a valid criterion? How do you calculate it? What are the costs associated with training personnel in the use of the new product? What about added maintenance costs? How do you pick the most cost-effective vendor? The most cost-effective product? The most cost-effective support?

Many of these questions will be answered in this volume. The list of contributing authors is impressive — and extensive — with some of the best expertise, both technical and economic, in the industry today. Some of the more business-specific issues you will have to address individually, and there is a plethora of consultants ready, willing, and able to assist you in the aspects of the technology business that you've chosen not to be in.

Technical

The chapters of this compendium address a large number of the types of server functions available on the market today. Some are mostly hardware, some software, some a reasonable combination of both. Within each type, there are many common issues: capacity, speed, scalability, cost, functions, features, support, maintenance, training, hardware/software, and appropriateness to your need. There are a few other issues, however, that also need addressing. There are two specific issues that stand out in this category: usability and compatibility.

Usability generally has two attributes: how easy a product is to use and how useful it is. If a product is not easy to use by the person needing it, it won't get used. That's a truism. Almost no one uses the clock on one's VCR. The manual is just too confusing, and the little pictures next to the buttons are incomprehensible (to me, anyway; all my VCRs keep flashing 12:00). True, the more functionality a product provides, the more buttons (menus, keystrokes, etc.) it must have. But to get the most use of a product, its basic functionality should be simple. If it gets too complex, people won't use it.

For example, the easiest CD-ROM server to use is one that simply adds a drive letter (K:, L:, and so on) for each CD on the server. The end user need only know that the Reference Library is on K:, and the Proposal Templates are on L:. If they want to access the Reference Library, they just double-click on the little box that says K:. If they have to go to a DOS prompt, type some commands, click multiple items, or hit a variety of unrelated keystrokes, many people will just ask someone else or not use the facility at all.

Some of the best fax servers hardly seem like fax servers at all, but printers. To send a newly created document to be faxed, all one has to do is print to a different printer. (The fax software loads as a printer driver program.) If one has to enter a new name/number, it is just typed in the "TO" box. In most cases, it will be added to the fax software phone book automatically. Most people with some computer experience already know how to change printers, whether on a network or stand-alone. Again, if someone has to do fancy fingerwork to fax something, the old fax machine that you never removed from the table by the water cooler will be used.

These and many other examples typify the variation in products and the importance of ease of use. The benefits of demonstrations or, better yet, evaluations cannot be overstated. You've got to see it, touch it, feel it, even try it before you buy it.

The other attribute of usability — how useful it is — is something that can usually be predetermined with a little homework. If your fax machine runs out of roll paper every other day, if its counter shows a high number of faxes sent daily, if you see lots of different people on line at the two fax machines every Friday afternoon, if your fax-line phone bills are approximating your rent bill — then you can probably safely assume a fax server will be very useful.

Compatibility

Compatibility is one of the most important issues in the world of technology today. It rears its ugly head in almost every aspect of the industry, almost daily. It shows up, like Murphy, at the most inopportune times, when least expected, at the minutest level, at the most important juncture in the life of your business. In other words, I can't express the importance of this issue enough.

Compatibility refers to how well a new product will work with the products you already have, or are purchasing at the same time, or plan to purchase. It means that there is a good probability that some feature of the new product will probably create a problem or problems, of varying degrees, with your existing environment. The new software makes a configuration change to your computer (without telling you) that impacts how some other program (that's already there) works. A new server defaults to a protocol other than the one your network uses (undocumented feature) that floods the network and is not accessible by any of your users. The addition of a CD-ROM server requires a setting in a configuration file that conflicts with the setting required by your network operating system. There are as many examples of these as there are Murphy's corollaries.

The resolution is, well, not quite so simple as it is straightforward — some combination of manufacturer's warranties, commitments in writing, competent consulting expertise, and, my favorite, not rocket science or bleeding edge. I do not recommend the purchase of any product new to the market. I never buy a release 1.0 of anything, regardless of who's using it or who has it (they're different), who's on the manufacturer's or reseller's client list, or what verbal promises are made. I do not need to expend blood, sweat, and tears trying to make something work. I want it to work right the first time, every time. The highest endorsement I can ever give to a manufacturer's product is that it works as advertised. Period.

SUMMARY

Deciding to install a server to share a service can be as simple as filling out a purchase order for a network engineer in a large company, or as confusing and difficult as an entrepreneur of a business about to purchase his sixth or seventh PC and dealing with some of the issues mentioned above. Proper identification and analysis of the business issues and costs can both save money and avoid mistakes and problems. One more time, even though skepticism abounds: there are a lot of competent and honest consultants out there who really want to help.

Ron Rosen is an Enterprise Consultant and Project Manager in the East Area of ENTEX Information Services, Inc. He manages large-scale networking projects for major corporate clients. He has been in the computer business for over 25 years, owned his own consulting firm, and worked for major corporations as a Project Manager, Network Manager, LAN/WAN Designer, troubleshooter, and technical consultant. He has published a number of articles and has made presentations to Networld shows.

Chapter 3
What Does a Server Do? A General Overview

Kristin Marks

INTRODUCTION

This chapter is for beginning network managers who want to make sure they have a grounding in the fundamental basics before they get lost in the alphabet soup of the more advanced technologies. Servers are considered the heart of any network. So before we get too far into all the different server issues, let's look at the very basic concepts that define what servers do and where servers fit into the larger network picture.

UNDERSTANDING NETWORKS

Networks exist to serve users, hence, the origin of the term *server* for a system that provides services. You have two jobs:

1. Learn the basic terms used in computer networking.
2. Understand the several possible network services along with how they can be implemented.

Let's start by defining just what *networking* is. At its most basic level, networking is sharing resources, which can include information or functional capabilities. Computer networking extends that sharing with communication tools that let our computers share resources.

Centralized Computing

At the very beginning (we're talking 1940s here), electronic computers were huge machines that used thousands of vacuum tubes (this was back in the dark ages before transistors and the large-scale integrated circuits we use today); a single computer system could consume the entire output of an electrical generating station. Governments, educational institutions,

0-8493-9823-1/00/$0.00+$.50

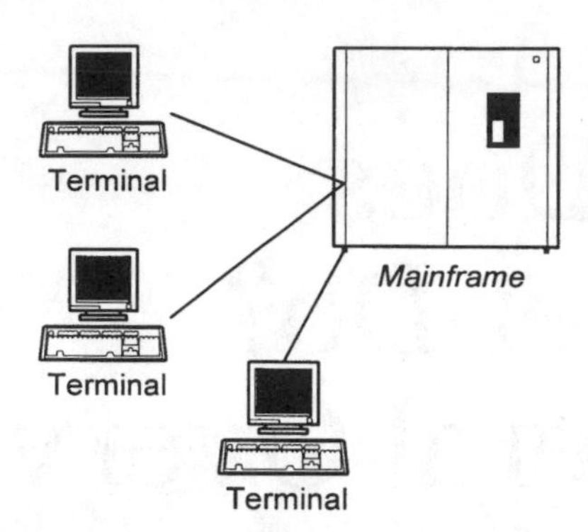

Exhibit 3.1 Mainframe and terminals.

and large corporations were the only places where these systems could be found, largely because no one else had the resources to build or buy them. By necessity, computing in its earliest form was centralized around these large "mainframes" used to store and organize information. Inputting data was done with *terminals* that featured input devices like keyboards. Multiple terminals could be connected to a single mainframe simultaneously by way of communication hardware.

In the *centralized computing* model in Exhibit 3.1, all data are managed by the mainframe, and all computing (or processing) is performed by the mainframe. The terminal shares no resources, but instead acts only as an input/output device. Computer networks actually began when mainframes were connected to each other in order to share information and resources.

Distributed Computing

Made possible by the introduction of personal computers in the mid-1970s, *distributed computing* features many smaller computers performing tasks that were once performed by a single mainframe computer system. In a distributed computing system, each separate personal computer can independently carry out smaller tasks without relying on a single central computer.

By connecting these personal computers together in computer networks, a distributed computing system (see Exhibit 3.2) provides many of the same functions and services once performed only by centralized mainframe computer systems.

Collaborative Computing

Collaborative computing has evolved largely from distributed computing and the capabilities of personal computer networks. Unlike the distributed

Exhibit 3.2 Distributed computing model with multiple personal computers connected together.

computing model where individual personal computers carry out independent tasks, collaborative computing involves two or more computers working together synergistically (there's a fifty-cent word) or cooperatively to perform processing tasks.

Today's computer networks integrate personal computers, mainframe computers, and a host of other computing and communication services — so many that it is almost impossible to describe any given network as centralized, distributed, or collaborative computing.

Today we most frequently classify computer networks based on the areas they service, as shown in Exhibit 3.3. Largely self-explanatory, the most commonly used network classifications and their accepted abbreviations are

- Local Area Network (LAN)
- Metropolitan Area Network (MAN)
- Wide Area Network (WAN)

Local Area Network. A *local area network*, or *LAN*, combines computer hardware and transmission media in a relatively small area, usually contained within one department, building, or campus, rarely exceeding a distance of tens of kilometers. LANs usually comprise only one transmission medium and are characterized by comparatively high speeds of data communication.

Metropolitan Area Network. A *metropolitan area network*, or *MAN*, is larger than a LAN, and — as its name implies — covers the area of a single, sometimes metaphorical city, rarely farther than one hundred kilometers. MANs frequently comprise a combination of different hardware and transmission media, are characterized by somewhat slower data communication

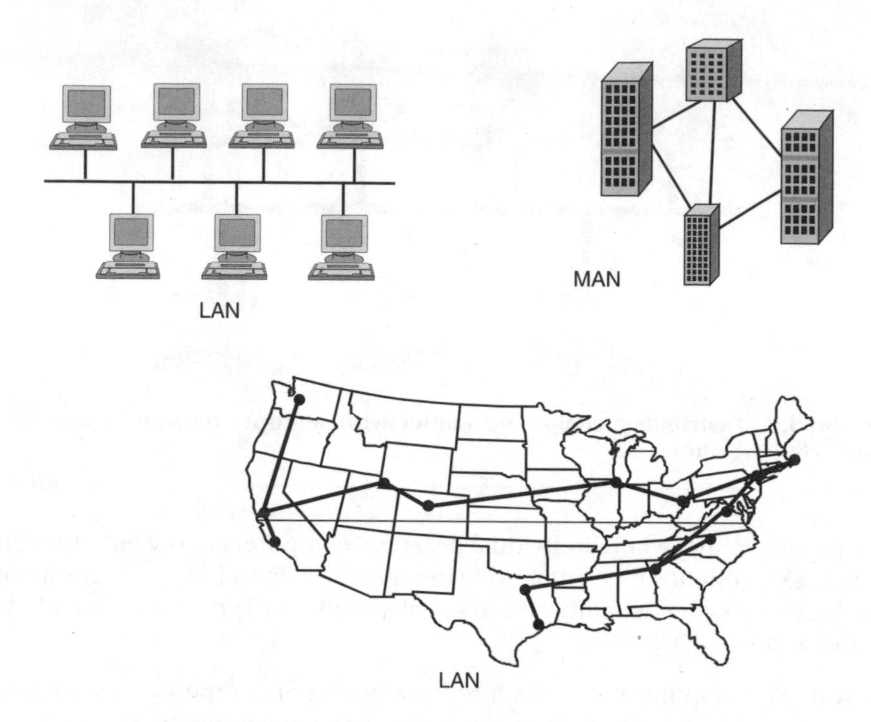

Exhibit 3.3 Comparison of LAN, MAN, and WAN.

rates than LANs, and are often used to provide connections between LANs from different sites.

Wide Area Network. A *wide area network*, or *WAN*, is any network larger than a MAN, and connects LANs together across the country or even around the world! Of the three network classifications, WANs are characterized by the slowest data communication rates and the largest distances.

An enterprise WAN network connects an entire organization, including the LANs at all its various sites. This term is used for large, widespread organizations such as corporations, universities, and government bodies.

As its name implies, a global network covers sites around the world. While not necessarily covering the entire world, a global network includes locations in multiple countries and connects networks of various organizations. Today, most people equate "*the* global network" with the Internet.

REQUIRED NETWORK ELEMENTS

Just as you can't make an omelet without breaking eggs, you can't have a network without certain key elements. They are

- A minimum of two users, or systems, with information to share
- A pathway that allows them to contact each other
- Rules or protocols that enable communication between them

Let's suppose you've been put in a room with people, every one of them from a different country and every one speaking a different language. You would certainly like to get to know the other people in the room, but you find it pretty rough going. Only two of the three key communication elements are present. Every one of you certainly has *information to share*, so the first element is present. You have a *pathway*, in that you can speak to and hear each other. The missing ingredient is a common language or *protocol* that allows communication to take place. Since none of you speaks the same language, it is virtually impossible for successful communication to take place!

Now let's suppose each of that group of people also speaks a second language, the same one. Once everyone realizes that no one else shares their native language, everyone tries their second language and communication begins. You've now fulfilled all three requirements for communication. Having agreed on the language, or protocol, you can now communicate. In fact, with all three of the required elements in place, you're networking!

In the example above, it's important to understand that having information to share and a pathway through which to share it is not enough. The pathway is there, but until you agree on the rules for communication, the language or protocol, no information is shared. Only after all three elements are in place can we share information.

Network Services

Using a combination of hardware and software, your individual workstation provides you with various *services*, including data storage and retrieval, printing and other input/output functions, communication functions like modem and fax services, and more. Much the same as your automobile provides the physical means for transportation, the physical construction of your PC provides the ability for these services to take place; the PC provides the framework.

But your automobile won't go far without a driver, will it? Your PC won't go far without software, either. That software includes an operating system and software application programs like word processing, spreadsheets, and so on.

Since a computer network is based on the Distributed Computing model, we can define *network services* as functions that are shared or shareable by way of the network. Just as your individual PC uses a combination of hardware and software to provide its services, so does the network.

A *service provider* is any combination of hardware and software that fulfills some particular job or function. Keep in mind that today's computers are able to perform many tasks at the same time, so our definition of a service provider is not necessarily a computer, but rather the combination of hardware and software required to perform the function or service.

The presence of services implies that someone or something will *use* those services, which brings us to our next definition — a *service requester.* And we'll use the term *entities* to generally describe a group of service providers and service requesters.

In today's computer industry, network service providers and service requesters fall into three types, based on the roles they perform on the network:

- Servers, whose role is limited to providing services
- Clients, whose role is limited to requesting services
- Peers, which may do both

Keep in mind the point made earlier; a service provider is not necessarily a computer. It's quite possible, though somewhat rare, for a single computer to act as a client, a server, and a peer, all at the same time. This assumes that the software provides all those abilities. This also means that it's software that defines a computer's role or roles as a client, server, or peer. Look out, here come more definitions.

In general, today's computer networks can often be described with one of these two classifications:

- Peer-to-peer
- Server-centric

Peer-to-Peer Networks. In a *peer-to-peer* network all entities are created equal and may be both service providers and service requesters. Peer-to-peer network software usually has no central point of control, relying on the peers to perform various and (or) similar functions for each other. Peer capabilities are built into Windows for WorkGroups, Windows 95, and Windows NT. Each individual system can act as a server. Most LAN administrators spend a great deal of time disabling peer services on desktop systems. The goal is to have the important stuff on a central system — the official server — so that basic administration tasks like backup can be controlled by the administrator.

Server-Centric Networks. The most popular and widely used network operating systems in today's computer market are server-centric. As of this writing, Novell's NetWare and Microsoft's NT Server are the leading examples of LAN-based server-centric network operating systems. Other server-centric operating systems include IBM's LAN Server, Banyan's VINES, and some versions of UNIX.

Transmission Media

Transmission media include a variety of physical cable technologies, but various wireless technologies are becoming more widely implemented. Transmission media can only guarantee that a pathway is available, much the same way a telephone line merely provides you with a channel through which you may converse with others — in order to communicate, you must send information across or through the pathway using a language understandable by both sender and receiver.

Protocols

Protocols are the rules of communication that must be mutually agreed upon by both sender and receiver, and may be as simple as just one rule or may be an extremely complex sets of rules. Modern spoken/written languages are good examples of complex protocols, involving a highly developed set of rules for vocabulary, spelling, syntax, and grammar.

In the "locked room" example we used earlier, once everyone agrees to use their commonly shared language you have established a communication protocol. A thorough discussion of computer network protocols is beyond the scope of this handbook.

NETWORK SERVICES

Network services, what we really care about in this book, cover a broad range of topics. Some of the most common network services are file, print, message, application, and database services. It is the provision of these services that makes a server and which the balance of this book addresses. We most often think of file servers, systems that provide file service, as *the server*. But that's not the only kind of server in the universe.

Common Network Services

Computer applications need certain resources in order to do the job for which they were designed. Those resources include data, actual computer processing, and input/output functions. But once you connect your personal computer to a computer network, many resources are available to you only through the network. Many applications have special network versions, or include the ability to take advantage of shared network services. Because many applications may not be aware of or able to access some network services, some network services are available to users by way of special "background" programs.

These “network services” programs are linked together into a single *network operating system*. Most of the desktop operating systems available today also include network services. Windows 95, for example, lets you share folders on your local drive with other network users. Network operating systems (NOS is the acronym) are designed from the ground up to provide various combinations of network services.

Local operating systems, also known as desktop operating systems, are the low-level computer programs that manage the resources of the individual computer (processor or CPU, storage, memory, printers, modems, and other peripherals). Some examples of local or desktop operating systems include DOS and comparable products like IBM's OS/2, UNIX, and Macintosh Finder. On the other hand, NOS is a specialized application designed for sharing services over a computer network. A NOS most often runs on a single computer, but can be spread out among various computers throughout the network. Novell's NetWare is far and away the most popular NOS in use today with over 50 percent market share, but other NOSs are also in widespread use, including IBM OS/2 WARP Server, Microsoft's NT Server, and Banyan's Vines.

The five most common network services are

- File services
- Print services
- Message services
- Application services
- Database services

File Services. *File services* allow you to store and retrieve files, and also allow you to manage those files by moving, copying, and deleting them. Well-implemented file services manage files across many different types of computer hardware and storage media. Some of the things network file services help you do are

- Move files from one location to another.
- Make efficient use of computer storage hardware.
- Create and manage copies of individual files.
- Make backups of critical data files.

All of this makes file management more efficient. Popular computer network operating systems make substantial improvements over the file services provided by local operating systems either by simply being faster, more efficient, or by adding more features and functions than provided by the local operating system.

Just as it's hard to imagine driving an automobile without fuel, it's hard to imagine a network operating system that does not provide file services.

It's easy to see why file services is one of the fundamental services offered by computer networks. We're going to break down in excruciating detail file service functions:

- File transfer
- File storage and data migration
- File update and synchronization
- File archival

File Transfer. In the early days of personal computers, about 15 years ago before networks, which might as well have been the Jurassic Age, in order to move data from one computer to another people would physically carry storage media (like diskettes) from computer to computer. This practice became affectionately known as "sneaker net" because network services were provided on foot. It's easy to see that this process is expensive and time consuming. It also can lead to reliability problems with the data involved.

Suppose both Jane and Bill "sneaker net" a copy of the same file from Tom's computer. Both Jane and Bill make changes to the data; at this point there would be three different copies of the same file — the original copy on Tom's computer, plus the two modified copies that Jane and Bill have.

Let's *really* mess things up by having Jane "sneaker net" her copy of the file back to Tom tomorrow morning. Look out! Here comes Bill with his modifications to the file tomorrow afternoon; Bill copies his modified version of the file back to Tom's machine, replacing both the original version of the file *and* Jane's modifications. Then Tom tries to fix what Bill has done to the file and it becomes so large it no longer fits on a floppy diskette.

Today's network operating systems provide functionality to prevent such reliability problems, using functions such as file sharing and file locking.

We can then define the network file service function *file transfer* as any service that stores, retrieves, or moves files for a network client. This provides much more efficient operation as organizations get larger and larger. It's easy to see how this principle applies in the case of today's business organizations, which may include offices, warehousing, and manufacturing facilities at widely scattered locations. If you work in such an organization and have to physically transport some critical data from one location to another, it can take a long time. To save yourself time and effort, you might hire an outside organization such as an overnight delivery service to transport the data, but this still can take an unacceptably long time.

By connecting all of your organization's sites to a computer network, you take advantage of network file services to electronically dispatch that critical data file to users at widely scattered multiple sites in a matter of seconds. While this newfound speed and power bring about greater efficiency of operation for many organizations, they also bring some headaches because it's now possible for users to gain access to data files and information that used to be inaccessible.

File services provided by network operating systems also allow organizations to place restrictions on that access. Some examples of information that might be placed under restricted access are personnel and payroll records, business plans and correspondence, proprietary or sensitive manufacturing data, and so on. Network operating systems implement such access restrictions by way of passwords and access rights. Most also encode or encrypt information so that it may only be read by those who are authorized to read it.

File Storage and Data Migration. The amount of computerized data has increased at an astronomical rate. Today we have a wealth of storage options, involving on-line, nearline, and off-line storage devices.

Storage technology utilizes a variety of magnetic and optical media, including disks, diskettes, and tapes. Each storage medium has specific characteristics that make it better for some purposes and functions than others. Today's network operating systems are able to provide management and control of storage on many different media types.

Each of the technologies that provide these storage options involves a combination of access speed, storage capacity, and price that determines where they can be best used. Let's start with the example of a hard disk drive that might provide a capacity of anywhere from 200 megabytes to 2 gigabytes and an access speed of a few milliseconds, for a cost ranging from a few hundred to more than a thousand dollars. With its relatively high speed and low price, a hard disk drive is a good example of technology that might be best used for on-line storage.

Consider a situation in which fairly quick but not instant access to reference information is required. An example would be a company's accounting data from the previous year. These data are no longer "active" because the year is over. It is not needed for day-to-day operations in the current year, but might be needed for comparison to the current year's operations. This type of information is best kept in nearline storage, which is not as quick but can still be accessed within a few seconds. Some examples of nearline storage systems include optical and tape "jukebox" systems that manage multiple optical disks or tapes in a single device. When information stored on a particular disk or tape is requested, the jukebox uses a mechanical system to load and access that particular disk or tape. The differentiating factor between nearline and off-line is that no "human" intervention is needed to access the data. No person has to load the CD-ROM or tape — the jukebox does that.

Older information such as business accounting data from years prior to last year may be needed for an audit, which requires access to the "archived" information. This older no-longer-in-use information can be stored off-line on high-capacity removable disks or tape. To access the information, we need to manually insert the appropriate disk or tape into the system to read it.

How does the information get from one type of storage to another? This is called *data migration*. As our data age and we no longer need them instantly at our fingertips, we can migrate or move the data — first from on-line to nearline, and then from nearline to off-line storage. You can set up the file services of many network operating systems to move or migrate data based on age or other criteria. This whole process is referred to as "hierarchical storage management," not to be confused with products of the same name from some backup vendors.

File-Update Synchronization. Business computing is becoming more and more mobile. The first *portable* computers were introduced back in the early 1980s; weighing in at 30 pounds or more, these monsters could more accurately have been called "transportable" computers. Today we use lap-top, notebook, and hand-held computers and personal data assistants (PDAs). It's common to see computer users relying on their portables almost anywhere in the world.

These mobile computer users need special support if they are to effectively use network services, largely because it's hard to be physically connected to a network while you're mobile. (The cable gets real tangled up going through highway interchanges.) Files that you would normally save on a network file server instead are saved to the portable computer's local hard disk. You need *file update synchronization.*

Assume you're the network manager for a medium-sized network consisting of fifty desktop workstations used by the office staff, plus twenty portable computers used by your outside salesforce to access the network. Individual sales representatives may only connect to the network once or twice a day, sending in orders, checking e-mail, etc.

Meanwhile, you (the network manager) have made a change in a crucial customer database file that's needed by almost everyone — especially the outside salesforce. Since the office staffers only uses the main copy of the database file stored on the network file server, they have no problem. The mobile users, however, due to the nature of their work, keep a separate copy of the database file on their portable computers; that copy is now outdated because of your change. File-update synchronization services see to it that all outside sales representatives receive an updated copy of the database file the next time they connect to the network.

File-update synchronization works by comparing time and date information of like files and using that information to decide which copy is most current. It may also keep track of what user has a specific file, and if transitional changes have been made. The file-update synchronization service will automatically update all copies of the file with the most recent and up-to-date version.

File Archival. If computer equipment never failed, we'd be living in a nearly perfect world. (OK, ending world hunger and war ranks up there, too.) But computers, like any mechanical device, fail. Offices can flood. World Trade Centers can be bombed. Electricity can fail at the most inopportune times. Acts of God do occur. The possibility of failure means we need to protect our information. One data protection technique is to create duplicates of the data on off-line storage media, such as tape. This process is called *file archival* or *file backup*. The duplicate or backup copy of the data can be restored in the event of an emergency. Remember to keep a set of data way off-line by storing it in a different geographical location.

A network administrator can use network interconnections to back up information stored on multiple network file servers, or even information stored on multiple workstations. This can be done from a single location using a specialized network backup system. Such systems include both hardware and software components, and provide substantial benefits in speed, storage capacity, and management.

Print Services. *Print services* are specialized applications that allow you to — you guessed it — print. Print services control printers and other paper-related equipment, like fax systems. Network print services handle print job requests, interpret printer commands and configuration information, manage print queues, and communicate with specialized network printers and fax equipment on behalf of network clients. You, the network user, are the client.

Some benefits of network print services include:

- Cutting the number of printers needed by an organization
- Locating printers where most convenient for the users
- Queuing print jobs to lower the amount of time your computer spends in the printing process
- Sharing specialized and expensive printers
- Streamlining the fax process (both sending and receiving) by way of the network — a specialized and fairly recent addition to the family of print services

Multiple Access from Limited Interfaces. Most printers have only one or two connection interfaces or ports. Some rare models may have more, but they are the exception rather than the rule. That means that no more than one or two computers can connect directly to a printer. Yet we are discussing interconnected computer networks, potentially consisting of many thousands of individual computers. Any or all of those computers may want to send data to a particular printer.

With network print services we only need one connection interface to connect a printer to a network. Many computers can send data to the printer through the network. This reduces the total number of printers needed for an organization. Instead of one printer for each computer, we can now share a few printers and place them in strategic or centralized locations. The direct result of needing fewer printers is financial savings, in purchase, maintenance, and management.

One of the reasons for the proliferation of local area networks is the Hewlett Packard LaserJet printer. When they first came out, everyone wanted their printouts to look like they came from the LaserJet. It was actually cheaper for some organizations to install a LAN to share a LaserJet than it was to buy LaserJets for every desk. Hard to imagine but true. It may be cheaper today for some companies to buy everyone a laser printer than to install a LAN, but that's another story.

Simultaneous Print Requests (Queuing). Sharing network print resources introduces a potential bottleneck. If all users send print requests directly to the shared network printer, simultaneous print requests could result in jumbled printer output, with multiple print requests mixed together. Since most printers aren't smart enough to tell the difference between multiple requests, the result would be useless, unintelligible "garbage," not to mention a waste of paper.

Imagine that Harry and Sally are two network users who send print requests to the network at the same moment. Harry's print request comes from his word processing application, while Sally prints a report from her database application. If both applications were able to send print requests directly to the printer, Harry's word processing document would most likely become mixed with Sally's database report.

Effective handling of simultaneous print requests is an important network print service solved by *queuing.* A queue is a waiting line. A print queue is a waiting line for print jobs on their way to the printer. When the desired printer is available and ready, the print requests are sent to the printer one at a time, thus eliminating the potential bottleneck of simultaneous print requests. Network operating systems provide a range of queuing features, including the ability to store, prioritize, reorder, and otherwise manage print jobs.

Since print queues can accept print information at network speeds (usually measured in millions, megas, of bits per second) instead of printer interface speeds (usually measured in a few thousand, kilos, bits per second), network print services make printing much more efficient. Since network clients can send print information to a print queue at network speeds, applications think they have printed and move on to other functions.

If you want to get really speedy, hook your printer up to the network cable directly instead of through one of the standard ports. It's like putting a network card inside your printer. Now you can move the queued print jobs from the network to the printer at network speed.

Network Fax Services. Network-based fax services are an outgrowth of print services and are being widely adopted by corporations. Traditional fax machines include an electronic scanner to create an electronic image of the document to be transmitted, a modem to digitally send or receive the electronic image, and a printer to print a hard copy of received fax transmissions. Use of traditional fax machines requires that the sender print an original document, and that the transmitted fax be printed at the receiving end.

Network fax systems save time in two ways: first, by eliminating the need to create an original document, and second by queuing faxes waiting for transmission. Instead of you waiting in line for the office fax machine to be available, your outgoing fax job can wait in the queue. In fact, many network fax services function as if they were network printers. Users send a fax by printing to a special printer driver from their application, and the fax software does the rest.

An additional feature of many network fax services is the ability to route electronic faxes to the intended recipient on the network. Direct-Inward-Dial, DID, is one such technology. The recipient can read the fax on a monitor or print it.

Message Services. *Message services* include storage, access, and delivery of various kinds of data, including text, binary, graphic, digitized video, and audio data. Sounds like file services, doesn't it? The difference is that message services refer to communication between different applications or network entities behind the scenes. Notification of new e-mail is a message service. It's not the e-mail message itself, just the alert that it's arrived. Message services include

- Passing electronic files from one user to another
- Integration of voice and e-mail systems
- Object-oriented software with objects distributed throughout the network
- Workflow routing of documents
- Keeping network information directories and phone books up to date

Electronic Mail. *Electronic mail*, which most of us call e-mail, is the electronic transfer of messages and message data between two or more network entities. The comparatively open nature of today's computer networks has led to an explosive growth in e-mail systems. Most e-mail messages are still text based — messages with quick queries and answers like "Do you want to have lunch?" However, you can get real fancy and send video and sound clips as e-mail messages. Video and sound take up lots more disk space than plain text. You can run out of network disk space because of cute e-mail messages.

Industry studies show that 50 percent or more of all network users will move at least once each calendar year. This raises large obstacles to keeping track of and managing name and address changes throughout the network.

As an example of the potential difficulties, imagine yourself as manager of a national postal system in a country where more than half of the citizens move every year. To add to your difficulties, many of your customers travel a lot, and your postal system is expected to deliver the mail to those travelers no matter where they may go.

E-mail has rapidly become an essential application in many organizations and is the most popular application installed on most networks. The popularity of public and private e-mail systems and bulletin boards has exploded exponentially. The biggest example is the public Internet. Private systems connected to the Internet include Prodigy, America On-Line, the Microsoft Network, and others.

Integrated E-mail and Voice Mail. One of the most interesting areas of emerging network technologies is the integration of voice and e-mail. Right now we have to check our voice mail and e-mail before we get all our messages. Voice mail is just a computer application. A computer running special software answers the phone and records digitally the message from your mother asking why you haven't called. Integrating voice mail computer systems with e-mail computer systems should be a simple matter. But standards have yet to take hold.

It is possible for an e-mail application to use speech analysis technology to convert audio voice mail messages to text information that can be displayed and read on a computer monitor. Conversely, similar technology can be used to read a text message and synthesize a voice playback. If you think it's great screening your incoming phone calls on your answering machine, wouldn't it be even cooler to screen incoming voice messages on your PC? Technology is under development that makes caller I.D. look primitive. What if the incoming phone call identified callers with not only their phone number but called up a picture of them and told you the last time they called, company name, etc. You could then accept or deny the call with a click and speak into your PC's microphone while their voice is projected from your PC's stereo quality speakers.

Object-Oriented Applications. *Object-oriented* applications combine small chunks of code and data to carry out elaborate functions. We discuss object-oriented applications here because message services provide communication between the chunks of code and data. These chunks are network entities. Network entities communicating qualifies under our definition of message services.

Object-oriented programming is the act of linking these chunks together into applications. Object-oriented programming increases the speed and efficiency of programmers by a factor of between 3 and 10 depending on whom you listen to. All you need to know is — it's faster.

Workgroup Applications. Workgroup applications are dependent on message services. The two most common types of workgroup applications are

- Workflow management
- Linked-object documents

Workflow management applications route information to appropriate users. The information can look and behave like a form to be filled out by multiple people or just be a document that is moved to the next editor on a list. (Remember Tom, Bill, and Jane all editing a single document and losing track of the latest version?)

A payroll process is an example of workflow management. In the old manual payroll system, employees filled out a paper time card. After the manager signed the time cards, they were sent to the payroll department. Once the time cards reached the payroll department, personnel entered information from the time cards, calculated pay, deductions, and benefits, and finally generated a paycheck. To verify reliability of information, each step of the process was audited. Most organizations forced the process to be done as quickly as possible in order to meet employee expectations (most of us like to be paid quickly!) and still comply with government regulations.

Here's where workflow management can help by intelligently routing electronic forms to the appropriate individuals. The time cards could be electronic forms. Each form may include data entered by other individuals, but the forms will also be customized in formats appropriate to each individual. We can blank out the dollar amounts on the screen to protect employee privacy when individuals who don't need to know how much each person makes access the forms. Since we're using our network computer system to transmit the data instead of a manual paper process, we can now move, store, manipulate, confirm, and present data to the users much more quickly than before.

Directory Services. For message services to send a message, they need to know the current address of the computer or other entity to which they are sending. Message services build and maintain directories. These directories are regularly distributed between and among message servers. Building and maintaining directories is called *directory services* or *directory synchronization*. The use of directory services is not limited to message services. Many network applications access directory service information in order to find and communicate with other applications.

When taking advantage of directory services, a computer user merely needs to provide the name of the intended recipient. Directory services will take care of the rest, determining the necessary location, address, syntax, format, and any other network routing or delivery information.

Application Services. *Application services* run a process at the request of a network client. This is not the same as file services. With application services more than one computer is required to complete a job. Client/server databases implement application services. The network client asks the database server to perform some function which is then returned to the client. Client/server databases manage the hardware and software to optimize the use of network resources. It is more efficient to perform the database query on the database server than it is to copy a really big database across the network cable to the client who then would have to sort through the data to get the answer to the query.

We'll talk about two other aspects often associated with application services:

- Server specialization
- Scalability and expansion

Server Specialization. The most significant benefit of application servers is that we can use specialized equipment to provide faster performance, higher reliability and integrity, and improved security. One of the biggest issues associated with application servers is how much processing should be done by the client, and how much should be shifted to a server. An application server is more than just another file server where you park all the applications. It is a souped up box that runs network software optimized for specific tasks. SQL Server is an example of specialized software designed to optimize the client/server relationship.

Say a user working at a personal computer needs to generate a graph that requires an extremely complex mathematical calculation. Even though the user has a very high-powered personal computer, it would take an unacceptably long time to perform the calculation. To speed up the process the user's PC sends the math segment of the calculation to a mainframe computer, which quickly performs the calculation and sends the results back to the user's PC. Now the user's personal computer only needs to gather the data and generate the graph, saving a substantial amount of time.

While it may not leap to mind, recent developments have seen another type of application server getting more attention, the private branch exchange (PBX). PBX systems are specialized computer systems that perform telephone switching services. Because they are computers, it turns out that PBX systems can be easily integrated into computer networks as we discussed in the messaging service section.

Scalability and Expansion. One of the hottest marketing terms in our industry is "scalability." This means that a set of application services can be run on more than one CPU. As the load increases, because of your company's terrific growth maybe, you can scale up to multiple CPUs, disseminating the load to maintain performance.

One current example of a specialized, scalable application service is Microsoft's SMS, Server Management Services, application not to be confused with Novell's SMS, which is a backup strategy. Server Management Services is a set of applications to help you manage your network. You can start off running it on a single PC. As your company grows, you can split portions off to other CPUs. The database part can live on one CPU, the communication module can live on another CPU, etc. This is scaling up. The same functions are being performed, but we're adding more horsepower to the server component to accomplish the task.

Database Services. Network *database services* are a specialized case of application services. We've just been talking about this — client/server databases. Certain design goals are implemented into these applications, like increasing transactions per second or reducing network traffic. Typically, the client will prepare requests and process responses, while the database server evaluates incoming requests and returns data. Network database services help centralize control of the data while providing quick access time and security. Two aspects of network database services are

- Coordination of distributed data
- Replication of data

Coordination of Distributed Data. An issue that frequently arises in large organizations is control of data, especially when data access is shared between multiple departments. Where are we to store the data? Who's responsible for backing it up? Network database services face these same problems. They also have solutions.

Network database services sometimes use a storage technique called *distributed data*. They divide control over pieces of the data among various computer systems. Even though the data are residing on more than one computer, it looks like it is one database. Database services coordinate the changing data across the disparate systems.

Replication of Data. When data are stored in a local database they can be retrieved much more quickly than if they are stored at a remote database. Consequently, users prefer to work on data stored locally. If there are multiple copies of a database, it is possible for users and organizations to make decisions using outdated or just plain incorrect data. *Replication* creates and synchronizes multiple copies of databases across the network.

There are two different methods of replication. In method one, all changes to the data are recorded in a master copy of the database. The master version then sends all changes to all the other copies. In the second method, every copy of the database is responsible for telling all other copies of the changes it knows about as well as integrating changes received from all the other copies.

SUMMARY

Networks have three common denominators: at least two users or systems with information to share, a pathway for sharing, and protocols for communication. All three elements can be broken down in excruciating detail — and have been in hundreds of books. This book concentrates on servers that provide a wide range of network services for users.

We've covered the generic definitions for a wide variety of network services. It's easy to see that there are several considerations involved in server-centric and peer-to-peer network services. It's also unfortunate that real-life considerations are rarely that easy. In many organizations, the organization's political structure may have more to do with file services decisions than technical considerations.

In a perfect world, strategies can be recommended for purely technical reasons. Clearly identifying the services you want your network to provide for your users will equip you with selection criteria for real world technologies. Good Luck.

Chapter 4
The Case for Mainframe Servers in the Enterprise

Brian Jeffery

An upgraded mainframe system equipped with new database and application capabilities can function as an enterprise server. This chapter explains techniques and proven solutions for making the transition from conventional mainframe computing to an enterprise server-based IS strategy.

INTRODUCTION

The mainframe continues to run core information systems without which most businesses could not function, providing high levels of availability, data integrity, and security. Mainframe-based computing retains its embedded strengths of economy of scale, robustness, and business-critical computing.

These systems, however, must also acquire new client/server capabilities. Central databases play a new role in concentrating, protecting, and providing access to all corporate data resources. Users employ PCs, workstations, and portable computers to access these data and to communicate via organizationwide network infrastructures. New development tools deliver high-quality, flexible applications in a fraction of the time previously required. In most cases, legacy data, applications, and skills carry over to the new IS environment. More important, there is no significant business disruption.

In most companies, the transition from conventional mainframe computing to an enterprise server-based IS strategy can be realized within 12 to 18 months. Key applications can be brought on-line even more rapidly. In an organization that already uses mainframe systems efficiently and employs modern hardware and software technologies, much of the infrastructure is already in place.

0-8493-9823-1/00/$0.00+$.50

Although costs will vary from company to company, such an IS strategy, if properly applied, can yield business gains and improvements in efficiency that more than justify the investments. Data center operating costs are frequently reduced by two to eight times. Applications development productivity can be increased up to eight times. PC/LAN costs can be lowered by 50 to 80 percent, and telecommunications costs by up to 50 percent.

ENTERPRISE SERVER-BASED STRATEGY IN PRACTICE

XYZ Company is a growing midsize retailer that has more than doubled its revenues during the last five years. Despite severe price-cutting and increased competition, gross margins have improved more than 30 percent over this period.

John is an XYZ customer. On XYZ's mainframe is a file on John, as well as files on more than five million other XYZ customers. The file contains details on all of John's purchases from the company over the past five years and more than 40 other categories of information. Because John is an XYZ preferred customer, he qualifies for discounts, special offers, and promotional rewards. This is why he buys from XYZ. Preferred customers are only 15 percent of XYZ's customer base, but account for more than 30 percent of sales and more than 40 percent of profits. XYZ's customer loyalty program is designed to keep existing preferred customers and find more of them.

When John makes a purchase, an hour later that purchase, along with hundreds of thousands of others, is being studied by XYZ's top managers. They view information displayed on a large-screen computer terminal showing precise sales patterns for every product line and department for each of the company's stores. If necessary, management can examine sales of any of the 12,000 inventory items the company carries, for each single checkout nationwide. They can examine not only the sale value, but also the profit contribution of each store, item, and customer.

Data are streamed upward from the point-of-sale systems in XYZ stores to a common customer database, where it is then analyzed automatically every hour for management by a highly sophisticated, custom-built command and control system. Exhibit 4.1 shows how a common customer database is the focal point of XYZ's entire IS infrastructure.

The XYZ Company profile is a composite of three of the fastest-growing retailers operating on three different continents. Similar examples can be found in a wide range of industries worldwide.

CONCENTRATING IS RESOURCES

In a large, diversified company, it may make sense to decentralize IS resources within the organization and subdivide responsibilities among

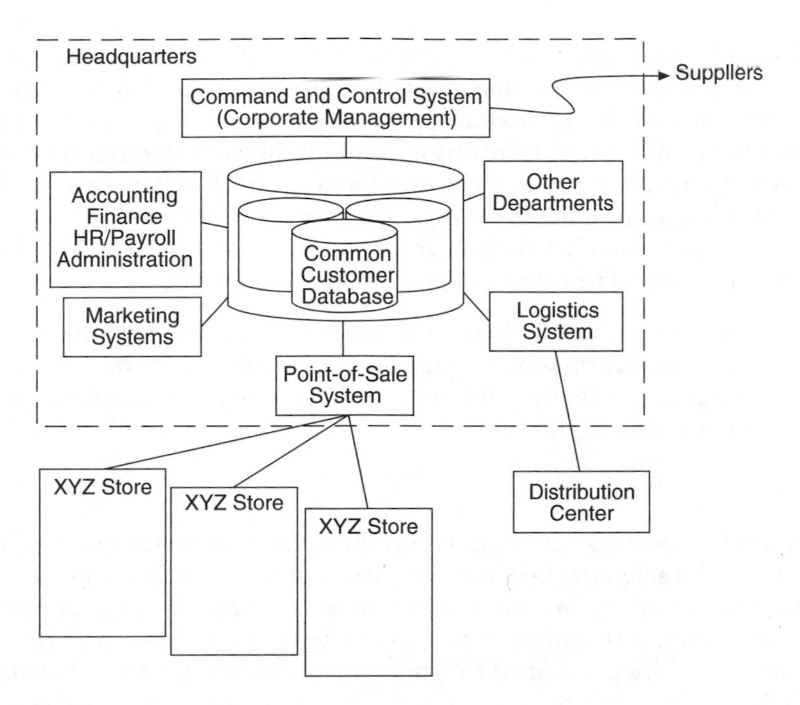

Exhibit 4.1 IS infrastructure built around a common customer database.

business units and divisions. By contrast, growing companies like XYZ target well-defined customer bases, are more focused in the products and services they supply, and usually operate within specific geographic markets. Management chains of command are relatively short, organizations are compact, and reporting structures are uncomplicated.

In this type of organization, there are major benefits in concentrating IS resources. This does not preclude the use of PCs, workstations, or distributed servers. However, all IS resources must work together with maximum effectiveness, for maximum business yield.

The deployment of all IS resources needs to be driven by the goal of improving competitive performance. Specifically, IS resources are focused on three key variables: information, speed, and cost.

Information

Leadership companies employ information technology to learn precisely who their customers are and how they behave, when and why their buying patterns change, and when and why they buy from competitors. This knowledge is used to acquire more customers, sell them more products and services, and retain their loyalty over time.

With a few modifications, the IS infrastructure also becomes the vehicle through which companies obtain valuable information about all the individuals, households, or businesses that should be their customers and about market trends, the actions of competitors, and all the other factors that affect what management reports to company shareholders. It also generates information continuously about the company's internal operations. That information is used aggressively to maximize the efficiency with which all financial, material, and human resources within the business are used.

The primary IS structure is the company's core databases that are the repository of its information resources. Applications generate and interpret information. Networks distribute it. PCs, workstations, and mobile computers provide access to it.

Speed

In competitive markets, leadership companies are able to respond quickly and intelligently because they are supported by IS infrastructures designed to allow key business decisions and operations to occur in real time. The speed with which strategic decisions are made in the executive suite and tasks are performed by front line employees provides a competitive edge.

There is not much point to having vast amounts of information if it takes too long to assimilate and act on it. In all companies, there is a path between the placing of a customer order and the final delivery of the product and service. That path should be as short and direct as possible. However, all too often that path goes astray: memos, forms, and printouts abound; one department after another checks, reviews, approves, signs off, and forwards reports to another department, where the process begins all over again.

The way in which IS is organized has a major impact on the business process. If applications and databases are fragmented because departments and workgroups work with separate, often incompatible computers, then data must be transferred between activities, functions, or operations through less than high-technology information-handling techniques. All of this slows the business.

The enterprise server-based strategy tackles these problems by:

- *Using precision information systems.* These systems interpret large volumes of data and present information in a clear, easy-to-understand form that can be acted on not only by management, but also by front-line employees, sales personnel, and administrative staff. Exhibit 4.2 provides a blueprint for this type of system. These new systems are a

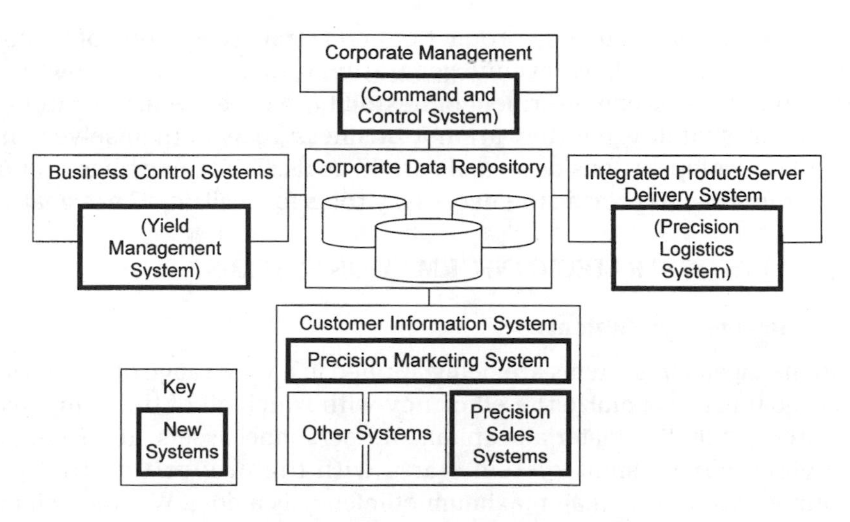

Exhibit 4.2 Blueprint for the IS strategy.

quantum leap beyond 1980s-style query and decision support tools running on PCs or departmental servers. As large-scale systems that operate companywide, the impact on bottom-line business performance is greater.

- *Developing integrated product/service delivery systems using flexible, highly functional software technologies.* These systems eliminate administrative inefficiencies, consolidate fragmented applications and databases into single, streamlined systems, and replace paper-based procedures with electronic information-handling. Such systems cut across traditional functional boundaries. Delivery of all products and services is accelerated, allowing the company to respond effectively and rapidly to changing market and competitive conditions.
- *Developing a command and control system.* Such a system assists the company's management team in determining market strategies, monitoring business performance, and initiating change. It is highly customized, using advanced technology to adapt to the planning needs, thinking processes, and work patterns of management.

Cost

Analyzing cost structures and projecting costs more effectively increases the company's control of its financial resources while improvements in throughput, resulting from streamlined delivery processes, mean that staff can be reduced and overhead costs cut.

Cost reduction is not a by-product of an IS strategy. It is one of its core mandates. Goals such as revenue generation or market share growth, or intangibles such as empowerment or flexibility, may be useful indicators of a company's vitality, but they are not business goals in themselves. In a leadership company, business goals are IS goals. Technology has no purpose other than to provide the necessary tools for realizing these goals.

NEW TYPES OF STRATEGIC INFORMATION SYSTEMS

Yield Management Systems

Yield management systems are being applied in a wide range of industries. Their goal is to maximize the efficiency with which all of the company's resources, including material, capital, and personnel assets, are used. Use of a yield management system starts with the proposition that any resource used at less than maximum efficiency is a cost. When efficiency improves, costs are reduced and profitability increases.

Precision Cost Analysis. Cost analysis in many organizations is handled in a piecemeal way. PC-based spreadsheets are usually employed in individual departments. A yield management system integrates and automates all procedures for cost accounting, analysis, and projection, while delivering radically higher levels of analytical power and statistical precision.

Cost analysis thus becomes a continuous, automatic process. Management can review cost patterns and identify emerging patterns and exploit opportunities weekly, daily, or even hourly. Targeting improvements in efficiency becomes a normal part of business operations rather than an occasional event.

Functional Areas. A company's yield management system may be used within several different functional areas. It can, for example, act as an overlay to an integrated product or service delivery system, such as a precision logistics system.

The system provides special value in cross-functional analyses. For example, a retail company uses a yield management system to realize major savings in procurement. Mining data on companywide external purchases allows details (shown in Exhibit 4.3) to be collected and analyzed for all stores and departments.

The identification of cost anomalies in particular is targeted. The company had initially carried more than 12,000 items in inventory. After analyzing each category, quantifying costs, and determining the effect on sales of changing stocking mixes, the total was reduced to less than 8,000. Costs were reduced by more than 15 percent without any reduction in sales.

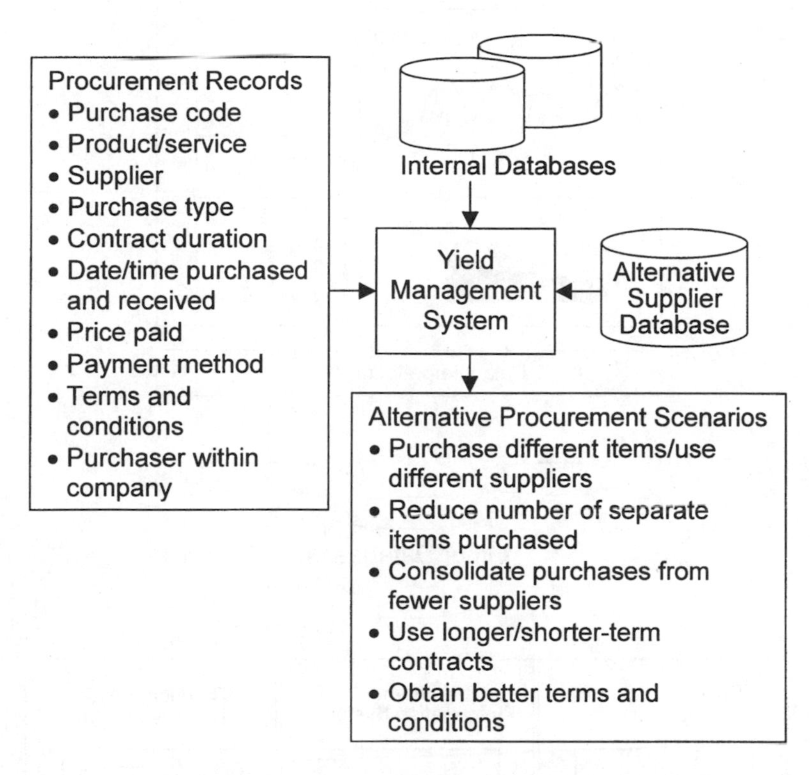

Exhibit 4.3 Yield management application.

Command and Control Systems

In large companies, it may be logical for corporate executives to avoid the complexities of day-to-day operations and focus on questions of policy and strategy. In growth companies like retailer XYZ, decisions about inventory, pricing, and specific marketing programs are central to maintaining competitiveness. Corporate management is directly involved in these decisions.

Command and control systems are designed to enhance the value of the skills of management principals by supplying them with comprehensive, timely, accurate information. Management can monitor the overall picture, make strategic decisions, identify critical situations, and when necessary, intervene at the tactical level. If this type of system is properly designed and implemented, it maximizes the top managers' role while allowing line managers to continue to deal flexibly with local business conditions.

Technology Design. An effective command and control system does not require that managers operate the system. Usually, technical design,

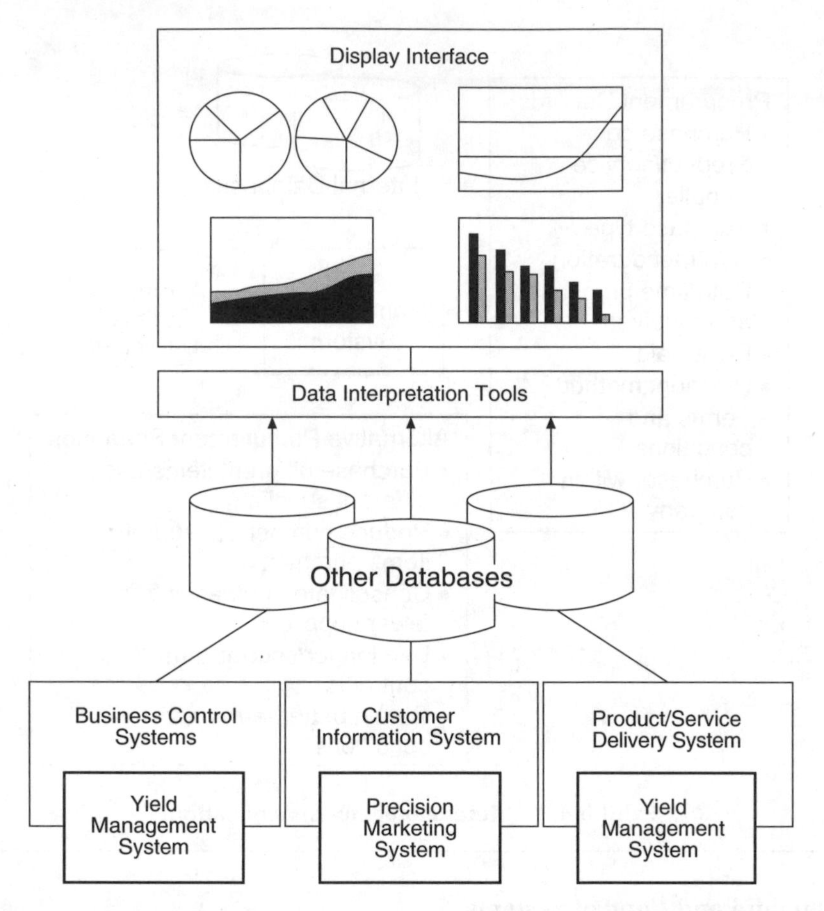

Exhibit 4.4 Technology components of command and control systems.

implementation, and operation are handled by IS professionals associated with either the executive staff or the IS organization.

The system should learn the way management principles work, not vice versa. It can and should be heavily customized to the needs, work patterns, and thinking processes of these individuals. In terms of technology, an effective command and control system, illustrated in Exhibit 4.4, consists of three primary components:

- *Display interface.* Display mechanisms may include high-resolution color graphics, presentation of data in 3D form, moving images, and large-scale projection screens. The result is similar to a business presentation, except that precision is much greater and the recipient depends less on the presenter for the interpretation of information.

- *Database infrastructure.* In a company using consolidated databases, virtually all of the electronic data that exist within the company is immediately available to management. If high levels of computerization are employed in business operations, there will be few (if any) significant items of information that are not available.
- *Data interpretation tools.* To reduce the complexities that managers must deal with, mechanisms must be in place not only to mine large volumes of data, but also to render the data into clear, easily understandable form. Conventional executive information systems (EISs) generally have not met expectations. PC-based tools may be appropriate for individual or departmental applications, but they are underpowered when it comes to dealing with data volumes and interpretation tasks needed to support strategic decision making. Large-scale tools for identifying, querying, and interpreting data resources perform better.

UPGRADING THE MAINFRAME: TECHNICAL IMPLEMENTATION

The technical infrastructure for this IS strategy is based on upgrading a mainframe system with new capabilities that enable it to function as an enterprise server.

The enterprise server is the cornerstone of this strategy. It concentrates key IS resources to better support distributed and remote users. It also increases the volume, quality, and accessibility of information for all users; improves availability of application, data, and network resources; and maximizes the cost-effectiveness with which all IS resources are used.

Enterprise Server Basics

An enterprise server combines the following distinct and, until recently, separate computing technology streams.

Mainframe Computing. The mainframe usually supports business control systems (i.e., accounting, finance, asset management, personnel administration, and payroll) and the core business-critical systems without which the company could not function. The mainframe's strengths are its robustness; its ability to handle large-scale, organizationwide workloads; its ability to provide high levels of availability and data integrity; and its capability to manage all system, data, and network resources effectively in performing these tasks. Mainframes also leverage economies of scale, benefits of concentrated resources, and consistencies of architecture that are inherent to the central IS environment.

Client/Server Computing. In practice, client/server computing is not a single phenomenon; it involves several different types of applications, including text processing, electronic mail, and personal productivity tools

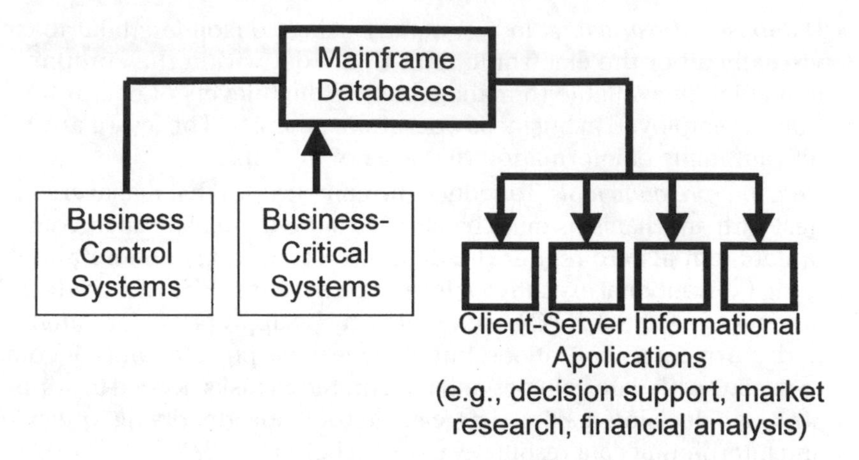

Exhibit 4.5 Mainframe and client/server synergy.

used on a PC or workstation. This category of applications is useful and fairly easy to implement in an end-user environment. From a business standpoint, the most valuable client/server applications are those that access, manipulate, and process data — for example, decision support, market research, financial analysis, human resources, and planning applications.

Consolidating Mainframe and Client/Server Technologies. In most organizations, mainframe databases are the primary source of data used by all such applications. The data are generated by the production business control and business-critical applications that run on mainframes. Thus, there is important synergy between mainframe and client/server computing (see Exhibit 4.5).

This synergy can be exploited to its full potential. By consolidating key databases and implementing reliable, organizationwide network infrastructures, all data can be made accessible at the workstation level.

In combining these technology streams, most growing companies are faced with the choice of upgrading a mainframe to act as a server or trying to equip a UNIX server to act as a mainframe. Upgrading the mainframe is usually both less expensive and less disruptive.

Moreover, it is substantially less difficult to provide client/server capability to a mainframe than it is to provide mainframe-class robustness to a UNIX server. The strengths of the mainframe environment have developed over more than 30 years. Few users have been prepared to entrust genuinely business-critical applications to UNIX servers.

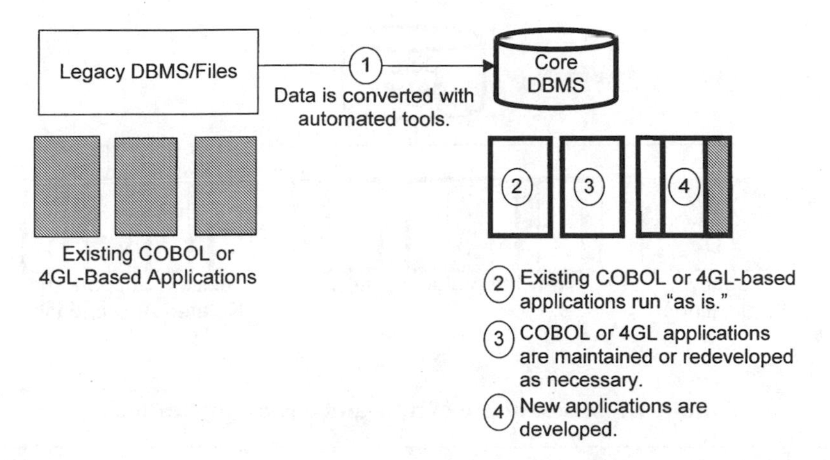

Exhibit 4.6 Migration from legacy mainframe environment to new core DBMS.

The Implementation Plan. The steps for upgrading the mainframe to act as a server involve:

- The migration to a core database management system (DBMS).
- The selection of tools to deliver new applications.
- The rebuilding of data center operations.
- The integration of PC/LAN clusters.
- The upgrade of the network infrastructure.

Migrating to a Core DBMS. The DBMS must be capable of handling On-Line Transaction Processing and batch workloads required for business control and business-critical applications, as well as the new query-intensive workloads generated by organizationwide client/server computing. Multiple databases, such as a common customer database and common operational database, are created within the core DBMS, which also ensures the accessibility, currency, integrity, security, and recoverability of all corporate data.

It may be necessary to replace older hierarchical database and file structures. This is, however, a relatively fast and easy process, especially if automated tools are used. Legacy data and applications can be converted to run on the new core DBMS. Existing applications written in Common Business Oriented Language and leading fourth-generation languages (4GLs) can be ported in this manner without the necessity for major rewrites. Exhibit 4.6 illustrates the migration from a legacy mainframe environment to a new core DBMS.

Selecting Tools for Building New Applications. Conventional COBOL- and 4GL-based techniques are often too slow for new applications and do not

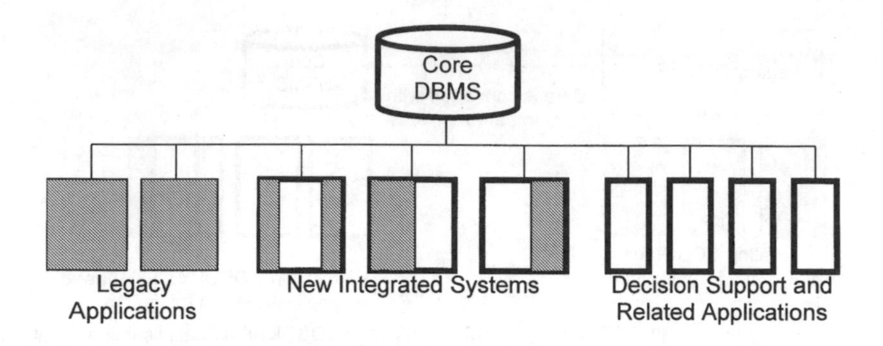

Exhibit 4.7 Coexistence of new and legacy applications.

provide adequate levels of integration and flexibility. Conversely, lightweight client/server development tools for PCs and small servers may not be able to handle the size and functional requirements of value-added solutions.

Packaged software can be used. However, most independent software vendor offerings address only standardized accounting or human resources requirements. They do not usually provide a direct competitive edge for companies that use them and are difficult to customize.

The tools to use to develop new high-value applications are latest-generation Computer-Aided Software Engineering, Rapid Application Development, and object-oriented development tools for the mainframe environment. Such tools deliver high-quality, large-scale applications that fully support client/server capabilities, including Graphical User Interface (GUIs). Moreover, they work in a fraction of the time required with conventional techniques.

For existing applications, several techniques can be used to make them more flexible and user friendly. For example, PC-based graphical user interfaces can typically be added to legacy applications without making major changes in platforms or software architectures. Similarly, existing applications can be redeveloped and maintained using new PC-based COBOL visual programming tools.

Once the core DBMS is in place, considerable flexibility is possible. New systems can coexist with legacy applications and light-duty database query, decision support, and related tools can also be employed (see Exhibit 4.7).

Rebuilding Data Center Operations. Another part of the implementation plan involves rebuilding data center operations using modern, efficient

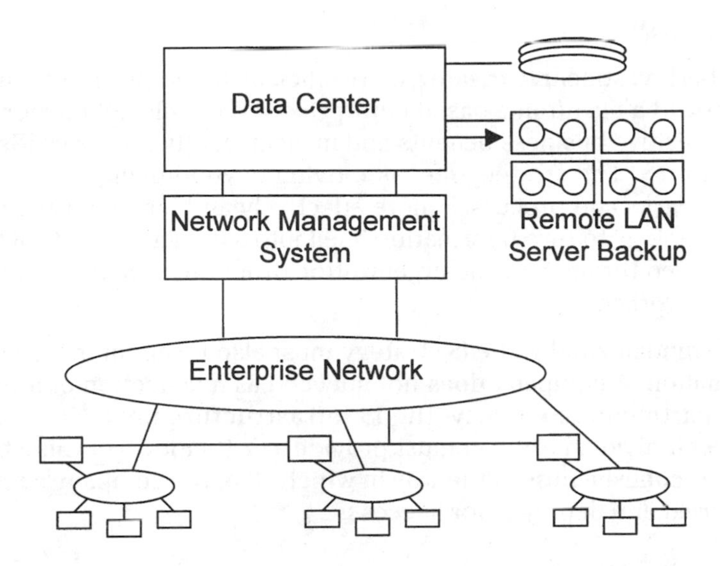

Exhibit 4.8 Enterprise server integration of PC/LAN clusters.

hardware and software technologies, automation tools, and improved management practices. The result is high levels of capacity utilization, increased performance, reduced operating costs, and minimal outages or production job failures.

Integrating PC/LAN Clusters. To more effectively support distributed PC users within the organization, local area networks (LANs) are interconnected and managed from a central point. Tools are used to regularly back up data on LAN servers to data center storage, thus ensuring that departmental data are properly managed and protected and are accessible to other users within the organization. These initiatives, illustrated in Exhibit 4.8, significantly improve the quality and availability of LAN infrastructures and reduce support costs.

Upgrading the Network Infrastructure. The network infrastructure should be functional enough to allow information to flow freely throughout the organization. Increased use of client/server computing usually increases network loading, which may mean that more bandwidth is needed at both the local and wide area network levels. In addition, wireless technologies may be used to support mobile sales, logistics, and other applications. Networking for remote computing is typically simpler and less expensive to implement than for client/server solutions built around traditional PCs and LANs.

CONCLUSION

The initial investment to create a more efficient, functional IS infrastructure built around a mainframe-based enterprise server will yield major payoffs both in positive business benefits and in significantly higher yields from IS expenditures. The strategy means moving beyond using IS to automate business tasks and processes more effectively and instead using information technology to turn information itself into a dynamic new tool that can be employed throughout the organization, from the executive suite to the front line worker.

More fundamentally, the IS strategy must also be about integration and coordination. A company does not succeed as a collection of individuals and departments. Similarly, the IS infrastructure must do more than empower individual users. It must provide a further level of value by creating a new cohesiveness in the way in which all of the company's resources are focused and deployed for success.

Section II
Server Functionality

In the first section in this handbook we primarily focused our attention upon the roles of the server, examining the use of PCs to mainframes as data repositories in an organization. Although it was difficult to avoid describing server functionality in our discussion of the roles of the server, detailed examination of their functionality was deferred until this section. Chapters included in this section provide detailed information covering a variety of technical aspects that govern the overall functionality of servers as well as the functions they perform, with the latter important when considering their installation and operation.

In Chapter 5, the first chapter in this section, Howard Marks provides us with his insight into changing technology. In his chapter, entitled "Server Issues and Trends 2000," Marks identifies a dozen trends in the server market we should be aware of and how those trends can affect the operation of server-based products.

The second chapter in this section, "Specifying Servers Based on Needs and Growth," also authored by Marks, examines the components of a server as a mechanism to specify a product. Marks addresses the features that are important to consider and how each feature is suited to different types of servers. Through a detailed overview of the components of a server, this chapter provides a comprehensive series of general specifications that can be selected as a basis for vendor negotiations, or which can even be used for developing a formally issued request for proposals (RFP). As you read this chapter, you will note that it is important to consider the server as an entity, to include its disk subsystem and network interface, as well as to plan for capacity enhancements and backup due to the critical nature of information placed on this device. Thus, this chapter also examines several issues beyond hardware that should be considered when preparing for your next server acquisition.

Many of us are aware of problems in support, training, and the cost of equipment as computers are distributed to the desktop user. Unfortunately, many organizations, without performing an analysis, assume that servers should also be fruitful and multiply. By this trite statement I refer to the propensity of organizations to use a separate server for most server-related functions as well as to distribute servers, literally, close to the user population without considering the potential benefits of consolidation. In Chapter 7, "Consolidated Server Overview," John Lillywhite focuses his attention on the advantages and economics associated with consolidated versus separate servers. This chapter also examines the hidden and incremental costs associated with distributed and consolidated server operations, and concludes by discussing how a consolidated server platform can impact the existing infrastructure as well as those of other departments and branch offices.

The old Boy Scout adage "proper planning prevents poor performance" is appropriate to many occupations and situations. It is especially true once you order a server and face the installation process that requires the computer to be placed into an operational environment that is both accessible and relatively free from harm. In Chapter 8, "Server Installation Considerations," Howard Marks provides us with several lists of problems associated with commonly used methods of server placement and also addresses the need for a UPS and environmental control and also the advantages of rack-mounted equipment.

Although not often recognized until too late, the management of distributed systems can be a complex process that requires a considerable amount of planning to achieve the results we all strive for — quality of service at minimal cost. In Chapter 9, "The Business of Managing Distributed Systems," Richard Ross addresses the concerns of IS department managers. He also proposes a model of service delivery for the distributed computing environment.

In Chapter 10, Jack T. Marchewka discusses how to mitigate problems associated with the design and use of information systems. The title of his chapter, "Including End Users in the Planning of Integrated Systems Development," literally describes one of the most common failures associated with poorly developed systems — the failure to involve the end user. In his chapter Marchewka describes the benefits of user involvement and presents a mechanism for facilitating a working user-developer relationship.

In Chapter 11, Steven P. Craig describes the process of business recovery planning for LANs with respect to network components. The server is a critical part of every network, and Craig's chapter, "Business Continuity in the Distributed Environment," provides the foundation that is critical for developing a plan to restore server functionality and to support your organization's mission in the event that the unthinkable becomes a reality.

In Chapter 12, "Developing Client/Server RDBMS Applications Using Java Servlets and JDBC" which concludes this section, Jonathan Held provides a detailed examination of the functionality of servers. This chapter guides you, hands on, through the development of server software that interacts with a back-end database, and returns data to a client's browser. In this chapter Held walks us through the steps necessary to create this application and support the previously described operations through the use of Java and JDBC. Thus, we conclude this section on server functionality with a detailed chapter which provides hands-on information concerning the manner by which a server can be configured to interact with a back-end database.

Chapter 5
Server Issues and Trends, 2000

Howard Marks

Like the rest of the computing industry the server market is changing at what is apparently an ever-increasing rate. Fortunately this change can be understood, planned for, and, most significantly, taken advantage of by clever network managers. The key is understanding both what your medium-term, 12 to 18 months out, needs are going to be and what is coming down the pipe from the industry so you can try to match them up.

We don't pretend to have a good long-term crystal ball so we won't try to tell you what's coming up for the turn of the millennium or later, but we have got a pretty good idea of what's coming down the pike for 2000 and beyond. While some of the trends we've identified here are just the PC server manifestations of broader trends that also effect the midrange, UNIX, and mainframe arenas, this chapter will look primarily at how these trends affect the Intel processor server market.

In addition to the general trend in the PC industry for the first-tier vendors (Compaq, IBM, HP, and Dell) to get bigger at the expense of second-tier vendors like AST and NEC, we've identified 12 trends in the server market for the near future. They are

1. 12-way Intel multiprocessor systems
2. New processors including Deschutes
3. 64-Bit processors including IA-64 Merced
4. OS trends and updates
5. Web management
6. I_2O
7. Hot pluggable PCI
8. Clustering
9. Thin client support
10. Storage area networks and fiber channel drive arrays

0-8493-9823-1/00/$0.00+$.50

12-WAY INTEL MULTIPROCESSOR SYSTEMS

Give us more processors! As Microsoft and Intel attempt to position their products as true enterprise platforms systems supporting up to four Pentium Pro processors using Intel's standard support chipset and SHV (Standard High Volume) motherboard, they just don't have the horsepower to compete with either traditional midrange systems or RISC-based UNIX machines.

Microsoft's answer is Windows NT Enterprise Edition, which adds support for two of this year's server hardware trends more-than-4-way multiprocessing and clustering. Hardware vendor solutions range from Compaq skipping Intel's chipsets and designing their own 4-way systems with larger processor cache to servers supporting even more processors. The trend started early in 1997 with 6-way servers from ALR, now a division of Gateway, and followed with 10-way servers from Unisys and a raft of announcements of 8-way SMP servers from almost all the first-tier server vendors.

As we enter the new millennium 12-way SMP product announcements could be viewed in trade publications and a logical expansion to 16-way products can be expected.

This is all part of the industry's continuing race to bigger and faster systems. It's turned into a great race to build the biggest multiprocessor servers. In the UNIX market, Sun Microsystems introduced servers with up to 64 UltraSparc processors.

Taking a quick look at the servers announced by major vendors, there are several designs with vendors signing on as OEMs for the various camps. Each of these designs works hard to solve the limitation in the Pentium Pro bus architecture, which provides for only four processors.

One of the first to ship is NCR's OctaScale motherboard used in their WorldMark 4380 server. This design links two 4-processor boards with 200-MHz Pentium Pro processors and 512 KB or 1 MB of cache per processor. DEC's Digital Server 9100 server also uses this design.

Another early shipper is Axil Computer's Northbridge NX801, which uses a similar architecture that bridges two 4-processor Pentium Pro busses through a custom set of bridge ASICs. Data General's AV8600 is based on the Axil design and motherboard.

The most powerful server, and the Windows NT winner on almost every benchmark to date, is Unisys Corp's Aquanta XR/6, which supports up to 10 processors on two processor cards that each have up to 3 MB of level 3 cache. Unisys provides a custom Windows NT Hardware Abstraction Layer (HAL) to support the additional processors.

The big winner seems to be Corollary Inc., whose Profusion architecture was so attractive to vendors, including Compaq, who will be offering it as an upgrade to their Proliant 7000, Data General, and Hitachi, that Intel acquired the company. Profusion uses three Pentium Pro system busses. Two are dedicated to processors, with the third bridged to the PCI I/O bus.

NEW 32-BIT INTEL PROCESSORS INCLUDING DESCHUTES

In 1997, server vendors faced an interesting conundrum. Intel's latest and fastest processor, the Pentium II, was available at speeds up to 300 MHz, but Intel, in designing the new Slot 1 interface for this processor, only provided support for single- and dual-processor configurations. Vendors wishing to build enterprise-class servers had to choose between building dual-processor 300-MHz Pentium II systems or using a larger number of older Pentium Pro processors which top out at 200 MHz.

The second generation of Pentium II processors, developed by Intel under the code name Deschutes, differs from the original Pentium II in several ways. The most significant is that the Level II cache in the card package runs at the same speed as the core processor, while first-generation Pentium II processors ran the Level II cache at half speed. The other major change is in the new Slot 2 connector that brings out the full Pentium Pro bus to support systems with more than two processors.

Announced for late 1998 are Deschutes processors that have 100-MHz external busses, which speed up server I/O somewhat, and extended MMX processors code named Katmai that have enhanced memory bandwidth. Additional instructions also speed up floating point processing.

In March 1999 Intel announced their most recent edition to their Pentium processor product line. Called the Pentium III, this new processor was initially available in 450- and 500-MHz versions, and a family of faster processors can be expected to be announced through 2000. In addition to an enhanced processor the Pentium III includes 70 new processor instructions developed to facilitate 3D, animation, and imaging. The first two versions of the Pentium III product line include a 512K level 2 cashe and support a 100-MHz system bus. While the system bus will more than likely be supported in future models, it is reasonable to expect an expanded use of level 2 cashe to occur in future Pentium III products.

64-BIT PROCESSORS

Intel's crowning achievement, the IA-64 (Intel Architecture 64 bit) processor co-developed under the code name Merced by Intel and HP, won't be available until the beginning of the millenium. Expected to offer many times the raw performance of today's processors, this new 64-bit device is in some ways the Holy Grail of microcomputing. The two companies hope that they will finally be bringing the architectural features of mainframe central processors to the PC chip level.

The IA-64 is expected to debut at 600 MHz and advance to 1 GHz over time. The performance of Merced is expected to exceed a rating of 100, according to the widely used SPECfp95 standard, which rates the processor's performance in engineering and scientific applications. Currently, the Pentium Pro musters about a 6.70 rating.

In addition to a faster clock speed, the chip is expected to execute multiple instructions at one time. The more computer instructions a chip can process simultaneously, the better performance it offers — above and beyond the pure speed that measures how fast each instruction is processed. The chip is also expected to use a new approach to processing instructions called Very Long Instruction Word (VLIW), which packs many instructions together.

Merced processors will also be able to directly address 64 GB of memory, breaking the 4-GB limit on current Intel offerings.

OS TRENDS AND UPDATES

New versions of both major PC network operating systems, Windows NT/ Windows 2000 and NetWare, enhance server operations. Coincidentally, or not so coincidentally, both serve to strengthen their prior versions' respective weaknesses, bringing NetWare and Windows NT even closer together in capabilities.

NetWare 5.0, formerly code named Moab, at long last supports TCP/IP without resorting to the inefficient and inelegant technique of embedding IPX packets within IP packets as Novell's older NetWare IP did. NetWare 5.0, despite the name change back from IntraNetWare, further embraces and supports Internet standards including Dynamic Name Services (DNS), Dynamic Host Configuration Protocol (DHCP), and the Lightweight Directory Access Protocol (LDAP) with integration between Novell's NDS and DNS.

Just as significantly, NetWare 5.0 includes a Java Virtual Machine running at the kernel level. Server-based applications can be written in Java, where older versions of NetWare required some arcane knowledge to create NLMs. In another major change, Novell has discontinued the NetWare Web Server, replacing it with Netscape servers ported to NetWare through the Novell/Netscape Novonyx joint venture.

While Novell enhances its support for TCP/IP and server-based application development, Microsoft, which had the lead in these areas, is making long-overdue enhancements to Windows NT's directory services. Windows NT 5.0's (now to be renamed Windows 2000) Active Directory enhances the domain architecture, integrating it with DNS and vastly simplifying the process of defining trust relationships. Active directory supports LDAP,

X.500-style naming, a distributed file system that allows users to access data without knowing the physical location and implied trusts. It also supports multiple master replication, replacing the clunky PDC/BDC (Primary Domain Controller/Backup Domain Controller) arrangement of earlier Microsoft network operating systems.

Microsoft has committed to providing an early developer's release of 64-bit NT before the end of the year, even as the company struggles to get the third beta of NT 5.0 out the door.

Whether or not the new version of the operating system will be further delayed is anyone's guess; however, its capabilities when delivered will represent a significant enhancement over NT Version 4. For example, the high-end version of Windows 2000 Server will support up to 64 Gbytes of memory which should enable users to consider consolidating applications presently running on many individual NT servers onto a single or a few Windows 2000 servers.

Both vendors are also adding support for new hardware standards including I_2O (Intelligent I/O) and hot-swap PCI.

WEB MANAGEMENT

Over the past few years we've moved from an environment where network managers managed servers by sitting in front of them reading obscure messages from text-based screens to servers that not only track the state of their internal components but will automatically call for help if a power supply or drive in a RAID array fails or the internal temperature rises to unacceptable levels. These management tools, including Compaq's Insight Manager and Intel's LANdesk Server Manager Pro, have simplified network managers' lives significantly. This is especially true for those of us who need to manage servers in multiple locations.

The problem with these tools is that there is no easy way to view the status of multiple servers from different manufacturers or to view both hardware and operating systems data. Each vendor of operating systems, hardware, or management tools has a unique console program you have to use to access the data from their application.

One of the true benefits of the Internet explosion has been the development of a truly universal client console, the web browser. Vendors have started adding web-based management and administration tools to network management tools like Seagate's Manage Exec, operating systems like Windows NT, and even networking hardware like hubs and routers.

Now that most server operating systems include web servers, we expect this trend will only accelerate.

Compaq, Intel, BMC Software, Microsoft, and Cisco have banded together to form the Web-Based Management Initiative. This group promotes the use of two new management-related technologies to provide data modeling, manipulation, and communication capabilities recently outlined at a meeting of the Internet Engineering Task Force (IETF):

- HyperMedia Management Schema (HMMS), an extensible data model representing the managed environment
- HyperMedia Management Protocol (HMMP), a communication protocol embodying HMMS, to run over HTTP

The HyperMedia Management Protocol has been presented to the IETF and is currently under discussion. The HyperMedia Management Schema will be defined, maintained, and evolved by the Desktop Management Task Force (DMTF), pending its approval. The schema will be maintained on a public web site using specially constructed tools to ensure consistency and longevity of the data model.

For more information see http:\\wbem.frerange.com.

I_2O

In the early 1990s several vendors, led by NetFrame, now owned by Micron Electronics, and Tricord, now strictly in the services business, developed and marketed what were known as super servers. These systems leapfrogged the development of server and PC architecture standards to provide higher I/O performance than was available through the limited architectures of the time.

These servers surpassed the performance of standard servers of the day by using proprietary memory and bus architectures and by adding processors to disk controllers, network interfaces, and other I/O components. As the PCI bus increased the performance of standards-based servers, the higher prices and proprietary nature of these super servers were no longer justifiable.

A careful examination of today's 4- and 8-way SMP servers shows that for many applications they are often processing basic I/O interrupts because their I/O architecture requires that the main processors handle the I/O workload. In fact, the biggest difference between a high-end PC server and a mainframe computer of just a few years ago isn't in raw compute performance but in the ability to move data from one I/O device to another.

The solution to this problem is to add intelligence to I/O devices and I/O controllers that offloads the processing of interrupts from the main processors. This technique works for mainframe computers and worked technically for the super servers of the past. The key is to avoid the proprietary nature, and therefore higher costs, of the super servers of the past.

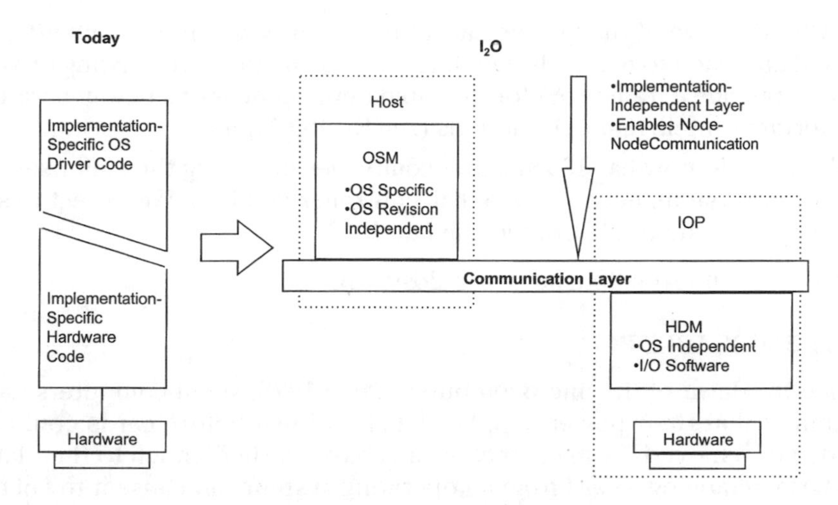

Exhibit 5.1 I_2O split driver model.

Learning from past mistakes, industry leaders, including 3Com Corp., Compaq Computer Corp., Hewlett-Packard Co., Intel Corp., Microsoft Corp., Novell Inc., and Symbios Logic Inc., formed the I_2O (Intelligent Input/Output) special interest group (I_2Osig) to design a new architecture for future systems.

The I_2O specification addresses two key problem areas in I/O processing:

- Performance hits caused by I/O interrupts to the CPU
- The necessity to create, test, and support unique drivers for every combination of I/O device and OS on the market

I_2O solves the first problem by offloading low-level interrupts to I/O processors (IOPs) designed specifically to handle I/O on each device. The I_2O architecture relieves the host of interrupt-intensive I/O tasks, greatly improving I/O performance in high-bandwidth applications such as networked video, groupware, and client/server processing.

It solves the second problem by using a "split driver" model for creating drivers that are portable across multiple OSs and host platforms. With the proliferation of network OSs (NOSs), most notably NetWare 4, Windows NT Server, and UNIXWare, the number of drivers that must be written, tested, integrated, and supported has escalated — one for every unique combination of OS and device. Through the split driver model, I_2O significantly decreases the number of drivers required: OS vendors write a single I_2O-ready driver for each class of device, such as disk adapter, and device manufacturers write a single I_2O-ready driver for each device, which will work for any OS that supports I_2O.

The objective of the I_2O specification is to provide an open, standards-based approach to driver design that is complementary to existing drivers and to provide a framework for the rapid development of a new generation of portable, intelligent I/O solutions (see Exhibit 5.1.)

The I_2O SIG now has 135 member companies including the major server, operating system, network, and disk controller vendors. We expect to see the first I_2O compatible products in late 1998.

For more information, see www.i2osig.org.

HOT PLUGGABLE PCI

Since the dawn of the microcomputer age in 1976, small computers have required that their power supplies be shut down before cards could be added or removed. As many technicians have learned, much to their later chagrin, removing a card from an operating system can cause it to fail the proverbial smoke test, destroying the card, motherboard, or both.

As PC servers are being used for more mission-critical applications, the downtime required to swap out or add network cards, disk controllers, or other devices has become increasingly problematic. A group of vendors, led by Compaq, HP, IBM, TI (Texas Instruments), Intel, and Cirrus Logic, has formed the PCI Hotplug workgroup as a subgroup of the PCI SIG to develop a standard for hot swappable PCI cards and systems. The workgroup is working with OS vendors to add their support.

The first hot-swappable systems shipped at the end of 1997. Operating system support and I/O controllers appeared in 1998. The Hot Swap workgroup merged with the I_2O SIG so future devices should support both sets of features.

CLUSTERING

Clustering, which we can think of as loosely coupled multiprocessing, is simply using several similar or identical boxes to act as one server. Clusters can be used to provide increased performance and increased fault tolerance, or availability. Incremental growth is accomplished by adding more nodes to the cluster. An operating system image and main memory address space exist within each server node, and there are various levels of coupling between the nodes to achieve greater performance.

Several different types of cluster systems have been available in the PC server market for a few years, primarily to provide greater availability. In order to understand clustered systems and their differences, we can divide clusters into three basic classes.

- Active/standby
- Active/active
- Concurrent access

Active/Standby Clustering

Active/standby clustering provides increased server availability but does not increase performance or scalability. Active/standby clustering has been available for several years from Vinca as their StandbyServer product for Windows NT or NetWare and is now available as Microsoft Cluster Server (codename Wolfpack phase 1), which is included in Windows NT Server Enterprise Edition.

Active/standby clustering is generally referred to as "failover clustering" and requires a relatively low-speed interconnect since the volume of communication between servers is fairly low. At most, a 100-Mbps interconnect is sufficient for failover clustering. One server actively runs the application while the other is ready to take over in the case of failure. While the mean time to repair the failed node may be two hours, failover time may be 10 to 20 minutes. This works for customers who run critical applications but can tolerate some short downtime and who require very cost-sensitive solutions.

Server availability is increased without requiring application modifications, but at the expense of redundant server components. Products range from those that provide a small increase in availability at relatively low cost, to products that deliver full fault tolerance at the expense of complete server redundancy. Some more sophisticated products allow one server to stand by for several active servers.

Novell's SFT can be thought of as falling somewhere in-between active/standby and active/active clustering. Like active/standby systems, there is no performance or load-balancing functionality, but because both servers do all the processing in parallel, the standby server takes over much faster than in a typical active/standby configuration.

Active/Active Clustering

Active/active clustering offers scalability, as well as availability, at a smaller cost increment per user than active/standby clustering. This may be done with or without load balancing. Here, the servers are described as loosely coupled.

Active/active clustering without load balancing occurs when both systems are running critical applications with their own resources. If one system fails, the users can switch to the remaining system, which takes over. There needs to be a fair amount of excess capacity on both systems, so that performance does not degrade excessively in case of failure.

While this is more expensive than an active/standby configuration, more users and applications are being supported. This is more desirable where there are two separate and distinct applications; that is, this configuration is not running the same application on multiple systems.

Active/active clustering with load balancing supports running one application on multiple systems. Users and data are partitioned, and, in case of the loss of a node, another node can pick up the users and keep them running. The load balancing is manual and requires that the data and users can be partitioned.

These systems must communicate to coordinate memory and disk access and to exchange data. This typically requires that shared SCSI disks be used in addition to a LAN-style interconnection. The next version of Microsoft Cluster Server Wolfpack Phase 2 will attempt to provide this capability.

Concurrent Access Clustering

Concurrent access clustering uses a very-high-speed interconnect to enable multiple instances of a service application to function as a single entity, operating against a single data set. This enables dynamic load balancing and dynamic user distribution in case of a node failure. Adding a machine to the cluster results in a near linear increase in performance, providing that the interconnect and disk-sharing systems are fast enough to support it, which in turn provides a high level of scalability.

This technology, which first appeared in Digital's VAXcluster over 10 years ago and exists in the midrange market, has not yet reached the PC server market.

THIN CLIENT SUPPORT

The microcomputer industry has always worked from the assumption that putting more computing power on users' desks will empower them to become more productive. Unfortunately, the race to build more and more powerful clients has also meant that a greater percentage of an organization's computing power is not in the clean, controlled environment of the glass house but out at the users' desks where it is difficult and expensive to maintain.

This problem came to a head in the past two years when The Gartner Group received a huge amount of press covering their Total Cost of Ownership (TCO) analysis that showed the cost of supporting a desktop PC in a typical organization exceeded three times its purchase price.

In response, vendors have developed a class of devices that provide a graphical user interface allowing users to run the kind of office automation applications that they've become accustomed to, while reducing the amount of maintenance required. These "thin clients" range from X-Terminals and Windows Terminals, both of which shift most or all of the computing load to the server, to Java-based Network Computers (NCs) and

NETPCs, which are more powerful computers in their own right. While these technologies have been discussed at great length in the press, they have yet to find acceptance in the lucrative corporate market.

FC-AL

Fiber Channel-Arbitrated Loop (FC-AL) is a subset of Fiber Channel technology. Using FC-AL, the storage subsystem delivers higher performance and is more cost efficient and more failure resistant than a storage subsystem implemented with parallel SCSI technology. There are numerous benefits of FC-AL.

- Delivery transfer rates of up to 100 MB/sec in each direction between nodes
- Support of up to 126 devices
- Support of SCSI protocol so most I/O drivers need only minor changes
- Highly sophisticated error-detection scheme

In a relatively low-cost manner, Fiber Channel can yield considerable benefits for users of storage subsystems: higher availability, higher scalability, higher performance, and very high capacity. These features are becoming crucial as companies increasingly rely on swift and continuous access to large volumes of data to support line-of-business applications.

SUMMARY

All these trends put together means that we'll have faster, smaller, more efficient, cheaper, easier-to-manage servers next year — just like *this* year we had them relative to *last* year. After all, if the automobile industry was like the computer industry, a Rolls Royce would have a top speed of 700 miles an hour, get 600 miles to the gallon, and cost 23 cents.

The differences between Intel servers is shrinking. Four or five years ago the super server products were very different from each other and from a Compaq server and other standard PC boxes running NOSes. Today those differences are smaller. An AST server based on Intel's SHV motherboard is very similar to an HP or Compaq server. Over the next two years, it will be very difficult to come up with true technical differentiation.

Chapter 6
Specifying Servers Based on Needs and Growth

Howard Marks

Most organizations' first LAN server was a desktop PC hidden under a secretary's desk, or away in an air-conditioning closet. The particulars of vendor, model, and even features were chosen based on the time-honored and completely unscientific process of buying whatever your network integrator wanted to sell that week.

The "server-of-the-month" club approach leaves you with huge expenses for support, spare parts, and overall management. If you're reading this chapter, you're probably looking for a better way.

There is no one, simple formula for specifying server configurations. File, database, web, and all servers have unique configuration requirements. In addition to each server's stated function, you also have to take into account your users, their locations, applications, and degree of computer use. Thirty salesmen who spend 60 percent of their time out of the office and use their computers primarily to cut and paste proposals together will, after all, place less strain on your servers than a single financial analyst slicing and dicing your databases to find the perfect customer for your company's new sky-blue pink nonstick widget spray.

This chapter will address the features that are currently being promoted by server vendors and how each is suited to different types of servers. The truth is no one understands the needs of your organization and users better than you do. By the end of the chapter, you should be able to get a good start on specifications for your own servers. You can then take these specs and use them as the basis for an RFP or a starting point for negotiations with your vendor.

0-8493-9823-1/00/$0.00+$.50

SERVER COMPONENTS AND SUBSYSTEMS

Processor(s)

The initial instinct of network managers is to buy the fastest processor their money can buy on the theory that the server is the busiest system on the network. In-depth analysis of the server's actual processing frequently reveals that the processor is idle 80 percent of the time. Getting a faster processor may not actually give your users a sense of improvement of their lot in life. It may be better to spend some of that processor money on more memory (RAM) or a faster disk subsystem.

The first thing you need to take into account when figuring out how much processor power your server is going to need is its role. File and print servers, especially those running Novell's NetWare, are almost never compute bound. Web servers, unless they host complex CGI scripts, are similarly primarily getting and delivering, as opposed to massaging, data so their performance is less dependent on their processor than disk and network subsystems. For these systems, we find the most cost-effective servers to be dual-processor-capable systems using the second- or third-fastest processor Intel is shipping at the time, with one processor installed. This configuration allows room for expansion should some more processor-intensive applications come along.

Database, messaging, and other application servers, on the other hand, do a lot more processing work. More processor power in these systems will have a significant impact on true performance. For these systems, actually coming up with a reasonable specification requires more extensive application profiling and testing — preferably real-world testing using your actual applications. You can start application profiling by using a metering program that tracks which applications are used by how many users over a period of time. You then need to correlate that to server statistics like processor utilization and disk I/O over the same time period. There are several products in these two market categories to choose from. Again we recommend buying a server with room for 50 percent processor expansion.

Memory

For most servers, regardless of role, adding additional memory will have a bigger effect on performance than any other change. NetWare servers cache large portions of their disk directories. Windows NT servers will swap data in memory to disk, thereby creating virtual memory as needed. Disk access is slower than memory access. Insufficient memory creates substantial performance loss with both these operating systems.

In addition to specifying enough memory, make sure to specify the type of *memory modules* used. Too often we've analyzed a client's server performance problem, ordered additional SIMMs (Single In-line Memory Module) or DIMMS (Dual In-line Memory Module), and found out too late that the vendor, used low-density modules in the initial configuration, filling all the sockets. For example, a server with six DIMM sockets and 192 MB of memory had six 32 MB DIMMS in the sockets, rather than three 64 MB DIMMS, which would have left us room for expansion without throwing the old memory away.

We recommend that you configure your server with half or less of its maximum capacity to allow for expansion and specify that the memory modules be of high enough density to leave free sockets for an additional 50 percent expansion. In the above example, with three 64 MB DIMMS we have three empty sockets that we can fill with another set of 64 MB modules.

Disk Subsystem

Most of us know that it's important to have enough disk capacity to hold all our users' data. That isn't all there is to specifying a disk subsystem for your server. You need to:

- Determine the rate of data growth to allow room for expansion
- Determine the appropriate RAID level for your applications
- Choose between internal and external drives
- Decide if you can cost justify hot-swappable drives

The most confusing choices are RAID level and array configurations. Since a separate chapter in this handbook discusses RAID levels and their application in detail, we're not going to analyze that aspect here. Do spend time looking at a variety of drive array products. Less common features, like dynamic array rebuild, let you replace a failed drive and have the array's fault tolerance restored without taking the server down. The less down time the better. Another feature, dynamic expansion, allows you to add capacity to a logical drive without backing up and restoring all the data. This means that if the server is running out of disk space you can add another drive without spending huge amounts of time backing up every bit, then adding the hardware and finally waiting for every bit to be restored.

The decision between internal and external arrays is also significant. Not only can you generally install more drives in an external array cabinet, but you can also use external drives in a shared disk cluster configuration like Microsoft Cluster Server formally code-named Wolfpack.

Network Interfaces

When specifying a server, you also need to specify the network interface. Because there will be significant traffic into and out of the server, make sure you get a speedy interface. Some network operating systems, including Novell's NetWare, will allow you to set up multiple connections to the same logical network to both provide fault tolerance and load balancing. For servers running these systems, we recommend two network interfaces, also known as network or LAN cards. If your network has multiple switches, each of the network interfaces should be connected to a different switch to keep the server on line in the event of a switch failure.

Other systems, like Windows NT running TCP/IP, run better if they have only one network connection. These systems should be given a single high-speed connection. Even a single 100-Mbps Ethernet provides as much bandwidth as most current PC servers need.

ADDITIONAL ISSUES

Aside from the standard components discussed above, you need to consider a few more issues when defining your next server purchase.

Standardization

Even though each user department within your company has different network and application needs, you should establish some set of standards across the board. Standardizing on a small number of server models from a single vendor makes it easier to manage the servers through a single console, allows you to keep spare and expansion parts like memory, disk controllers, network interfaces, and hot-swappable disk drives. Standardizing on a single vendor may also give you support contract negotiating clout.

Capacity

Your network will need to grow. No matter how much extra speed, memory, and disk space you invest in today, you will have more users, more applications, and more network traffic in the future. We can guarantee that tomorrow's applications will take more disk space, more memory, make bigger files, and take up more processor power. So you need to project what your needs are. It's much saner installing extra disk drives or memory modules during planned down time than having a hundred users breathing down your neck while you try to install during unplanned down time. Projecting your rate of growth is tricky. You'll need trend analysis tools and you may have to cobble a set together. Track disk space usage over time. Then find the average rate of increase for a given time period — financial quarters works for many environments that budget similarly. Inventory programs may be able to track some of this data for you. Follow the same process for network traffic, application usage, etc.

Reliability

Redundant Power Supplies. Given the high voltages and temperatures inside a typical computer power supply, it isn't surprising that power supplies are more likely to fail than just about any other electronic component in your server. Since a server without a working power supply is useless, high-end server vendors build their systems with multiple power supplies to allow the system to continue to operate when a single supply fails.

The simplest and most common way to provide power supply redundancy is to have two power supplies with automatic fallover. Each power supply is big enough to run the whole server and when the primary supply fails, the backup takes over.

A better system is dual load-sharing power supplies. In this system, the second power supply, rather than just sitting around waiting for the primary to fail, provides half the power to run the system under normal conditions. Load sharing is better because each power supply, running at just half its capacity or less, is less likely to fail or overheat in the first place.

As servers get bigger, with the corresponding increased power demands, building a single power supply that can handle the whole load starts to be very expensive. These systems can use an even more sophisticated technique called N+1, where a system that uses 100 watts of power will have five 25-watt power supplies.

If you're going to buy a system with redundant power supplies, you also want them to be hot-swappable so a failed supply can be replaced without taking the server down.

Battery Backed-Up Disk Cache. Today's RAID systems use intelligent caching controllers to improve performance. Most cache disk-writes as well as disk-reads. In this environment, the disk controller reports to the server processor when data are written to the memory in the disk controller rather than when they are later written to the disk. If the server crashes through a software or power failure and the disk controller doesn't have battery backup, the data still in the disk controller cache may be lost.

Manageability. Even the most reliable server will occasionally have a problem. Even it it's just a user who decides that everyone in his department should back up their workstations' hard disks to the server on Monday morning. So everyone in his department logs into the network on Monday morning, and all the data on their local hard drive starts winding its way through the network cable to your file server. Network throughput falls to its knees and the disk drive(s) on your server start chugging for dear life. Servers with the appropriate management features and software can let you know that these problems are developing before they become a true crisis.

Hardware Monitoring. Today's servers have multiple fault tolerant subsystems including power supplies, disks, and ECC memory. While these features improve system reliability, they can also hide component failures and lull you into a false sense of security. More than once we've visited client sites where a disk drive or power supply has failed and no one noticed.

Hardware monitoring features and software products like Compaq's Insight Manager, Intel's LANdesk Server Manager Pro, and Hewlett Packard's TopTools record these component failures and report them to a console machine across your network. In addition, they collect data from the temperature sensors and new-generation smart disk drives to alert you before something actually fails. This is proactive monitoring. With tools like these you just need to remember to check the console regularly.

OS Health Monitoring. Products like Seagate's Manage Exec, Intel's LANdesk Manager, and BMC Patrol keep track of operating system- and software-related server parameters like processor and memory utilization, free disk space, etc. For example, you can set a threshold that generates a warning when you fill 75 percent of your current disk space. Your trend analysis for capacity growth may have projected that you won't need more disk space until next year, but wouldn't it be nice to know if you're running out of capacity quicker than you projected before you get that annoying "out of disk space" message. Running one of these tools allows you to place your orders for server upgrades before your users start complaining about the system running slowly or being out of disk space.

HOW MUCH WILL WE CONSOLIDATE?

One early philosophical decision you're going to have to make is how much you want to consolidate multiple functions into a smaller number of servers. In some ways, the ideal network would have just one huge server that provided all the services the users needed. The administrators of such a network would have just one server to manage and could keep close watch on it via a staff of dedicated operators. With only one server, the organization could invest its money in providing that server with the right air and power conditioning equipment, physical security, high availability disk subsystems, and management tools. We could even use thin client technology to make the users' workstations easier to manage.

If you've been in the information technology field for more than a few years, the environment described above should sound familiar. It was the most common configuration for the first 20 years or so of computing. That single, consolidated server was called a mainframe computer and the thin clients, terminals.

We're all familiar with some of the disadvantages of the mainframe model. Leaving aside fascist MIS directors, bad user interfaces, and the other disadvantages of the traditional mainframe, a single server still has significant problems for all but the smallest organization. The most obvious is that a single server is also a single point of failure. Even the most reliable server will be down some of the time.

The other obvious disadvantage is the limited services a single system can offer. A typical network will need naming and other directory services, file service, print service, messaging, and other communications services. Putting all these functions on a single box is a recipe for disaster. You need to balance the benefits of having fewer servers to manage with the disadvantages of limited services running slowly.

GENERAL-PURPOSE COMPUTER-BASED OR BLACK BOX

For some services, especially those that provide communications and network services like routers and remote access servers, you need to choose between providing the service through software running on a general purpose computer like a PC or through a dedicated hardware server — the so-called black box.

PC-based servers are generally easier for network administrators who don't have a lot of telecom experience to manage. They're managed with the same tools that you use to run the basic file and other services and integrate into the network operating system security architecture. On the down side, PC servers often aren't as completely integrated into the telecom network and don't scale well to very large configurations. Black box servers are generally simpler, have fewer moving parts, and are therefore more reliable than PC servers.

DON'T TRY TO PLAN TOO FAR

One mistake we've seen several network managers make is to try to provide today for their needs five years from now. We've never seen anyone do it successfully and cost effectively. The truth is that between the increasingly rapid rate of performance and capacity advancement, the general decline in technology pricing over time, and the occasional rapid technology shift, like the mid-'90s' sudden shift to internet and web technologies, it just can't be done.

You're better off buying systems that meet today's needs and can be easily expanded to cover your projected needs for 18 months to two years. This generally means buying systems that can accept at least twice the processors, memory, and disk than you'll initially install them with. Worry more about individual server components and which set of components is more important for which type of server. Once you've identified these, you can use the RFP, Request for Proposal, chapter in this book to get the best prices and support.

Chapter 7
Consolidated Server Overview

John Lillywhite

During the past year, we have seen a dramatic increase in interest concerning the value of consolidating multiple smaller LAN infrastructure servers into centrally managed large or super servers. The reasons for this rise in interest are twofold. First, the number of servers and network change requests in many environments has become unmanageable, and many companies are looking to reduce this complexity via centralization. Second, as the price of servers and other LAN components has continued to escalate, the signature authority has moved away from departmental managers to more traditional IS areas of responsibility. These IS managers are used to a more centralized approach (whether dealing with hardware or people) and are looking for ways to reduce costs and improve service, while still reining in control of the distributed environment.

D. Capuccio
Gartner Group
Report on Consolidated Server Technology
15 February 1995

CURRENT SITUATION

Short-term benefits — long-term problems. The constant demand for more information or communications capabilities has placed a great demand upon IS management and network administrators. Squeezed from the field to provide more data faster, and from corporate finance to cut costs, IS management and network administration have been forced to implement less than ideal short-term solutions at the expense of higher hidden costs, increased network complexity, and reduced flexibility.

CONSOLIDATED SERVERS — A DEFINITION

Consolidated servers contain an array of independent, PC-based processing elements with system-level hardware and software management capabilities. The processing elements within the array are typically nothing more than "repackaged" fully functional, completely independent PCs with

0-8493-9823-1/00/$0.00+$.50

special circuitry to perform high levels of hardware management, monitoring, and control.

Most consolidated server technologies utilize modular components to simplify administration and provide expansion capabilities beyond that which is possible using traditional desktop PCs or higher-end symmetrical multiprocessing (SMP) servers.

Consolidated server technology attempts to combine the advantages of existing computer technologies into a single system while eliminating disadvantages (see Exhibit 7.1).

The market segment matrix in Exhibit 7.2 shows the ability of desktop PCs, SMP servers, and consolidated servers to satisfy a corporation's enterprise network requirements as a function of costs.

In most cases, consolidated servers provide solutions to systems-level requirements at a substantially lower cost. In many cases, network administrators are able to migrate to consolidated servers and completely bypass SMP server technology.

MIGRATION TO CONSOLIDATED SERVERS

As is normal from most migration models, computer users traditionally migrate to systems which provide more power, lower costs, increased productivity, etc. Migration to consolidated servers is likely to originate from the traditionally less managed, less productive desktop PC segment. Most corporations utilize standard desktop PCs as their entry platform for various network applications, including mainframe gateways, e-mail servers, print servers, communications servers, etc. As their numbers grow (typically over 10 to 20 PCs), desktop PC platforms as network application servers become unmanageable and expensive. (Costs will be discussed later.) Consolidated servers, using standard desktop PC technology, enable network administrators to port their existing applications and maintain their investment in software and administration tools.

CONSOLIDATED SERVER ADVANTAGES

In discussing the migration to consolidated server technology, it is assumed that one will migrate from traditional desktop PC platforms as discussed above. Therefore, only comparisons are made between consolidated servers and their desktop PC counterparts.

Increased Flexibility

Microsoft's NT, NetWare 4.x, SMP architectures, higher bandwidth WANs, and the Internet as a productivity tool have made it virtually impossible to predict with any accuracy what networks will look like in the next several

Exhibit 7.1 Product matrix: stand-alone PC servers and SMP platforms vs. consolidated servers.

	Characteristics	Vendors	Advantages	Disadvantages
Desktop PCs	Single processor, typically Intel-based. Single enclosure, typically several unused slots. Designed for desktop.	Dell, Compaq, HP, Gateway, Packard Bell, AST, etc.	Standards-based architecture. Huge variety of software. Huge installed base. Easy to use and understand. Low cost.	Unable to monitor functionality. Difficult to administer. Difficult to scale. Lack component redundancy. Not manageable. Large footprint. Difficult to upgrade processors.
File server PCs	Single processor enclosure, typically Intel-based. Designed for high-speed network and disk I/O. Used for network file server applications.	Compaq, HP, Dell, AST	Standards-based architecture. Wide range of NOS support. Easy to use and understand. Tuned for high-speed I/O. High levels of management tools. High data availability through component redundancy. Large installed base.	Too expensive for non-mission critical applications.
SMP servers	Multiple processor enclosure. Very high-speed I/O for network and file server applications.	Netframe, Tricord, Compaq, HP, AST, Dell	Tuned for very high-speed I/O. High levels of management tools. High data availability through component redundancy.	Limited NOS or application support. Lacks standards-based architecture. Expensive; questionable price/performance ratios.
Consolidated servers	Multiple independent processors within single enclosure. Typically, modular architecture to facilitate adaptability and manageability.	Compaq, HP, ChatCom, Cubix	Standards-based architecture. Huge variety of software. High levels of systems management tools. High data availability through modularity and component redundancy. Scalable. Low cost.	Small installed base. Not SMP-ready. Not suited for very high-speed processing.

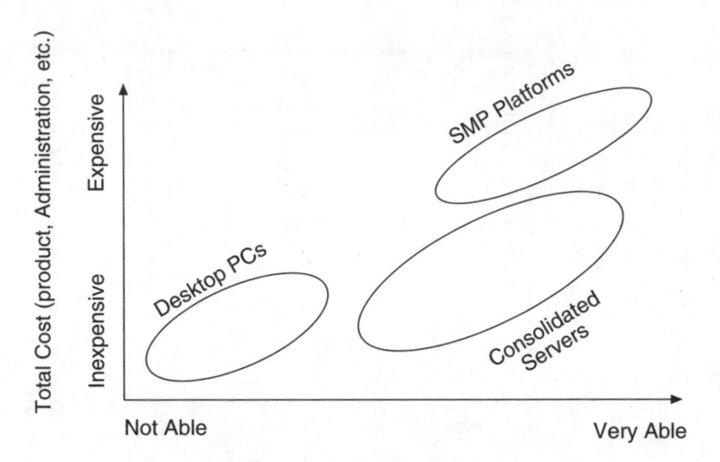

Exhibit 7.2 Platform's ability to satisfy IS/IT network system requirements.

years. This is making it difficult to make long-term, enterprise-wide computer purchases with confidence. Consolidated servers are designed to adapt to radical change. They use modular components — most hot-swappable — enabling administrators to modify networks with minimal effort or costs. Modifications or enhancements are made by replacing or upgrading a single processing module within the consolidated server.

Increased Data Availability

Consolidated server technology addresses data availability three ways: rapid replacement of failed components, component redundancy, and early fault detection and isolation.

Reduced Mean Time To Repair (MTTR). Hot-swappable, modular technology substantially reduces the time required to recover from hardware failures. All mission critical components (server modules, power supply modules, and hard drives) can be extracted and replaced with backup modules within minutes.

Component Redundancy. Consolidated servers include redundant power supplies and RAID storage configurations to ensure operation in the event of failure. (Processor module redundancy is generally not yet supported in consolidated server architectures. There are few operating systems today that support processor failover features.) Within the processor cards, however, more generic intraprocessor card fault tolerant features such as error correction code (ECC) memory are supported on most consolidated server processor modules.

Systems-Level Management. The ability to access all processor logic and hardware within a consolidated server provides the greatest advantages over stand-alone desktop PCs. Sensors located throughout the enclosure and embedded on each processor are capable of providing detailed operational status in ways not achievable with desktop PC technology.

Consolidated servers typically include software to monitor the functionality and health of the entire system. System temperature, power supply voltages, and cooling fan rotation are monitored and reported by most consolidated server vendors via SNMP management software. Additional features can potentially resolve catastrophic faults without human intervention. Most consolidated servers have the ability to detect "hung" processors, automatically cold-reset them, and report intervention activity.

SMP TECHNOLOGY VS. CONSOLIDATED SERVER TECHNOLOGY

SMP platforms include sophisticated component redundancy designs and systems management techniques to increase data availability. SMP systems are capable of running many different applications among their multiple loosely or closely coupled processing elements. But in the event of a failure of the total SMP system, all applications running on it become unusable. In a consolidated server architecture, applications maintain a one-to-one relationship between application and processor. A failed consolidated server processor will only impact a single application. In addition, since all the various applications on an SMP server run together under a single copy of the server operating system, these systems are susceptible to application interactions, where a combination of two or more applications on the same server results in instability.

CONSOLIDATED SERVERS VS. DISPARATE SERVERS — COST MODELING AND COMPARISONS

Capital Costs — Distributed vs. Consolidated Servers

In most instances, hardware costs for consolidated servers will remain at par to their distributed counterparts. Only substantial increases in disk subsystem and vendor maintenance costs will offset the reduction in costs of all other server hardware.

In a February 1995 consolidated server report, Gartner Group reports that server consolidation will reduce capital budgets by only 7.1 percent while the more traditional hidden costs (management and administration, tuning, monitoring, upgrading, etc.) will reduce by almost 44 percent.

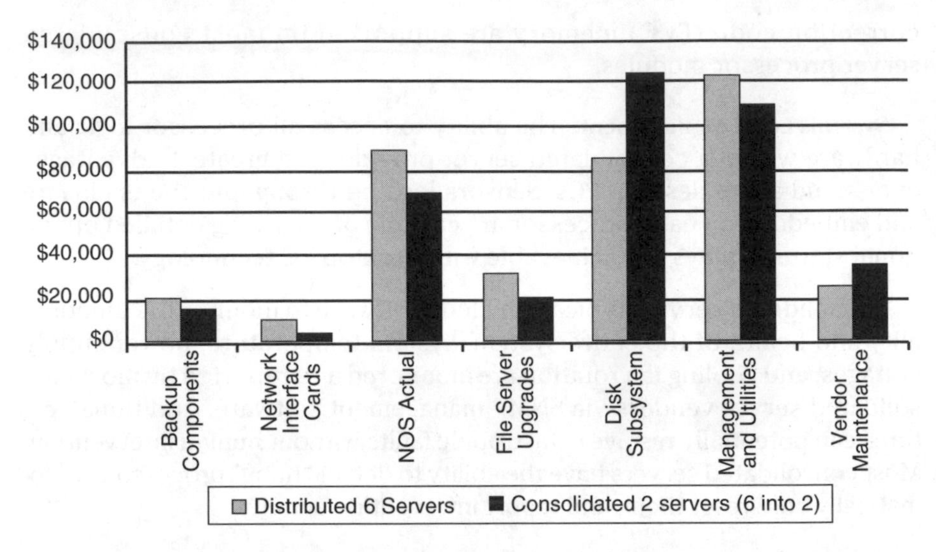

Exhibit 7.3 Overall component costs for two environments over a three-year time line.

Exhibit 7.3 reflects overall component costs for two environments over a three-year time line.

Hidden Costs — Distributed vs. Consolidated Servers

The problems associated with the deployment of standard PCs as specialty servers become apparent in medium and large organizations. Administration, maintenance, and troubleshooting equate to significant hidden costs. Exhibit 7.4 shows those costs that could be quantified in the same February 1995 Gartner Group report. Unknown, but expectedly substantial, costs are lost productivity costs resulting from interrupts and unplanned downtime while distributed servers are being serviced. The numbers represent costs over a three-year timeline.

Larger Incremental Costs with Every Distributed PC

Leading industry analysts have concluded that even minor movement toward a consolidated server architecture can substantially reduce operational costs and increase flexibility.

Research from IDC, Gartner Group, and corporate IS management shows that maintenance and support costs are not linear functions against the number of distributed servers in the computer room. Costs increase incrementally as the number of servers increases. In other words, the costs to

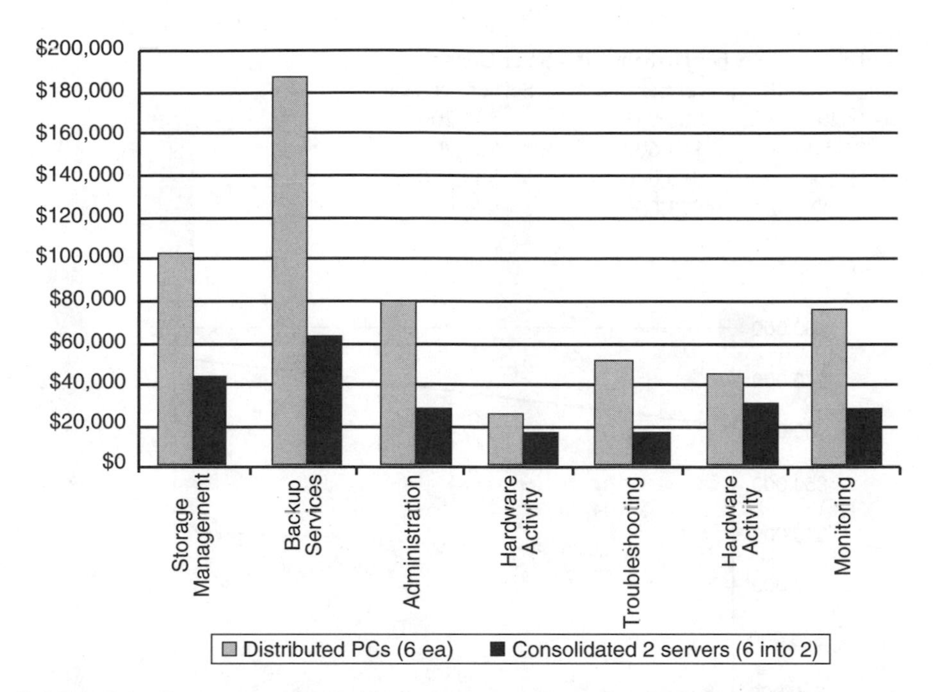

Exhibit 7.4 Costs associated with deployment of standard PCs as specialty servers.

administer an additional PC is greater than the one preceding it. The numbers in Exhibit 7.5 have been compiled independently through analyst research information and end user interviews.

What To Do Next

Assuming a network manager is sufficiently compelled to consider consolidated server technology, the next step is not to destroy the existing array of independent, distributed desktop computers and replace them with consolidated servers. An effective migration to consolidated server technology should simply mean that no more desktop computer enclosures are added to the computer room. Use consolidated server technology for future network expansion. Replace existing desktop computer platforms only when/if more power, increased management, etc. are required. In no time at all, however, network managers will find themselves replacing existing distributed desktop computers for the more robust, more manageable, modular consolidated server elements. The primary benefit of consolidated servers is their ability to scale well. The recommendation is to begin slowly and work up from there, accelerating the process as confidence with the technology increases.

# of Servers	$ per Distributed server per year	$ per Consolidated Server per year
<20	$60,833	$34,200
21-50	$64,600	$36,700
51-100	$69,000	$38,600
>100	$73,000	$40,000

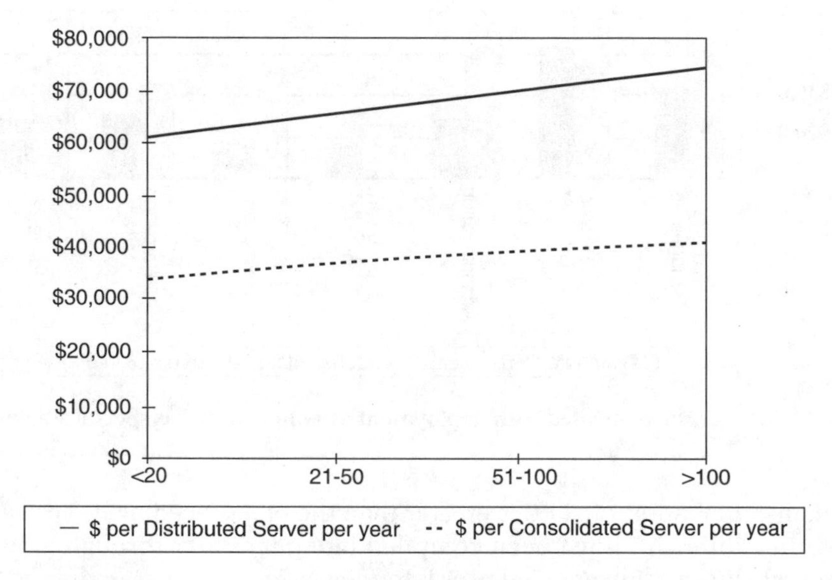

Exhibit 7.5 Costs for distributed servers vs. consolidated servers.

Plan for Expansion! Consider the Big Picture

A good question is to ask whether an existing computer room could adequately support the cooling and AC current distribution requirements of 10 or 20 additional distributed desktop PCs. Many of today's distributed desktop computer installations got to their current state one piece at a time and would be incapable of supporting the onslaught of a large number of computers. Rarely does a network administrator install dozens of independent, distributed computers all at once, and the state of the computer room shows this fragmented or sporadic growth.

Since consolidated server platforms can take on lives of their own, it's important to plan for sometimes rapid growth. Projects often grow beyond a network manager's expectations and can easily get twice as large in half the time anticipated. Consolidated server installations begin with at least half a dozen computers. Premises wiring, air conditioning and cooling, and

AC current capacity considerations are absolutely necessary prior to installation. Consolidated server modules may be more compact, but they still generate the same amount of heat and require the same cabling.

Purchasing and Approval Considerations

Although consolidated server technology may be functionally identical to its distributed desktop computer counterparts, its real advantages are total systems management and the associated long-term cost savings. This is an important consideration as network managers requisition this new technology, as it will initially carry a higher short-term price tag. This is a result of the managed, fault-tolerant enclosure into which network managers place the computer modules. Only one enclosure is required for many computer modules, which usually retail for between $4,000 and $6,000. Beyond the enclosure, prices of individual computer modules are basically at par with their desktop computer counterparts. All told, it's not unusual to see investments of above $100,000 for a full-blown consolidated server solution. These platforms usually include upwards of 50 consolidated server modules.

One of the significant benefits of distributed desktop computers is their availability from about a billion different vendors. A network manager's reliance on any vendor's desktop computer is virtually nil. Consolidated servers, on the other hand, can lock an IS department into a single vendor for future upgrades and enhancements. Since vendors' architectures vary considerably, it is impossible to interchange vendor computing modules. The advice, therefore, is to consider a consolidated server platform that minimizes the use of proprietary hardware (that is, standard motherboard technology, standard plug-in CPU card architectures, etc.). Beyond the hardware platform, the process of integrating software and third-party cards should be straightforward and identical to their distributed desktop counterparts.

Consolidated servers require a systems level architect mentality. Network managers must consider how the addition of a consolidated server platform is going to impact the existing infrastructure as well as those of other departments or branch offices. Further, network managers must consider how they will be able to show the cost- and time-saving advantages of this new platform. The best advice for those who understand the advantages of consolidated server technology, but are reluctant to move forward with this new technology, is to consult with end users and other systems-level architects on issues pertinent to their installation. Consolidated server vendors should be willing to provide lists of customers with which to consult. In the long run the consolidated server solution will be the most cost-effective solution for many environments.

John Lillywhite, currently Director of Marketing at ChatCom, Inc., has over 20 years technical and marketing experience in the computer, communications, and telephone industry. Prior to ChatCom, Lillywhite held executive marketing positions at CommVision and Cubix Corporation, both of which manufacture competitive products to ChatCom. This puts him in a highly regarded position of understanding consolidated server industry dynamics better than most. In the last half of the 1980s Lillywhite held engineering positions within AST Research, where he developed its mainframe and LAN connectivity products. Prior to AST Research, Lillywhite was among the staff at CXC and Anderson Jacobson who developed the first enterprise-wide voice data PBX switch.

Chapter 8
Server Installation Considerations

Howard Marks

If you have one server or a multiacre server farm, the issues you must face in protecting your server(s), and in reality protecting your data, are the same. The solutions are, however, somewhat different. This chapter, while exploring issues applicable to even the smallest installation, is designed to help those with growing server farms.

Too many information systems managers spend too many weeks evaluating the technical fine points of their new multiprocessor, RAID equipped servers, and less than 15 minutes thinking about where they're going to put it. We've seen servers under a secretary's or network manager's desk, in the back of a coat closet, and even hidden on a shelf in the janitor's closet.

Just in case you hadn't figured it out, none of these are really suitable places for servers. Each fails to provide a server with the few things it really needs to run long and reliably.

Placing a server under a users' desk exposes the server to:

- Dust
- Temperature swings during nights and weekends
- Static due to low humidity in winter
- Power loss due to wire kicking, cleaning staff, etc.
- Power disturbance due to vacuum, hot-pot, etc.
- Spilled lunches
- Theft
- Accidental turn off

The back of a coat closet adds:

- Poor air flow
- Lint
- Water exposure from wet coats

0-8493-9823-1/00/$0.00+$.50

The back of a janitorial closet adds the possibility of cleaners spilling into the case.

Servers need plenty of care if they are to perform as expected. Some of these requirements are

- Assured supply of clean power
- Adequate connection to the network
- A free flow of air of reasonable temperature and humidity
- Protection from static electric shocks
- Protection from fire and smoke
- Protection from curious, inept, and (or) malicious hands

Servers that don't get all of these things suffer from premature hardware failures, random reboots, and scrambled data. If you're having difficulty wrangling office real estate for a server room from management, you can always strike the fear of God into them by whispering the word "hacker."

Leaving servers and their keyboards exposed to any casual passer-by also exposes all your company's confidential data. Don't let the argument "no one here would do that" delude you into complacency. Destruction isn't always predicated by malicious intent. The other popular argument for not giving your servers a room of their own is that "no one even knows how to hack here — we can barely manage to use our email system." This is very weak indeed, as any child knows that rebooting the system may get them a screen they understand.

PROVIDING CLEAN POWER

Most network designers recognize the need for power protection early. They know that even if the power from their local utility company is 99.98 percent reliable like the utility company claims, the power will be out on average for 105 minutes a year.

The first thing most IS departments do is make sure that their servers are always provided with an uninterruptible power supply. Many, however, work under the simplistic delusion that simply plugging their server into a UPS is a permanent solution to all their power problems. Unfortunately for them, there is more to protecting a server from power problems than just going down to Akbar and Jeff's Computer Hut and buying a UPS at random.

While any UPS will protect your server from a blackout, only a small percentage of power problems that can cause problems for your server are total power failures. In addition to power outages, you want to protect your servers from:

- Spikes and surges
- Short to very short increases in voltage; can be caused by lighting, large motors stopping, etc.
- Sags and brownouts
- Short- to long-term low voltage; can be caused by copiers or other devices sucking large quantities of juice from the power lines
- Wave form defects caused by other computers, industrial equipment, etc.
- Noise caused by office and industrial equipment, distant lighting, etc.
- Frequency variations
- Unusual fluctuction in utility power; commonly found in backup generator power as load changes, especially at startup

It's the power problems you can't see, when the lights are on, that can wreak havoc on your server and consequently your data.

TYPES OF UPSES

There are three types of devices commonly sold in the server market as an uninterruptible power supply. The simplest is a backup power supply. In the simplest case, a backup UPS supplies line power to your server unless the line voltage falls below some predetermined level. When this happens, the backup power supply switches to battery power. This very basic device really protects your server only against blackouts. Most commercial backup power supplies also include surge suppressors and noise filers improving their overall performance.

A step up from a backup system is a line interactive UPS. These systems add voltage regulation to the backup technology, bucking, or boosting, the output voltage as the input voltage varies. If you're in an area that suffers from summer utility blackouts or brownouts, you should get at least a line interactive UPS.

On-line UPSes run your server(s) on battery power under normal conditions, switching to line power only in the event of a problem with the battery system. On-line UPSes provide protection against all power problems and are the products of choice if you have especially dirty power like an industrial plant or rural site at the end of a long power line. On the down side, on-line UPSes cost more to purchase and use more power than backup or line interactive devices.

Room vs. Individual UPSes

This is an easy decision for those of you who have one or two PC servers. Just get a stand-alone UPS for each server. If, however, you have a server farm and are planning a server room, there are several advantages to using a single large UPS:

- Greater electrical efficiency
- Manual power transfer switch for maintenance
- Lower cost for same technology and capacity
- Battery expansion options for extended run time
- Automatic battery testing

Beyond the UPS

Most users specify a UPS to provide roughly 15 minutes of backup power to allow a server to perform an orderly shutdown. In order for this to be effective, you need a connection between the server and the UPS so the server can know that the line power has failed. The server operating system can then perform an orderly shutdown.

If, however, you need to be able to not just shut the server down cleanly but need to continue operating, you'll need a backup generator using natural gas, propane, or diesel fuel. Backup generators should always be equipped with a manual test transfer switch so you can start it up once a month to make sure it's still working. You should also use an on-line UPS between the backup generator and your server farm to clean up the generator's output before it gets to your delicate server. Backup generators are notorious for providing "dirty," or uneven, power that servers just can't handle.

CONNECTING TO THE NETWORK

In most networks, 70 to 90 percent of the network traffic is to or from one or more servers. This means that the server's connection to the network is a potential bottleneck. While the detailed design of a high performance network is beyond the scope of this chapter, common practice today is to provide 100-Mbps connections to servers. Obviously, that connection cable should be well marked and very well connected — locked down if possible, with backup cables at the ready if the primary fails.

You'll need to check the fire codes for your area and make sure that any cable that runs through ceiling ducts or under raised floors complies. This generally means that you need to use Plenum cable in air ducts like ceilings and raised floors. Non-Plenum cable releases poisonous gas when it catches fire.

It will also be useful to your sanity to color code your cabling. The simplest way to do this is to order cable with colored jackets. Then document which color cable is for which connection.

ENVIRONMENTAL CONTROL

While today's servers, even UNIX-based midrange servers, are much less picky about temperature and humidity than the water-cooled mainframes of the 1950s and 1960s, they do need somewhat more care than the average office environment provides. The typical office is heated and cooled to make people comfortable during the hours that people are working. A window office in a modern skyscraper can frequently reach over 100 degrees on a summer Sunday afternoon and 50 degrees in the wee hours of a winter's morning.

Even top-quality servers don't like that kind of treatment. At a bare minimum you should make sure that your server room has 24-hour heat and air conditioning sufficient to remove all the heat generated by the equipment in the room. If your servers are in a modern building with central air, you'll need to add a supplemental air conditioner in the server room.

Sophisticated computer rooms use specialized environmental control systems, and if you have more than $250,000 invested in server hardware, you should, too. These systems, built by vendors like Liebert, not only cool, and as a by-product dehumidify, the air like standard air conditioners but also heat and humidify if needed. In addition, computer room air conditioning systems have alarms to let you know when the temperature or humidity is outside the acceptable range and can even send an SNMP alert to your network management console so your operators two floors, or 500 miles, away will know something is wrong before systems start to fail due to heat prostration.

We all know that running computers in a rain-forest-like 100-percent humidity environment is a bad idea as contacts will corrode, condensation inside the computer will cause short circuits, and the system will generally just rot. The worst case we've seen was a computer system with mold growing on its power supply, which caused an arc and then started a fire.

Low humidity is also an issue as it leads to static problems. When we think of static, most of us, especially those from northern climes, think of it as a problem of users walking across a carpeted area and discharging themselves into a computer system, giving it the high-tech version of electroshock therapy.

The best cure for this problem in the office environment is the strategic use of grounded, conductive surfaces to draw the static electric charge from people before they can contact a computer. Floor mats, conductive carpets, and desk pads, all with wires leading to an electrical ground, are the tools of the trade. In the computer room we want a grounded floor with a hard, conductive surface. Carpet is to be avoided at all costs because it is difficult to keep clean and vacuuming spreads dust through the room.

FIRE PROTECTION

Like any valuable asset, computers, especially large concentrations of computers like a server farm, need to be protected from the risk of fire. The problem is that the cheapest and most common fire protection system, the water sprinkler, is itself dangerous to computer systems. Since water sprinkler systems occasionally leak, or trigger inappropriately, using water sprinklers in a computer room is a bad idea.

The traditional solution was to equip computer rooms with a Halon® gas fire suppression system. Halon is a very effective fire extinguishing agent and relatively nontoxic. Unfortunately, it is a chlorofluorocarbon, closely related to Freon, which is quite harmful to the ozone layer and no longer commercially available.

Current computer room fire systems use carbon dioxide, nitrogen, or proprietary gas mixtures like Fenwal's FM2000.

TO RACK OR NOT TO RACK

As your server farm grows beyond five or six servers, you have to start thinking about how you're going to manage to house all of your servers. There are basically three common solutions.

The first is to pile multiple servers on tables, steel shelves, and (or) baker's racks. While initially cost effective, this quickly becomes difficult to manage, leads to complicated and often tangled cable configurations, and is hard to maintain.

The second choice is to use standard tower or desktop cabinet servers and install them in storage racks that have built-in power distribution bars, slide-out keyboard drawers, slide-out server shelves and optional keyboard, video, and mouse switches so you can use a single keyboard, monitor, and mouse to control 4 to 16 servers. These systems from vendors like Wright Line and Hergo make for a neater, higher-density installation.

If you have a large number of servers or need to expand and maintain servers in place, the best solution is to use 19"-wide industry standard rack mount servers and racks. These racks, used in the electronics industry for decades, allow servers to be mounted on ball bearing pull-out slides so you can add cards without moving the server and unplugging all its cables. A single rack taking just 3 to 4 square feet of floor space can hold four or five servers, a UPS, and all the other associated hardware. Rack mount servers cost a little more but are well worth it for a large, highly reliable environment.

THE DOOR

Having wrested space for a server room by pointing out the security risks, you can't neglect securing the door. Obviously, not every Tom, Dick, and

Harriet should be allowed unlimited access to the server room. At the same time, you will need to get in occasionally to troubleshoot and perform regular maintenance. It never fails to amaze us that some companies require you to have a key or combination code for the bathroom but not for the server room.

The most common security choices are keys, combination locks, and key cards. Keys are the least practical choice because they get lost. When someone leaves the company either by choice, design, or request, the key may not be recovered and calling a locksmith to regain access is an unwieldy process.

Combinations are slightly better because they can be more easily changed in case of personnel changes and they can't get lost. Card systems are best, however, for a couple of reasons. The first is that individual cards can be selectively deactivated when someone leaves the company. The second is because you can restrict cards to certain hours or days for additional security. The third is that they allow you an audit trail of which card, and when, provided access to the server room. All the data for key cards are, after all, tracked by computer. If you have a serious security breach you can cross-reference the time of the breach with the access card.

CONCLUSION

As the information processing industry shifts from mainframes to distributed systems, we shouldn't forget the lessons learned in the past about protecting system integrity. Servers, while they have many similarities to desktop PCs, are valuable resources used by many that need greater protection. If you have a large investment in server resources, they need to be provided with clean, cool, well-powered, and secure environments to work in.

Chapter 9
The Business of Managing Distributed Systems

Richard Ross

Many of the top concerns of managers in IS departments relate directly to the issues of distributing information technology to end users. The explosive rate at which information technology has found its way into the front office, combined with the lack of control by the IS organization (ostensibly the group chartered with managing the corporation's IT investment), has left many IS managers at a loss as to how they should respond. The following issues are of special concern:

- Where should increasingly scarce people and monetary resources be invested?
- What skills will be required to implement and support the new environment?
- How fast should the transition from a centralized computing environment to a distributed computing environment occur?
- What will be the long-term impact of actions taken today to meet short-term needs?
- What will be the overall ability of the central IS group to deliver new standards of service created by changing user expectations in a distributed computing environment?

In large companies during the past decade, the rule of thumb for technology investment has been that the opportunity cost to the business unit of not being able to respond to market needs will always outweigh the savings accruing from constraining technology deployment. This has resulted in a plethora of diverse and incomparable systems, often supported by independent IS organizations.

0-8493-9823-1/00/$0.00+$.50

In turn, these developments have brought to light another, even greater, risk — that the opportunity cost to the corporation of not being able to act as a single entity will outweigh the benefit of local flexibility.

For example, a global retailer faces conflicts if it has sales and marketing organizations in many countries. To meet local market needs, each country has its own management structure with independent manufacturing, distribution, and systems organizations. The result might be that the company's supply chain becomes clogged — raw materials sit in warehouses in one country while factories in another go idle; finished goods pile up in one country while store shelves are empty in others; costs rise as the number of basic patterns proliferate. Perhaps most important, the incompatibility of the systems may prevent management from gaining an understanding of the problem and from being able to pull it all together at the points of maximum leverage while leaving the marketing and sales functions a degree of freedom.

Another example comes from a financial service firm. The rush to place technology into the hands of traders has resulted in a total inability to effectively manage risk across the firm or to perform single-point client service or multiproduct portfolio management.

WANTED — A NEW FRAMEWORK FOR MANAGING

A distributed computing environment cannot be managed according to the lessons learned during the last 20 years of centralized computing. The distributed computing environment is largely a result of the loss of control by the central IS group because of its inability to deliver appropriate levels of service to the business units. Arguments about the ever-declining cost of desktop technology are well and good, but the fact of the matter is that managing and digesting technology is not the job function of users. If central IS could have met their needs, it is possible users would have been more inclined to forgo managing their own systems.

It is not just the technology that is at fault. Centralized computing skills themselves are not fully applicable to a distributed computing environment. The underlying factors governing risk, cost, and quality of service have changed. IS departments need a new framework, one that helps them to balance the opportunity cost to the business unit against that to the company, while optimizing overall service delivery.

DEFINING THE PROBLEM: A MODEL FOR DCE SERVICE DELIVERY

To help IS managers get a grip on the problem, this chapter proposes a model of service delivery for the distributed computing environment (DCE). This model focuses on three factors that have the most important influence on service as well as on the needs of the business units vs. the corporation — risk, cost, and quality (see Exhibit 9.1). Each factor is analyzed to understand its

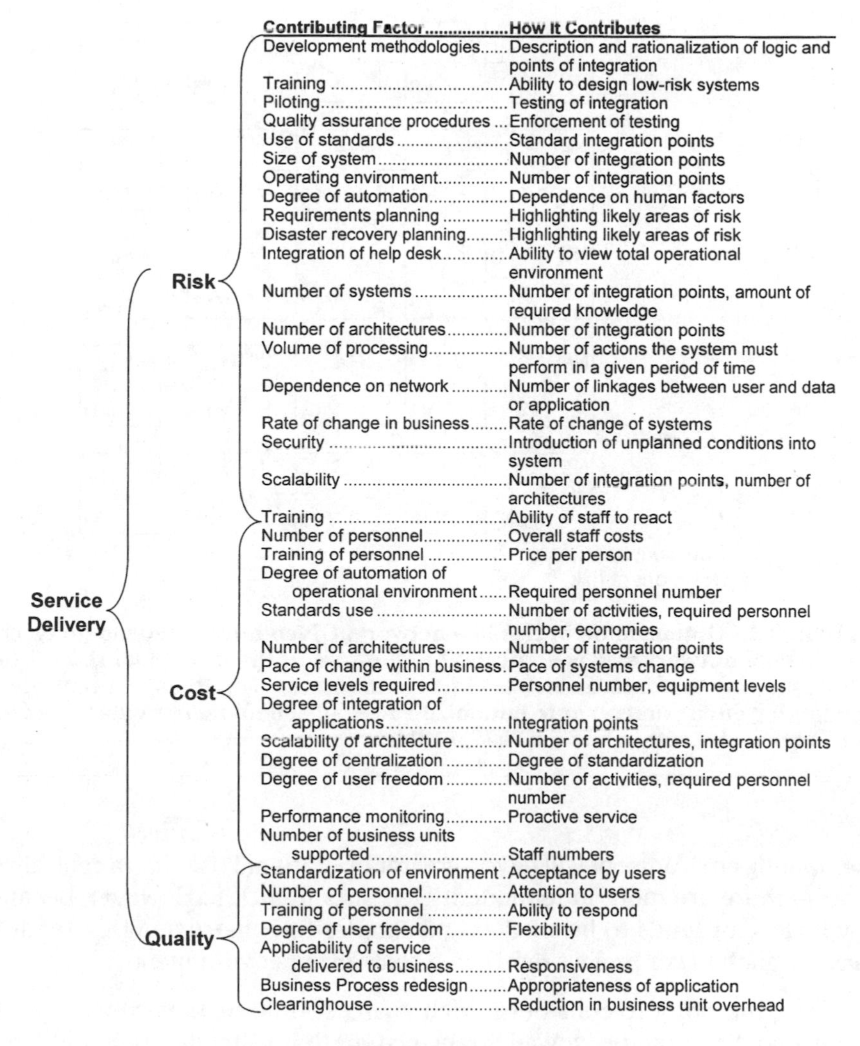

Contributing Factor.................How It Contributes
Service Delivery
Risk
Development methodologies......Description and rationalization of logic and points of integration
Training ....Ability to design low-risk systems
Piloting.......................................Testing of integration
Quality assurance procedures ...Enforcement of testing
Use of standards ....Standard integration points
Size of system ....Number of integration points
Operating environment...............Number of integration points
Degree of automation.................Dependence on human factors
Requirements planning ....Highlighting likely areas of risk
Disaster recovery planning.........Highlighting likely areas of risk
Integration of help desk..............Ability to view total operational environment
Number of systems ....Number of integration points, amount of required knowledge
Number of architectures.............Number of integration points
Volume of processing.................Number of actions the system must perform in a given period of time
Dependence on network ....Number of linkages between user and data or application
Rate of change in business........Rate of change of systems
Security ....Introduction of unplanned conditions into system
Scalability ....Number of integration points, number of architectures
Training ....Ability of staff to react
Cost
Number of personnel..................Overall staff costs
Training of personnel ....Price per person
Degree of automation of operational environment......Required personnel number
Standards use ....Number of activities, required personnel number, economies
Number of architectures.............Number of integration points
Pace of change within business.Pace of systems change
Service levels required...............Personnel number, equipment levels
Degree of integration of applications.........................Integration points
Scalability of architecture ....Number of architectures, integration points
Degree of centralization ....Degree of standardization
Degree of user freedom ....Number of activities, required personnel number
Performance monitoring.............Proactive service
Number of business units supported.............................Staff numbers
Quality
Standardization of environment .Acceptance by users
Number of personnel..................Attention to users
Training of personnel ....Ability to respond
Degree of user freedom ....Flexibility
Applicability of service delivered to business...........Responsiveness
Business Process redesign........Appropriateness of application
Clearinghouse ....Reduction in business unit overhead

Exhibit 9.1 A model of service delivery.

cause and then to determine how best to reduce it (in the case of risk and cost) or increase it (as in quality).

Risk

Risk in any systems architecture is due primarily to the number of independent elements in the architecture (see Exhibit 9.2). Each element carries its own risk, say for failure, and this is compounded by the risk associated with the interface between each element. This is the reason that a distributed

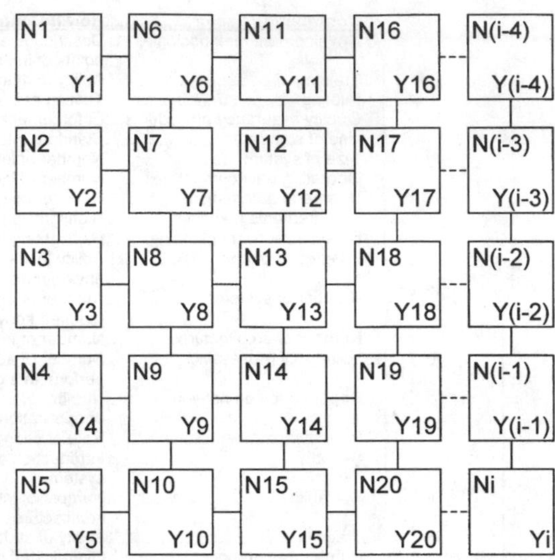

N = Component number
Y = Component risk

Exhibit 9.2 Optimization of risk in a network. Given fully independent components, total network risk is equivalent to the sum of the individual risks, 1 to i. Thus, the way to minimize risk is either to minimize (i.e., to have a centralized computing environment) or to minimize Y for each component by standardizing on components with minimum risk profiles.

---

computing environment will have a greater operational risk than a centralized one — there are more independent elements in a DCE. However, because each element tends to be smaller and simpler to construct, a DCE tends to have a much lower project risk than a centralized environment.

Thus, the point to consider in rightsizing should be how soon a system is needed. For example, a Wall Street system that is needed right away and has a useful competitive life of only a few years would be best built in a distributed computing environment to ensure that it gets on-line quickly. Conversely, a manufacturing system that is not needed right away but will remain in service for years is probably better suited for centralization.

One other difference between a distributed environment and a centralized environment is the impact of a particular risk. Even though a DCE is much more likely to have a system component failure, each component controls such a small portion of the overall system that the potential impact of any one failure is greatly reduced. This is important to take into account when performing disaster planning for the new environment.

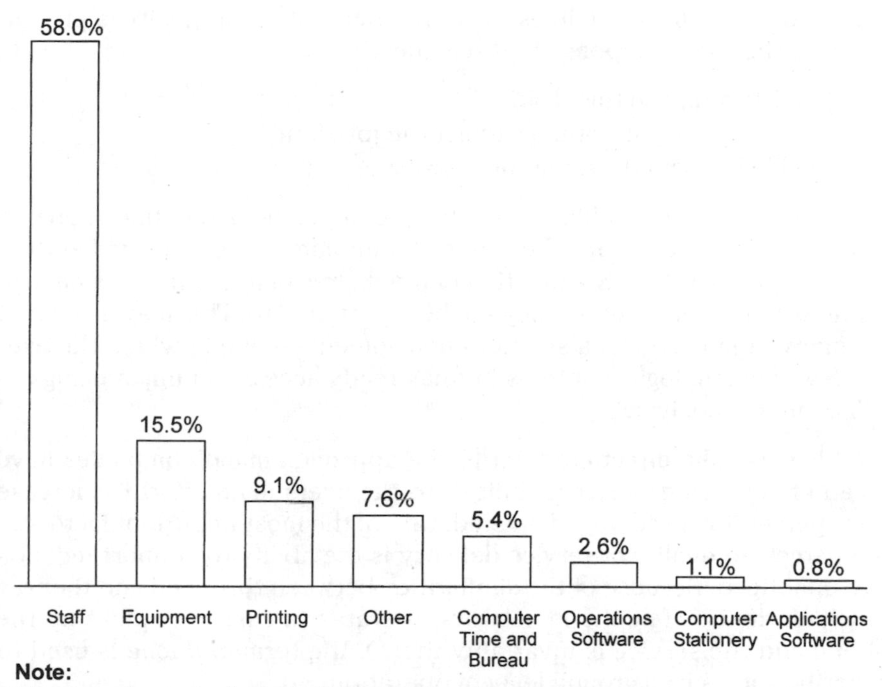

Exhibit 9.3 Cost profile of the IS function.

Cost

Cost is largely a function of staff levels (see Exhibit 9.3). As the need for services increases, the number of staff members invariably increases as well. People are flexible and can provide a level of service far beyond that of automation. Particularly in a dynamic environment, in which the needs for response are ill defined and can change from moment to moment, people are the only solution.

Unfortunately, staff is usually viewed as a variable cost, to be cut when the need for budget reductions arises. This results in a decrease in service delivered that is often disproportionately larger than the savings incurred through staff reductions.

Perceived Quality

Quality is a subjective judgment, impossible to quantify but the factor most directly related to the users' perception of service where information technology is concerned. In essence, the perception of quality is proportional to the users' response to three questions:

- Can I accomplish this task?
- Am I able to try new things to get the job done?
- Am I being paid the attention I deserve?

Except for the sheer ability to get the job done, the perception of quality is not necessarily a factor of how much technology a user is provided with. Instead, quality is largely a function of the degree of freedom users have to try new things and whether they are being listened to. This may mean that for many organizations, a simpler environment — one in which the user has fewer technological options but has ready access to human beings — will be more satisfying.

This view is in direct contrast to the approach many companies have taken of delivering more capabilities to the users in an effort to increase their perception of service delivered. One of the most important factors in the perceived quality of service delivery is the ability to support technology unnoticed. Because of the similarities between this need and the way in which the U.S. telephone network operates (a customer picks up the phone and the service is invariably there), the term *dial tone* is used to describe such a background level of operation.

"Dial Tone" IS Service. One problem with highly functional IS environments is that users must think about them to use them. This is not the case with the telephone system, which operates so dependably that people and businesses have integrated it into their routine working practices and use it without much conscious effort. The phone companies maintain this level of usefulness by clearly separating additional features from basic service and letting the customer add each new feature as the customer desires.

Contrast this with the typical business system that represents an attempt to deliver a package of functions on day one and to continually increase its delivered functionality. The impact on users is that they are forced to continually adapt to changes, are not allowed to merge the use of the system into the background, and must continually stop delivering on their jobs just to cope with the technology. This coping might be as simple as looking something up in a manual or changing a printer cartridge, or it may mean not working at all while the system is rebooted.

COMPLEXITY: THE BARRIER TO SERVICE DELIVERY

In general, the basic driver to each of the three service factors is complexity. Complexity increases risk by increasing the number of interfaces between system elements as well as the number of elements themselves. It increases cost by increasing the need for staff as the only way to deal with ill-defined environments. Finally, it affects quality by making it harder to provide those services that users base their perception of quality on (i.e., dial tone and personal attention), in response to which even more staff are added.

This is the paradoxical environment in which IS operates. To improve the quality of service, they find themselves increasing the risk and cost of the operation. Improved application delivery cycles result in more systems to manage. End-user development tools and business unit-led development increase the number of architectures and data formats. Increasing access to corporate data through networks increases the number of interfaces.

Conversely, trying to improve the risk and cost aspects, typically through standardization of the environment, usually results in decreased levels of service delivered because of the constraints placed on user freedom. This paradox did not exist in the good old days of centralized computing, when the IS organization dictated the service level.

FIVE PIECES OF ADVICE FOR MANAGING DISTRIBUTED COMPUTING

The measure of success in a distributed computing environment is therefore the ability to deliver service through optimizing for the factors of risk, cost, and quality while meeting the needs of both the business units and the corporation. It sounds like a tall order, but it is not impossible. There are five key practices involved in corporate information processing:

- Manage tightly, but control loosely.
- Organize to provide service on three levels.
- Choose one good standard — even a single bad one is better than none or many good ones.
- Integrate data at the front end — do not homogenize on the back end.
- Minimize the use of predetermined architectures.

Manage Tightly, Control Loosely

A distributed computing environment should be flexible and allow users to be independent, but at the same time it should be based on a solid foundation of rules and training and backed up with timely and appropriate levels of support.

In the distributed computing environment, users must react to the best of their abilities to events moment by moment. IS can support users in a way that allows them to react appropriately, or can make them stop and call for a different type of service while the customer gets more and more frustrated.

BPR and Metrics. Two keys to enabling distributed management are business process redesign (BPR) and metrics. BPR gets the business system working first, highlights the critical areas requiring support, builds consensus between the users and the IS organization as to the required level of support, and reduces the sheer number of variables that must be managed at any one time. In essence, applying BPR first allows a company to step back and get used to the new environment.

Without a good set of metrics, there is no way to tell how effective IS management has been or where effort needs to be applied moment to moment. The metrics required to manage a distributed computing environment are different from those IS is used to. With central computing, IS basically accepted that it would be unable to determine the actual support delivered to any one business. Because centralized computing environments are so large and take so long to implement, their costs and performance are spread over many functions. For this reason, indirect measurements were adopted when speaking of central systems, measures such as availability and throughput.

But these indirect measurements do not tell the real story of how much benefit a business might derive from its investment in a system. With distributed computing, it is possible to allocate expenses and effort not only to a given business unit but to an individual business function as well. IS must take advantage of this capability by moving away from the old measurements of computing performance and refocusing on business metrics, such as return on investment.

In essence, managing a distributed computing environment means balancing the need of the business units to operate independently while protecting corporate synergy. Metrics can help do this by providing a clear basis for discussion. Appropriately applied standards can help, too. For example, studies show that in a large system, as much as 90 percent of the code can be reused. This means that a business unit could reduce its coding effort by some theoretical amount up to 90 percent — a forceful incentive to comply with such standards as code libraries and object-oriented programming.

Pricing. For those users who remain recalcitrant in the face of productivity gains, there remains the use of pricing to influence behavior. But using pricing to justify the existence of an internal supplier is useless. Instead, pricing should be used as a tool to encourage users to indulge in behavior that supports the strategic direction of the company.

For example, an organization used to allow any word processing package that the users desired. It then reduced the number of packages it would support to two, but still allowed the use of any package. This resulted in an incurred cost to the IS organization due to help desk calls and training problems. The organization eventually settled on one package as a standard, gave it free to all users, and eliminated support for any other package. The acceptance of this standard package by users was high, reducing help calls and the need for human intervention. Moreover, the company was able to negotiate an 80 percent discount over the street price from the vendor, further reducing the cost.

This was clearly a benefit to everyone concerned, even if it did not add up from a transfer pricing point of view. That is, individual business units may have suffered from a constrained choice of word processor or may have had to pay more, even with the discount, but overall the corporation did better than it otherwise would have. In addition to achieving a significant cost savings, the company was able to drastically reduce the complexity of its office automation environment, thus allowing it to deliver better levels of service.

Organize to Provide Service on Three Levels

The historical IS shop exists as a single organization to provide service to all users. Very large or progressive companies have developed a two-dimensional delivery system: part of the organization delivers business-focused service (particularly applications development), and the rest acts as a generic utility.

Distributed computing environments require a three-dimensional service delivery organization. These services are

- Dial tone, or overseeing the technology infrastructure
- Business-focused or value-added services, or ensuring that the available technology resources are delivered and used in a way that maximizes the benefit to the business unit
- Service that maximizes leverage between each business unit and the corporation

Dial tone IS services lend themselves to automation and outsourcing. They are complex, to the degree that they cannot be well managed or maintained by human activity alone. They must be stable, as this is the need of users of these services. In addition, they are nonstrategic to the business and lend themselves to economies of scale and hence are susceptible to outsourcing (see Exhibits 9.4 and 9.5).

Value-added services should occur at the operations as well as at the development level. For example, business unit managers are responsible for overseeing the development of applications and really understanding the business. This concept should be extended to operational areas, such as training, maintenance, and the help desk. When these resources are placed in the business unit, they will be better positioned to work with the users to support their business instead of making the users take time out to deal with the technology.

Common Operational Problem	Responsiveness to Automation
Equipment hangs	●
Network contention	◑
Software upgrades	●
Equipment upgrades	○
Disaster recovery	●
Backups	●
Quality assurance of new applications	●
Equipment faults (e.g., print cartridge replacement, disk crash)	○
Operator error (e.g., forgotten password, kick out plug)	○
Operator error (e.g., not understanding how to work application)	●

Responsiveness
High ●
Medium ◑
Low ○

SOURCE: Interviews and Decision Strategies Group analysis

Exhibit 9.4 Responsiveness of operations to automation.

People do not learn in batches. They learn incrementally and retain information pertinent to the job at hand. This is why users need someone to show them what they need to know to accomplish the job at hand. Help desks should be distributed to the business units so that the staff can interact with the users and solve problems proactively, before performance is compromised, and accumulate better information to feed back to the central operation.

The third level of service — providing maximum leverage between the business unit and the corporation — is perhaps the most difficult to maintain and represents the greatest change in the way IS does business today. Currently, the staff members in charge of the activities that leverage across all business units are the most removed from those businesses. Functions such

Dialtone Function	Applicability
Equipment maintenance	●
Trouble calls	●
Help Desk	●
Installations	●
Moves and changes	●
Billing	◑
Accounting	◑
Service level contracting	○
Procurement	○
Management	○

Responsiveness
High ●
Medium ◑
Low ○

SOURCE: Interviews and Decision Strategies Group analysis

Exhibit 9.5 Applicability of outsourcing to dialtone.

as strategic planning, test beds, low-level coding, and code library development tend to be staffed by excellent technical people with little or no business knowledge. IS managers must turn this situation around and recruit senior staff with knowledge of the business functions, business process redesign, and corporate planning. These skills are needed to take the best of each business unit, combine it into a central core, and deliver it back to the businesses.

Choose One Standard

If the key to managing a distributed computing environment is to reduce complexity, then implementing a standard is the thing to do. Moreover, the benefits to be achieved from even a bad standard, if it helps to reduce complexity, will outweigh the risk incurred from possibly picking the wrong standard.

The message is clear: there is more to be gained from taking inappropriate action now than from waiting to take perfect action later.

It should be clear that information technology is moving more and more toward commodity status. The differences between one platform and another will disappear over time. Even if IS picks a truly bad standard, it will likely merge with the winner in the next few years, with little loss of investment. More important, the users are able to get on with their work. In addition, it is easier to move from one standard to the eventual winner than from many.

Even if a company picks the winner, there is no guarantee that it will not suffer a discontinuity. IBM made its customers migrate from the 360 to the 370 architecture. Microsoft is moving from DOS to Windows to Windows NT. UNIX is still trying to decide which version it wants to be. The only thing certain about information technology is the pace of change, so there is little use in waiting for things to quiet down before making a move.

Integrate Data at a Front End

At the very core of a company's survival is the ability to access data as needed. Companies have been trying for decades to find some way to create a single data model that standardizes the way it stores data and thus allows for access by any system.

The truth of the matter is that for any sufficiently large company (i.e., one with more than one product in one market), data standardization is unrealistic. Different market centers track the same data in different ways. Different systems require different data formats. New technologies require data to be stated in new ways. Trying to standardize the storage of data means ignoring these facts of life to an unreasonable extent.

To try to produce a single model of all corporate data is impossible and meaningless. The models invariably grow to be so large that they cannot be implemented. They impose a degree of integration on the data that current systems technology cannot support, rendering the largest relational database inoperative. And they are static, becoming obsolete shortly after implementation, necessitating constant reworking. Moreover, monolithic data models represent unacceptable project risk.

The standardization approach also ignores the fact that businesses have 20 to 30 years' worth of data already. Are they to go back and recreate all this to satisfy future needs? Probably not. Such a project would immobilize the business and the creation of future systems for years to come.

Systems designed to integrate and reconcile data from multiple sources, presenting a single image to the front end, intrinsically support the client/

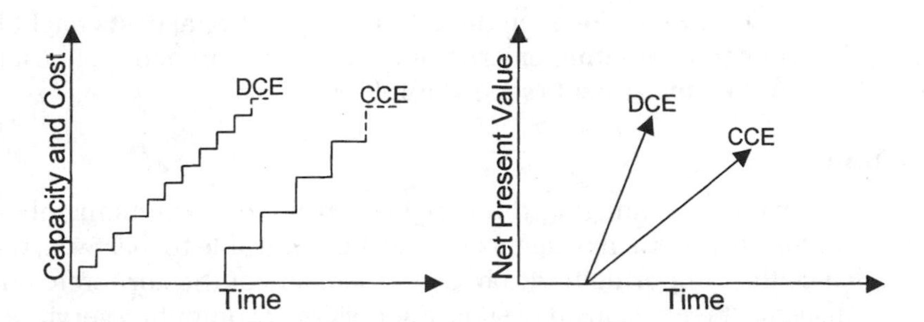

Exhibit 9.6 Net present value of distributed vs. centralized computing. Distributed computing environment (DCE) has a higher net present value because its capacity can be used sooner relative to its marginal costs when compared with the centralized computing environment (CCE). SOURCE: Decision Strategies Group.

server model of distributed computing and build flexibility into future applications. They allow data to be stored in many forms, each optimized for the application at hand. More important, they allow a company to access its data on an as-needed basis. These integration systems are an important component to successfully managing future growth.

Less Architecture Is More

To overdesign a systems architecture is to overly constrain the organization. Most architecture arises as a function of rightsizing of applications on the basis of where the data must be stored and used. Understanding this helps the IS manager size the network and associated support infrastructure.

The management of risk and impact also drives architecture by forcing redundancy of systems and, in some cases, mandating the placement of data repositories regardless of user preferences. Assessing project vs. operational risk helps to determine whether a system is built for central or distributed use.

Economics. This view is one in which the business needs drive the shape of the architecture. It results in a dynamic interconnection of systems that respond flexibly to business needs. Under a centralized computing environment, it was impractical to employ such an approach. It took so long and cost so much to implement a system that investment had to come before the business needs. This necessitated preplanning an architecture as an investment guide.

The economics of distributed computing are different. Systems cost much less and can be quickly implemented. This means that their use can be responsive to business needs instead of anticipatory. It also results in a greater net present value, for even though their operational costs might be higher, distributed computing environments are more immediately useful for a given level of investment (see Exhibit 9.6).

SUMMARY

This framework for managing the new computing environment relates directly to the business. If managers of IS functions are able to master it, they can enhance their opportunities to become members of the corporate business management team instead of simply suppliers of computing services.

Success in a distributed computing environment requires a serious culture shift for IS managers. They must loosen up their management styles, learning to decentralize daily control of operations. They must provide direction to staff members so that they can recognize synergies among business units. Some jobs that were viewed as low-level support activities (e.g., value-added services such as help desk and printer maintenance) must be recognized as key to user productivity and distributed. Others, viewed as senior technical positions (e.g., dial tone functions such as network management and installations), might best be outsourced, freeing scarce IS resources.

Most important, IS must understand the shift in power away from themselves and toward users. The IS organization is no longer the main provider of services; it now must find a role for itself as a manager of synergy, becoming a facilitator to the business units as they learn to manage their own newfound capabilities.

Chapter 10
Including End Users in the Planning of Integrated Systems Development

Jack T. Marchewka

Despite continued improvements in information technology, information systems still experience serious problems and even failures because of "people problems" associated with the design and use of application systems. Symptoms of such problems and failures include nonacceptance by the users, cost/schedule overruns, high maintenance costs, and opportunity costs associated with unrealized benefits. This chapter offers suggestions on how to mitigate these problems by improving mutual understanding between end users and systems developers.

INTRODUCTION

The traditional approach to information systems development (ISD) assumes that the process is both rational and systematic. Developers are expected to analyze a set of well-defined organizational problems and then develop and implement an information system (IS). This, however, is not always the case.

The full extent of IS problems and failures may not be known, as most organizations are less than willing to report these problems for competitive reasons. However, a report by the Index Group indicates that 25 of 30 strategic system implementations studied were deemed failures, with only five systems meeting their intended objectives. Moreover, it has been suggested that at least half of all IS projects do not meet their original objectives.[1]

0-8493-9823-1/00/$0.00+$.50

Previous research suggests that a lack of cooperation and ineffective communication between end users and system developers are underlying reasons for these IS problems and failures. Typically, the user — an expert in some area of the organization — is inexperienced in integrated systems development, while the developer, who is generally a skilled technician, is unacquainted with the rules and policies of the business. In addition, these individuals have different backgrounds, attitudes, perceptions, values, and knowledge bases. These differences may be so fundamental that each party perceives the other as speaking a foreign language. Consequently, users and developers experience a communication gap, which is a major reason why information requirements are not properly defined and implemented in the information system.

Furthermore, differences in goals contribute to a breakdown of cooperation between the two groups. For example, the user is more interested in how information technology can solve a particular business problem, whereas the system developer is more interested in the technical elegance of the application system.

On the other hand, users attempt to increase application system functionality by asking for changes to the system or for additional features that were not defined in the original requirements specifications. However, the developer may be under pressure to limit such functionality to minimize development costs or to ensure that the project remains on schedule.

Subsequently, users and developers perceive each other as being uncooperative, and integrated systems development becomes an "us vs. them" situation. This leads to communication problems that inhibit the user from learning about the potential uses and benefits of the technology from the developer. The developer, on the other hand, may be limited in learning about the users' functional and task requirements. As a result, a system is built that does not fit the users' needs, which, in turn, increases the potential for problems or failure. Participation in the integrated systems development process requires a major investment of the users' time that diverts them from their normal organizational activities and responsibilities. An ineffective use of this time is a waste of an organizational resource that increases the cost of the application system.

The next section examines the conventional wisdom of user involvement. It appears that empirical evidence to support the traditional notion that user involvement leads to IS success is not clear-cut. Subsequently, this section suggests that it is not a question of whether to involve the user but rather a question of how or why the user should be involved in the integrated systems development process. In the next section, a framework for improving cooperation, communication, and mutual understanding is described. Within the context of this framework, the following section provides a classification of user and developer relationships. This is followed by a section that describes how to assess, structure, and monitor the social relationship between users and developers.

USER INVOLVEMENT AND COMMON WISDOM

The idea that user involvement is critical to the successful development of an information system is almost an axiom in practice; however, some attempts to validate this idea scientifically have reported findings to the contrary.[2] Given the potential for communication problems and differences in goals between users and developers, it is not surprising, for example, that a survey of senior systems analysts reported that they did not perceive user involvement as being critical to information systems development.

Moreover, a few studies report very limited effects of user involvement on system success and suggest that the usual relationship between system developers and users could be described as one in which the IS professionals are in charge and users play a more passive role.[3]

There are several reasons why involving the user does not necessarily guarantee the success of an information system. These include:

- If users are given a chance to participate in information systems development, they sometimes try to change the original design in ways that favor their political interests over the political interests of other managers, users, or system developers. Thus, the potential for conflict and communication problems increases.
- Users feel that their involvement lacks any true potential to affect the development of the information system. Consequently, individuals resist change because they feel that they are excluded from the decision-making process.

Despite equivocal results, it is difficult to conceive how an organization could develop a successful information system without any user involvement. It therefore may not be a question whether to involve the user, but how or why the user should be involved. More specifically, there are three basic reasons for involving users in the design process:[4]

- To provide a means to get them to "buy in" and subsequently reduce resistance to change.
- To develop more realistic expectations concerning the information technology's capabilities and limitations.
- To incorporate user knowledge and expertise into the system. Users most likely know their jobs better than anyone else and therefore provide the obvious expertise or knowledge needed for improved system quality.

While the more traditional integrated systems development methods view users as passive sources of information, the user should be viewed as a central actor who participates actively and effectively in system development. Users must learn how the technology can be used to support their processes, whereas the system developer must learn about the business processes in order to develop a system that meets user needs. To learn from each other, users and developers must communicate effectively and develop a mutual understanding between them. This leads to improved definition of system requirements and increased acceptance, as the user and developer co-determine the use and impact of the technology.

COOPERATION, COMMUNICATION, AND MUTUAL UNDERSTANDING

Earlier research suggests that effective communication can improve the integrated systems development process and is an important element in the development of mutual understanding between users and system developers. Mutual understanding provides a sense of purpose to the integrated systems development process. This requires that users and developers perceive themselves as working toward the same goal and being able to understand the intentions and actions of the other.[5]

To improve communication and mutual understanding requires increased cooperation between users and developers. As a result, many of the inherent differences between these individuals are mitigated and the communications gap bridged; however, the balance of influence and their goals affects how they communicate and cooperate when developing information systems.

The Balance of Influence

In systems development, an individual possesses a certain degree of influence over others by having a particular knowledge or expertise. This knowledge or expertise provides the potential to influence those who have lesser knowledge. For example, a system developer uses his or her technical knowledge to influence the design of the information system. If the user has little or no knowledge of the technology, the system developer, by possessing the technical knowledge needed to build the application system, has a high degree of influence over the user. On the other hand, the user has a high degree of influence over the system developer if the user possesses knowledge of the domain needed to build the application system. By carefully employing their knowledge or expertise, the user and the developer cultivate a dependency relationship. Subsequently, the balance of influence between the user and developer determines how these individuals communicate with each other and how each individual tries to influence the other.

Reconciling the Goals Between the User and Developer

Even though users and developers may work for the same organization, they do not always share the same goals. More specifically, the nature of the development process creates situations in which the user and developer have different goals and objectives. For example, the developer may be more interested in making sure that the IS project is completed on time and within budget. Very often the developer has several projects to complete, and cost/schedule overruns on one project may divert precious, finite resources from other projects.

Users, on the other hand, are more interested in the functionality of the system. After all, they must live with it. A competitive situation arises if increasing the system's functionality forces the system to go over schedule or over budget or if staying on schedule or within budget limits the system's functionality.

In 1949, Morton Deutsch presented a theory of cooperation that suggests cooperation arises when individuals have goals linked in a way that everyone sinks or swims together. On the other hand, a competitive situation arises when one individual swims while the other sinks.[6] This idea has been applied to the area of information systems development to provide insight as to how goals might affect the relationship between users and developers.[7]

Cooperation arises when individuals perceive the attainment of their goals as being positively related (i.e., reaching one's goals assists other people in attaining their goals). Cooperation, however, does not necessarily mean that individuals share the same goals, only that each individual will (or will not) attain their goals together. Here the individuals either sink or swim together.

The opposite holds true for competition. In competition, individuals perceive their goals as being negatively related (i.e., attainment of one's goals inhibits other people from reaching their goals). In this case, some must sink if another swims.

Cooperation can lead to greater productivity by allowing for more substitutability (i.e., permitting someone's actions to be substituted for one's own), thus allowing for more division of labor, specialization of roles, and efficient use of personnel and resources. Cooperative participants use their individual talents and skills collectively when solving a problem becomes a collaborative effort. Conflicts can be positive when disagreements are limited to a specific scope, and influence tends to be more persuasive in nature.

Cooperation also facilitates more trust and open communication. In addition, individuals are more easily influenced in a cooperative situation

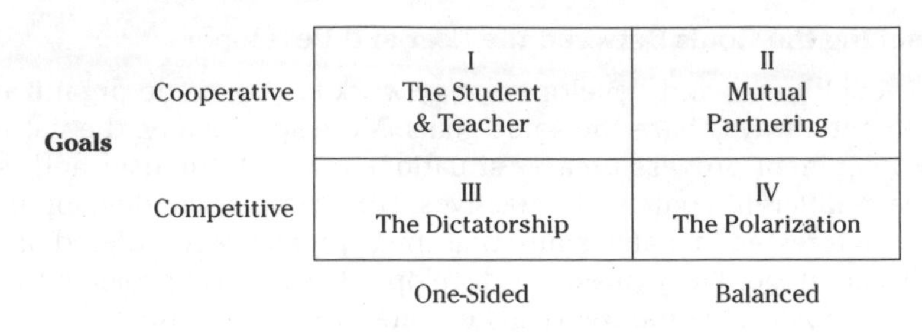

Exhibit 10.1 Classification of user/developer relationships.

than in a competitive one. Communication difficulties are reduced when persuasion rather than coercion is used to settle differences of viewpoints. Honest and open communication of important information exemplifies a cooperative situation. Competition, on the other hand, is characterized by a lack of communication or misleading communication.

The competitive process also encourages one party to enhance its power while attempting to reduce the legitimacy of the other party's interests. Conflict is negative when discussions include a general scope of issues that tend to increase each party's motivation and emotional involvement in the situation. Defeat for either party may be less desirable or more humiliating than both parties losing. In addition, influence tends to be more coercive in nature. Competitive individuals tend to be more suspicious, hostile, and ready to exploit or reject the other party's needs and requests. The cooperative process supports trust, congenial relations, and willingness to help the other party's needs and requests. In general, the cooperative process encourages a convergence of similar values and beliefs. The competitive process has just the opposite effect.

A CLASSIFICATION OF USER AND DEVELOPER RELATIONS

Exhibit 10.1 provides a classification scheme for viewing potential user/developer relationships based on the interdependency of goals and their balance of influence. Classification of relationships clarifies the social process of user involvement (i.e., how the user currently is involved or how the user should be involved) in integrated systems development.

Quadrant I: The Student and Teacher

In this quadrant, the balance of influence is one-sided; however, the goals between the user and the developer are positively related. Subsequently, this relationship resembles a teacher/student relationship for two reasons.

First, because the balance of power is one-sided, the more experienced or knowledgeable individual most likely leads the integrated systems development process. Because they both perceive their goals as being positively related, the less-experienced individual most likely follows the advice of the more influential individual.

The second reason has to do with a one-way model of learning. If the more influential individual leads the integrated systems development process, he or she has more to offer in terms of being able to share his or her knowledge or expertise than the less-experienced individual. As a result, learning generally takes place in one direction as in a typical teacher/student relationship.

An example of this type of relationship is an experienced developer teamed with a novice user. The users' limited knowledge or experience may make it difficult to specify their requirements. The developer may then attempt to control or lead the integrated systems development process, in which users contribute to the best of their knowledge or expertise.

Since these individuals perceive their goals as being positively related, the potential for resistance may be low. The user may view the development process as an opportunity to learn about the technology from the developer and may be easily influenced by the developer. An information system may be developed and viewed as successful; however, as the user becomes more experienced and familiar with the system, he or she may begin to request changes. Subsequently, these requests may result in higher maintenance costs later on.

Quadrant II: Mutual Partnering

In this quadrant the user and developer share the same degree of influence and have positively related goals. Here users play a more active role in the integrated systems development process than a novice, as their knowledge and level of expertise are greater. Because the developer also is experienced and knowledgeable in integrated systems development, the potential for a two-way model of learning exists.

Users, for example, learn how the technology supports their needs, whereas the developer learns about the business processes. Because the goals of these individuals are positively related, the potential for resistance may be low. Subsequently, a two-way model of learning suggests a higher degree of mutual learning and understanding where a system may be built successfully with lower maintenance costs later on.

Quadrant III: The Dictatorship

In the third quadrant the individuals exhibit more of a dictatorial relationship, in which the individual with the greater potential to influence leads the integrated systems development process. Resistance is high, because the goals of these individuals are negatively related. If the developer has the greater potential to influence the user, for example, he or she may view the user as a passive source of information. Users may perceive themselves as lacking any real chance to participate and then subsequently offer a high degree of resistance. The developer may build a system that fits the developer's perception of user needs or wants, to attain his or her goals. As a result, a system is developed that does not meet the initial requirements of the user and ultimately exhibits high maintenance costs. The system might be characterized as a technical success but an organizational failure.

On the other hand, if the user has the greater potential to influence the developer, the user tries to increase, for example, the functionality of the system to attain his or her goals. As a result, the developer offers passive resistance when asked to comply with the users' requests to minimize what the developer perceives as a losing situation. Conflicts may be settled through coercion with limited learning occurring between these individuals.

Quadrant IV: The Polarization

The fourth quadrant suggests a situation in which both the user and developer have an equal balance of influence but negatively related goals. Mutual learning is limited or exists only to the degree needed by an individual to attain his or her goals. Settlement of conflicts is achieved through political means, and a high degree of resistance results if one side perceives itself as being on the losing side. Conflicts increase each individual's motivation and emotional involvement in the situation, making defeat less desirable or more humiliating than both parties losing. These individuals may become more suspicious, hostile, and ready to exploit or reject the other party's needs or requests. Subsequently, this type of relationship may potentially be the most destructive and could lead to the abandonment of the IT project if neither side is willing to capitulate.

STRUCTURING THE USER–DEVELOPER RELATIONSHIP

The framework presented in the previous section may be used to assess and structure the relationship between users and developers. A three-step process is now presented to assess, structure, and then monitor the social relationship between these individuals.

Assessment

Assessment of the user–developer relationship is useful for choosing project participants as well as for gaining insight during the project if problems begin to arise. Using the framework presented in the previous section, a manager may begin by determining the potential balances of influence. Examples of factors that affect the balance of influence for both developers and users include:

- The level of technical knowledge
- The level of domain knowledge (i.e., knowledge of the business processes or functions that are the core of the IS project)
- The years of experience in the company or industry
- The prior involvement in system development projects
- The rank or position within the organization
- The reputation (i.e., how the individual's level of competency is perceived by others in the organization)

Other factors relevant to the project or organization can and should be used to assess the balance of influence among individuals. Subsequently, the manager should begin to get a clearer picture as to whether the balance of influence will be one-sided or balanced.

The next step is to assess individuals' goals. An easy way to make this assessment is to ask each individual to list or identify factors that are important to him or her. Examples include:

- What do you have to gain should the project succeed?
- What do you have to lose should the project fail?
- How would you determine whether the project is a success or failure?

After having both users and developers list what is important to them, a manager can compare these items to determine whether the individuals have potentially conflicting goals. End users and developers need not have exactly the same items of interest; these items need only not cause a win/lose situation. Asking the individuals to list how they would determine whether the project is a success or failure may uncover other potentially conflicting goals. For example, a manager may discover that users value functionality over cost, whereas the developer is concerned with ensuring that the project is developed by a specific date.

Structuring

A manager has several alternatives that can alter the balances of influence. The first alternative — choosing project participants — has the greatest impact. In an ideal situation, a manager can choose from a pool of personnel that includes both users and developers with varying degrees of skill, expertise, and knowledge. Unfortunately, if the pool is small, the number of possible combinations is reduced. As a result, training may prove to be a valuable tool for users becoming more knowledgeable about technology and developers becoming more knowledgeable about the business processes and functions.

As suggested in the framework presented, the goals of these individuals may be the most important factor in improving the social process of system development. For example, involving novice or less experienced individuals on a project is desirable when the project participants perceive their goals as positively related; however, serious communication problems arise if these same participants have negatively related goals.

To increase cooperation, then, the goals of the development team should be structured so that the individuals' goals are positively related. This may be accomplished in a number of ways.

Make Project Team Members Equally and Jointly Accountable. This may be in terms of a bonus or merit system under which each of the project team members is equally and jointly accountable for the success or failure of the information system. In other words, both the users and developers should be concerned with such issues as the functionality of the system and cost/schedule overruns.

Goals Should Be Made Explicit. Each individual involved in the development of the information system should have a clear, consistent perception that his or her goals are related in such a way that all sink or swim together. It is important not only that users and developers be held accountable using the same reward or merit system but also that they are aware that each is held accountable in the same way.

Management's Actions Must Reinforce Project Team Goals. It is important that management not allow project team members the opportunity to point fingers or assign blame. Subsequently, the actions of management must be consistent with the goals of the project team members. This is the most difficult challenge of all, because a change in values, attitudes, and possibly culture is required. For example, if goals are to be positively related, there can be no "us vs. them" ideology. Instead, both users and developers should see themselves as part of the same team.

Monitoring

The goals and perceptions of the individuals may change over the course of the project. Just as the allocation of time and resources must be monitored during a project, the social process between the project participants should be monitored as well. Monitoring should be continual during the project to identify any problems or negative conflict before they adversely affect the project. Similar to assessment, a manager may want to look for warning signs. Examples of warning signs include:

Finger Pointing When Problems Arise. Project team members should fix the problem, not assign the blame.

Negative Conflict. Individuals focus on petty issues that do not move them closer to their goals but only serve one party at the expense of the other. However, users and developers should agree to disagree. Conflict can be positive, especially when developing innovative approaches or refining new ideas.

Lack of Participation or Interest. All members of the project should be involved actively. However, even when all members perceive themselves as having a cooperative relationship, some individuals may become less involved. This may occur when the balance of influence is one-sided. Too often systems developers take control of the IS project and attempt to act in the best interest of the users. Although the developers may mean well, they may attempt to develop a system that the user never asked for and does not want.

Assessment, structuring, and monitoring should be a cycle. If specific problems are identified, a manager should assess the balance of influence and goals of the project team members. Changes can be made to alter or fine-tune the balances of influence or goals among the members. By managing this social process between users and developers, a manager increases the likelihood that systems that meet the objectives originally envisioned are developed on time and within budget.

CONCLUSION

This chapter suggests that even though users and developers may work for the same organization, they do not always share the same goals. Subsequently, problems in integrated systems development arise, especially when the user is more interested in functionality and the developer is more interested in maintaining cost/time schedules.

Cooperation facilitates improved communication and leads to greater productivity, because individuals perceive their goals as being positively related. In addition, the goals of the individuals provide some insight as to how each party uses its influence in the development of an information system. This idea was presented through a classification of user and developer relationships that considered their interdependency of goals and their balance of influence. Using this framework, a manager can assess, structure, and monitor the social relationship between users and developers. Managing this social relationship may result in more systems being developed within budget and on schedule and that meet the needs of the user.

Jack T. Marchewka is an assistant professor of management information systems at Northern Illinois University, DeKalb.

References

1. M. Keil, *Escalating Commitment: A Theoretical Base for Explaining IS Failure* (unpublished paper, Georgia State University, 1993).
2. H. Barki and J. Hartwick, Measuring user participation, user involvement, and user attitude, *MIS Quarterly* 18, 1 (March 1994), pp. 59-82.
3. J.T. Marchewka and K. Kumar, The social process of user and system developer interaction: a conceptual model (*Proceedings of the Decision Sciences Institute,* Honolulu, Hawaii, November 1994), pp. 964-966.
4. J.D. McKeen, T. Guimaraes, and J.C. Wetherbe, The relationship between user participation and user satisfaction: an investigation of four contingency factors, *MIS Quarterly* 8, 4 (December 1994), pp. 427-451.
5. M. Tan, Establishing mutual understanding in systems design, *Journal of Management Information Systems* 10, 4 (Spring 1994), pp.159-182.
6. M. Deutsch, Theory of cooperation and competition, *Human Relations* 2 (1949), pp. 129-152.
7. J.T. Marchewka, Goal interdependence and distributions of power: stimuli for communication in information systems development (*Proceedings of the Decision Sciences Institute,* Boston, MA, November 1995).

Chapter 11
Business Continuity in the Distributed Environment

Steven P. Craig

This chapter describes the process of business recovery planning specifically for local area networks (LANs) and the components that comprise the LAN. These procedures can be applied to companies of any size and for a recovery scope ranging from operational to catastrophic events.

INTRODUCTION

Today's organizations, in their efforts to reduce costs, are streamlining layers of management, while implementing more complex matrices of control and reporting. Distributed systems have facilitated the reshaping of these organizations by moving the control of information closer to its source, the end user. In this transition, however, secure management of that information has been placed at risk. Information technology departments must protect the traditional system environment within the computer room plus develop policies, standards, and guidelines for the security and protection of the company's distributed information base. Further, the information technology staff must communicate these standards to all users to enforce a strong baseline of controls.

In these distributed environments, information technology personnel are often asked to develop systems recovery plans outside the context of an overall business recovery scheme. Recoverability of systems, however, should be viewed as only one part of business recovery. Information systems, in and of themselves, are not the lifeblood of a company; inventory, assets, processes, and people are all essential factors that must be considered in the business continuation design. The success of business continuity planning rests on a company's ability to integrate systems recovery in the greater overall planning effort.

0-8493-9823-1/00/$0.00+$.50

Business Recovery Planning — The Process

Distinctive areas must be addressed in the formulation of a company's disaster recovery plan, and attention to these areas should follow the steps of the scientific method: a statement of the problem, the development of a hypothesis, and the testing of the hypothesis. Like any scientific process, the development of the disaster recovery plan is iterative. The testing phase of this process is essential because it reveals whether the plan is viable. Moreover, it is imperative that the plan and its assumptions be tested on an ongoing, routine basis. The most important distinction that marks disaster recovery planning is what is at stake — the survival of the business.

The phases of a disaster recovery plan process are

- Awareness and discovery
- Risk assessment
- Mitigation
- Preparation
- Testing
- Response and recovery

Recovery planners should adapt these phases to a company's specific needs and requirements. Some of the phases may be combined, for example, depending on the size of the company and the extent of exposures to risk. It is crucial, however, that each phase be included in the formation of a recovery plan.

Awareness and Discovery. Awareness begins when a recovery planning team can identify both possible threats and plausible threats to business operations. The more pressing issue for an organization in terms of business recovery planning is that of plausible threats. These threats must be evaluated by recovery planners, and their planning efforts, in turn, will depend on these criteria:

- The business of the company
- The area of the country in which the company is located
- The company's existing security measures
- The level of adherence to existing policies and procedures
- Management's commitment to existing policies and procedures

Awareness also implies educating all employees on existing risk exposures and briefing them on what measures have been taken to minimize those exposures. Each employee's individual role in complying with these measures should be addressed at this early stage.

In terms of systems and information, the awareness phase includes determining what exposures exist that are specific to information systems, what information is vital to the organization, and what information is proprietary and confidential. Answering these questions will help planners determine when an interruption will be catastrophic as opposed to operational. For example, in an educational environment, a system that is down for two or three days may not be considered catastrophic, whereas in a process control environment (e.g., chemicals or electronics), just a few minutes of downtime may be considered this.

Discovery is the process in which planners must determine, based on their awareness of plausible threats, which specific operations would be affected by existing exposures. They must consider what measures are currently in place or could be put in place to minimize or, ideally, remove these exposures.

Risk Assessment. Risk assessment is a decision process that weighs the cost of implementing preventive measures against the risk of loss from not implementing them. There are many qualitative and quantitative approaches to risk analysis. Typically, two major cost factors arise for the systems environment. The first is the loss incurred from a cease in business operations due to system downtime. The second is the replacement cost of equipment.

The potential for significant revenue loss when systems are down for an extended period of time is readily understood in today's business environment, because the majority of businesses rely exclusively on systems for much of their information needs. However, the cost of replacing systems and information in the event of catastrophic loss is often grossly underrated. Major organizations, when queried on insurance coverage for systems, come up with some surprising answers. Typically, organizations have coverage for mainframes and midrange systems and for the software for these environments. The workstations and the network servers, however, are often deemed not valuable enough to insure. Coverage for the information itself is usually neglected as well, despite the fact that the major replacement cost for a company in crisis is the recreation of its information database.

Notably, the personal computer, regardless of how it is configured or networked, is usually perceived as a stand-alone unit from the risk assessment point of view. Even companies that have retired their mainframes and embraced an extensive client/server architecture, and that fully comprehend the impact of the loss of its use, erroneously consider only the replacement cost of the unit rather than of the distributed system as the basis of risk.

Risk assessment is the control point of the recovery planning process. The amount of exposure a company believes it has, or is willing to accept, determines how much effort the company will expend on this process. Simply put, a company with no plan is fully exposed to catastrophic loss. Companies developing plans must approach risk assumption by identifying their worst-case scenario and then deciding how much they will spend to offset that scenario through mitigation, contingency plans, and training. Risk assessment is the phase required to formulate a company's management perspective, which in turn supports the goal of developing and maintaining a companywide contingency plan.

Mitigation. The primary objectives of mitigation are to lessen risk exposures and to minimize possible losses. History provides several lessons in this area. For example, since the underground floods of 1992, companies in Chicago think twice before installing data centers in the basements of buildings. Bracing key computer equipment and office furniture has become popular in California because of potential injuries to personnel and the threat of loss of assets from earthquakes. Forward-thinking companies in the South and southern Atlantic states are installing systems far from the exterior of buildings because of the potential damage from hurricanes.

Although it is a simple exercise to make a backup copy of key data and systems, it is difficult to enforce this activity in a distributed systems environment. As systems have been distributed and the end user empowered, the regimen of daily or periodic backups has been adversely affected. In other words, the end user has been empowered with tools but has not been educated about, or held responsible for, the security measures that are required for those tools. One company, a leader in the optical disk drive market, performs daily backups of its accounting and manufacturing systems to optical disk (using its own product), but never rotates the media and has never considered storing the backup off-site. Any event affecting the hardware (e.g., fire, theft, or earthquake) could therefore destroy the sole backup and the means of business recovery for this premier company. Mitigation efforts must counter such oversights.

Preparation. The preparation phase of the disaster planning process delineates what specific actions must be taken should a disaster occur. Based on an understanding of plausible threats, planners must determine who will take what action if a disaster occurs. Alternates should be identified for key staff members who may have been injured as a result of the event. A location for temporary operations should be established in case the company's building is inaccessible after a disaster, and the equipment, supplies, and company records that will be required at this site should be identified. Preparation may include establishing a hot site for systems and telecommunications. Off-hours or emergency telephone numbers should be kept for all vendors and services providers that may need to be contacted.

Moreover, the contingency plans must be clearly documented and communicated to all personnel.

Testing. The testing phase proves the viability of the planning efforts. The recovery planner must determine, during testing, whether there are invalid assumptions and inadequate solutions in the company's plan. It is important to remember that organizations are not static and that an ever-changing business environment requires a reasonable frequency of testing. Recovery planners must repeat this phase of the plan until they are comfortable with the results and sure that the plan will work in a time of crisis.

Response and Recovery. This final phase of the contingency plan is one that organizations hope never to have to employ. Preparing for actual response and recovery includes identifying individuals and training them to take part in emergency response in terms of assessment of damage, cleanup, restoration, alternate site start-up, emergency operations duties, and any other activities that managing the crisis might demand.

Every phase of the planning process, prior to this phase, is based on normalcy. The planning effort is based on what is perceived to be plausible. Responses are developed to cover plausible crises and are done so under rational conditions. However, dealing with a catastrophic crisis is not a normal part of an employee's work day, and the recovery team must be tested under more realistic conditions to gauge how they will perform under stress and where lapses in response might occur. Ideally, recovery planners should stage tests that involve role playing to give their team members a sense of what they may be exposed to in a time of crisis.

DEPARTMENTAL PLANNING

Often, consultants are asked to help a company develop its business resumption plan and to focus only on the systems environment to reduce the overall cost of planning efforts. Often, companies take action on planning as the result of an information systems audit and thus focus solely on systems exposure and audit compliance. These companies erroneously view disaster recovery as an expense rather than as an investment in business continuity.

A plan that addresses data integrity and systems survivability is certainly a sound place to begin, but there are many other factors to consider in recovery planning. Depending on the nature of the business, for example, telecommunications availability may be much more important than systems availability. In a manufacturing environment, if the building and equipment are damaged in a disaster, getting the systems up and running may not necessarily be a top priority.

A company's business continuation plan should be a compilation of individual department plans. It is essential that each department identify its processes and prioritize those processes in terms of recovery. Company-wide operating and recovery priorities can then be established by the company's management based on the input supplied by the departments. Information technology, as a service department to all other departments, will be better equipped to plan recovery capacity and required system availability based on this detailed knowledge of departmental recovery priorities.

Information Technology's Role

Information technology personnel should not be responsible for creating individual department plans, but they should take a leadership role in the plan development. Information technology generally has the best appreciation and understanding of information flow throughout the organization. Its staff, therefore, are in the best position to identify and assess the following areas.

Interdepartmental Dependencies. It is common for conflicts in priorities to arise between a company's overall recovery plan and its departmental plans. This conflict occurs because departments tend to develop plans on their own without considering other departments. One department may downplay the generation of certain information because that information has little importance to its operations, but the same information might be vitally important to the operations of another department. Information technology departments can usually identify these discrepancies in priorities by carefully reviewing each department's plan.

External Dependencies. During the discovery process, recovery planners should determine with what outside services end-user departments are linked. End-user departments often think of external services as being outside the scope of their recovery planning efforts, despite the fact that dedicated or unique hardware and software are required to use the outside services. At a minimum, departmental plans must include the emergency contact numbers for these outside services and any company account codes that permit linkage to the service from a recovery location. Recovery planners should also assess the outside service providers' contingency plans for assisting the company in its recovery efforts.

Internal and External Exposures. Stand-alone systems acquired by departments for a special purpose are often not linked to a company's networks. Consequently, they are often overlooked in terms of data security practices.

For example, a mortgage company funded all of its loans via wire transfer from one of three stand-alone systems. This service was one of the key operations of the company. Each system was equipped with a modem and a uniquely serialized encryption card for access to the wire service. However, these systems were not maintained by the information technology department, no data or system backups were maintained by the end-user department, and each system was tied to a distinct phone line. Any mishap involving these three systems could have potentially put this department several days, if not weeks, in arrears in funding its loans. Under catastrophic conditions, a replacement encryption card and linkage establishment would have taken as much as a month to acquire.

As a result of this discovery, the company identified a secondary site and filed a standby encryption card, an associated alternate phone line, and a disaster recovery action plan with the wire service. This one discovery, and its resolution, more than justified the expense of the entire planning effort.

During the discovery process, the recovery planner identified another external exposure for the same company. This exposure related to power and the requirements of the company's uninterruptible power supply (UPS). The line of questioning dealt with the sufficiency of battery backup capacity and whether an external generator should be considered in case of a prolonged power interruption. An assumption had been made by the company that, in the event of an areawide disaster, power would be restored within 24 hours. The company had eight hours of battery capacity that would suffice for its main operational shift. Although the county's power utility company had a policy of restoring power on a priority basis for the large employers of the county, the company was actually based in a special district and acquired its power from the city, not the county. Therefore, it would have power restored only after all the emergency services and city agencies were restored to full power. Moreover, no one could pinpoint how long this restoration period would be. To mitigate this exposure, the company added an external generator to its UPS system.

Apprise Management of Risks and Mitigation Costs. As an information technology department identifies various risks, it is the department's responsibility to make management aware of them. This responsibility covers all security issues — system survivability issues (i.e., disaster recovery), confidentiality, and system integrity issues.

In today's downsized environments, many information technology departments have to manage increasingly more complex systems with fewer personnel. Because of these organizational challenges, it is more important for the information technology staff involved in the planning process to present management with clear proposals for risk mitigation. Advocating comprehensive planning and security measures and following through with management to see that they are implemented will ensure that a depleted information technology staff is not caught off-guard in the event of disaster.

Policies. To implement a system or data safeguard strategy, planners must first develop a policy, or standard operating procedure, that explains why the safeguard should be established and how it will be implemented. The planners should then get approval for this policy from management.

In the process of putting together a disaster recovery plan for a community college's central computing operations, one recovery planner discovered that numerous departments had isolated themselves from the networks supported by the information technology group. These departments believed that the servers were always crashing, which had been a cause for concern in years past, and they chose to separate themselves from the servers for what they considered to be safer conditions. These departments, which included accounting, processed everything locally on hard drives with no backups whatsoever. Needless to say, a fire or similar disaster in the accounting department would severely disrupt, if not suspend, the college's operations.

The recovery planner addressed this problem with a fundamental method of distributed system security: distribute the responsibility of data integrity along the channels of distributed system capability. A college policy statement on data integrity was developed and issued to this effect. The policy outlined end-user security responsibilities, as well as those of the department administrators.

Establish Recovery Capability. Based on departmental input and a company's established priorities, the information technology department must design an intermediate system configuration that is adequately sized to permit the company's recovery immediately following the disaster. Initially, this configuration, whether it is local, at an alternate company site, or at a hot site, must sustain the highest-priority applications yet be adaptable to addressing other priorities. These added needs will arise depending on how long it takes to reoccupy the company's facilities and fully restore all operations to normal. For example, planners must decide that the key client/server applications are critical to company operations, whereas office automation tools are not.

Restore Full Operational Access. The information technology department's plan should also address the move back from an alternate site and the resources that will be required to restore and resume full operations. Depending on the size of the enterprise and the plausible disaster, this could include a huge number of end-user workstations. At the very least, this step is as complex as a company's move to a new location.

PLANNING FOR THE DISTRIBUTED ENVIRONMENT

First and foremost, planners in a distributed environment must define the scope of their project. Determining the extent of recovery is the first step. For example, will the plan focus on just the servers or on the entire enterprise's systems and data? The scope of recovery, the departmental and company priorities, and recovery plan funding will delimit the planner's options. The following discussion outlines the basics of recovery planning regardless of budget considerations.

Protecting the LAN

Computer rooms are built to provide both special environmental conditions and security control. Environmental conditions include air conditioning, fire-rated walls, dry sprinkler systems, special fire abatement systems (e.g., Halon, FM-200), raised flooring, cable chase-ways, equipment racking, equipment bracing, power conditioning, and continuous power (UPS) systems. Control includes a variety of factors: access, external security, and internal security. All these aspects of protection are built-in benefits of the computer room. Today, however, company facilities are distributed and open; servers and network equipment can be found on desktops in open areas, on carts with wheels, and in communications closets that are unlocked or have no conditioned power. Just about anything and everything important to the company is on these servers or accessible through them.

Internal Environmental Factors. A computer room is a viable security option, though there are some subtleties to designing one specifically for a client/server environment. If the equipment is to be rack mounted, racking can be suspended from the ceiling, which yields clearance from the floor and avoids possible water damage. Notably, the cooling aspects of a raised floor design, plus its ability to hide a morass of cabling, are no longer needed in a distributed environment.

Conditioned power requirements have inadvertently modified computer room designs as well. If an existing computer room has a shunt trip by the exit but small stand-alone battery backup units are placed on servers, planners must review the computer room emergency shutdown procedures. The function of the shunt trip was originally to kill all power in the room so that, if operational personnel had to leave in a hurry, they would be able to come back later and reset systems in a controlled sequence. Now, when there are individual battery backup units that sustain the equipment in the room, the equipment will continue to run after the shunt is thrown. Rewiring the room for all wall circuits to run off the master UPS, in proper sequence with the shunt trip, should resolve this conflict.

Room placement within the greater facility is also a consideration. When designing a room from scratch, planners should identify an area with structural integrity, avoid windows, and eliminate overhead plumbing.

Alternate fire suppression systems are still a viable protection strategy for expensive electronics and the operational, on-site tape backups within a room. If these systems are beyond the company's budget, planners might consider multiple computer rooms (companies with a multiple-building campus environment or multiple locations can readily adapt these as a recovery strategy) with sprinklers and some tarpaulins handy to protect the equipment from incidental water damage (e.g., a broken sprinkler pipe). A data safe may also be a worthwhile investment for the backup media maintained on site. However, if the company uses a safe, its personnel must be trained to keep it closed. In eight out of ten site visits where a data safe is used, the door is kept ajar (purely as a convenience). The safe only protects the company's media when it is sealed. If the standard practice is to keep it closed, personnel will not have to remember to shut it as they evacuate the computer room under the stress of an emergency.

If the company occupies several floors within a building and maintains communication equipment (e.g., servers, hubs, or modems) within closets, the closets should be treated as miniature computer rooms. The doors to the closets should be locked, and the closets should be equipped with power conditioning and adequate ventilation.

Physical Security. The other priority addressed by a properly secured computer room is control: control of access to the equipment, cabling, and backup media. Servers out in the open are prime targets for mishaps ranging from innocent tampering to outright theft. A thief who steals a server gets away not only with an expensive piece of equipment but with a wealth of information that may be prove much more valuable and marketable than the equipment itself.

The college satellite campus discussed earlier had no backup of the information contained within its network. The recovery planner explained to the campus administration, which kept its servers out in the open in its administration office area (a temporary trailer), that a simple theft of the $2,000 equipment would challenge its ability to continue operations. All student records, transcripts, course catalogs, instructor directories, and financial aid records were maintained on the servers. With no backup to rely on and its primary source of information evaporated, the campus administration would be faced with literally thousands of hours of effort to reconstruct its information base.

Property Management. Knowing what and where the organization's computer assets (i.e., hardware, software, and information) are at any moment is critical to recovery efforts. The information technology department must be aware not only of the assets within the computer room but of every workstation used throughout the organization: whether it is connected to a network (including portables); what its specific configuration is; what software resides on it; and what job function it supports. This knowledge is achievable if all hardware and software acquisitions and installations are run through the IT department, if the company's policies and procedures support information technology's control (i.e., all departments and all personnel willingly adhere to the policies and procedures), and if the department's property management inventory is properly maintained. Size is also a factor here. If the information technology department manages an organization with a single server and 50 workstations, the task may not be too large; however, if it supports several servers and several hundred workstations, the amount of effort involved is considerable.

Data Integrity. Information, if lost or destroyed, is the one aspect of a company's systems that cannot be replaced simply by ordering another copy or another component. The company may have insurance, hot-site agreements, or quick-replacement arrangements for hardware and global license agreements for software, but its data integrity process is entirely in the hands of its information technology specialists. The information technology specialist and the disaster recovery planner are the individuals who must ensure that the company's information will be recoverable. Based on the initial risk assessment phase, planners can determine just how extensive the data integrity program should be. The program should include appropriate policies and education addressing frequency of backups, storage locations, retention schedules, and the periodic verification that the backups are being done correctly. If the planning process has just begun, data integrity should be the first area on which planners focus their attention. None of the other strategies they implement will count if no means of recovering vital data exists.

Network Recovery Strategies

The information technology specialist's prime objective with respect to systems contingency planning is system survivability. In other words, provisions must be in place, albeit in a limited capacity, that will support the company's system needs for priority processing through the first few hours immediately following a disaster.

Fault Tolerance vs. Redundancy. To a degree, information technology specialists are striving for what is called fault tolerance of the company's critical systems. Fault tolerance means that no single point of failure will stop the system. Fault tolerance is often built in as part of the operational component design of a system. Redundancy, or duplication of key components, is the basis of fault tolerance. When fault tolerance cannot be built in, a quick replacement or repair program should be devised. Moving to an alternate site (i.e., a hot site) is one quick replacement strategy.

Alternate Sites and System Sizing. Once the recovery planner fully understands the company's priorities, the planner can size the amount of system capacity required to support those priorities in the first few hours, days, and weeks following a disaster. When planning for a recovery site or establishing a contract with a hot-site service provider, the information technology specialist must size the immediate recovery capacity. This is extremely important, because most hot-site service providers will not allow a company to modify its requirements once it has declared a disaster.

The good news with respect to distributed systems is that hot-site service providers offer options for recovery. These options often include offering the use of their recovery center, bringing self-contained vans to the company's facility (equipped with the company's own required server configuration), or shipping replacement equipment for anything that has been lost.

Adequate Backups with Secure Off-Site Storage. This process must be based on established company policies that identify vital information and detail how its integrity will be managed. The work flow of the company and the volatility of its information base dictates the frequency of backups. At a minimum, backup should occur daily for servers and weekly or monthly for key files of individual workstations.

Planners must decide when and how often to take backups off-site. Depending on a company's budget, off-site could be the building next door, a bank safety deposit box, the network administrator's house, the branch office across town, or a secure media vault at a storage facility maintained by an off-site media storage company. Once the company meets the objective of separating the backup copy of vital data from its source, it must address the accessibility of the off-site copy.

The security of the company's information is of vital concern. The planner must know where the information is to be kept and about possible exposure risks during transit. Some off-site storage companies intentionally use unmarked, nondescript vehicles to transport a company's backup tapes to and from storage. These companies know that this information is valuable and that its transport and storage place should not be advertised.

Adequate LAN Administration. Keeping track of everything the company owns — its hardware, software, and information bases — is fundamental to a company's recovery effort. The best aid in this area is a solid audit application that is run periodically on all workstations. This procedure assists the information technology specialist in maintaining an accurate inventory across the enterprise and provides a tool for monitoring software acquisitions and hardware configuration modifications. The inventory is extremely beneficial for insurance loss purposes. It also provides the technology specialist with accurate records for license compliance and application revision maintenance.

Personnel. Systems personnel are too often overlooked in systems recovery planning. Are there adequate systems personnel to handle the complexities of response and recovery? What if a key individual is affected by the same catastrophic event that destroys the systems? This event could cause a single point of failure.

An option available to the planner is to propose an emergency outsourcing contract. A qualified systems engineer hired to assist on a key project that never seems to get completed (e.g., the network system documentation) may be a cost-effective security measure. Once that project is completed to satisfaction, the company can consider structuring a contractual arrangement that, for example, retains the engineer for one to three days a month to continue to work on documentation and other special projects, as well as cover for staff vacations and sick days, and guarantees that the engineer will be available on an as-needed basis should the company experience an emergency. The advantage of this concept is that the company maintains effective outsourced personnel who are well-versed in the company's systems if the company needs to rely on them during an emergency.

TESTING

The success of a business recovery plan depends on testing its assumptions and solutions. Testing and training keep the plan up to date and maintain the viability of full recovery.

Tests can be conducted in a variety of ways, from reading through the plan and thinking through the outcome to full parallel system testing, or setting up operations at a hot site or alternate location and having the users run operations remotely. The full parallel system test generally verifies that the hot-site equipment and remote linkages work, but it does not necessarily test the feasibility of the user departments' plans. Full parallel testing is also generally staged within a limited amount of time, which trains staff to get things done correctly under time constraints.

Advantages of the Distributed Environment for Testing

Because of their size and modularity, distributed client/server systems provide a readily available, modifiable, and affordable system setup for testing. They allow for a testing concept called cycle testing.

Cycle testing is similar to cycle counting, a process used in manufacturing whereby inventory is categorized by value and counted several times a year rather than in a one-time physical inventory. With cycle counting, inventory is counted year long, with portions of the inventory being selected to be counted either on a random basis or on a preselected basis. Inventory is further classified into categories so that the more expensive or critical inventory items are counted more frequently and the less expensive items less frequently. The result is the same as taking a one-time physical inventory in that, by the end of a calendar year, all the inventory has been counted. The cycle counting method has several advantages:

- Operations do not have to be completely shut down while the inventory is being taken.
- Counts are not taken under time pressure, which results in more accurate counts.
- Errors in inventories are discovered and corrected as part of the continuous process.

The advantages of cycle testing are similar to those of cycle counting. Response and recovery plan tests can be staged with small, manageable groups so they are not disruptive to company operations. Tests can be staged by a small team of facilitators and observers on a continual basis. Tests can be staged and debriefings held without time pressure, allowing the participants the time to understand their roles and the planners the time to evaluate team response to the test scenarios and to make necessary corrections to the plan. Any inconsistencies or omissions in a department's plan can be discovered and resolved immediately among the working participants.

Just as more critical inventory items can be accounted for on a more frequent basis, so can the crucial components required for business recovery (i.e., systems and telecommunications). With the widespread use of LANs and client/server systems, information systems departments have the opportunity to work with other departments in testing their plans.

CONCLUSION

Developing a business recovery plan is not a one-time, static task. It is a process that requires the commitment and cooperation of the entire company. To perpetuate the process, business recovery planning must be a company-stipulated policy in addition to being a company-sponsored goal. Organizations must actively maintain and test plans, training their employees to respond in a crisis. The primary objective in developing a business resumption plan is to preserve the survivability of the business.

An organization's business resumption plan is an orchestrated collection of departmental responses and recovery plans. The information technology department is typically in the best position to facilitate other departments' plan development and can be particularly helpful in identifying the organization's interdepartmental information dependencies and external dependencies for information access and exchange.

A few protective security measures should be fundamental to the information technology department's plan, no matter what the scope of plausible disasters. From operational mishaps to areawide disasters, recovery planners should ensure that the information technology department's plan addresses:

- An adequate backup methodology with off-site storage
- Sufficient physical security mechanisms for the servers and key network components
- Sufficient logical security measures for the organization's information assets
- Adequate LAN/WAN administration, including up-to-date inventories of equipment and software

Finally, in support of an organization's goal to have its business resumption planning process in place to facilitate a quick response to a crisis, the plan must be sufficiently and repeatedly tested, and the key team members sufficiently trained. When testing is routine, it becomes the feedback step that keeps the plan current, the response and recovery strategies properly aligned, and the responsible team members ready to respond. Testing is the key to plan viability and thus to the ultimate survival of the business.

Chapter 12
Developing Client/Server RDBMS Applications Using Java Servlets and JDBC

Jonathan Held

Client/server computing is by no means a novel concept; it has been around nearly as long as the computer. What is new, however, is how the rise of the World Wide Web (circa 1992) impacted this computing concept. Client/server computing, given this venue, has reached new ground and its popularity is indelibly tied to the astounding success that the Internet has seen. What makes the Web so attractive, in part, is the price — client software is free. Using Netscape's Communicator or Microsoft's Internet Explorer (or any other capable browser), one can get a multitude of information on virtually any subject. The information has to be stored somewhere, and, in most cases, it is kept in a relational database management system (RDBMS), with which the browser (translate as client) interacts.

What you'll need:

- Some prior knowledge of the Java programming language and Structured Query Language (SQL)
- Java Development Kit (JDK) 1.2
- Microsoft Access (MSACCESS)
- Sun's Servlet Software Development Kit (SDK)
- Web server software

0-8493-9823-1/00/$0.00+$.50

If you think Web-based databases haven't caught on, you might want to reconsider. Consider the Web search sites (Lycos, Yahoo, Excite, Metacrawler, Webcrawler, or Hotbot, to name a few); where do you think the "hits" come from?

If you're as much of an Internet junkie as I am, you may even go so far as to check online to see what movies are playing in your local area. Two online sites offer such information: http://www.movielink.com and http://www.moviefinder.com. I enter my zip code, click the mouse a couple of times, and I know what movies are playing at what theaters and their show times. Why pick up the phone, call the theater, and get a recording that you can barely hear? If you would rather stay at home and park yourself in front of the couch with a bag of potato chips, try http://www.tvguide.com and you can choose the television listings available by cable company. So, if you were purchasing the Sunday paper just for the *TV Week* magazine that came with it, save yourself some money and cancel your subscription.

These examples all have several things in common. The first is that the Web browser is the client application. As a developer, you can now breathe a sigh of relief knowing that you can completely concentrate your programming efforts on the server-side interface to the data repository.

So how does it all work? Well, the short (and extremely simplified) answer is that the client, you and your browser, initiate a process that somehow interacts with the back-end database. This process is also responsible for returning content back to the browser, although what it returns may vary on what action was being performed. If you are merely submitting personal information about yourself or making an entry into a guest book, the response might simply consist of a confirmation that the information was successfully entered into the database.

As you can probably well imagine, there are a number of technologies available today that would allow us to accomplish such tasks. We could opt to adopt Common Gateway Interface (CGI) scripts, but this option is replete with security risks, making it an unattractive solution to even experienced programmers. Active Server Pages (ASP), a Microsoft technology designed to operate in conjunction with that company's Internet Information Server (IIS) 4.0, is another possibility, but it locks us into an operating system and a Web server that our Internet service provider (ISP) might not be using. Of course, there are a number of other options available, but perhaps one of the better but less explored ones is made possible by Java servlets and JDBC™.

THE JAVA INCENTIVE

There are two key requirements for database programmers:

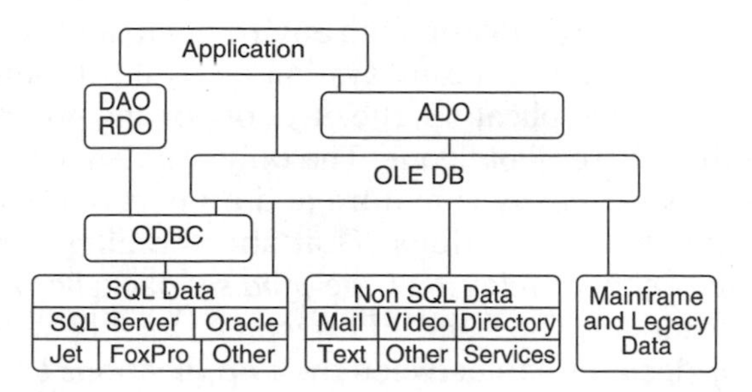

Exhibit 12.1 Comparison of ADO, DAO, and RDO.

- They must have intimate knowledge of the language construct used to manipulate databases.
- They need to be cognizant of what means are available for invoking these constructs from external applications.

Of course, the syntax for performing the former task is accomplished by a data query language that is now universal across different computer systems — SQL. SQL is neither difficult to learn nor use; rather, it is the means of using SQL in programs that, until recently, presented the greater challenge.

At first, many database applications were developed by making Open Database Connectivity (ODBC) Application Programming Interface (API) calls. But despite all that Microsoft's ODBC allowed you to do, it wasn't without its own problems. Chief among these were the following:

- ODBC was written exclusively in the C programming language, so there was no concept of objects or methods. The logical organization that is intrinsic to Object Oriented Programming (OOP) was nowhere to be found, resulting in a great deal of frustration when you were trying to find the right procedure or function to call.
- The API was extremely large, hard to follow, and required a fair amount of knowledge on the part of the programmer.

These shortcomings were noted and Microsoft proceeded to create several object models that programmers could use instead. These new collections of objects and methods were ODBC wrappers; they encapsulated calls into the ODBC API and hid the implementation details from the programmer. They exist today in the form of Data Access Objects (DAO), Remote Data Objects (RDO), and the more recent ActiveX Data Objects (ADO), as illustrated in Exhibit 12.1.

Then came Sun Microsystems and the rise of Java. Java made many new promises, but what made it so attractive was that it was designed to offer a secure (or more secure) programming environment and could run on any platform regardless of the operating system being used. Now, if one could create a Java database application, the days of porting programs from one machine to another were all but gone. The only problem was that Java, like all new things, was extremely immature and no part of the core language had database-enabled applications. That shortcoming was noticed and fixed with the subsequent release of the *java.sql* package, which contains the JDBC object model. The JDBC API became the mechanism by which programmers bridged the gap between their applications and databases. It defines a number of Java classes that allow programmers to issue SQL statements and process the results, dynamically learn about a database's schema etc. It is by far one of the easier to understand object models, and it is nearly effortless to incorporate it into Java applications.

THE PROJECT

So what is it that we're going to set out to do? Let's suppose we wanted to create a fortune/quotation server that interacts with a Microsoft Access database and returns an entry and five lucky, random numbers back to the client's browser. We're going to create this application and support this functionality using Java and JDBC, but one more thing is needed that requires us to make some development decisions.

We could create an applet that is downloaded by the browser and provides a user interface enabling information retrieval from the database. However, this solution has some notable drawbacks. First and foremost, to use an applet and interact with a database requires a JDBC driver. There are many types of commercially available drivers, but they are prohibitively expensive and a project of this scope does not justify the expense. Another disadvantage to using these drivers is that they typically consist of a large number of class files. The more files that the browser has to download over a slow connection, the more irate clients will get at using the system, eventually abandoning it if it becomes to burdensome (i.e., time-consuming) to use. We could opt to use Sun Microsystem's JDBC-ODBC bridge, which is free, but it is not thread-safe. And unfortunately, incorporating this driver into an applet requires that we take some additional steps to make it a trusted component. So now we have to explore how we can manipulate the browser's built-in security manager so it works, and this is far more trouble than it's worth for our simple task.

A final disadvantage of using applets is that they can only make connections back to the machine from which they were downloaded. This means that if we use a JDBC driver, the database it communicates with must be co-located with the Web server. It is possible to use a proxy server to circumvent this restriction, but, short of doing this, we should see if an easier solution exists (after all, why make more work for ourselves than is necessary?)

The solution we'll use that enables us to get around all of these potential pitfalls is the Java servlet. The servlet concept was first introduced in April of 1997, in conjunction with the first all-Java Web server. Servlets are protocol and platform independent server-side components. You can think of them as an applet for a server. They are almost identical to their CGI counterparts, and they can do anything that CGI can do. But servlets differ in several ways: they are easier to write than CGI programs/scripts written in C++ or PERL, and they are noticeably faster and much safer. There are four important reasons why we'll turn our attention to the servlet solution:

- *Performance:* Servlets do not require a new process for each request (CGI does, and if a server fails to load-balance or put a limit on the number of concurrent requests, it can easily be brought to its knees). The servlet *init()* method allows programmers to perform resource-intensive operations common to all servlet invocations once at startup. For example, by having the *init()* method establish a database connection, this process can be done once. Consequently, the slowest performance occurs the very first time the servlet is executed; subsequent invocations occur much more rapidly.
- *Portability:* Because Java is platform independent, so are servlets. We can move our compiled servlet code from one machine to another without having to recompile, and we can use our code with many different types of Web servers.
- *Security:* Servlets have the Java advantage — memory access and strong typing violations are simply not possible. By default, all servlets are untrusted components and they are not allowed to perform operations such as accessing network services or local files unless they are digitally signed and accorded more freedom by manipulating Java's security manager.
- *Flexibility:* Although servlets are written in Java, their clients can be written in any programming language. Servlets can be written as clients to other services that are written in any programming language. For example, we can use them with JDBC to contact a RDBMS. They can process data submitted via an HTML form, allow collaboration between people by synchronizing requests to support systems, such as online conferencing, and pass requests to other servlets to load-balance the amount of work that a system or servlet is performing.

With all these good things going for us, we should be convinced that servlets are a viable option for our project. The only part that remains now is to put this thing together, but that is where the fun begins.

THE BACK-END DATABASE

Creating the database for this project was by no means a difficult process, but it was time-consuming to populate it with 700 fortunes/quotations.

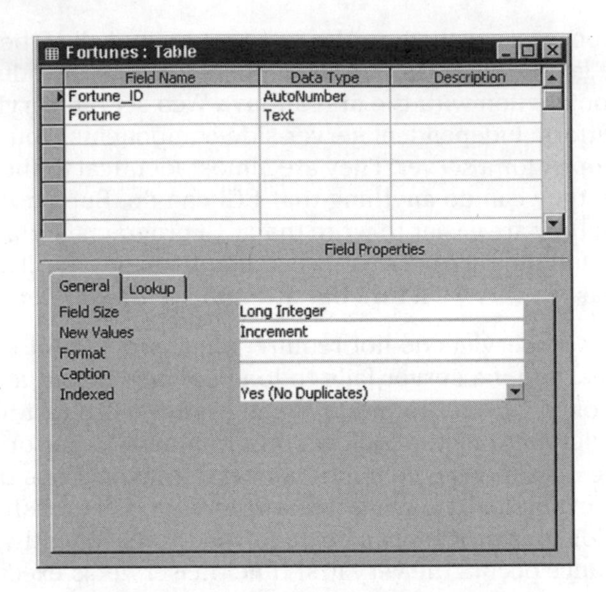

Exhibit 12.2 The *Fortunes* table schema.

Fortunately, should you decide to put this project together on your own personal computer, you can just download the Microsoft Access database. To give you an appreciation of what was done and how, we'll briefly outline the database's schema, and how we used and configured the control panel ODBC applet.

The fortune database has only one table. This table has two fields:

- a *Fortune_ID*, and
- a *Fortune*.

The *Fortune_ID* is a unique, self-generated autonumber that is indexed and serves as the table's primary key. The *Fortune*, as you might expect, is a text entry of up to 200 characters that holds all the wise adages we'll be delivering to the client. Exhibit 12.2 is a screen capture of the database design view as it appears in Microsoft Access, and Exhibit 12.3 shows you the datasheet view.

We now need to decide where to place the Access database in the file system. Because we intend to use our database from the web, We might be inclined at first to move it to where all our other Web files are located. A better solution, though, is to place the *.mdb* Access file in a directory named *Internet Databases* (or whatever name you choose) that resides entirely elsewhere. This is a good practice in general, especially for security reasons (we don't want someone downloading our database, do we?).

Fortunes : Table

Fortune_ID	Fortune
93	It is unwise to be too sure of one's own wisdom. It is healthy to be re
94	In the attitude od silence the soul finds the path in an clearer light, a
95	Adaptability is not initation. It means power of resistance and assum
96	It is the quality of our work which will please God and not the quanti
97	Honest differences are often a healthy sign of progress
98	A keen sense of humor helps us to overlook the unbecoming, under
99	Hot heads and cold hearts never solved anything.
100	Most of us follow our conscience as we follow a wheelbarrow. We p
101	The test of a preacher is that his congregation hgoes away saying,
102	Comfort and prosperty have never enriched the world as much as a
103	Abstaining is favorable both to the head and the pocket
104	Fame usually comes to those who are thinking of something else.
105	Common sense is very uncommon
106	Journalism will kill you, but it will keep you alive while you're at it
107	Fame is vapor, popularity an accident, riches take wings. Only one

Record: 1 of 700

Exhibit 12.3 The *Fortunes* table datasheet view.

Exhibit 12.4 The ODBC control panel applet (also listed as ODBC Data Sources [32 bit] in Windows 98).

To do this, create your directory. Once this is completed, open the Windows Control Panel and double click on the ODBC icon as shown in Exhibit 12.4.

This should display a tabbed dialog box appropriately titled *Data Source Administrator.* We use this program to inform the system of data source names and locations; we'll use it in our servlet programs to refer to the database we wish to manipulate with SQL statements. Once you have placed the fortune database in a directory, select the *System DSN* tab and click *Add.* You'll be prompted for the type of driver for which you want to set up a data source. Since we're using an Access database, we want the Microsoft Access driver. Click *Finish*, and you should then be directed to a new dialog titled *ODBC Microsoft Access Setup*. Here, there are two pieces of information which we have to provide:

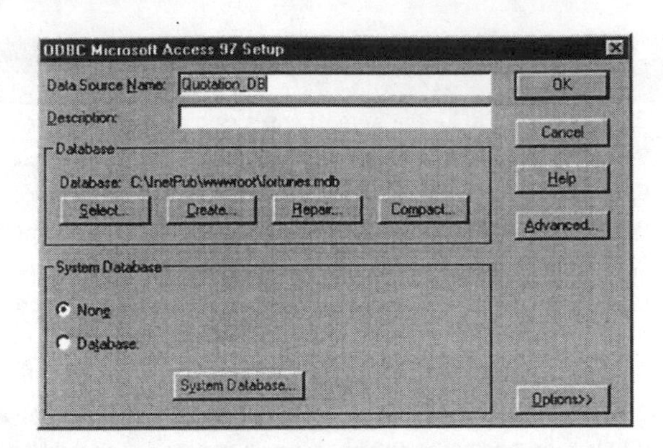

Exhibit 12.5 Configuring the ODBC data source.

- The name of the data source
- The location of the database

In Exhibit 12.5 above, the data source name is *Quotation_DB*, and it is located on the C drive, in the *wwwroot* subdirectory of the *InetPub* directory. You indicate this path by clicking the *Select* button and manually locating the *.mdb* file. With this step of the process successfully completed, you are one third of the way in creating the client/server application.

THE HTML FRONT END

We now need to provide a Web interface through which the client will interact with the database we have set up. The easiest and quickest way to do this is by using a form on an HTML page. Forms enable page authors, such as us, a means of requesting and then processing user input. Every form is submitted to the server via a method specified by the *ACTION* attribute. This attribute can have one of two values:

- *GET*: This operation sends name/value pairs for each form element appended to the end of the URL. Each name/value pair is encoded and separated by an ampersand before being sent to the server.
- *POST*: Data are transmitted to the server via the standard-input, i.e., via HyperText Transfer Protocol (HTTP) headers. Information can be posted only to files that are executable, such as CGI scripts.

To demonstrate how this works, we'll create two forms — one that uses the *GET* method to get a fortune/quotation and five lucky numbers, and one that uses the *POST* method to search the database for a particular keyword. The HTML source code is displayed in Exhibit 12.6, and Exhibit 12.7 illustrates what you should see in your browser.

```
<html>
<head>
<meta http-equiv="Content-Type"
content="text/html; charset=iso-8859-1">
<title>So you want a fortune?</title>
</head>
<body bgcolor="#000080">
<CENTER><font color="#FFFFFF" size="6">700 Quotations/Fortunes as of
   10/19/98!!!</font></p></CENTER><BR>
<form action="127.0.0.1:8080/servlet/FortuneClientServlet" method="GET">
<CENTER><font color="#FF0000" size="5"><strong>So you want a
   fortune/quotation, huh? Don't we all... <br>
We got good ones and bad ones, so take a chance and grab one (or
   many)...</strong></font></CENTER><BR>
<CENTER>
<input type="submit" name="B1" value="I'm daring enough to push this
   button!">
</CENTER>
</form>

<form action="127.0.0.1:8080/servlet/QuoteSearch" method="POST">
<table border="0" width="100%">
   <tr>
      <td><CENTER><font color="#FFFF00" size="5"><strong>ADDED 10/20/98:
         SEARCH THE QUOTATION DATABASE BY KEYWORD!!!!<br>
      </strong></font><font color="#FF00FF" size="3"><strong>(Be patient,
         as the search may take some time.)</strong></font></CENTER>
      </td>
   </tr>
   <tr>
      <td><table border="0" width="100%">
            <tr>
               <td><CENTER><font color="#FF8040" size="5">Text you want to
                  search for:</font></p>
               </td></CENTER>
               <td><input type="text" size="38" name="keyword">
               </td>
            </tr>
         </table>
      </td>
   </tr>
   <tr>
      <td><CENTER><input type="submit" name="B1" value="Search!"></CENTER>
      </td>
   </tr>
</table>
</form>
</body>
</html>
```

Exhibit 12.6 Raw HTML source code.

THE MIDDLEWARE

In a two-tier client/server system, the business logic is either contained in a user interface like an applet, or it resides within the database on the server (e.g., a set of stored procedures). Alternatively, it can be in both locations. Two-tier systems are slightly more complex to manage because they are not as modular as systems which successfully separate the application

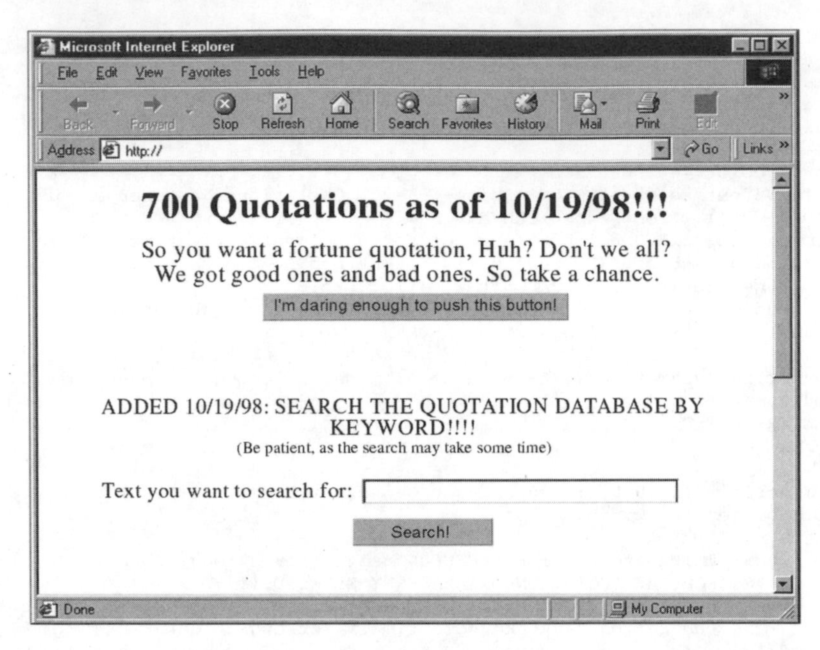

Exhibit 12.7 Visual representation of the HTML displayed in Exhibit 12.6.

---

and business logic from each other and the data. Our servlet project is a three-tier example that does just this. Here, the application logic (user interface) we don't even need to worry about — Microsoft, Netscape, and others have done the work for us. The servlet is the business logic which is going to mediate access between the client and the RDBMS. The servlet can be considered middleware, a vague term that refers to all the software needed to support interactions between clients and servers.

The first thing we're going to do before we even write a servlet, however, is to concentrate on the Fortune/Quotation server. The code for this project component is shown in Exhibit 12.8.

How Does the Server Work?

Let's examine the code and get a general idea of what's going on here, and how this component can be used in conjunction with servlets to complement our project. First, notice that the *FortuneServer* class is a subclass of *Thread*. This means that it has all of the *Thread* methods and data members, and the methods will remain as written unless we explicitly override

```
import java.net.*;
import java.io.*;
import java.sql.*;
import RequestProcessor;
import WriteToFile;

/**
 * The FortuneServer object binds to port 8888 and waits for clients to
 * connect. When it receives a connection, it interprets this as a request
 * for a fortune and starts a  RequestProcessor thread to handle the
 * request.
 * Created October 15, 1998.
 * @author Jonathan S. Held
 * @version 1.0
 * @see RequestProcessor
 */
public class FortuneServer extends Thread {

   java.net.ServerSocket fortuneSocket = null;
   java.net.Socket clientSocket = null;
   java.lang.String url = "jdbc:odbc:Quotation_DB";
   java.sql.Connection con = null;
   java.sql.Statement stmt = null;
   static long numberOfRequests = 0;
   final int DB_SIZE = 700, DATA_DUMP = 50;
   static int queries[];

   /**
    * Class constructor
    * Creates a socket on port 8888 and binds to it. Attempts to load the
    * Sun bridge driver which is used to talk to the MSAccess database.
    * Enters into the log file fortune.log the date on which the log file
    * entries that follow were created.
    * @param none
    * @exception ClassNotFoundException thrown if the FortuneServer is
    * unable to load the Sun Jdbc-Odbc bridge driver
    * @exception SQLException thrown if the database url is unaccessible
    * @exception IOException thrown if unable to bind to port 8888 (e.g.,
    * the port is already in use by another process)
    */
   FortuneServer(){
      try {
         queries = new int[DB_SIZE];
         fortuneSocket = new ServerSocket(8888);
         System.runFinalizersOnExit(true);
         System.out.println("Fortune server successfully bound to port
            8888.");

         try {
            Class.forName("sun.jdbc.odbc.JdbcOdbcDriver");
            con = DriverManager.getConnection(url, "sa", "");
            stmt = con.createStatement();
            System.out.println("Established connection to database.");
            System.out.println("Awaiting client requests...");
            java.util.Calendar ts = java.util.Calendar.getInstance();
            java.lang.String info = new String("Log file created on " +
               ts.getTime().toString());
            (new WriteToFile(info)).start();
```

Exhibit 12.8 The *FortuneServer* Java code.

```
        }
        catch (java.lang.ClassNotFoundException e1) {
           System.err.println(e1.toString());
        }
        catch (java.sql.SQLException e2){
           System.err.println(e2.toString());
        }
     }
     catch (java.io.IOException e3){
        System.err.println("Unable to bind to port 8888.");
        System.err.println(e3.toString());
        System.err.println("Hit any key to continue.");
        try {
           System.in.read();
        }
        catch (java.io.IOException e4){
           System.out.println(e4.toString());
        }
     }
  }//end FortuneServer() constructor

  /**
   * Uses the socket.accept() method to wait for an incoming request. The
   * server indicates how many requests it has processed, determines if
   * it needs to dump statistical information to the log file (currently
   * done after every 50 requests), and then starts a new
   * RequestProcessor thread to handle the request. The RequestProcessor
   * object is passed the client's socket information as well as a JDBC
   * statement object that is used to query the MSAccess database.
   * This method is run in a while(true) loop and can only be terminated
   * by system shutdown or CTRL-C.
   * @param none
   * @see RequestProcessor
   * @exception IOException thrown if unable to accept incoming client
   * requests
   * @return none
   */
  private void runServer(){
     while (true){
        try {
           clientSocket = fortuneSocket.accept();
           System.out.println("Processing request number " +
              (++numberOfRequests));
           if (numberOfRequests % DATA_DUMP == 0)
              writeStatistics();
     (new RequestProcessor(clientSocket, stmt)).start();
        }
        catch (java.io.IOException e){
           System.out.println("Unable to fulfill fortune request.");
           System.out.println(e.toString());
        }
     }
  }//end runServer()

  /**
   * Creates a new FortuneServer object and calls the thread's start
   * method. @param args[] a series of command line arguments stored in
   * array; not used.
   * @exception none
```

Exhibit 12.8 (continued)

```
     * @return none
     */
   public static void main(String args[]){
      //start a new FortuneServer
      (new FortuneServer()).start();
   }//end main()

   /**
    * Called when the thread is started; calls the private utility method
    * runServer
    * @param none
    * @return void
    */
   public void run(){
      runServer();
   }//end run()

   /**
    * responsible for creating a new WriteToFile object and writing
    * information to the  logfile fortune.log.
    * @param none
    * @see WriteToFile
    * @return void
    */
   private void writeStatistics(){
      java.lang.StringBuffer statistics = new StringBuffer("Data Dump for
         " + Long.toString(numberOfRequests) + " requests: ");
      for (int ix=0; ix < DB_SIZE; ix++){
         statistics.append(Integer.toString(queries[ix]) + " ");
         if ((ix !=0) && (ix % 25 == 0))
            statistics.append(" | BREAK | ");
      }
      (new WriteToFile(statistics.toString())).start();
   }//end writeStatistics()
}//end class FortuneServer
```

Exhibit 12.8 (continued)

---

them by redefining their behavior. The server is going to be a multi-threaded process so it can capably handle many concurrent requests.

The *FortuneServer* begins by executing code contained in its *main()* method. It is here that we simply create a new *FortuneServer*, and then start the thread that the application just spawned. We should briefly look at the class constructor to see what happens when we create a new *FortuneServer* object. Here, the variable *queries*, a 700-element integer array, is created and its contents are initialized to 0. We're going to use this variable to keep track of how many times a particular fortune was displayed. In this manner, we can examine our logfile later to determine if we're really getting a random, distributed return of fortunes. Once the array has been initialized, we need to get the server to bind to a port. We do this by creating a new *ServerSocket* called *fortuneSocket* and binding it to port 8888. If all is successful, you should see the message "Fortune server successfully bound to port 8888" when you run the program.

Of course, the next important step the server needs to make is to connect to the database. We could leave this task to the servlet, and do it once and only once in its *init()* method; however, it's just as appropriate for the *FortuneServer* to do this job on its own. This is exactly what happens in the *try/catch* block that follows. We load the *sun.jdbc.odbc.JdbcOdbcDriver* and then use a JDBC *Connection* object to connect to our remote data source. Notice that we specify what data source we want to use with a string. In our example, the string is set to "jdbc:odbc:Quotation_DB", where *jdbc* is the protocol, *odbc* is the subprotocol, and *Quotation_DB* is the name of the data source. Because the server is going to run on the same machine as the data source, there is no need for a host name or Internet Protocol (IP) address to let the application know where the database is. If this were not the case, i.e., there was physical separation between the server and the database, you would need to use a different driver and syntax.

This brings us to the *run()* method, where most threads contain the specialized code they are going to perform during their lifetime. Our *run()* method is called *runServer()*, which waits for a client to connect. The *fortuneSocket accept()* method is a blocking call which keeps the program waiting here until that connection is made. Once a client binds to the port the server is listening on, another message appears that indicates what request number is being processed. A data dump of the *queries* variable into our logfile occurs every 50 requests (by making a call to *writeStatistics()*), and execution continues by turning over control to the *RequestProcessor* component. This allows the server to continue its job of waiting for requests, while some other part of the system processes the actual request and responds to the client. The *RequestProcessor* code is shown in Exhibit 12.9.

What does the RequestProcessor do? The *RequestProcessor* is itself a thread, and the server spawns a new *RequestProcessor* thread for each new client request. Notice that this class does not have a *main()* method; rather, the object's *start()* method is called and control is eventually routed to the *run()* method. When one of these objects is created, two vitally important pieces of information are needed — the client's socket and an initialized JDBC *Statement* object. We retain the information about the client because it is to this port number that we are going to transfer information. The *Statement* object is initialized from the *Connection* object, so whenever we perform SQL operations (which is why we want it), the *Statement* object inherently knows what data source it is tied to.

The SQL statement we're going to use is

```
"SELECT * FROM Fortunes WHERE Fortune_ID = " + random
```

```
import java.net.*;
import java.io.*;
import java.sql.*;
import FortuneServer;
import java.util.Random;


/**
 * The RequestProcessor object is used by the FortuneServer to handle
 * client requests. This thread is created when the server needs to get a
 * quotation or fortune from the MSAccess database, generate five lucky
 * numbers, and send the information back to the FortuneClientServlet.
 * Created October 15, 1998.
 * @author Jonathan S. Held
 * @version 1.0
 * @see FortuneClientServlet
 */
public class RequestProcessor extends Thread {

   java.net.Socket cs = null;
   java.sql.Statement statement = null;
   final int MAX_FORTUNES = 700;
   final int LUCKY_NUMBERS = 5;
   final int LOTTERY_NUMBER_MAX_VALUE = 50;

   /**
    * Class constructor
    * @param clientSocket the socket the client attached from
    * @exception statement a JDBC Statement object associated with a
    * database connection; these parameters are passed from the
    * FortuneServer at the time a new RequestProcessor object is created
    */
   RequestProcessor(java.net.Socket clientSocket, java.sql.Statement
      stmt){
      cs = clientSocket;
      statement = stmt;
   }

    /**
    * Called when the RequestProcessor thread is started; run generates a
    * random number, selects the quotation from the database based on this
    * number, then makes creates random numbers; this information is sent
    * back to the FortuneClientServlet, which will then process it and
    * send it back to the client's browser.
    * @param none
    * @return void
    * @exception IOException thrown if an outputstream cannot be created
    * to the client @exception SQLException thrown if an SQL error occurs
    * when trying to query the database
    */
   public void run(){

      try {
         Random generator = new Random();
         int random = Math.abs(generator.nextInt() % MAX_FORTUNES) + 1;
         int num[] = new int[LUCKY_NUMBERS];
         java.lang.String query = new String("SELECT * FROM Fortunes WHERE
            Fortune_ID = " + random);
         FortuneServer.queries[random-1] += 1;
```

Exhibit 12.9 The *RequestProcessor* Java code.

```
      java.lang.String response = null;
      java.sql.ResultSet rs = statement.executeQuery(query);
      while (rs.next()){
         rs.getInt(1);
         response = new String(rs.getString(2));
         response += "<BR><BR><font color='#004080'>Your lucky numbers
            are: </font>";

         for (int ix=0; ix<LUCKY_NUMBERS; ix++){
            int number = Math.abs(generator.nextInt() %
               LOTTERY_NUMBER_MAX_VALUE) + 1;

            if (ix !=0){
               boolean check = true;
               while (check){
                  for (int jx=0; jx <= ix; jx++){
                     if (num[jx] == number)
                        number = Math.abs(generator.nextInt() %
                           LOTTERY_NUMBER_MAX_VALUE) + 1;
                     else {
                        check = false;
                        num[ix] = number;
                     }
                  }
               }
            }
            else num[ix] = number;
         }
         response += "<font color='#FF0000'>" + num[0] + ", " + num[1]
            + ", " + num[2] + ", " + num[3] + ", " + num[4] +
            "</font>";
         if (response != null){ break; }
      }
      java.io.BufferedWriter out = new java.io.BufferedWriter(new
      OutputStreamWriter(cs.getOutputStream()));
      out.write(response, 0, response.length());
      out.flush();
      out.close();
      cs.close();
   }
   catch (java.io.IOException e1){
      e1.printStackTrace();
   }
   catch (java.sql.SQLException e2){
      System.out.println(e2.toString());
   }
 }//end run()
}//end class RequestProcessor
```

Exhibit 12.9 (continued)

This object's *run()* method generates a random number which corresponds to the fortune/quotation we are going to return. The SQL statement is executed by using a *ResultSet* object. The net effect of the line that reads

rs = statement.executeQuery(query)

is to execute the SQL string specified by the variable *query* and to return a reference of the results back to the *ResultSet* object that invoked the method. In this case, we expect to get only one tuple (or row) back from the database. The *getXXX()* methods of the rs object allow us to pick off the values contained in each column (or field). Without any real reason, we make a call to rs.getInt(1) to illustrate how to retrieve the *Fortune_ID* number. It is the next part that we make use of — rs.getString(2) returns the text of the fortune/quotation to the *response* string. To this, we append our five lucky numbers (which includes a little algorithm for ensuring all numbers are unique), and generate some HTML code that is sent back to a servlet via a *BufferedWriter* object.

The only part that remains is somehow to tie the browser and the server together. We do this with the *FortuneClientServlet.* This component will be invoked by the HTML form and will connect to the server on the client's behalf. Once this is done, all of the actions that were described above take place. Let's turn our attention to this project's centerpiece, the *FortuneClientServlet* code (see Exhibit 12.10), as without it we would be unable to make any of this happen.

Creating the Client Servlet

The *FortuneClientServlet* is a subclass of *HttpServlet.* It contains one and only one method *-doGet()* — that redefines the behavior the superclass provided. When we click the button "I'm daring enough to push this button," on the HTML form, a program called *servletrunner* (part of the servlet SDK) is executing on the target machine, takes the form request and any information the form contains, and acts as a proxy by directing it to the appropriate servlet. Our *FortuneClientServlet* gets called, and code execution begins in the method *doGet()* — *doPost()* if this were a *POST* action. Notice that the *FortuneClientServlet* attaches to the port the server is listening to; the server delegates the task of getting a fortune to the *RequestProcessor,* and this last component returns the fortune to the servlet. The servlet has initiated a chain of events that effectively limits its participation in this system to receiving a fortune, then forwarding it to the client that requested it. The culmination of this part of the project is shown in Exhibit 12.11.

Searching the Database

Surely one of the more popular tasks today is being able to perform searches against databases. For that reason, we've developed a *QuoteSearch* servlet. The client can enter a keyword, then exhaustively search all 700 fortunes/quotations, and, if the keyword is found, the entry is returned. This servlet is no more difficult to develop than the former; however, it does illustrate some things we haven't talked about, e.g., how do we capture form input from a servlet and how do we use the *init()* method to our benefit? Before we continue, take some time to examine the code in Exhibit 12.12.

```
import java.io.*;
import java.net.*;
import javax.servlet.*;
import javax.servlet.http.*;
import WriteToFile;

/**
 * FortuneClientServlet creates a new socket and attaches to the
 * FortuneServer object. The connection to the fortune server generates a
 * request for a fortune, and FortuneClientServlet waits until its request
 * has been fulfilled before returning the fortune and five lucky numbers
 * to the client that invoked it. Please note that this is not like a
 * regular object (there is no constructor). Creation of the
 * FortuneClientServlet is done by the servletrunner utility program,
 * which is part of the Servlet Software Development Kit (SDK).
 * Created October 15, 1998.
 * For more information, please see <a href="http://jserv.java.sun.com/
 * products/java-
 * server/servlets/index.html">the Servlet SDK.</a>
 * @author Jonathan S. Held
 * @version 1.0
 */

public class FortuneClientServlet extends HttpServlet
{

   /**
    * doGet() - Overridden from HttpServlet to handle GET operations.
    * @param request HttpServlet request object encapsulating
    * communication from the client
    * @param response HttpServletResponse object encapsulating means of
    * communicating from the server back to the client
    * @return void
    * @exception IOException thrown if the servlet cannot create a socket
    * to the server on port 8888
    * @exception ServletException handled by the superclass
    * This method implements a GET operation called from an HTML form's
    * ACTION URL. HTML is sent back to the client via the response object.
    */
   public void doGet (HttpServletRequest request, HttpServletResponse
      response) throws ServletException, IOException
   {
         java.lang.String fortune = new String();
         java.io.PrintWriter out;
         String title = "Your lucky fortune/quotation...";
         response.setContentType("text/html");
         out = response.getWriter();
         out.println("<HTML><HEAD><TITLE>");
         out.println(title);
         out.println("</TITLE></HEAD><BODY>");
         out.println("<body bgcolor='#FFFF00'>");
   try {
      java.net.Socket socket = new Socket("127.0.0.1", 8888);
      java.io.BufferedReader in = new BufferedReader(new
      InputStreamReader(socket.getInputStream()));

         for (int ch = in.read(); ch > 0; ch = in.read())
         fortune += (char)(ch);

         socket.close();
```

Exhibit 12.10 The *FortuneClientServlet* code.

```
      }
      catch (java.io.IOException e){}

      out.println("<CENTER><font color='#000000'><H1><B><I>" + fortune +
         "</I></B></H1></font><BR></CENTER>");
      out.println("</BODY></HTML>");
      out.close();

      java.util.Calendar ts = java.util.Calendar.getInstance();
      java.lang.String info = "On " + ts.getTime().toString() + " received
         request from " + request.getRemoteAddr();
      System.out.println(info);
      (new WriteToFile(info)).start();
   }//end doGet()
}//end class FortuneClientServlet
```

Exhibit 12.10 (continued)

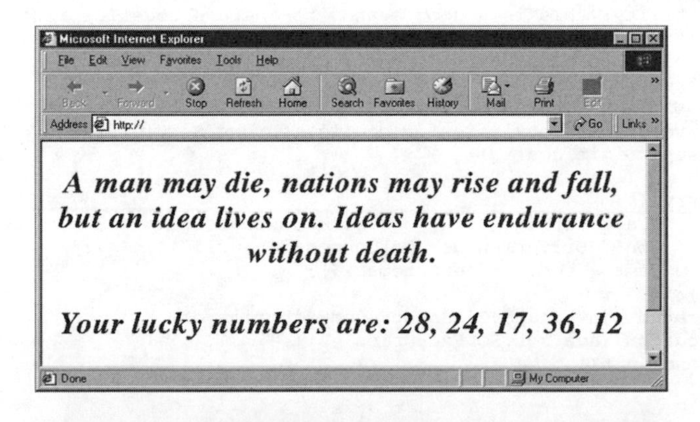

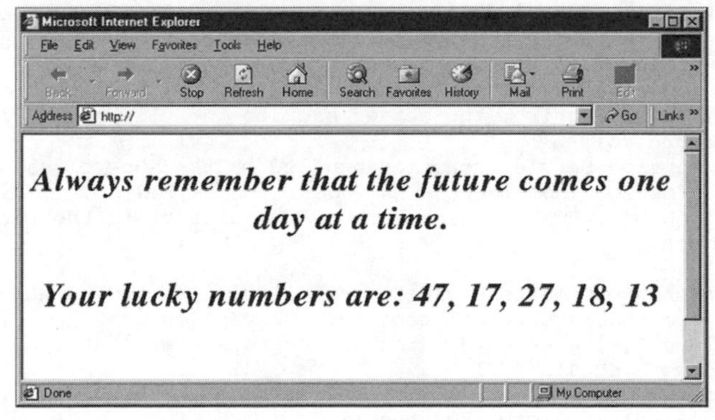

Exhibit 12.11 Random fortunes and quotations as seen by the client.

```
import java.io.*;
import javax.servlet.*;
import javax.servlet.http.*;
import java.sql.*;

/**
 * QuoteSearch is a Java servlet created to allow a client to search the
 * database for a keyword.
 * Created October 15, 1998.
 * For more information, please see <a href="http://jserv.java.sun.com/
 * products/java-server/servlets/index.html">the Servlet SDK.</a>
 * @author Jonathan S. Held
 * @version 1.0
 */
public class QuoteSearch extends HttpServlet
{
   static java.sql.Connection con;
   static java.sql.Statement stmt;
   static final java.lang.String url = "jdbc:odbc:Quotation_DB";
   static final int INITIAL_SIZE = 20;

      /**
       * init() - Servlet method invoked only once by the servletrunner
       * utility; this is a good method to include code for resource-
       * intensive operations, such as connecting to a database
       * @param response ServletConfig object
       * @return void
       */
   public void init(ServletConfig config) throws ServletException {
         super.init(config);

         try {
            Class.forName("sun.jdbc.odbc.JdbcOdbcDriver");
            con = DriverManager.getConnection(url, "", "");
            stmt = con.createStatement();
         }
         catch (java.lang.ClassNotFoundException e1) { }
         catch (java.sql.SQLException e2){ }
      }//end init()

   /**
       * doPost() - Overridden from HttpServlet to handle POST operations.
       * @param request HttpServlet request object encapsulating
       * communication from the client
       * @param response HttpServletResponse object encapsulating means of
       * communicating from the server back to the client
       * @return void
       * @exception ServletException handled by the superclass
       * This method implements a GET operation called from an HTML form's
       * ACTION URL. HTML is sent back to the client via the response
       * object.
       */
   public void doPost (HttpServletRequest request, HttpServletResponse
      response) throws ServletException, IOException {
      java.lang.String keyword = request.getParameter("keyword");
      if (keyword.equals(""))
         return;
      else goFindIt(keyword, response);
   }
```

Exhibit 12.12 The *QuoteSearch Servlet* code.

```
/**
 * goFindIt() - Searches for a keyword in a fortune/quotation.
 * Returns the fortune/quotation with the keyword highlighted to the
 * client.
 * @param response whatToFind a string representing the keyword to
 * find
 * @param response HttpServletResponse object encapsulating means of
 * communicating from the server back to the client
 * @return void
 */
private void goFindIt(java.lang.String whatToFind, HttpServletResponse
   response)
{
   java.lang.String query = "SELECT Fortune FROM Fortunes";
   int number_found = 0, total_quotes = 0;
   java.io.PrintWriter out;
   java.lang.String title = "Matches...";

   try {
      response.setContentType("text/html");
      out = response.getWriter();
      out.println("<HTML><HEAD><TITLE>");
      out.println(title);
      out.println("</TITLE></HEAD><BODY>");
      out.println("<body bgcolor='#800000'><font color='#00FF00'
         size='5'>");
      out.println("<H1><I>Searching... Matches appear
         below:</I></H1>");
      out.flush();
      java.sql.ResultSet rs = stmt.executeQuery(query);
      while (rs.next()){
         java.lang.String quote = rs.getString(1);
         total_quotes++;

         if (inQuote(whatToFind, quote)){
            number_found++;

            int index =
               quote.toLowerCase().indexOf(whatToFind.toLowerCase());

            out.print("<img src='http://127.0.0.1/images/speaking.gif'
               width='25' height='25'>");

            for (int ix=0; ix < index; ix++)
               out.print(quote.charAt(ix));
            out.print("<B><I><font color='#FFFF00'>");

            int match_length = whatToFind.length();
            for (int jx=index; jx<index+match_length; jx++)
               out.print(quote.charAt(jx));
            out.print("</font></B></I>");

            int start = index+whatToFind.length(), end =
               quote.length();
            for (int kx=start; kx < end; kx++)
               out.print(quote.charAt(kx));
            out.println("<BR><BR>");
            out.flush();
         }
      }
```

Exhibit 12.12 (continued)

```
        out.println("</font><font color='#FF0080' size='4'>");
        out.println("Number of quotations is " + total_quotes + "<BR>");
        if (number_found == 0)
           out.println("Sorry... Your keyword was not found in any " +
              "quotations/fortunes.");
        else
           out.println("Your query resulted in " + number_found + "
              matches.");
        rs.close();
        out.println("</font></BODY></HTML>");
        out.close();
     }
     catch (java.io.IOException e) { }
     catch (java.sql.SQLException e) { }

  }

  /**
      * inQuote() - Returns a boolean value indicating whether the
      * keyword being looked for is anywhere in the fortune/quotation;
      * this is a case insensitive search
      * @param lookingFor the keyword string
      * @param quote the text to be searched
      * @return boolean indicating whether lookingFor is in the quote or
      * not
      */
  private boolean inQuote(java.lang.String lookingFor, java.lang.String
     quote)
  {
     boolean found = false;
     if (quote.toLowerCase().indexOf(lookingFor.toLowerCase()) != -1)
        found = true;
        return found;
  }
}
```

Exhibit 12.12 (continued)

Much of the code we see here should look familiar — the process of connecting to the database and working with SQL statements remains the same. We perform the initial resource-intensive operation of connecting to the database only once — in the *init()* method. The *servletrunner* proxy, which listens for servlet requests, ensures that each servlet's *init()* is executed just once.

After the client enters the keyword and clicks the Submit button, a *POST* operation is performed. For this reason, we override the *doPost()* method and tailor our response to the client's action with any code we place in here. Notice that we have an *HttpServletRequest* and an *HttpServletResponse* object. These objects contain a number of methods that allow us to learn, respectively, information about the request that was generated (such as where it came from, information that was passed in the request via HTTP headers, etc.) and a means for responding to the request as we see fit.

We use the *HttpServletRequest* method *getParameter()* to retrieve values from forms. This method takes a string that represents the name we assigned to the HTML text control. If the client tries to submit the form without entering a keyword, which we explicitly check for, no action is taken (although a white screen will appear). We could later customize this servlet to return an error message, if we were so inclined. If a keyword is entered, we make a call to *goFindIt()*, which requires two parameters: the keyword being searched for and the *HttpServletResponse* object which is used to communicate back with the client.

Some HTML is immediately generated and sent back to the client, so when you run this servlet you'll get a maroon screen that informs you a search is in process. All quotations are retrieved from the database, and *inQuote()* determines if the keyword is found. If it is, the quotation is returned (with the keyword portion highlighted in yellow), and the search process goes on until the entire database is examined. Meanwhile, the client gets the perception that the page is still loading. When the servlet is done executing, some summary statistics are returned. I promise not to scrutinize the code any further (since you can examine it as well as I can). Suffice it to say that this search is slow and could be significantly improved in a couple of ways: if a keyword appears as part of word, the keyword portion is highlighted; if it appears twice in a fortune, only the first occurrence is highlighted. These are areas for improvement that I'll leave as an exercise for the reader. Exhibit 12.13 shows two screen captures of what you should expect the *QuoteSearch* servlet to return.

QUICK SETUP

Installation of the servlet SDK will create a JSDK2.0 directory and subdirectories for documentation (*doc*), executable programs (*bin*), library files (*lib*), and source code (*src*). You'll find the *servletrunner* utility in the *bin* directory. You configure this program, i.e., associate a servlet name and its compiled class file, by modifying the *servlet.properties* file in a text editor. Examples of how to use this file are illustrated in Exhibit 12.14.

Writing your own servlets requires two more things: all your programs must import the *javax.servlet* and *javax.servlet.http* packages, and you must start the *servletrunner* utility after you've edited the *servlet.properties* file. The easiest way to import the packages into your programs is by modifying your *CLASSPATH* setting as follows:

```
SET CLASSPATH = %CLASSPATH%;C:\jsdk2.0\lib\jsdk.jar
```

This will allow you to use the *javac* compiler without error, and the only thing left to do is to start the *servletrunner* utility. You can do this by simply typing the name of the program at a DOS command prompt, or you can

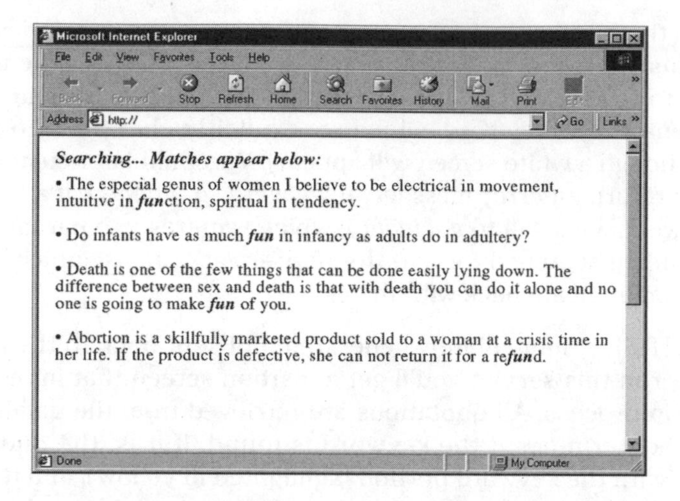

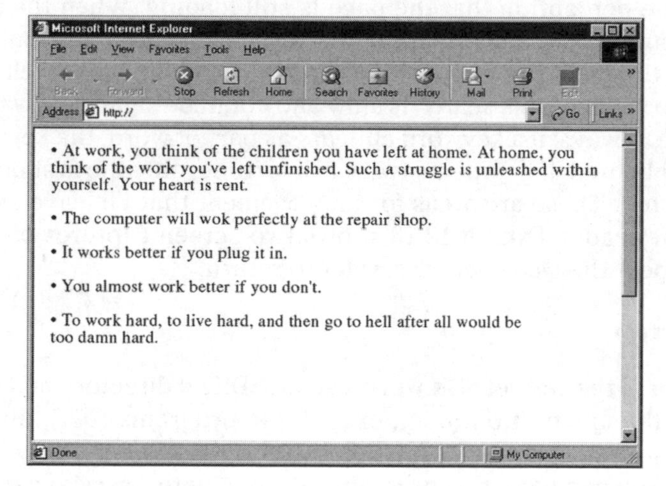

Exhibit 12.13 ***QuoteSearch Servlet* results for the keywords "fun" and "work."**

---

append a number of parameters to customize its configuration. Exhibit 12.15 shows you what command-line parameters are available.

CONCLUDING REMARKS

Servlets are a useful extension to the Java programming language that have almost the identical functionality and utility of CGI programs, but unlike the latter, they are not as prone to security risks and are much easier to write. This chapter has demonstrated how you can rapidly develop and deploy a three-tier client/server RDBMS application using this technology.

```
# @(#)servlets.properties 1.86 97/11/14
#
# Servlets Properties
#
# servlet.<name>.code=class name (foo or foo.class)
# servlet.<name>.initArgs=comma-delimited list of {name, value} pairs
#       that can be accessed by the servlet using the
#       servlet API calls
#

# simple servlet
servlet.simpleservlet.code=SimpleServlet

# survey servlet
servlet.survey.code=SurveyServlet
servlet.survey.initArgs=\
   resultsDir=/tmp

servlet.FortuneClientServlet.code = FortuneClientServlet

servlet.QuoteSearch.code = QuoteSearch
```

Exhibit 12.14 The *servlet.properties* file.

```
C:\JSDK2.0\bin>servletrunner /? >t
Usage: servletrunner [options]
Options:
   -p port          the port number to listen on
   -b backlogthe    listen backlog
   -m max           maximum number of connection handlers
   -t timeout       connection timeout in milliseconds
   -d dir           servlet directory
   -s filename      servlet property file name
java.exe: No error
```

Exhibit 12.15 *Servletrunner* command-line parameters.

If you have specific questions that you cannot find answers to after consulting the resources listed below, you may contact the author via e-mail at jsheld@hotmail.com.

Resources

1. JDK1.2: http://www.javasoft.com/products/jdk/1.2
2. Servlet 2.1 SDK: http://www.javasoft.com/products/servlet/index.html
3. Servlet 2.1 API: http://www.javasoft.com/products/servlet/2.1/html/servletapiTOC.fm.html
4. JDBC™ 2.0 API: http://www.javasoft.com/products/jdbc/jdbcse2.html
   All Java code (source and compiled class files) and the MSAccess database for the project described in this chapter can be obtained from this magazine's Web site.

Section III
Hardware

A server is similar to other types of computers in that its operational capability is governed by hardware and software. In this section we turn our attention to the former, focusing our attention on the server's hardware components that govern its operational capability. To do so, the following chapters examine memory, processors, data bus structures, the use of redundant arrays of inexpensive disks, and two specific types of servers.

In Chapter 13 David Hanson provides us with an overview of different types of memory and the advantages and disadvantages associated with each. In his chapter entitled "Server Memory Issues," Hanson addresses how different types of memory affect the performance of different types of servers. In doing so, his chapter tackles such questions as

- Will doubling memory be more of a performance boost than adding another processor?
- Will increasing the amount of RAM boost cache size and deliver better performance?
- At what point will upping RAM be less effective than adding more processors?

Continuing our focus upon hardware, A. Padgett Peterson examines the two types of processors used to provide the computational platform upon which servers are based. In Chapter 14 titled "CPUs," Peterson explains the role of RISC and CISC processors as well as the relationship of the computer's memory bus, cache, and the use of speed accelerators to enhance overall processor performance.

Chapter 15 continues our examination of key hardware components of a server. The chapter entitled "Server Data Bus Structures," also authored by A. Padgett Peterson, provides us with an overview of the evolution of the computer bus and describes its role in enhancing network, disk, and other peripheral operations. Because bus width and speed will govern I/O operations, this chapter provides information that can be of considerable value when selecting an appropriate server platform.

Because the storage subsystem of a server represents one of its most important hardware components, Chapter 16, "The RAID Advantage" by Tyson Heyn, focuses on this critical area. This chapter reviews the operation and utilization of different RAID levels, explains how this storage system is interfaced to a computer, and, through the use of a series of exhibits, makes us aware of how we can obtain an appropriate level of data storage redundancy through the selection of an appropriate storage subsystem.

The two concluding chapters in this section discuss hardware by examining the entire server. In Chapter 17, titled "Multiprocessing Servers," Bill Wong examines the rationale for servers' supporting multiple processors and describes and discusses products from Intel, Sun, and IBM. Wong describes several types of multiprocessor systems to include RISC-based products that provide up to 46G floating point operations per second (FLOPS).

Chapter 18 concludes this section with another chapter authored by A. Padgett Peterson entitled "Super Servers — One Step Beyond." This chapter first examines a core set of requirements that can justify the use of a super server. It then describes the technology which defines this high-end server and makes us aware of the special software that may be necessary to optimize the performance of the server's hardware components.

Chapter 13
Server Memory Issues

M. David Hanson

The World Wide Web has grown from the small ARPAnet network of the 1970s, devoted to document-sharing among a handful of universities and governmental agencies to today's large *network of networks* serving millions of computers and users. It is, at the same time, a huge repository of information and a marketplace for commercial and financial transactions.

The paradigm shift on the Web has been to change the role of Web servers, from serving flat HTML files, to hosting Web applications that serve live, dynamic content. In effect, Web servers are now high-powered applications servers. Client/server computing has taken on a whole new life.

As a result, servers have evolved to the point where they are, in many cases, more powerful than mainframes and can scale up to 64 or more processors, and over 1-gigabyte (G) of memory. The issue to be addressed in this chapter is how to optimize the server hardware configuration in general, and the RAM requirements in particular. This chapter addresses the issue of how memory affects server performance, including such questions as

- Will doubling memory be more of a performance boost than adding another processor?
- Will increasing the amount of RAM boost cache size and deliver better performance?
- At what point will increasing the RAM be less effective than adding more processors?

The following environments will be considered:

- File-server environment, measuring how well a server handles I/O requests and how much throughput it provides to clients
- Application-server environment, in which the server has the added responsibility of hosting the applications the client machines, are running (transactions/sec)
- HTTP/Web environment, in which the server handles requests of both static and dynamic content (requests/sec, throughput)

0-8493-9823-1/00/$0.00+$.50

WHAT IS RAM?

The CPU (such as a Pentium) is the heart of the client/server environment. It is where data are processed and program instructions are interpreted. Integrated with the CPU is the system's main memory, the random access memory we call RAM. Together, these two components make up the core of the client/server machine; components such as hard disks, controllers, and video cards are peripheral to this central activity, and are therefore known as peripherals. The CPU uses its RAM as a storage area for data, calculation results, and program instructions, drawing on this storage as necessary to perform the tasks required by programs. In order to store data and draw from the data store, the CPU specifies the memory address of the required information. The address bus allows the CPU to send the address to RAM, and the data bus allows the actual data to transfer to the CPU.

The term *bus* itself refers to the connection between the two devices that allows them to communicate. An important measurement of RAM performance is access time, the amount of time that passes between the instant the CPU issues an instruction to RAM to read a particular piece of data from a particular address and the moment the CPU actually receives the data.

Today's RAM chips typically have a 60-nanosecond (ns) access time, which means it takes 60 ns (a nanosecond is a billionth of a second) to perform this round trip function. This access time is much faster than that of the 100- to 120-ns chips of a few years ago, but it's still much slower than the ideal access time of zero, which would be realizable if the CPU itself stored all the data.

To speed things further, the CPU has access to cache memory (usually referred to as the *cache*). At 20 ns or better, cache memory is faster than main memory, but systems contain less of it than main memory (cache memory is expensive), and therefore only select data — the data the CPU will probably need next — is placed inside it. The cache controller handles the selection.

Memory chips function by storing electronic charges. The chips are made of a capacitor and a transistor, with the capacitor storing the charge and the transistor turning the charge on or off. With RAM chips, the system can alter the on or off state of the charges, but with ROM (read-only memory) chips the charges are either permanently on or permanently off. This chapter deals only with RAM.

All RAM technologies emphasize speed and attempt to offer more of it without an increase in cost. But CPU technology keeps getting faster, and memory technology must keep pace — hence the need for different types of RAM (more on that later). In general, all computers, whether clients or servers, deal with information in two ways: as memory (or storage) and as logic (or processing). So, to summarize, there are two kinds of active memory:

- ROM (read only memory) holds the basic routines for communicating with most of the devices used by the PC. It is a memory chip located on the motherboard inside the PC's system unit and cannot be modified.
- RAM (random access memory) *temporarily* holds the programs and data while the computer is working with them. Think of RAM as the computer's short-term memory.

RAM is a series of microchips (integrated circuits called SIMM or DIMM) on the motherboard or on special circuit cards plugged into the motherboard. These chips contain millions of tiny "memory" transistors and capacitors, which in general tend to handle information more slowly than so-called *logic* transistors, which are used in the CPU. The instructions that tell the PC what to do and the information it processes are kept in RAM while it is on, so the more RAM installed, the more the PC can do at one time. This is a very important consideration when the PC is functioning as a server, serving thousands of clients.

A computer uses RAM to hold temporary instructions and data needed to complete tasks. This enables the computer's CPU to access instructions and data stored in memory very quickly. A good example of this is when the CPU loads an application program — such as a word processor or page layout program — into memory, thereby allowing the application program to run as quickly as possible. In practical terms, this means more work can be done with less time spent waiting for the computer to perform tasks.

Entering a command from the keyboard results in data being copied from a storage device (such as a hard disk drive or CD-ROM drive) into RAM, which can provide data to the CPU more quickly than storage devices.

When the computer is turned on, the operating system and other files are loaded into RAM, usually from the hard disk. RAM can be compared to a person's short-term memory and the hard disk to the long-term memory. The short-term memory focuses on work at hand, but can keep only so many facts in view at one time. If short-term memory fills up, the brain sometimes is able to refresh it from facts stored in long-term memory. A computer also works this way. If RAM fills up, the processor needs to continually go to the hard disk to overlay old data in RAM with new, slowing down the computer's operation. Unlike the hard disk, which can theoretically fill up and put the server out of business, so to speak, RAM never runs out of memory. It keeps operating, but much more slowly than desirable.

RAM is physically smaller than the hard disk and holds much less data. A typical client computer may come with *32 million* bytes of RAM and a *4 billion*-byte hard disk. A server can be much larger, up to gigabytes of RAM or more. RAM comes in the form of discrete (meaning separate) microchips and also in the form of modules that plug into slots in the computer's motherboard. These slots connect through a bus or set of electrical paths to the processor. The hard drive, on the other hand, stores data on a magnetized surface that looks like a stack of CD-ROMs.

Having more RAM reduces the number of times that the computer processor has to read data from the hard disk, an operation that takes much longer than reading data from RAM. (RAM access time is in nanoseconds; hard disk access time is in milliseconds.)

RAM is called *random access* because any storage location can be accessed directly. Originally, the term distinguished regular core memory from offline memory, such as magnetic tape in which an item of data could be accessed only by starting from the beginning of the tape and finding an address sequentially. Perhaps it should have been called "nonsequential memory" because RAM access is hardly random. RAM is organized and controlled in a way that enables data to be stored and retrieved directly to specific locations.

ROM is a more expensive kind of memory that retains data even when the computer is turned off. Every client and server computer comes with a small amount of ROM that holds just enough programming so that the operating system can be loaded into RAM each time the computer is turned on.

In general, RAM is much like an arrangement of post-office boxes in which each box can hold a 0 or a 1. Each box has a unique address (see Exhibit 13.1) that can be found by counting across columns and then counting down by row. In RAM, this set of post-office boxes is known as an array and each box is a cell. To find the contents of a box (cell), the RAM controller sends the column/row address down a very thin electrical line etched into the chip. There is an address line for each row and each column in the set of boxes. If data are being read, the bits that are read flow back on a separate data line.

In describing a RAM chip or module (see Exhibit 13.2), a notation such as 256 K × 16 means 256 thousand columns of cells standing 16 rows deep. An 8-MB module of dynamic RAM contains 8 million capacitors and 8 million transistors and the paths that connect them. In the most common form of RAM, dynamic RAM, each cell has a charge or lack of charge held in something similar to an electrical capacitor. A transistor acts as a gate in determining whether the value in the capacitor can be read or written. In static RAM, instead of a capacitor-held charge, the transistor itself is a positional "flip-flop" switch, with one position meaning 1 and the other position meaning 0.

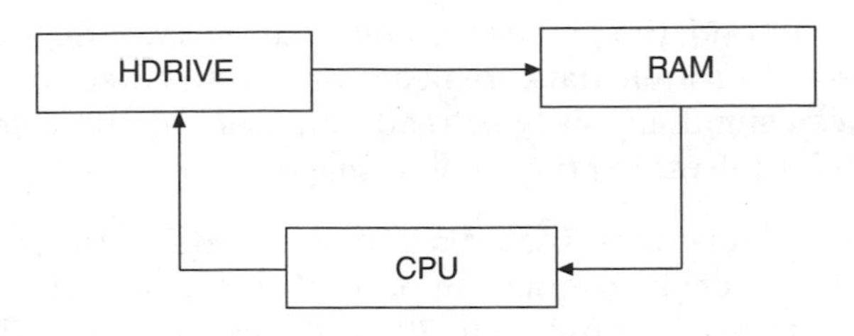

Exhibit 13.1 Unique addresses.

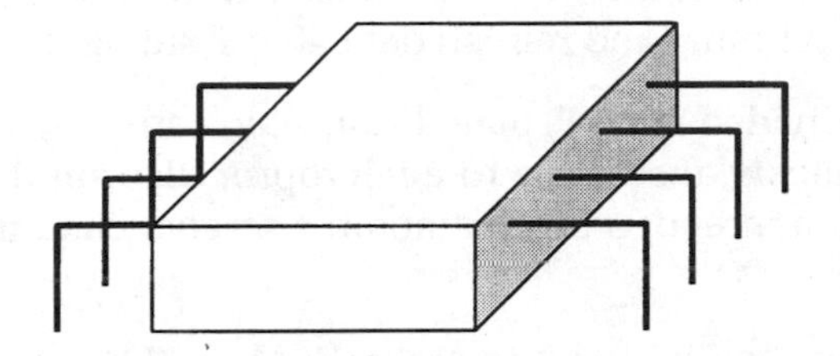

Exhibit 13.2 DIP integrated circuit.

Externally, RAM is a chip that comes embedded in a personal computer motherboard with a variable amount of additional modules plugged into motherboard sockets. To add memory, more RAM modules are added in a prescribed configuration. These are single in-line memory modules (SIMMs) or dual in-line memory modules (DIMMs). Since DIMMs have a 64-bit pin connection, they can replace two 36-bit (32-bits plus 4 parity bits) SIMMs when synchronous DRAM is used.

When the processor or CPU gets the next instruction it is to perform, the instruction may contain the address of some memory or RAM location from which data are to be read (brought to the processor for further processing). This address is sent to the RAM controller. The RAM controller organizes the request and sends it down the appropriate address lines. Transistors along the lines "open up" the cells so that each capacitor value can be read. A capacitor with a charge over a certain voltage level represents the binary value of 1 and a capacitor with less than that charge represents a 0.

For dynamic RAM, before a capacitor is read, it must be power-refreshed to ensure that the value read is valid. Depending on the type of RAM, the entire line of data where the specific address happens to be located may be read or, in some RAM types, a unit of data called a page is read. The data that are read are transmitted along the data lines to the processor's nearby data buffer, known as level 1 cache, and another copy may be held in level 2 cache. For video RAM, the process is similar to DRAM except that, in some forms of video RAM, while data are being written to video RAM by the processor, data can simultaneously be read from RAM by the video controller (for example, for refreshing the display image).

The amount of time that RAM takes to write data or to read it once the request has been received from the processor is called the access time. Typical access times vary from 9 to 70 ns, depending on the kind of RAM. Although fewer nanoseconds is better, user-perceived performance is based on coordinating access times with the computer's clock cycles. Access time consists of latency and transfer time. Latency is the time to coordinate signal timing and refresh data after reading it.

RAM can be divided into (1) main RAM, which stores every kind of data and makes it quickly accessible to a microprocessor and (2) video RAM, which stores data intended for the monitor screen, enabling images to be displayed faster.

Main RAM can be divided into static RAM (SRAM) and dynamic RAM (DRAM). Static RAM is more expensive, requires four times the amount of space for a given amount of data than dynamic RAM, but, unlike dynamic RAM, does not need to be power-refreshed and is therefore faster to access. One source gives a typical access time as 25 ns in contrast to a typical access time of 60 ns for dynamic RAM. (More recent advances in dynamic RAM have improved access time.)

Static RAM is used mainly for the level 1 and level 2 caches that the microprocessor checks before looking in dynamic RAM (DRAM). DRAM uses a kind of capacitor that needs frequent power refreshing to retain its charge. Because reading a DRAM discharges its contents, a power refresh is required after each read. DRAM must also be refreshed about every 15 microseconds just to maintain the charge that holds its contents. DRAM is the least expensive kind of RAM.

Before newer forms of DRAM, fast page mode DRAM (FPM DRAM) was the most common kind of DRAM in personal computers. Page mode DRAM essentially accesses a row of RAM without having to continually respecify the row. A row access strobe (RAS) signal is held active while the column access strobe (CAS) signal changes to read a sequence of contiguous cells. This reduces access time and lowers power requirements. Clock timings for FPM DRAM are typically 6-3-3-3 (meaning three clock cycles for access setup, and three clock cycles for the first and each of three successive accesses based on the initial setup).

Enhanced DRAM (EDRAM) is the combination of SRAM and DRAM in a single package that is usually used for a level 2 cache. Typically, 256-byte static RAM is included along with the dynamic RAM. Data are read first from the faster (typically 15 ns) SRAM; if no data are found there, it is read from thc DRAM, typically at 35 ns.

Extended data output RAM (EDO RAM) or extended data output dynamic RAM (EDO DRAM) is up to 25 percent faster than standard DRAM and reduces the need for level 2 cache.

Synchronous DRAM (SDRAM) is a generic name for various kinds of DRAM that are synchronized with the clock speed for which the microprocessor is optimized. This tends to increase the number of instructions that the processor can perform in a given time. The speed of SDRAM is rated in MHz rather than in nanoseconds. This makes it easier to compare bus speed and RAM chip speed. You can convert the RAM clock speed to nanoseconds by dividing the chip speed by 1billion ns (which is one second). For example, an 83 MHz RAM would be equivalent to 12 ns.

PC100 SDRAM is SDRAM that states that it meets the PC100 specification from Intel. Intel created the specification to enable RAM manufacturers to make chips that would work with Intel's i440BX processor chip set. The i440BX was designed to achieve a 100-MHz system bus speed. Ideally, PC100 SDRAM would work at the 100-MHz speed, using a 4-1-1-1 access cycle. It's reported that PC100 SDRAM will improve performance by 10 to 15 percent in an Intel Socket 7 system (but not in a Pentium II because its L2 cache speed runs at only half of processor speed).

Double data rate SDRAM can theoretically improve RAM speed to at least 200 MHz. It activates output on both the rising and falling edge of the system clock rather than on just the rising edge, potentially doubling output. It's expected that a number of Socket 7 chip set makers will support this form of SDRAM.

SyncLink DRAM is, along with direct RAM bus DRAM (DRDRAM), a protocol-based approach in which all signals to RAM are on the same line (rather than having separate CAS, RAS, address, and data lines). Since access time does not depend on synchronizing operations on multiple lines, SLDRAM promises RAM speed up to 800 MHz. Like double data rate SDRAM, SLDRAM can operate at twice the system clock rate. SyncLink is an open industry standard that is expected to compete and perhaps prevail over direct RAM bus DRAM.

Parity RAM sends an extra data bit to check for errors in transmission, making a total of nine bits instead of eight. With new technology and better quality motherboards, these error detection bits are no longer needed. While most Pentium motherboards using 72-pin SIMMS no longer require the use of parity memory, most will work with it. However, motherboards that require parity memory will not work without it. Logic Parity SIMMS use a cheaper circuit to mimic the response of true parity SIMMS during system

Exhibit 13.3 The RAM table.

RAM technology	Application and location	Access speed range	Ports	Characteristics
SRAM	L1/L2 Cache	Fast	One	Always charged
BSRAM	L2 Cache	Fast	One	SRAM in burst mode
DRAM	Main Mem	Slow	One	Generic term for any dynamic RAM
FPM DRAM	Main Mem	Slow	One	Was most common
EDO DRAM	Main Mem	15 percent >FPM	One	Overlapping reads
BEDO DRAM	Main Mem	>EDO 4-1-1-1 @66	One	Not widely supported
EDRAM	L2 Cache			
NVRAM	Modem	Fast	One	Battery powered
SDRAM	Main Mem		One	Generic term for DRAMs With synch interface
JEDEC SDRAM	Main Mem	Fast	One	Dual bank/burst mode Most common form
PC100 SDRAM	Main Mem	100 MHz 4-1-1-1	One	Intel spec for i440bx
DDR DRAM	Main Mem	<200 MHz	One	2 * PC100
ESDRAM	Main Mem	100 MHz	Two	2 * SDRAM
SLDRAM	Main Mem	Fastest 200 MHz+	One	Uses packets
DRDRAM	Main Mem	800 MHz 16 bit	One	

boot. To avoid problems it is advised to purchase true parity SIMMS. See Exhibit 13.3 for a summary of the above.

IMPACT OF RAM ON SERVER PERFORMANCE

With the demand on memory that NT and MS BackOffice place, especially SQL, a robust RAM environment is absolutely essential. This section discusses the performance benefits of more memory in Windows NT-based servers. Specific issues include the following:

- How much more work can a server do with each memory configuration?
- When is it better to add memory to a system than to add processors?
- How many users can a system support with additional memory?

These questions have to be answered within the context of a typical server configuration. In addition, the following questions have to addressed. Answers for these questions, relevant to a *typical* server, are given in parentheses.

- What operating system does the server use? (Windows NT, UNIX, Novell)
- At what internal clock speed do the processors in the server operate? (333 MHz)
- How many processors are installed in the server? (more than 4)

- How much disk storage is maintained on the server? (16 or more gigabytes)
- Which functions does the server perform? (application, communication, remote access, e-mail, Internet/intranet, file server, multimedia, print, database?)
- Typically, how many users are logged onto the server at the same time? (1000)
- How many print jobs are run on this server in an average day? (1000 or more)
- How critical to the corporate mission is the server's response time? (very critical)
- How tolerant are the users of delays in server processing? (Not at all!)
- What is the file working size? (622 MB)
- What is the directory size? (50,000 to 400,000 entries)

Summary

- Estimated memory requirement: 768 to 1024 MB.
- Migrating to Windows NT 5.0, processing or analyzing large databases, and utilizing the server for particularly memory-intensive programs will increase the required memory to 1152 to 1408 MB.

Web server results:

- Double the memory, and cut server response time by more than half.
- A minor investment in memory will produce enormous gains in server productivity (by over 500 percent).
- Memory plays a key role in performance increase, at times taking on a greater importance than the CPU.

Application server results:

- Increasing memory in an application server significantly improves the application performance, up to 3000 percent.
- Doubling the memory dramatically increases, 3 to 10 times, the number of clients supported on an application server.
- The number of clients supported on a server is directly related to the amount of RAM installed in the system.

Directory server results:

- Doubling the memory dramatically boosts directory server performance by an average of 1000 percent.
- Increasing memory provides a dramatic performance increase for clients accessing requests from the server.
- These results are a valuable guideline for IT managers in determining their server memory needs.

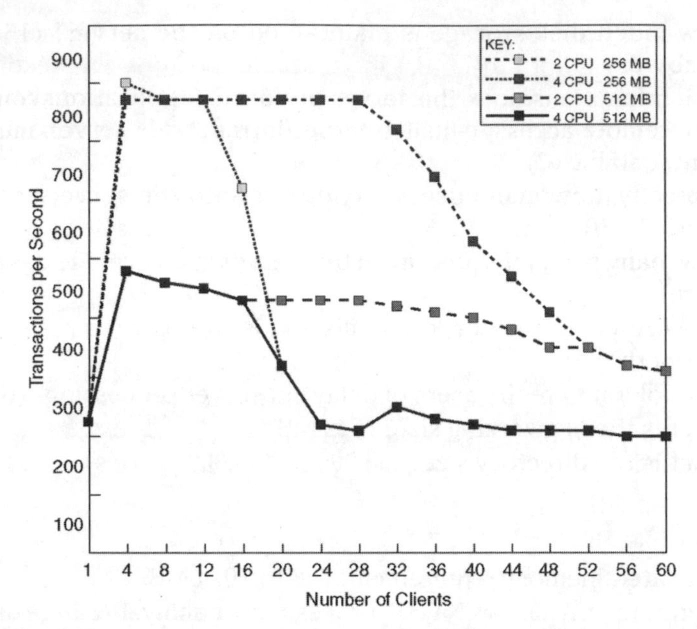

Exhibit 13.4 Higher transactions-per-second numbers indicate better performance and scalability.

Other Considerations

Obviously, there is a complex and mostly empirical relationship between all the parameters listed above. Here are some additional thoughts:

One CPU is not enough! To take full advantage of increased cache (see below), a server needs more processing power. Adding processors as RAM is added will significantly boost performance; a single CPU is simply not sufficient for most application servers. For example, increasing RAM by as little as 128 MB and adding a second processor can increase peak performance by as much as 90 percent. And going from two processors to four can improve performance by yet another 50 percent.

In fact, increasing the number of processors has a much greater effect on performance than simply adding RAM. There was no real gain when upgrading a two-processor server from 256 to 512 MB of RAM. Although the extra RAM effectively doubled the number of clients the server could handle before the cache was depleted, performance actually began to slip because of the heightened disk activity. After increasing the number of processors to four, however, performance rose by approximately 70 percent over that of the two-processor configuration. Exhibit 13.4 shows how important memory and CPUs are to performance. As with all servers, once the cache is exhausted, performance will drop as the disk takes on the overflow.

The bottom line is that as the number of clients accessing an application server increases, both RAM and CPUs should be added to accommodate them and take full advantage of the additional cache. If the disk becomes a problem, simply adding processors will not boost performance unless RAM is added to take some of the burden off the disk. Similarly, file and Web servers benefit from the right mix and quantities of memory and CPUs.

Cache Memory

Cache memory is a special high-speed memory designed to accelerate processing of memory instructions by the CPU. The CPU can access instructions and data located in cache memory much faster than instructions and data in main memory. For example, on a typical 100-MHz system board, it takes the CPU as much as 180 ns to obtain information from main memory, compared to just 45 ns from cache memory. Therefore, the more instructions and data the CPU can access directly from cache memory the faster the computer can run.

Types of cache memory include primary cache (also known as level 1 (L1) cache) and secondary cache (also known as level 2 (L2) cache). Cache can also be referred to as internal or external. Internal cache is built into the computer's CPU, and external cache is located outside the CPU. Primary cache is the cache located closest to the CPU. Usually, primary cache is internal to the CPU, and secondary cache is external. Some early personal computers have CPU chips that don't contain internal cache. In these cases the external cache, if present, would actually be the primary cache. Earlier we used the analogy of a room with a work table and a set of file cabinets to understand the relationship between main memory and a computer's hard disk. If memory is like the work table that holds the files you're immediately working on, making them easy to reach, then cache memory is like a bulletin board that holds the papers to which you refer most often. When you need the information on the bulletin board, you simply glance up and there it is.

Cache memory is like a bulletin board that makes the work at the memory "work table" go even faster.

You can also think of cache memory as a worker's tool belt that holds the tools and parts needed most often. In this analogy, main memory is similar to a portable toolbox and the hard disk is like a large utility truck or a workshop. The "brain" of a cache memory system is called the cache memory controller. When a cache memory controller retrieves an instruction from main memory, it also takes back the next several instructions to cache. This occurs because there is a high likelihood that the adjacent instruction will also be needed. This increases the chance that the CPU will find the instruction it needs in cache memory, thereby enabling the computer to run faster.

Numbering Plan

In this section we discuss the binary numbering system that forms the basis of computing and how memory modules are designed to work within the binary system.

All computers speak a language made up of only two numerals: 0 and 1. This two-numeral form of communication is called machine language; the numerals combine to form binary numbers. Machine language uses binary numbers to form instructions for the chips that drive computing devices — such as computers, printers, and hard disk drives.

A bit is the smallest unit used by a computer and can be either a 1 or a 0. A byte consists of 8 bits (more about bytes later). Because binary numbers consist of only 1s and 0s, binary number values appear different from the decimal values we use in everyday life. For example, in the decimal system the number 1 followed by two 0s (100) represents a value of one hundred. In binary, however, the same number combination — 100 — represents a value of four.

The process of counting in binary isn't all that different from the decimal system. As you count in decimals, when you reach 9, the count resets to 0 and a 1 shifts into the tens column. When you count in binary, the process is very similar. However, because there are only two numerals, the resetting and shifting of digits happens more quickly.

Take a look at the table in Exhibit 13.5. On the top are the decimal numbers 0 through 15; on the bottom are the binary equivalents. Going back to how machine language uses 1s and 0s, remember that each digit in a binary number represents 1 bit. Machine language thinks of each bit as either on or off. A bit with a value of 1 is considered on, and a bit with a value of 0 is off. Therefore, determining the value of a binary number is simply a matter of adding up the columns that are "turned on," in other words, wherever a 1 appears. (This on/off idea comes in handy later.) In the decimal numbering system, each column (ones, tens, hundreds, and so on) has ten times the value of the previous column. In the binary numbering system, however, each column has a value twice that of the previous column (ones, twos, fours, eights, sixteens, and so on).

Decimal and Binary Number Comparison

Exhibit 13.6 compares the same number combination (111) in each of the two systems. In the decimal numbering system, the number 111 represents the addition of 100+10+1. In binary, 111 represents a binary value of 7 because it's the result of adding 4+2+1. Because your computer understands binary values consisting of 1s and 0s, a binary value exists for every possible keyboard character. The most widely accepted standardized system for numbering

Exhibit 13.5 Decimal numbers and their binary equivalents.

Decimal															
Tens	0	0	0	0	0	0	0	0	0	1	1	1	1	1	1
Ones	1	2	3	4	5	6	7	8	9	0	1	2	3	4	5
Binary	=	=	=	=	=	=	=	=	=	=	=	=	=	=	=
Eights	0	0	0	0	0	0	0	0	1	1	1	1	1	1	1
Fours	0	0	0	0	1	1	1	1	0	0	0	0	1	1	1
Twos	0	0	1	1	0	0	1	1	0	0	1	1	0	0	1
Ones	1	0	1	0	1	0	1	0	1	0	1	0	1	0	1

Exhibit 13.6
Decimal and binary number comparison.

Sample Number: 111		
Decimal		
Hundreds	1	100
Tens	1	10
Ones	1	1
Total		**= 111**
Binary		
Fours	1	4
Twos	1	2
Ones	1	1
Total		**= 7**

keyboard characters is called the ASCII system. (ASCII is pronounced "askee" and stands for American Standard Code for Information Interchange.)

CPU AND MEMORY REQUIREMENTS

A computer's CPU (central processing unit) processes data in 8-bit chunks. Those chunks, as already learned, are commonly referred to as bytes. Because a byte is the fundamental unit of processing, the maximum number of bytes it can process at any given time often describes the CPU's processing power. For example, the most powerful Pentium and PowerPC microprocessors currently are 64-bit CPUs, which means they can simultaneously process 64 bits, or 8 bytes, at a time. Each transaction between the CPU and memory is called a bus cycle. The number of data bits a CPU can transfer during a single bus cycle affects a computer's performance and dictates what type of memory the computer requires. Most desktop computers use either 72- or 30-pin SIMMs. A 30-pin SIMM supports 8 data bits; a 72-pin SIMM supports 32 data bits.

NOTE: With most computer models, mixing different-capacity SIMMs within the same bank prevents the computer from accurately detecting the amount of available memory. This causes one of two things to occur: the computer will not boot, or the computer will boot but will not recognize or use some of the memory in the bank. For example, if a bank had three 1-MB SIMMs and one 4-MB SIMM, it would recognize them all as 1-MB SIMMs.

30-pin SIMMs

Let's look at the example of a CPU that supports 32 data bits. If the computer's system board has 30-pin SIMM sockets, each of which supports 8 data bits, you'll need four 30-pin SIMMs to supply 32 bits. (This is a common configuration for systems that use 30-pin SIMMs.) The memory configuration on such a system is typically divided into two memory banks — bank zero and bank one. Each memory bank consists of four 30-pin SIMM sockets. The CPU addresses memory one bank at a time.

72-pin SIMMs

The 72-pin SIMM was developed to satisfy the ever-expanding memory requirements of desktop computers. One 72-pin SIMM supports 32 data bits, which is four times the number of data bits supported by a single 30-pin SIMM. If you have a 32-bit CPU — such as an Intel 486 or Motorola's 68040 — you need only one 72-pin SIMM per bank to provide the CPU with 32 data bits. As we saw in the previous section, that same CPU would require four 30-pin SIMMs per bank to get its 32 data bits. See Exhibit 13.4.

DIMM Memory

Dual in-line memory modules, or DIMMs, closely resemble SIMM-type memory. Like SIMMs, most DIMMs install vertically into expansion sockets. The principal difference between the two is that on a SIMM, opposing pins on either side of the board are joined to form one electrical contact; on a DIMM, opposing pins remain electrically isolated to form two separate contacts. DIMMs are often used in computer configurations that support a 64-bit or wider memory bus. In many cases, these computer configurations are based on powerful 64-bit processors like Intel's Pentium or IBM's PowerPC processors. For example, Kingston's KTM40P/8 DIMM module used in IBM's PowerPC 40P RISC 6000 computer is a 168-pin DIMM.

The full size, 168-pin DIMM supports 64-bit transfers without being twice the size of the 72-pin SIMM, which supports only 32-bit transfers.

Memory Banks

Computers have memory arranged in what are called memory banks. The number of memory banks and their specific configurations vary from one computer to another, determined by the computer's CPU and how it receives information. The needs of the CPU determine the number of memory sockets required in a bank. While we can't look at every possible memory configuration, we can look at a system for depicting memory configuration requirements called the bank schema. A bank schema is a diagram of rows and columns that shows the number of memory sockets in a system. This visual display is a theoretical bank layout and not an actual system board layout; it is designed to help you quickly determine the configuration requirements when adding memory modules.

MEMORIES OF THE FUTURE

The memory business is poised to enter the next century with the DRAM market in perpetual evolutionary transition. DRAMs are migrating their interfaces to SDRAM, and then again to RAM bus DRAMs. One thing is certain about the memory market of the future: it will change.

The Rambus DRAM, DDR DRAM, SDRAM, EDO DRAM, and FPM DRAM either has taken over or will take over from its predecessor. This does not mean that FPM and EDO are going away. FPM and EDO revenue was still projected to be 40 percent of the market in 1998. Gradually, these types will decline as DRAM vendors raise prices on them to encourage designs of current types of DRAM. Embedded DRAM at the smaller density requirements is also expected to help eliminate the need for the smaller density, older type DRAMs.

The 128-megabit DRAM began shipping in the third quarter of 1998, and 256-megabit DRAMs were shipping in prototype quantities. The DDR DRAM began shipping in the second half of 1998. Engineering samples were supplied to PC OEMs for the SLDRAM in April of 1998. Engineering samples of the DRDRAM were also made available in 1998. In 1999, the market will begin to determine which of the new DRAMs will be the volume winner.

The revenue mix between FPM, EDO and SDRAM, SDRAM II and Direct RDRAM or SyncLink DRAM (SLDRAM) through the year 2001 is projected to show that FPM DRAM will continue to decline. As systems implementing the FPM DRAM are redesigned, SDRAM will be the DRAM of choice. Many workstation and server manufacturers will continue to use FPM and EDO DRAM. However, manufacturers will try to convert OEMs from FPM DRAM so they do not have to manufacture as many types of DRAM.

DDRDRAM and DRDRAM or SLDRAM are projected to begin to ship in production quantities in 1999. Quantities are small in this forecast period for the revolutionary DRAMs.

The conventional wisdom is that the DDR DRAM will be the transition DRAM between SDRAM and DRDRAM/SLDRAM. The DRDRAM and SLDRAM

Exhibit 13.7 RAM prices.

32MB of RAM	Low	High	Average
8×32, 60 ns, EDO SIMMs	$35	$85	$63
4×64, 10 ns, SDRAM DIMMs	$39	$59	$48
4×72, 8 ns, PC100 SDRAM DIMMs (2)	$57	$70	$63
64MB of RAM	**Low**	**High**	**Average**
16×32, 60 ns, EDO SIMMs	$90	$159	$120
8×64, 10 ns, SDRAM DIMMs	$80	$149	$104
8×72, 8 ns, PC100 SDRAM DIMMs (2)	$97	$150	$130
128MB of RAM	**Low**	**High**	**Average**
32×32, 60 ns, EDO SIMMs	$187	$286	$235
16×64, 10 ns, SDRAM DIMMs	$159	$199	$178
16×72, 8 ns, PC100 SDRAM DIMMs (2)	$189	$277	$236

may share the market. The final decision has not been made as to which will be the main memory of the future. With changes in buying patterns and the popularity of the $1000 or less PC, there will potentially be more than one type of DRAM consumed in volume by PCs. Presumably the same forecast will hold for servers.

Intel has endorsed the DRDRAM, but DRAM vendors continue to work on DDR DRAM, DRDRAM, and SLDRAM.

Main memory for computers continues to be the major consumer of memory. Therefore, the type of DRAM used for main memory will be the highest volume and largest revenue generator. The FPM DRAM, EDO DRAM, and SDRAM are currently the highest volume products. The high volume DRAM densities for the forecast period of 1998 to 2002 are 16, 64, 128, and 256 megabit. The 128-megabit DRAM began shipping in the second half of 1998, but did not rival the 64-megabit quantities. The predominant organizations are ×4, ×8 and ×16. Some ×32 organization DRAM is available, but not in high volume at this time. (See Exhibit 13.7.)

Chapter 14
CPUs

A. Padgett Peterson

PROCESSOR TYPES: CISC VS. RISC

The two basic divisions between processors today are those between CISC and RISC processors. CISC, or Complex Instruction Set Computers, are the traditional form in which the manufacturer decides the use of the processor, designs the microcode for all of the instructions the processor will use, and implements that microcode on the chip itself, optimizing the data paths for speed. The most common CPUs today, the Intel 80x86 and Motorola 680x0 series, are examples of CISC processors.

With a CISC processor, compilers and assemblers produce code which runs for the most part directly on the processor. In turn, the CISC processor offer a very rich instruction set with provision for such things as floating point arithmetic, character and string manipulations, and an array of I/O manipulations.

Beginning in the late 1970s with bit slice processors, such as the AMD 2901, a different path was formed with the RISC or Reduced Instruction Set Processor. In this case the on-board microcode is limited to a very small set of instructions, typically only involving simple register and memory operations. The advantage of RISC is a very small microcode store and very short data paths which give the RISC processors a decided speed advantage over their CISC relatives.

For example, RISC processors are now capable of clock speeds in the 300- to 900-MHz range, while their CISC relations are running at 100 to 500 MHz.

As usual, vaporware is rampant as marketeers promise 750 MHz Intel chips and 1,000 MHz Alphas. While tomorrow may well bring such marvels, the current foundry capability (the companies that make the wafer dies from which chips are constructed) will probably make speeds not much faster than today's processors.

While such numbers look impressive and are often used to promote models, it is important to remember that true machine throughput depends on a balance of all the components. In an application-sensitive

0-8493-9823-1/00/$0.00+$.50

firewall or a graphical design application, a very-high-speed CPU may be important, while for most file server applications, the load is not very heavy.

For this reason, while a well-designed server may use multiple CPUs, the purpose will be more reliability in the same way RAIDs are used rather than for pure throughput.

However, this is not without cost when real work must be done, as it typically takes a RISC processor many more instructions to do the same task that a CISC computer can do with only one. An old joke in the computer industry is that the ultimate RISC does nothing at all but does it very fast.

In terms of a typical programming mix of human–computer activities the RISC unit will be somewhat slower than the equivalent CISC. The reason is that while a full-featured operating system can be layered directly on the CISC, often it is necessary to augment the RISC instruction set through a layer of software called an emulator.

This is best shown by the original "Power PC," which was able to emulate the Macintosh/680x0 instruction set very well but was noticeably slower when called upon for Windows/Intel duties.

However, the real power of the RISC can be found when the full set of instructions is not generally needed, such as for server operations, provided the software is written directly for the RISC set without requiring an intermediate emulation layer. In this case the speed advantage, particularly for logical operations, can make up for the fact that more instructions must be executed to accomplish the same task.

For this reason, the later Intel processors beginning with the 486 series retain the full CISC instruction set. However, the most commonly used instructions, such as MOV, were optimized to be executable in a single clock cycle. This characteristic was carried forward in the Pentium and Pentium Pro series, which used even greater optimization for an overall reduction in clock cycles per instruction. For this reason, the Pentium 90 and the 486DX5-133 are approximately equal in practical speed.

MEMORY SPEED

Any discussion of CPU speed must be related to the speed at which the memory can deliver instructions to the processor. Optimally, the processor would request a memory location and on the same cycle it would be able to access that information. In practice this does not work that way for a number of reasons.

The first is that a digital designer must consider that while the memory address may not be available until near the end of the requesting clock cycle, the data must be stable at the beginning of the cycle at which it is expected.

Second, gate delays in the associated circuitry take time to translate the address request from the processor to the memory device assigned to that location. Third is the reaction time of the memory device itself.

For the first PCs of 1981 with a 4.77-MHz clock cycle, 200 ns memory was adequate for a three-cycle operation. On the first clock, the address was provided to the memory circuitry; the second was allowed to provide stable information on the data bus; and on the third clock the data would be read in.

This three-clock cycle was the basis for all PC-type computers that came later. When computers are set up today, this is the same baseline, so if the user selects 1 wait state for memory, this means that four cycles are used instead of three.

As processor speeds went up, memory access times had to decrease since for a 33-MHz bus clock this provided only 30 ns per clock or 90 ns total for no-wait-state operation. Given a 20- to 30-ns gate delay, 80 ns memory often required an additional wait state to operate. The fact that the 40-ns cycle allowed by a 25-MHz clock provided the same 120 ns with three clocks that a 33-MHz unit needed four to achieve meant that for 80-ns memory there was little real performance difference between a 25- and a 33-MHz computer.

MEMORY BUSSES

Another contributing factor is the bus width or the amount of data that can be transferred at one time. While the Motorola 680x0 series of processors have always had a 32-bit bus width allowing four bytes of data to be received in parallel, the Intel-based processors have gone through a series of stages to get there.

Beginning with an 8-bit data bus width for the 8088 to make it compatible with 8080 and Z-80 machines, the IBM PCs and their first clones were handily outperformed by machines from those companies that designed their systems around the companion 8086 processor. The only difference was a 16-bit data bus width.

With the PC-AT and its 80286 processor, the standard bus width became 16 bits and accounted for most of the performance increase since clock speed was increased only slightly to 6 MHz.

The introduction of the 386 made Windows possible by curing a problem in the 286 — the inability to switch from virtual to real mode without rebooting. It also increased the bus width to 32 bits, achieving for the first time parity with the 680x0 used in the Macintosh. In an attempt to use up stocks of 16-bit bus motherboards originally intended for AT class machines, Intel also produced the 386SX, a 16-bit version and hence considerably slower than the DX.

With the introduction of the 486, things became even more confusing when both the SX and DX versions were 32-bit but were soon joined by the 486SLC from other vendors, which had a 16-bit data bus. Fortunately, few were sold.

The final entry (so far) are the Pentium and Pentium Pro processors. Built with superscalar architecture using submicron design rules, these super microprocessors utilize a 64-bit data path enabling access of 8 bytes at a time. To fully utilize this, memory chip carriers have transformed from 30-pin SIMMs to 72-pin units in the last few years.

The latest addition to the Pentium product line, the Pentium III, was introduced in March 1999. Two models of this new processor were initially offered, operating at 450 and 500 MHz. Both include over 70 new instructions that facilitate 3D, multimedia, and parallel processor operations.

RISC computers have always enjoyed an advantage in this area with the first units having a 32-bit bus width and the latest Alpha chips accessing a 256-bit-wide memory bus. Combined with 500- to 600-MHz clock speeds, the latest RISC stations enjoy a clear performance advantage.

CACHE

Cache memory is an attempt to increase the processing speed of a computer by reducing the need for access to slow main memory by adding a relatively small amount of very fast memory in a separate area. Whenever memory is accessed, it is also written to this store. On the next access, if it is still in the cache area, it is retrieved from there with no wait states involved. That cache memory typically has a 15- to 20-ns access time, while main memory typically is of 60 to 80 ns access time, is indicative of the speed required for next cycle access at a 33-MHz bus speed.

Introduced on some 386-class machines, such as those from Zenith Data Systems, the first caches were fully populated at 64K of RAM and materially increased effective performance in the same way that disk caches in memory had had the same effect some years earlier.

Experiments showed that performance increased dramatically at first as more RAM was added, then slowed after 128K with a maximum "bang for the buck" at about 512K. Further increases, while assisting in some operations, were hampered by the increased activity necessary to determine if the just-requested segment was already in cache.

The second cache was introduced with the introduction of the 486-class processor, which had 8K of onboard RAM. Located on the CPU itself, the amount was limited but was as fast as the processor. Since now there were two types of cache available, designations were needed, with onboard cache being referred to as level 1 or simply "L1" cache, while separate RAM cache on the motherboard became known as level 2 or "L2."

Unfortunately, while companies like Texas Instruments experimented with increased L1 cache, up to 16K, others marketed products with as little as 1K, which had minimal performance effect. When combined with a "486" designated device, which was really a 386 architecture with a few added instructions, this led to "486" machines which were slower than the equivalent 386. This was further confused by a series of "486SLC" devices designed for a 16-bit data path to take advantage of leftover 386SX motherboards. Combined with a ROM change, these systems reported themselves to be 486 boards — just very slow ones.

In contrast, several vendors packaged TI 486 chips onto special substrates, allowing them to be used on true 386DX motherboards. Coupled with a driver program to enable the L1 cache, when used on a motherboard already equipped with L2 cache, speeds very close to true 486 boards can be obtained.

At the present time there are a few chipsets available that permit "586" chips to be used on older 486 motherboards with a concurrent performance improvement. Unfortunately, such hybrids typically use an ISA or VESA peripheral bus instead of the high-speed PCI bus (see next chapter), limiting the effective throughput. One major reason is that the true Pentium uses a 64-bit direct access to L2 cache, doubling the effective cache throughput, and incidentally larger effective cache sizes (512 Kb is effective), though the speed improvement with a larger cache is not linear — going from 256K to 512K provides a typical speed increase of only 2 to 3 percent. 486 chips and motherboards do not support this, though some "dual use" boards have been built.

If cache is an advantage to CISC-based systems, it can be of even greater assistance to RISC systems since they typically require more instructions to perform the same operation. In this case, if the entire program to be executed can be loaded in cache, then the only need for main memory is to load the program initially and retain data. In this case the complete operation can be from cache, while main memory takes the place of disk storage. For this reason and due to the drastic drop in memory prices in 1996, very large cache systems are becoming popular, with cache approaching early disk drive capacity becoming common.

Again for RISC processors, the rules are different. Where Intel-based motherboards are often limited to 1 Mb of cache (super servers may provide more), with a RISC machine the cache is typically extended to three levels. The current DEC Alpha Station 500s have 8 Mbytes of L3 cache as standard and most are shipping with 256 Mb of memory.

SPEED DOUBLERS AND TRIPLERS

Given sufficient cache for most operations, it is possible for the processor to operate independently of the system within which it is operating. At the same time, ever-decreasing prices for peripherals are driven by the stable bus speeds. In this case the basic clock speed for the PC has been hardware limited to about 33 MHz.

To achieve increased internal speeds, processors now have internal clock circuits which are synchronized to the external hardware clocks, but which operate at two or three times the hardware clock rates. The first of these raised 25 MHz clocks to 50 MHz and 33 MHz clocks to 66 MHz.

The numbering system used by "doublers" bears some examination as it is not outwardly clear why a "486DX4-100" has a bus speed of 33 MHz rather than 25. The reason is that when doublers were first introduced, it was not clear yet what the designs would allow. Accordingly, speed points were set at 2x, 2.5x, 3x, and 4x clock and numbered 2, 3, 4, and 5. Hence, a "486DX3-80" would use a 33-MHz system clock, while a "486DX2-80" would require a 40-MHz clock. Just for a frame of reference given a 33-MHz system clock, the advertised speeds would be as follows: 2-66 MHz, 3-80 MHz, 4-100 MHz, and 5-133 MHz.

All of this was made obsolete by the Pentium processors, which are advertised solely on their internal speed. Here once again the early practice of overdriving a chip is becoming common with the DS2-50 chip often found in 33 (66) MHz machines. This is popular since often the elaborate cooling towers/fans which are used to cool the latest high-speed chips obscure the manufacturers markings.

Note that this was primarily a way to allow the millions of existing motherboards to receive an apparent speed improvement with little cost to the consumer. So long as they are able to operate out of L1 and L2 cache, performance will improve, but when required to access main memory or I/O, the limit is that of the original design, the processor spends much of its time waiting for the rest of the machine to catch up.

In general RISC machines do not use doublers, preferring to require an upgrade of all of the supporting systems when a processor speed advance is made. Since consumer price pressure does not exist for most RISCs, being high-end items, this is possible today.

HEAT

Very dense chips such as CPUs have always run hot — often near the 85°C (185°F) limit of most plastic cases. With the introduction of the very large 486 series, this became a real problem. The first 25-MHz chips were marginal.

Once the 33-MHz threshold was reached it became critical. To combat this elaborate convection, cooling towers and fans were attached.

Unfortunately, the plastic cases are not very good heat conductors, and it was critical that thermally conductive grease be applied between the tower and the CPU or often the CPU heated up even faster.

With the introduction of the 3.3V CPUs replacing the earlier 5V units, much of the problem evaporated since these dissipate much less heat than the earlier designs. When replacing a CPU it is essential to know the voltage the board is designed for (some accommodate both) since use of a 3.3V chip in a 5V socket is a quick way to achieve maximum smoke.

BIOS

The final element that affects processing speed is the BIOS or Basic Input Output System. Essentially an operating system on a chip, the BIOS is what makes an Intel box "100 percent compatible" (with the original IBM specification).

The problem is not the memory structure of the BIOS; it uses ROM or Read Only Memory, which can be faster than the common DRAM (Dynamic Random Access Memory) used for the main memory (RAM) or even the faster Static RAM memory (SRAM). Instead the bottleneck is usually the 8-bit data bus used to connect the CPU to the BIOS ROM.

One typical way to correct this is to copy the ROM memory into RAM on the 32-bit data bus, which has much faster access time. This is called "shadow RAM."

However, this is unnecessary when a true 32-bit operating system is used, since all I/O is through RAM anyway and the BIOS is mapped out. When server operating systems such as UNIX® or Novell Netware® are used, a small RAM savings (64K) can be achieved by turning off all ROM shadowing. This mechanism is a legacy from the first 8088-based Intel machines which is also maintained in the interest of minimum price and compatibility with REAL mode, another 8088 requirement. RISC machines have no such limitations.

PIPELINING

The final element that must be considered relating to processor speed is the pipeline depth. 680x0 series CPUs have always been pipelined, though the current Pentium CPUs have a similar depth. RISC processors rely heavily on pipelining for burst speed.

Essentially an "L0" cache, the pipeline works best with interleaved memory. This is a technique whereby more than one memory location is accessed on a single clock. With interleaving, when memory is accessed, the surrounding memory locations are also activated and read into the pipeline. That way when one instruction completes, the next has already been loaded so that no memory wait cycle is required.

Early CPUs had very short pipes but the current Pentium series chips often incorporate dual pipe of 50 instructions. The problem is that pipes work best with in-line code. Every time a branch or jump is taken, the pipeline must be flushed and refilled. Programs with many JMPs or CALLs then must often flush and refill. Taken to extremes the pipelined system can be slower than a system with none at all. Hence, it is very important to utilize a compiler optimized for the processor.

A poor compiler can lengthen run times by 20 to 30 percent over a well-optimized one particularly if a very large pipe is involved.

For RISC machines, the pipelining is a critical element and care is usually taken to avoid branching which would flush the pipe. Part of the speed improvements found on most RISC platforms result from the fact that the compilers optimize for speed rather than size. In turn this makes cache size even more critical. Many RISC machines are designed today to operate programs entirely out of cache if possible.

BALANCE

As can be seen from the above, it is essential that the system be properly balanced for each of its components. Memory must be properly matched to bus speed (as seen above, with slow memory, slowing the clock to reduce total wait states can improve throughput). Better yet of course is to use fast memory in the first place.

While a 33-MHz memory bus system can run with 80 ns RAM, extra wait states will be required. Increasing the memory speed to 70 or 60 ns can allow a reliable return to "zero-wait-state" operation (as we have seen, zero really means three).

The next consideration is cache. More is generally better, though the performance increase approaches zero once the actual program length is reached. For CISC processors, 256K is a good place to start, while a RISC machine may benefit from a megabyte or more.

L1 cache is another case of "if enough is good, more is better." Generally 8K is a minimum and 16K is better. Below 8K, performance drops off quickly so that a 1K L1 cache is almost useless.

Again the problem is that too much can hurt. If the cache is very large, the system may spend longer deciding a particular range is in cache than it would to retrieve it in the first place. While improvement will vary with instruction mix, the numbers above have proven to be good for today's systems.

For the RISC machines "the rules are different" with very large caches (4 to 8 Mbytes) common. This can be done effectively because the design is optimized for maximum use rather than cost sensitivity and if the entire program can be encapsulated inside the cache, then there is no need for disk access once loaded.

Thus it is essential to properly balance the components in a system and then tune to the application mix. If one area cannot meet the system throughput design requirements, the system will be slow. Too high performance devices will not impact throughput, but will generally impact cost. Balance is vital.

Chapter 15
Server Data Bus Structures

A. Padgett Peterson

One of the determining features of any processor is the bus speed, since any operation is limited by the slowest portion. In the previous chapter we discussed the relative operations of various processors and their memory subsections. Here we will discuss the relationship between the processor speeds and the bus speeds.

Toward the end of the chapter, nonstandard structures such as the crowbars used in parallel processing operations will be described.

DATA BUSSES

In modern systems while memory busses are tightly incorporated into the system design and are related to the processor in use, data busses used for peripherals are much more flexible and the current PCI bus is defined as processor-independent.

The importance of the bus speeds is paramount in a well-designed system since all I/O is via the system bus which is shared around the peripherals: keyboard, display, disk drives, and network. Since buses are typically parallel operations similar to memory busses, there are two elements to be considered: bus (clock) speed and bus width. In both cases, more is better.

What this means is that for an 8-bit-wide (one byte) bus running at 10 MHz, the theoretical maximum data rate is 10 Mbyte/s (8 bits * 10,000,000/8 bits per byte). However, this assumes that the bus is simply a pipe with as much data as it can process on one end and unlimited capability to process on the other. This theoretical maximum is also sometimes known as "burst" or "streaming" mode.

In practice, real-world throughputs are somewhat less, about two thirds maximum for well-organized and balanced systems, one half or less for poorly assembled systems such as a Pentium platform with a 16-bit ISA

0-8493-9823-1/00/$0.00+$.50

video card and complex graphical displays. It must be remembered that any system will be limited by the speed of its slowest component.

Thus when a 32-bit PCI bus is said to have a 133-Mbytes/s throughput, that is a theoretical maximum and indicates a 33-MHz clock speed (133*8/32) with no wait states. Real-world factors, such as the commands necessary to initiate the transfer or to acknowledge receipt or for error correction, will all consume cycles. If the processor itself with inherent wait states must be involved (as opposed to direct DMA transfers), the effective throughput will be even less.

While mainframes typically have 64- or 128-bit bus widths, the systems typically used for microcomputer-based servers are smaller. Further, while some common computer bus structures, such as VME or Q-Bus, are both wide and fast, these are rarely used for servers. Today the standards are VLB (VESA Local Bus) for 486-class machines, and PCI (peripheral component interconnect) for Intel Pentium and follow on processors, Power PC and Dec Alpha systems.

Evolution of Data Busses

The busses used by the IBM PC are an excellent starting point for understanding the evolution of bus structures. The first was the 8-bit PC-Bus introduced with the IBM PC, which runs at the same 4.77 MHz as the 8088-based PC. This gives a maximum theoretical throughput of about 5 Mbyte/s. The continuing popularity of the 3Com 3C503 Ethernet card for low-performance applications shows that this was and often still is adequate for low-performance applications and can typically provide effective net speeds around 70 Kbytes/second transfer rates. In this case the address bus had a width of 20 bits and a granularity of 1 byte, limiting access to 1 Mbyte. Since the 8088 only had 16-bit registers, a system of 64K "segments" had to be used. On the other hand, the memory bus was only 8 bits wide, so data had to be accessed 1 byte at a time.

The next change came in 1985 with the introduction of the PC-AT (advanced technology) and the 16-bit ISA architecture. Doubling the memory bus width to 16 bits and increasing the bus speed (originally to 6 MHz but quickly expanded to 8 MHz and often stretched to 10 MHz). Effectively quadrupling the theoretical bandwidth, two additional cards/vendors, the WD (Western Digital) 8003/SMC (Standard Microsystems Corp.) Elite and the NE (Novell Ethernet) 2000, became leaders. Transfer rates of 160 to 280 Kbytes/second over networks became common.

For marketing reasons, the ISA bus had to maintain commonality with the earlier 8-bit bus so a dual connection was used — for 8-bit boards, the original connector was maintained with a 4.77-MHz bus speed and 8-bit-wide connection. For the new 16-bit boards, a second connector was placed in-line allowing 16-bit width and 8 MHz (often pushed to 10 MHz) data rate. The design limit for the ISA bus was 8.77 Mbytes/s throughput.

IBM attempted to improve again with the 1987 MCA (microchannel architecture) bus. Thirty-two bit and 10 MHz, the very high card I/O possibly compared to the 8 MHz ISA and made OS/2 a common choice for servers for many years, even though the proprietary nature of the bus led to high card prices, which in turn kept the PS/2 from becoming any more than a market niche. MCA did away with the familiar dual-edge connector sockets of the AT, replacing them with finer (.050" wide rather than .100") contacts.

Few vendors took part since the market was small and MCA used an early (and user-hostile) version of what we know now as "Plug-N'-Play," requiring a special setup disk for each machine and each card. Some vendors responded with dual-use cards — ISA connection on one side and MCA on the other. Others built adapter cards but none were very successful.

One element introduced with the PS/2 was widely accepted — the VGA graphics structure — but this and the miniature keyboard/mouse connectors are about the only elements that remain today.

To counter the MCA bus, a group of vendors led by Compaq supported the EISA (extended ISA) bus — an extension of the ISA bus which promised 33-Mbyte/s transfer rates and compatibility with older 286 machines (though the 33-MHz rate required burst capability). Squabbles among the vendors kept EISA out of all but niche markets, primarily OEM (such as Compaq) components. One problem was that while ISA cards could be used in EISA slots, the EISA standard changed the pin assignments so that while to the user the sockets looked the same, they were actually different and incompatible.

The final answer which eliminated the PS/2 from serious consideration in the marketplace (though the flat memory model of OS/2 maintained their use as servers) was the VESA VL bus, which allowed a 32-bit width and even higher speed (160-Mbytes/s transfer rate in version 2) than the MCA bus. Such cards were designed to be able to be placed in a standard 16-bit ISA bus slot as well, although the speed then dropped to that of a normal ISA card.

The VESA portion required yet another connector to be placed behind the original PC bus connector and the AT bus connector for a total of three connectors in line. In an age of ever-reducing motherboard size, this made for a very long footprint, and so most motherboards only had three VESA sockets since video, disk, and network cards were the only uses felt to require such speed.

Also often known as "Local-Bus," "VESA Local Bus," or "VLB" in the beginning there were two competing standards.

Attempts have been made to extend the life of the VLB bus structure with expansion to a theoretical 160-Mbytes/s data rate through use of a 40-MHz clock. But other than the AMD 486DX-40, TI 486DX2-80, and AMD 486DX4-133, no processors are available for this use —which also requires very high-speed memory to avoid the inevitable wait states.

Next to appear was the PCMCIA (PC-Card) standard for notebook devices. Though rarely seen on desktop machines, the PCMCIA bus promised 33-MHz clock speeds. The early implementations ran somewhat slower. In the future, the use of PCMCIA capable slots may appear on the desktop to foster the use of smart cards for on-line electronic commerce expected to explode in 1999. The PCMCIA bus (now known as "PC-CARD") is really a parallel development and is unlikely to impact servers.

The final link is the PCI bus introduced with the Pentium processors. This utilizes a fine contact specialized connector much like the MCA but providing both 32- and 64-bit access sections and peak transfer rates of 133 Mbytes/s (32 bit) or 266 Mbytes/second (64 bit). The current Dec Alpha RISC processors use the PCI bus structure with some enhancements. At present, the only 64-bit peripheral cards are for disk access (SCSI-2).

(*Note:* The access speeds such as 133 Mbytes/s are the maximum theoretical throughput and are presented for establishing relationships between bus structures. Actual throughput in terms of real-world performance is often about half the theoretical capability.)

BUS LOADS

As mentioned earlier, the network card is not the only device requiring access. While the display and keyboard require some service, these are demand devices and once operational on a server their loads are minimal. Of course if the server is also used as a graphical workstation, this will change drastically, which is the reason a server should always be dedicated to that task unless the network loads are very light.

The other device requiring service is the disk storage subsystem, which also passes data. In fact the I/O requirements are almost exactly the same as that of the network since most data passes between these two devices, with the processor acting as an intermediary.

Better disk subsystems will have large caches to minimize processor loads. However, data passing from the network to the disk, or the disk to the network, must pass via the processor along the bus. Thus each byte must take at least two clocks to get from input device to processor, then another two clocks to get from processor to output device. Often this latency is what determines actual throughput.

Finally, just because the bus can support a high data rate does not mean that the system will operate at that speed; many factors determine this. For instance, if an ISA card such as the SMC "Ultra" is placed on a PCI bus machine, the speed will be that of the lowest component, in this case the 16-bit 8-MHz PC-AT bus and not the 33-MHz, 32-bit PCI bus. While a lower rated device will often work on a higher rated system, typically the speed will suffer over an integrated device.

For today, a Pentium-based system with 32-bit PCI bus structure will be adequate for low to medium loaded networks, while for heavy loads, a RISC-based system such as the Dec Alpha with a 64-bit PCI bus would be preferred.

PERIPHERALS

Once the basic architecture has been determined, the peripherals must be considered. Oddly enough, what is visible is not that important, particularly in a server environment. For example, consider a 6x CD-ROM with a theoretical capability of 900(6x150) Kb/s. I have seen many people dissatisfied with such devices when the real culprit is an 8-bit controller card that limits the effective throughput to around 400 Kb/s.

The opposite is true of the human interface elements (keyboard, pointing device), printers, and even modems because these devices are incapable of and do not need very high-speed access. For these devices the original 8-bit PC bus is sufficient.

Really high-speed operations must, by their nature, be contained within the computer itself. Primarily, because for nanosecond response times, lead latencies (6.67 ns/ft of lead length is a good rule of thumb) become significant. Accordingly, the most critical elements must be chosen to apply a balanced load on the system when operating at its design point. For a server, these elements are CPU and memory (covered earlier), disk drive subsystem, network interface, and display controller.

DISK DRIVES

The disk drive structure needs to have the highest performance capability in a server. Until recently, there was little difference in disks since the limiting function was typically the bus and processor speed. Since 1990, the pressure has been on the disk drives, with caching controllers, extended IDE and SCSI-2 specifications, and higher rotational speeds by drives all increasing performance capability.

The basic limitation of any drive is the physical: the rotation speed of the spindle, the numbers of bytes per sector and real sectors per track, and the positioning rate of the heads. Note that the caveat "real" with respect to sectors per track appears because often a logical translation is made to comply with older BIOS limitations. Hence, the number entered into a CMOS table may bear little relation to the actual physical layout.

Historical Origins

The first drives for PCs used the ST-506 (Seagate Technology) standard for drive manufacture. This required a separate drive and controller. The long lead lengths slowed effective throughput and such devices were limited to effective transfer rates of about 300 Kbytes/second — about the same as a 2x CD-ROM today. The theoretical transfer rate for an MFM drive with seventeen 512-byte sectors per track rotating at 3600 rpm was about 510 Kbytes/s but was approachable only with an interleave of 1:1. Any deviation slowed the rate down. The 3:1 interleave optimally used on an XT-class machine (we have seen many with 6:1) was induced by the electronics/data bus. (I have successfully run an original full-height 10-Mb drive (type 1) at 1:1 with a 386 and a good controller with a throughput of over 300 Kbytes/s.) While adequate for personal computers, this was not sufficient for any except the most basic servers.

The next adoption was ESDI, or enhanced small device interface, which introduced the concept of "on-disk" controllers. This allowed a theoretical transfer rate increase to 3 Mbytes/second, which was never approached in practice but looked good in advertising.

In the latter 1980s, SCSI (small computer system interface) devices also became popular. This was surprising since the cost was generally higher than for a similar ST-506 device and since the 8-bit data bus inherent in the initial standard limited the initial devices to a peak of 4 Mbytes/second.

Still this was not bad for the time and had a definite advantage in the PC environment. Where the standard Intel-based PC could only support two disks, the SCSI bus could support up to eight devices of any kind — disk, CD-ROM, etc. When coupled with FAST-SCSI adapters having theoretical data rates of 10 Mbytes/s (real world, about 5) performance was more than adequate. This has now been incorporated into SCSI-II.

In addition, since controllers had their own BIOSes to manage the disks, the TYPE function, which limited available disk sizes in many PCs, did not apply. For this reason, the first very large disks were primarily SCSI.

Today, SCSI-II and III standards have extended the capabilities to 16- and 32-bit bus widths for even greater theoretical capability. Currently SCSI is the drive mechanism of choice in high-performance RISC workstations with 64-bit data widths and extended PCI bus structures allowing RAID configured systems to exceed 120-Mbytes/s real transfer rates. Theoretical maximum is over 260 Mbytes/s so some improvements are still possible without major change. This is not to be found in CISC-based workstations as yet, and there is question whether such a workstation would be able to handle such a large throughput anyway.

IDE

For smaller systems and lower-performance servers (though the gap is narrowing,) IDE (intelligent drive electronics a.k.a. ATA or AT attachment) drives extended the ESDI promise by moving all electronics onto the drive and using the controller card merely to establish port addresses and connect the drive to the system bus.

The promise of the IDE standard was in its 16-bit data width and the speeds inherent in having the drive electronics as close as possible to the heads. A large number of sectors per track (often between 50 and 80) increases disk speeds as do rotational speeds of up to 7200 rpm (most currently run at 4800 rpm). Coupled with 1:1 interleaving, this allows very high throughput, usually limited by the bus structure to effectively around 1 Mbytes/second. Later local bus controllers utilizing the EIDE (extended IDE) standard raised this to about 3 Mbytes/s — good enough for workstations and low-end servers but not for truly high-performance systems.

The latest update to the EIDE specification, known as Ultra DMA or UDMA, boosts performance to match midrange SCSI drives with a burst transfer rate of 33 Mbytes/s. Even so, the extensions allow very low drive prices, currently around $50/gigabyte, and as such are very attractive to the consumer market since a good SCSI system also requires a high-performance controller often costing more than the drive it supports. (E)IDE has no such limitation, and with BIOSes supporting large disks and LBA (large block access — essentially a streaming mode), to those upgrading to such a system, the throughput seems very good.

The bottom line is that for a very fast RISC-based server in a high load condition, there is no substitute for fast wide (64-bit) SCSI drives. For lesser applications, particularly with an Intel base, EIDE drives will perform satisfactorily at a slightly lower cost.

RAID

The final note on drives involves both speed and error correction. RAID (redundant arrays of inexpensive drives) offers one solution that can enhance reliability, improve throughput, and allow hot swapping of defective drives without having to bring the system down. For standard server operation, Level 5 or Level 6 is generally the best choice if RAID is desired.

NETWORK CARDS

Until recently, the standard was Ethernet using thick coax (aui adapter required), thin coax (RG-58 cable with BNC connections), or hub-based twisted pair using 6 pin RJ-45 connectors (4 pin RJ-11 connectors a.k.a. "telephone jacks" have also been used but this is nonstandard) at 10-Mbits/second rate or under 2 Mbytes/s. As seen, this level of performance can be handled by a standard ISA bus structure. In fact it was common to connect a high-performance server to multiple networks simultaneously since the CPU and the disk structure were only lightly loaded.

Today fast-Ethernet, FDDI (fiber distributed data interface), and ATM (asynchronous transfer mode) are pushing network speed past 100 Mbits/s or 12 Mbytes/second. To support these speeds, the current servers are being stressed to keep up, particularly under heavy loads.

Ethernet is a serial network. Only one packet can be received at a time, so each must be handled in turn. For these systems, multiple CPUs and arrays of RAIDs are becoming common both for fault tolerance and improved throughput. This will be covered in more depth in the chapter on super servers. However, the bottom line is that new and emerging network systems will be able to fully oust the throughput capability of a powerful server and are no longer the bottleneck they once were.

DISPLAYS

Unless the server is to be used as a workstation, there is little need to invest heavily in display technology. While 4 Mb of display memory coupled with high-performance graphics processors can provide impressive displays in 64,000,000 colors, there is no need for this in a server display. A common VGA video card with 1 Mb of memory will generally be sufficient even if complex graphics are utilized to support load displays.

CD-ROMS

The addition of one or more CD-ROM devices will not place a significant load on the system even in the current 10X (1.5 Mbytes/s) systems since such high throughputs will rarely be sustainable by the hardware. Like a floppy disk, CD-ROM usage will differ depending on the installation; the hardware performance level is not yet sufficient to be a strain on any modern server.

Keyboard and pointing device (mouse). Since these operate at human speed, virtually any device will do since speed/throughput is not an issue.

CONCLUSION

The main I/O will always be the bottleneck of any server installation. The main factors are disk access times and throughput and network capability. While important, CPU speed is generally not the limiting factor. In a server environment, the first task is the design of the disk farm, with throughput a major factor. Overall system speed and network load capability as well as necessary bus widths and speed will fall out from this. Servers, being by nature serial devices, will be limited by the least capable device.

Chapter 16
The RAID Advantage

Tyson Heyn

INTRODUCTION

Electronic data processing evolved from virtually nothing 50 years ago to its virtual omnipresence in the industrialized societies of the world today. The technologies that have been harnessed to manipulate data converted to its lowest common denominators (zeros and ones) have made nothing short of a huge impact on the lives of people throughout the world. Digitized information, or data, are being used to enable everything from live conversations between continents via satellite to the advancement of scientific discoveries and research, to controlling the temperatures of different rooms in a home. The recently emerged raft of on-line services provides not only the links to communicate with personal computers, but provides access to oceans of information to navigate, capture, and use by anyone with a computer. Businesses like banks and credit card companies use massive computing systems to provide everyday conveniences such as easier and faster access to money, in turn making it easier to bill or manage accounts. Even supermarkets and retail department stores are using powerful data-intensive information systems to do everything from managing inventories to monitoring consumer spending habits. The applications list goes on and on; everyone in virtually all walks of life is exposed in some manner or form to the impact of the ongoing revolution we call the Information Age.

The engines behind this revolution, of course, are computers. Today's Pentium-class personal computers, RISC workstations, minicomputers, supercomputers, and even (still!) mainframes provide the power that drives this infinite mass of data we rely upon to make everything from bank transactions to the purchase of groceries as easy as possible. The flow of data between computers, whether networked or linked via on-line services or the Internet, has become nothing less than a raging flood.

This astounding volume of data being transmitted between systems today has created an obvious need for data management. As a result, more and more servers — whether they are PCs, UNIX workstations, minicomputers or supercomputers — have assumed the role of information or data

0-8493-9823-1/00/$0.00+$.50

Exhibit 16.1 A simple RAID subsystem.

Disk 1	Disk 2	Disk 3	Disk 4
Block 1	Block 2	Block 3	Block 4
Block 5	Block 6	Block 7	Block 8
Block 9	Block 10	Block 11	Block 12
Block 13	Block 14	Block 15	Block 16
Block 17	Block 18	Block 19	Block 20

traffic cops. The number of networked or connectable systems is increasing by leaps and bounds as well, thanks to the widespread adoption of the client/server computing model, the boom in home computer use and the rise of Internet access service providers.

Hard disk storage plays an important role in enabling improvements to networked systems, because the vast and growing ocean of data has to reside somewhere. It also has to be readily accessible, placing a demand upon storage system manufacturers to not only provide high-capacity products, but products that can access data as fast as possible and to as many people at the same time as possible. Such storage also has to be secure, placing an importance on reliability features that best ensure that data will never be lost or otherwise rendered inaccessible to network system users.

RAID: THE SOLUTION TO SERVER GRIDLOCK AND DATA INTEGRITY

The solution to providing access to many gigabytes of data to users fast and reliably has been to assemble a number of drives in a gang or array of disks. These are known as RAID subsystems, which stands for redundant arrays of independent disks. Simple RAID subsystems (as shown in Exhibit 16.1) are basically a clutch of up to five or six disk drives assembled in a cabinet that are all connected to a single controller board. The RAID controller orchestrates read and write activities in the same way a controller for a single disk drive does, and treats the array as if it were in fact a single or virtual drive. RAID management software that resides in the host system provides the means to manage data to be stored on the RAID subsystem.

RAID Elements

Despite its multidrive configuration, RAID subsystems disk drives remain hidden from users; the subsystem itself is the virtual drive, though it can be as large as 1,000 Gbytes. The phantom virtual drive is created at a lower level within the host operating system through the RAID management software. Not only does the software set up the system to address the RAID unit as if it were a single drive, it allows the subsystem to be configured in ways that best suit the general needs of the host system.

RAID subsystems can be optimized for performance, the highest capacity, fault tolerance, or a combination of two or three of the above. Different so-called RAID levels have been defined and standardized in accordance with those general optimization parameters. There are six such standardized levels of RAID, called RAID 0, 1, 2, 3, 4, or 5, depending on performance, redundancy, and other attributes required by the host system. The RAID software is what is used to configure the desired RAID level of features in an array described in more detail below.

The RAID controller board is the hardware element that serves as the backbone for the array of disks: it not only relays the input/output (I/O) commands to specific drives in the array, but provides the physical link to each of the independent drives so they may easily be removed or replaced. The controller also serves to monitor the health or integrity of each drive in the array to anticipate the need to move data should it be placed in jeopardy by a faulty or failing disk drive (a feature known as fault tolerance).

The Array of RAID Levels

The RAID 0 through 5 standards offer users and system administrators a host of configuration options. These options allow the arrays to be tailored to their application environments. Each of the various configurations listed below focus on maximizing the abilities of an array in one or more of the following areas: capacity, data availability, performance, and fault tolerance.

RAID Level 0. An array configured to RAID Level 0 is an array optimized for performance, but at the expense of fault tolerance or data integrity. RAID Level 0 is achieved through a method known as striping. The collection of drives (or virtual drive) in a RAID Level 0 array has data laid down in such a way that it is organized in stripes across the multiple drives. A typical array can contain any number of stripes, usually in multiples of the number of drives present in the array. As an example, imagine a four-drive array configured with 12 stripes (four stripes of designated space per drive). Stripes 0, 1, 2, and 3 would be located on corresponding hard drives 0, 1, 2, and 3. Stripe 4, however, appears on a segment of drive 0 in a different location than Stripe 0; stripes 5 through 7 appear accordingly on drives 1, 2, and 3. The remaining four stripes are allocated in the same even fashion across the same drives such that data would be organized in the manner depicted in Exhibit 16.2.

In Exhibit 16.2, a RAID Level 0 configuration, a virtual drive is comprised of several stripes of information. Each consecutive stripe is located on the next drive in the chain, evenly distributed over the number of drives in the array.

Practically any number of stripes can be created on a given RAID subsystem for any number of drives. Two hundred stripes on two disk drives

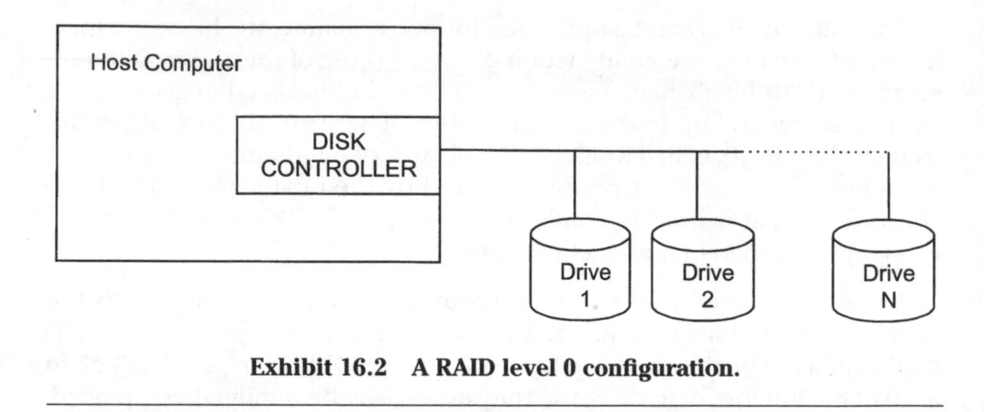

Exhibit 16.2 A RAID level 0 configuration.

are just as feasible as 50 stripes across 50 hard drives. Most RAID subsystems, however, tend to have between 3 and 10 stripes.

The reason RAID 0 is a performance-enhancing configuration is that striping enables the array to access data from multiple drives at the same time. In other words, since the data are spread out across a number of drives in the array, they can be accessed faster because they are not bottled up on a single drive. This is especially beneficial for retrieving very large files, since they can be spread out effectively across multiple drives and accessed as if it were the size of any of the fragments it is organized into on the data stripes.

The downside to RAID Level 0 configurations is that it sacrifices fault tolerance, raising the risk of data loss because no room is made available to store redundant data. If one of the drives in the RAID 0 fails for any reason, there is no way of retrieving the lost data as can be done in other RAID implementations described below.

RAID Level 1. The RAID Level 1 configuration employs what is known as disk mirroring, and is done to ensure data reliability or a high degree of fault tolerance. RAID 1 also enhances read performance, but the improved performance and fault tolerance come at the expense of available capacity in the drives used.

In a RAID Level 1 configuration, the RAID management software instructs the subsystems controller to store data redundantly across a number of the drives (mirrored set) in the array. In other words, the same data are copied and stored on different disks (or mirrored) to ensure that, should a drive fail, the data are available somewhere else within the array. In fact, all but one of the drives in a mirrored set could fail and the data stored to the RAID 1 subsystem would remain intact. A RAID Level 1 configuration can consist of multiple mirrored sets, whereby each mirrored set can be a different capacity. Usually the drives making up a mirrored set are

Exhibit 16.3 A RAID level 1 subsystem.

Disk 1	Disk 2	Disk 3	Disk 4
Block 1	Block 1	Block 6	Block 6
Block 2	Block 2	Block 7	Block 7
Block 3	Block 3	Block 8	Block 8
Block 4	Block 4	Block 9	Block 9
Block 5	Block 5	Block 10	Block 10

of the same capacity. If drives within a mirrored set are of different capacities, the capacity of a mirrored set within the RAID 1 subsystem is limited to the capacity of the smallest-capacity drive in the set, hence the sacrifice of available capacity across multiple drives.

The read performance gain can be realized if the redundant data are distributed evenly on all of the drives of a mirrored set within the subsystem. The number of read requests and total wait state times both drop significantly — inversely proportional to the number of hard drives in the RAID in fact. To illustrate, suppose three read requests are made to the RAID Level 1 subsystem (see Exhibit 16.3). The first request looks for data in the first block of the virtual drive; the second request goes to block 0; and the third seeks from block 2. The host–resident RAID management software can assign each read request to an individual drive. Each request is then sent to the various drives, and now — rather than having to handle the flow of each data stream one at a time — the controller can send three data streams almost simultaneously, which in turn reduces system overhead.

In Exhibit 16.3, a RAID Level 1 subsystem provides high data reliability by replicating (or mirroring) data between physical hard drives. In addition, I/O performance is boosted as the RAID management software allocates simultaneous read requests between several drives.

RAID Level 2. RAID Level 2 is rarely used in commercial applications, but is another means of ensuring data are protected in the event drives in the subsystem incur problems or otherwise fail. This level builds fault tolerance around Hamming error correction code (ECC), which is often used in modems and solid-state memory devices as a means of maintaining data integrity. ECC tabulates the numerical values of data stored on specific blocks in the virtual drive using a special formula that yields what is known as a checksum. The checksum is then appended to the end of the data block for verification of data integrity when needed.

As data get read back from the drive, ECC tabulations are again computed, and specific data block checksums are read and compared against the most recent tabulations. If the numbers match, the data are intact; if there is a discrepancy, the lost data can be recalculated using the first or earlier checksum as a reference point.

Example of Error Correction Coding. Here is an example of one method of error correction coding. Suppose the phrase being stored is HELLOTHERE. The checksum is computed for every 10 bytes of data.

Data being stored:	H	E	L	L	O	T	H	E	R	E	
Numerical representation:	72	69	76	76	79	84	72	69	82	69	
Checksum formula:	x 1	x 2	x 3	x 4	x 5	x 6	x 7	x 8	x 9	x 10	
Multiplied out:	72	138	228	304	395	504	504	414	738	690	
[Check]sum of all values:	72	+138	+228	+304	+395	+504	+504	+414	+738	+690	= 3987

So, the data are stored on the drive as: 72 69 76 76 79 84 72 69 82 69 3987

As the data are read back from the drive, the same calculations with the data segment are made. The newly computed checksum is compared against the previously stored checksum, thus verifying data integrity.

This form of ECC is actually different from the ECC technologies employed within the drives themselves. The topological formats for storing data in a RAID Level 2 array is somewhat limited, however, compared to the capabilities of other RAID implementations, which is why it is not used all that often in commercial applications.

RAID Level 3. This RAID level is really an adaptation of RAID Level 0 that sacrifices some capacity for the same number of drives, but achieves a high level of data integrity or fault tolerance. It takes advantage of RAID Level 0's data striping methods, except that data are striped across all but one of the drives in the array. This drive is used to store parity information that is used to maintain data integrity across all drives in the subsystem. The parity drive itself is divided up into stripes, and each parity drive stripe is used to store parity information for the corresponding data stripes dispersed throughout the array. This method achieves very high data transfer performance by reading from or writing to all of the drives in parallel or simultaneously but retains the means to reconstruct data if a given drive fails, maintaining data integrity for the system (see Exhibit 16.4). RAID Level 3 is an excellent configuration for moving very large sequential files in a timely manner.

In Exhibit 16.4, a RAID Level 3 configuration is very similar to a RAID Level 0 configuration in its utilization of data stripes dispersed over a series of hard drives to store data. In addition to these data stripes, a special drive is configured to hold parity information used to maintain data integrity throughout the RAID subsystem.

The stripes of parity information stored on the dedicated drive are calculated using the Exclusive OR function. Exclusive OR is a logical function

Exhibit 16.4 A RAID level 3 configuration.

Disk 1	Disk 2	Disk 3	Disk 4	Disk 5
Bit/Byte 1	Bit/Byte 2	Bit/Byte 3	Bit/Byte 4	Parity
Bit/Byte 5	Bit/Byte 6	Bit/Byte 7	Bit/Byte 8	Parity
Bit/Byte 9	Bit/Byte 10	Bit/Byte 11	Bit/Byte 12	Parity
Bit/Byte 13	Bit/Byte 14	Bit/Byte 15	Bit/Byte 16	Parity
Bit/Byte 17	Bit/Byte 18	Bit/Byte 19	Bit/Byte 20	Parity

Exhibit 16.5 Standard OR function: group A group B.

Group A	Group B	Result
0	0	0
1	0	1
0	1	1
1	1	1

between the two series that carries most of the same attributes as the conventional OR function. The difference occurs when the two bits in the function are both non-zero: in Exclusive OR, the result of the function is zero, whereas with conventional OR it would be one, as described in Exhibit 16.5.

By using Exclusive OR with a series of data stripes in the RAID, any lost data can easily be recovered. Should a drive in the array fail, the missing information can be determined in a manner similar to solving for a single variable in an equation (for example, solving for x in the equation, $4 + x = 7$). Similarly, in an Exclusive OR operation, it would be an equation like $1\ x = 1$. Thanks to Exclusive OR, there is always only one possible solution (in this case, 0), which provides a complete error recovery algorithm in a minimum amount of storage space.

RAID Level 4. This level of RAID is similar in concept to RAID Level 3, but emphasizes performance for different applications, e.g., Database TP vs. large sequential files. Another difference between the two is that RAID Level 4 has a larger stripe depth, usually of two blocks, which allows the RAID management software to operate the disks much more independently than RAID Level 3 (which controls the disks in unison). This essentially replaces the high data throughput capability of RAID Level 3 with faster data access in read-intensive applications.

RAID Level 4 builds on RAID Level 3 technology by configuring parity stripes to store data stripes in a non-consecutive fashion. This enables independent disk management, ideal for multiple-read-intensive environments.

Exhibit 16. 6 RAID level 5 overcomes the RAID level 4 write bottleneck.

Disk 1	Disk 2	Disk 3	Disk 4
Parity (0, 1, 2)	Block 0	Block 1	Block 2
Block 3	Parity (3, 4, 5)	Block 4	Block 5
Block 6	Block 7	Parity (6, 7, 8)	Block 8
Block 9	Block 10	Block 11	Parity (9, 10, 11)

A shortcoming of RAID level 4 is rooted in an inherent bottleneck on the parity drive. As data gets written to the array, the parity encoding scheme tends to be more tedious in write activities than with other RAID topologies. This more or less relegates RAID Level 4 to read-intensive applications with little need for similar write performance. As a consequence, like its Level 3 cousin, it doesn't see much common use in commercial applications either.

RAID Level 5. This is the last of the most common RAID levels in use and is probably the most frequently implemented. RAID Level 5 minimizes the write bottlenecks of RAID Level 4 by distributing parity stripes over a series of hard drives. In doing so it provides relief to the concentration of write activity on a single drive, which in turn enhances overall system performance (see Exhibit 16.6).

In Exhibit 16.6, RAID Level 5 overcomes RAID Level 4's write bottleneck by distributing parity stripes over two or more drives within the system. This better allocates write activity over the RAID drive members, thus enhancing system performance.

The way RAID Level 5 reduces parity write bottlenecks is relatively simple. Instead of allowing any one drive in the array to assume the risk of a bottleneck, all of the drives in the array assume write activity responsibilities. The distribution frees up the concentration on a single drive, improving overall subsystem throughput.

RAID Level 5's parity encoding scheme is the same as Levels 3 and 4; it maintains the system's ability to recover any lost data should a single drive fail. This can happen as long as no parity stripe on an individual drive stores the information of a data stripe on the same drive. In other words, the parity information for any data stripe must always be located on a drive other than the one on which the data resides.

Other RAID levels. Other, less-common RAID levels have been developed as custom solutions by independent vendors (they are not established standards):

Exhibit 16.7 RAID configuration options.

RAID Level	Capacity	Data Availability	Data Throughput	Data Integrity
0	High	Read/Write High	High I/O Transfer Rate	
1		Read/Write High		Mirrored
2	High		High I/O Transfer Rate	ECC
3	High		High I/O Transfer Rate	Parity
4	High	Read High		Parity
5	High	Read/Write High		Parity
6		Read/Write High		Double Parity
10		Read/Write High	High I/O Transfer Rate	Mirrored
53			High I/O Transfer Rate	Parity

- RAID Level 6, which emphasizes ultra-high data integrity by writing two sets of parity data, providing redundancy for drive failures.
- RAID Level 10 (also known as RAID Level 0 & 1), which focuses on high I/O performance and very high data integrity
- RAID Level 53, which combines RAID Level 0 and 3 for uniform read and write performance

Tailor-made RAID. Perhaps RAID technology's biggest advantage is the sheer number of possible adaptations available to users and systems designers. RAID offers the ability to customize an array subsystem to the requirements of its environment and the applications demanded of it. RAID's inherent variety of configuration options provides several ways in which to satisfy specific application requirements (see Exhibit 16.7).

Customization, however, doesn't stop with a RAID level. Drive models, capacities, and performance levels have to be factored in as well as what connectivity options that are available.

INTERFACE OPTIONS

Differential SCSI (small computer systems interface), for example, allows a subsystem to be cabled as far as 18 feet from a host with no degradation to the data signal. Fast/Wide SCSI, another interface option, can be combined with differential SCSI or employed by itself; it essentially doubles the 10 Mbyte/s throughput of Fast SCSI, enabling data rates of up to 20 Mbytes/s. The newest parallel SCSI interface option is UltraSCSI, a 40 Mbyte/s interface standard.

An emerging new serial interface standard known as fibre channel-arbitrated loop (FC-AL) is yet another interface option for RAID subsystems, and is the most powerful of them all. FC-AL is capable of up to 200 Mbyte/s data throughputs (dual loop configurations), while allowing RAID subsystems or other connected peripherals to be placed as far as 10 kilometers from the host. It also enables easy connection of up to 126 disk drives on a single controller (compared to seven devices with conventional SCSI!). The potential impact of FC-AL alone will undoubtedly be enormous on the evolution of RAID subsystems. FC-AL can be operated in either single or dual loop configurations. The dual loop allows another level of redundancy by allowing two separate data paths for all attached devices.

SCA: CLEANING UP THE CABLE MESS

Many of these interface options, including serial FC-AL and parallel UltraSCSI, support the SCSI single connector attachment (SCA) standard. SCA is an elegant means of eliminating the miles of wiring involved with connecting several drives via conventional backplane architectures. Before SCA, conventional connections involved two cables per drive: one for power and the other for data transmission. Arrays with more than a few drives would amass a lot of spaghetti at the rear of the rack, and especially large arrays would have an unwieldy mess of wire to connect the drives. SCA, however, allows for drives to be plugged directly into a backplane without cables. It not only rids subsystems of the mass of cabling previously required, but facilitates hot plugging (removal or insertion of a drive while the subsystem is on line) and improves the reliability of the system as a whole because of the substantially reduced number of connections.

Chapter 17 Multiprocessing Servers

Bill Wong

How many processors does your PC use? If you answered one, you are wrong. Today's PC is a multiprocessing wonder on a small scale. The keyboard chip has its own processor as do many video cards and most SCSI controllers. These form a loosely coupled, asymmetrical multiprocessing (AMP) system, but the symmetrical multiprocessing (SMP) systems are what typically come to mind when multiprocessing is mentioned in casual conversation. SMP uses multiple, identical processors typically tied together with a shared memory subsystem. The processors are used interchangeably, distributing the workload, and thereby providing better overall performance compared to a single processor.

The first step is moving from a single processor to a dual processor system. Dual processor servers were the first on the PC scene, with major vendors like ALR and Compaq leading the way. Moving from one to two is not too difficult. Increasing memory bandwidth and coordinating two processors is easy compared to supporting a dozen processors. The speed advantage gained by adding another processor is not 100 percent but typically 80 or 90 percent. The hard part with dual processor systems is whether the trade-off is better with two processors or with a faster single processor. This type of trade-off is less of an issue as the number of processors grows.

The hard part of growing the number of processors is bandwidth, performance, and reliability. Multiple processors need access to memory and disks. The bandwidth to these devices limits how well the processors will work together. Bandwidth in turn controls overall system performance. The goal is to increase performance as much as possible. Reliability is important because a failed processor has the potential of bringing the entire system to a standstill. Robust hardware and software are equally important to maintaining a highly reliable system.

The typical, low-end SMP system has two to six processors. Dual processors are often put on a single motherboard. Systems with more processors

0-8493-9823-1/00/$0.00+$.50

use multiple cards with one or two processors on each. A high-speed memory bus is the typical processor interconnection mechanism. Each processor uses a cache to maintain performance, and each cache must remain in sync.

Increasing the number of processors to dozens, or hundreds, requires even more complex connection mechanisms such as crossbar memory switches and hierarchical memory. Memory subsystem design is difficult, but providing access to large amounts of information from hard disks, tape, and optical disks is even more difficult.

Clustering is an alternative and is also complementary to more tightly coupled multiprocessing systems. Single processor and multiprocessor systems can be clustered together using high-speed, but typically not memory-speed, connections. Clusters provide coarse multiprocessing as well as high reliability solutions.

RISC processors found their way into multiprocessing systems quickly. They have a long track record in large systems. Of course, just as the Intel processors have overtaken the desktop, so too are they invading the multiprocessor and clustering arena. You are equally apt to find multiple Pentium Pro chips in a system as you might find Sparc, MIPS, Alpha, or Power PC chips.

Two to four processor systems are slowly finding their way to the desktop, but multiprocessing systems are typically found as servers on a network. So, are today's file servers turning into multiheaded wonders?

DOES IT WORK? WHERE?

Actually, if multiprocessing didn't work, there wouldn't be so many companies supplying multiprocessing solutions. The number of products indicates that it works exceptionally well. Multiprocessing systems can support two general types of applications. The first type (S) of application is one designed to run on a single processor system. This includes multithreaded applications. The second type (M) is an application that is designed to run on multiple processors. SMP systems can improve the performance of both kinds of applications but in different ways. Multiple processors can help S applications run faster by running more of them at the same time. Run only a single S application with multiple processors and you will not see much of an improvement. The best example of this type of improvement is on a dual processor development PC. The development environment can continue to run at full speed while a compiler runs on the second processor.

Multiprocessing demand is up because network servers are moving from basic file and print services to application servers. SMP systems can improve file and print service performance, but AMP systems often work better because additional I/O processors improve file and print service performance.

Database and web servers benefit from SMP support because database processing can be easily split among similar processors. Database access can also be partitioned based on the requests coming in allowing multiple processors to be used effectively. Dozens or even hundreds of processors can be used in this type of environment.

THE INTEL REALM

Intel's (www.intel.com) multiprocessing specification (MP Spec) was drawn up at the start of the Pentium era, which is quickly being eclipsed by the Pentium Pro. Still the MP Spec was an important starting point for multiprocessing in the Intel processor realm because prior to this point the multiprocessing systems required customized versions of operating systems such as UNIX and OS/2. The MP Spec included specifications for the programmable interrupt controller (PIC) architecture and the multiprocessor BIOS. It allowed Intel processor compatible operating systems to be implemented so they would run on any MP PC with little or no reconfiguration.

The Pentium Pro has moved to the forefront of Intel-based multiprocessing support because of its built-in cache and multiprocessing support. Intel has developed the 82450 PCI chip set family to work with multiple Pentium Pro processors with minimal support overhead. It supports up to four Pentium Pro processors. It is found on dual and quad multiprocessor motherboards. The 82450 provides support for multiple PCI buses as well as making the combination ideal for network servers that require PCI adapters for disk subsystems and network adapters. The 82450 actually comes in two versions. The 82450GX is designed for SMP servers, while the 82450KX is designed for dual processor workstations. Intel calls the 82450 a glueless multiprocessing solution because it incorporates most of the multiprocessor support not already included in the Pentium Pro, such as arbitration logic and memory controller.

Two early multiprocessing leaders, Tricord and NetFrame, are still plugging away with Pentium and Pentium Pro-based systems based on their own architectures. Tricord's (www.tricord.com) 267 MB/sec Power Bus works with up to eight processors.

NetFrame's (www.netframe.com) 8500 Server splits things up with up to four independent multiprocessor parallel server architecture (MPSA) buses providing an aggregate throughput of 100 MB/second. Multiple processors can be connected to each bus including up to eight I/O server modules, thereby mixing SMP and AMP into one high-performance package.

The whole collection is called a ClusterServer and it can support up to sixteen Ethernet, eight Token Ring, or four FDDI ports.

Tricord and NetFrame do more than just add processors and mix. Fault tolerance, an attribute that used to be limited to high-end applications, is a hallmark and now available at a reasonable cost for medium and large application environments.

At the low end, multiprocessing Intel platforms are popping up all over the place. Compaq (www.compaq.com), ALR (www.alr.com), and HP (www.hp.com) have moved from supplying a few small, multiprocessing systems to making multiprocessing systems the center of their Intel server offerings. Now just about every Intel PC vendor has an SMP multiprocessing solution.

NOT SO RISCY BUSINESS

RISC multiprocessing systems have been around almost as long as RISC processors. RISC vendors like HP and Sun (www.sun.com) provide multiprocessing support in all but their low-end desktop solutions, with dozens or hundreds of processors finding a home in a single server.

The PA-RISC chip is the basis for HP's RISC products. It complements HP's Intel processor based systems, which it also provides in a multiprocessing package, but the PA-RISC systems make the Intel solutions look pale in terms of performance and expandability. HP's X-Class Exemplar Technical Server series is based on HP's popular PA-8000 systems. The X-Class supports up to 64 processors that provide up to 46.08 GFLOPS. FLOP is floating point operation per second. GFLOP is one billion flop. With performance like this it is no wonder that large database server sites and web sites fit nicely into an X-Class system.

Sun took the risk out of RISC with its SPARC processor. The SPARC chip has undergone numerous upgrades. Up to fourteen 167-MHz 64-bit UltraSPARC I chips can be found in Sun's Ultra Enterprise 5000 systems. The 5000 utilizes a 2.6-GB/s, packet-switched Gigaplane bus to feed the SPARC chips and up to fourteen 200-MB/s I/O channels. The Gigaplane is actually designed for up to 30 processors and 30 I/O channels for really large servers. Sun's Ultra Port Architecture (UPA) bus ties the I/O processors to SBus I/O peripherals.

The Solaris operating system, Sun's UNIX variant, ties all the UltraSPARC chips together. It provides a time-tested environment for applications designed to take advantage of the multiprocessing environment.

HP and Sun are not the only multiprocessing solution providers. While Intel Pentium Pro's have taken a lot of the limelight, there are still a number of RISC system vendors, and every one of them has a multiprocessing solution in his bag of products. This includes IBM's Power PC, which is showing up in a variety of places.

Intel and Microsoft Windows NT has generated a lot of interest but much of this is at the low end of the multiprocessing spectrum, which is also the high end of the network server arena. At this point, the combination doesn't hold a candle to the high-end RISC solutions, which is why the only place to look for large and robust solutions is with the likes of Sun and HP.

BLUE IRON

IBM (www.ibm.com) makes a variety of multiprocessing PC servers that are comparable to the products described in the last two sections. IBM is also known for three product lines that also have multiprocessing support: the AS/400, the RS/6000, and the S/390. The AS/400 and RS/6000 are often thought of as minicomputers but they are effectively in the realm of the PC and RISC solutions both in cost and construction. The S/390 is considered big iron but it too has shrunk from the giant, air-conditioned shrines. Big iron is still big, but it often makes a better network server solution than a multiprocessing PC alone.

The AS/400 uses an AMP approach. Multiple file server I/O processor (FSIOP) boards can be plugged into an AS/400 chassis. FSIOP boards come in different forms, but a typical FSIOP runs an Intel processor. It is also home to a special version of IBM's Warp Server. Warp Server provides a gateway to an Ethernet or Token Ring network and the server utilizes the AS/400 for storage as well as providing a gateway to the AS/400's main processor and its applications. The AS/400 may not be known as a multiprocessing solution, but it is — and an effective one at that.

The RS/6000 uses IBM's Power PC RISC processor that typically runs AIX, IBM's UNIX-variant. Multiple processor RS/6000 systems are common and AIX is no stranger to multiprocessing either. In fact, the RS/6000 series is the IBM solution that most people think of when you mention multiprocessing. A variety of software products take advantage of the RS/6000's multiprocessing support including transaction processing systems such as Tuxedo.

S/390 is one of the granddaddies of the mainframe world where multiple processing abounds. SMP and AMP appears throughout. Multiple main processors are augmented by multiple I/O and communication processors. Disk farms and hierarchical storage management (HSM) systems proliferate in large networks. Clustering is also a common trait in large computational environments. Network gateways come in a variety of forms, which also complement the mixed AMP/SMP environment.

Why would anyone want to invest in more big iron? Well, multiprocessing was in mainframes well before it was ever considered for PCs. Multiprocessing hardware support is complex, but at this point the hardware is the best understood and most reliable portion of PC, mini, or mainframe multiprocessing solutions. The place where mainframes actually have an edge is in the software. IBM has years of multiprocessor experience tucked under its belt and it shows.

Big iron isn't dead; it is actually thriving.

PEANUT CLUSTERS

Multiprocessing systems are tightly coupled, and all processors normally reside in the same box. Clustered systems spread the processors out a bit. Like tightly coupled multiprocessing systems, clustered systems have a variety of connection methods. The idea behind clustering is the same as multiprocessing. Increase the number of processors and you can increase the performance and reliability. The main difference between tightly coupled multiprocessing systems and clustered systems is the speed of interaction and the level of integration. Clustered systems typically incorporate software redundancy. Detecting and superceding a failed processor can take seconds, which is great for most high availability environments.

All major mainframe, mini, and high-end PC server vendors have cluster support in one form or another. Sun, IBM, and Digital (www.digital.com) all have clustering technology, but hardware support alone is not sufficient. Software, especially middleware, is important. Middleware is the key because it provides both coordination and error recovery. Like tightly coupled multiprocessing, clusters will support applications written for a single processor, but effective use of the cluster is made when the applications are written to take advantage of the middleware.

Microsoft (www.microsoft.com) has been working with Windows NT system vendors to bring clustering to Windows NT. The project is code named Wolfpack. It is both a hardware and software specification with software support coming from Microsoft as well. Wolfpack is exciting because of the interest in Windows NT, but if you need solutions today, Wolfpack may be a bit early for you. Wolfpack promises to lower the cost of clustering, but, as anyone who has installed Windows NT knows, the resources and horsepower needed to run Windows NT is comparable to UNIX, which is where clustering has an exceptional track record.

IT'S THE BLADES

Remember the razor company motto: give away the razor and sell the blades. Multiprocessing systems are not in the same class as razors but the idea still applies. The starting cost for a multiple processing system is typically higher than a single processing system, but you can minimize the starting cost to customers and make money with the add-ons.

Cost cutting on single processor file and print servers is common, but multiprocessing servers tend to have a higher reliability requirement. Processors are not the only thing that tend to multiply in a multiple processor system. Power supplies, disk and tape drives show up with hot swapping and arrays being very common. A system still has a single case, but they are typically larger and more robust. Environmental monitoring assists in proactive problem determination.

Hard disks and memory are incremental growth areas with multiprocessing systems. Multiple processors can provide performance and reliability gains but only if they have sufficient resources. It is often more efficient to add more memory than to add another processor. In fact, adding another processor often leads to a need for a lot more memory.

The need for supplies also grows with a multiprocessing system. More tapes and optical disks are needed to back up larger disk arrays. Information on high-reliability systems is typically more valuable so complete or incremental backups are apt to be done on a regular basis.

SUMMARY

Multiprocessing solutions are great for VARs because they have more peripherals, more memory, and more support software than single processor solutions. Multiprocessing solutions also provide an ideal upgrade path for customers, which are a future source of revenue for VARs.

The trick, of course, is to make multiprocessing more effective than single processor solutions in addition to making them more powerful. A single point of management and redundant hardware and software support are definite benefits of a multiprocessing system.

Multiprocessing systems have been around for ages, but low-cost multiprocessing systems are bringing benefits to more networks than ever before. Medium to large networks will soon find multiprocessing servers to be the norm.

Tracking all the multiprocessing solutions is a full-time job, so if your job is selling, installing, or supporting you will want to narrow your focus. If you plan on having clusters or multiprocessor systems in your network, then you will want to keep them as similar as possible or spend a lot of time learning how to support them. Multiprocessing solutions are usually worth the investment, when they work.

One key point to remember about multiprocessing systems is that the choice of components is critical. Just plugging in a few processors does not make a fast or robust system. All parts come into play, including the memory subsystem and peripheral adapter subsystem, not to mention support hardware like UPS and environment monitoring. The best way to provide multiprocessing solutions is to know what combinations will be the most effective and the most cost effective.

Chapter 18
Super Servers — One Step Beyond

A. Padgett Peterson

The concept of a super server is that of "more." Whatever the current state of the art is, a super server is more/faster/better.

Being more than the best, SS technology is essentially price insensitive and low quantity. Designed for the site which must do more than is available from standard technology, exactly what a super server is becomes a moving target, always in advance of current offerings.

Occasionally, advances developed for SS technology will become standard, such as RAID. More often the technology will be obviated by advances in another area with the risk that the unit will become unmaintainable.

One problem is that it is almost impossible to tell in advance which will succeed and which will not; all are plausible and effective within their area. The question is more one of economics: if the mechanism becomes standard so that volume increases and prices drop, then success is likely.

ESDI (Enhanced Small Device Interface) appeared to have such promise and was widely used for servers, since considerably higher throughput was available than for standard ST-506 drives. However, technology leapfrogged directly to IDE, which was a logical extension of the ESDI concept, leaving ESDI in a backwater.

The promise of the latest technology is compelling, particularly when contemplating replacement of an older system. Putting emotion aside, three questions need to be asked first:

1. Is this really necessary?
2. Could the same effect be achieved with multiple servers? (This often is a better solution when multiple protocols/tasks must be supported simultaneously.)
3. Is there cost/benefit justification?

0-8493-9823-1/00/$0.00+$.50

Very important is the realization that technology is doubling throughput about every 18 months. A viable option may be to simply wait for a few months. Such delay is particularly apt when technology is undergoing a shift such as the current move to web sites and access.

So before deciding that a super server is necessary, take a moment to assess the real need as opposed to the possible drawbacks:

1. Maintainability
2. Upgradabiity
3. Sole source for parts
4. Higher price (initial and ongoing)
5. Special training for administrators

(Reliability is not mentioned since often a reason for going to a super server is for redundancy and fault tolerance.)

On the other hand, "standard" servers are generally designed to meet a cost point and that in turn generally allows for single point failures which could remove the server from use for an extended period. A corporate mail server serving over 3,000 employees with no fault tolerance/mirroring could leave an organization open to a charge of culpable negligence.

SUPER SERVER TECHNOLOGY

Some elements have been staples of super servers: the first and most common element is RAID disk storage, usually providing very high burst throughput (real-world use expectation should be about half what is quoted). More important is the fault tolerance level — for critical operations dual-fail-operational with hot-swapping should be a minimum (Level 6) with a hybrid mechanism where safety considerations are involved.

SMP (symmetric multiprocessing) is another staple of super server technology. Again the promise is generally increased throughput through task designation. Here there are two mechanisms used, the SS may use either or both.

The first is multitasking, where the programs are divided into modules and a semaphore mechanism is used to detect task completion on the part of a processor and to assign the next task on a stack to it. This mechanism is inherently scalable as the number of processors is independent of the program. The scheduler never needs to be concerned about which processor is assigned to which task. Concern needs to be given to fault tolerance to detect when a task is not being completed and a handshake is necessary to be sure a processor is really available.

However, such a system is inherently fault tolerant, as on failure of a CPU; the tasks are merely reassigned to a different processor and the system continues at a reduced throughput capacity. In general it is a good idea to size the system so that full expected throughput can be maintained with a single failure and safe operation maintained on dual-fail.

The second mechanism used on critical systems is mirroring. As in RAID 1, this calls for two fully independent systems to be performing the same tasks, each fully capable of maintaining the system load. In this case throughput increase is not observed, since both are running identical operations. However, in the event of failure at any point, operation is maintained.

While such systems may be found in a single system, look for duality in every element, including power supplies. At the highest level, each element may support either path so that if there is a failure in power supply 1 and CPU 2, power supply 2 will support CPU 1 to maintain operations. Such an installation is dependent on the switching/selection mechanism, which may take the form of a third independent element.

In a truly high-end system, both mirroring and SMP may be found in a single unit, with multiple peripheral busses and RAID disk farms, extending the RAID philosophy to every element of the server.

Here there is a caveat with the instability that is present at the end of the 20th century: at one time companies could be expected to support their wares nearly indefinitely, but since World War II this practice has been slowly disappearing. By 1970 the average support period was 10 years. Ten years later the norm was five. Today, many companies' life spans are less.

So for the risk–adverse environment where continuity and reliability are paramount, it is important to select a company that appears stable.

If analysis justifies the extra cost of the super server as opposed to multiple standard servers, analysis of the vendor should be of the same importance. Further investment in a service pack of expected replacement and unique spare parts is a good investment. Next, a good practice is that if a part fails, buy two replacements — one to replace the failed part and a second to have on hand. Since the investment in a super server is not trivial, investment for spares should be considered as well.

A final note: in general super servers will require special software to optimize their use (often standard software will degrade performance). When new versions appear, if the system is running smoothly it is often a good idea to delay installing the upgrade for a few months after release to see if any problems are detected. Unfortunately, another facet of the 1990s is that what manufacturers release as a final version often has many of the characteristics of beta software. For this reason upgrades are not necessarily needed unless there is a compelling reason to do so.

Section IV
Network Operating Systems

Two of the major mechanisms used to categorize the operation and use of servers are their hardware platform and network operating system (NOS). The previous section in this handbook examined the key hardware components that define the processing and storage capability of servers. In this section we change our orientation from hardware to software, examining the use of different operation systems.

Until recently, any discussion of network operating systems was primarily a choice between NetWare and Windows NT. While both network operating systems continue to dominate the server market, a viable newcomer that is gaining the attention of industry, academia, and government is Linux. Recognizing the fact that the dynamic duo of network operating systems may evolve into a troika, this section includes chapters focused on all three operating systems.

In Chapter 19, Dennis Dillman provides us with detailed information covering one of the most successful network operating systems in history, "NetWare 3.X." Although many people consider NetWare 3.X to be dated and it is, it remains a very popular operating system with hundreds of thousands of sites still using this reliable NOS. Dillman first provides us with a detailed overview of the architecture of NetWare 3 to include the manner by which the NOS is initiated and its requirement to use a predefined directory structure. He then reviews client software required to interoperate with this NOS and utilities that facilitate managing of the operating system.

A second chapter covering NetWare turns our attention to the support of the TCP/IP protocol. In Chapter 20, "intraNetWare Overview," James E. Gaskin provides us with an examination of the features incorporated into the first version of NetWare developed to support the TCP/IP protocol stack.

Recognizing the important role of the Novell Director Services (NDS), the third chapter in this section turns our attention to this topic. In the chapter "Migrating to intraNetWare 4.X and Novell Directory Services,"

Eric Stral provides us with a detailed overview of NDS, including its benefits, structure of its network-wide database, types of objects supported by the database, and its directory tree structure. Using the preceding information as a base, Stral illustrates the process involved in designing a Novell directory services tree and describes how to plan and implement an NDS partition and describes a replication strategy. He also describes and discusses the advantages and disadvantages of four methods associated with migrating to NetWare 4.1.

Commencing with Chapter 22 we begin a series of chapters covering Microsoft's Windows NT network operating system. In that chapter, titled "Windows NT Architecture," Gil Held provides an overview of the basic structure of this NOS. The second chapter covering the NT operating system details the differences and similarities between the workstation and server versions of the operating system. In this chapter, which is Chapter 23 and entitled "Windows NT Workstation vs. Server 4.0," Stuart Miller provides us with a description of the critical differences between the two versions of this NOS, including memory and storage requirements. Although Windows NT Workstation does not support fault tolerances, Miller reviews the manner by which NT Server provides such support which supplies additional insight into the rationale for using the more costly version of this NOS even when you have only a handful of users that need to access a common series of programs and data files.

In concluding our series of Windows NT-oriented chapters, Gil Held, in Chapter 24 titled "Evaluating Client/Server Operating Systems: Focus on Windows NT," looks at some of the key features of both Windows NT Workstation and Windows NT server, including built-in security, performance monitoring capability, and network support.

In Chapter 25, "A Look at Linux," Daniel Carrere provides us with a detailed overview of the Linux operating system. He introduces us to the technology of this relatively new NOS, provides a list of major business applications that run under it, and through a series of referenced screen images describes such important operations as an NFS mount, the use of SAMBA, and Apache server, as well as the support of FTP. In concluding this introduction to Linux, Carrere provides a list of references for additional information. This resource list covers Linux Web sites and addresses of several journals and organizations that promote this operating system as well as a list of books on this topic.

"With Linux and Big Brother Watching Over Your Network, You Don't Have to Look Over Your Shoulder (Or Your Budget)," Chapter 26, continues our exploration of Linux. This chapter, also written by Daniel Carrere, introduces us to the use of the Big Brother UNIX Network monitor to monitor services offered by the Linux operating system. After describing the core components of Big Brother, Carrere provides us with the details of Linux service that can be monitored also and illustrates, through a series of narrated screen images, how different services are monitored.

In the first eight chapters in this section we examined the characteristics of specific network operating systems. In Chapter 27 Howard Marks addresses a question many readers may have concerning the inevitability of change. In his chapter, entitled "Changing Server Operating Systems," Howard first looks at the issues involved in managing a mixed NetWare and Windows NT network and then carefully guides us through the migration process associated with moving from NetWare to Windows NT.

In concluding this section, Scott Koegler makes sure we also are informed about the potential of UNIX. In his chapter "UNIX as an Application Server," Koegler Scott tells us about the resurgence of interest in this operating system and its ability to supplement other operating systems due to its support of TCP/IP and its ability to provide network printing and file services and to function as a high-capacity Web server.

Chapter 19
NetWare 3.x

Dennis Dillman

INTRODUCTION

NetWare 3 was one of the most successful network operating systems (NOSs) in history, especially in terms of market share. This chapter will discuss NetWare 3's history, some reasons it is still a significant force in the market, its architecture, what files are needed to bring up a NetWare 3 server, what software is needed to connect to and manage a NetWare 3 server, and other optional software that extends and enhances the capabilities of a NetWare 3 server.

BRIEF HISTORY

NetWare was originally released in 1983 as a solution for networking IBM PCs and clones. Shortly thereafter Novell released NetWare 2. There were two main problems with NetWare 2. These were a lack of scalability (it could only access 16 megabytes of RAM) and problems adding services to the server (the server had to be brought down in order to add or modify services like network printing).

NetWare 3 was released in 1989 and addressed both problems. It can access up to 4 gigabytes of RAM, and additional services can be added easily without bringing the server down or interrupting service to end users. The last release of NetWare 3 was version 3.12. Version 3.12 will be the last revision of the product.

NetWare 4 was released in 1993 and addressed the concerns and needs of managers of large networks. It is currently the flagship product of Novell and is the product receiving nearly all the development efforts of the company. Novell fully expects NetWare 4 to replace NetWare 3 eventually.

SO, WHY BOTHER WITH NETWARE 3?

NetWare 3 Is neither Obsolete nor Irrelevant

The conventional wisdom, that is to say the wisdom of salespeople, is that NetWare 3 is an obsolete solution and that shiny new solutions should be implemented in its place. The conventional wisdom is wrong. The truth is

0-8493-9823-1/00/$0.00+$.50

that NetWare 3 continues to sell well and that there are a large number of sites that are well served by NetWare 3. It is fair to say that NetWare 3 is no longer an example of cutting edge technology, and it is probably not the best solution for extensive network installations. But for many small and medium-sized installations, NetWare 3 may still be the best choice for a NOS. This is because NetWare 3 has three benefits that no other NOS has: a massive installed base, extremely low hardware requirements, and a huge number of readily available and inexpensive support mechanisms.

NetWare 3 Has a Massive Installed Base. At its peak popularity NetWare 3 was running on 80 percent of PC-based servers worldwide. Studies indicate that NetWare 3 still runs on 40 percent of PC-based servers worldwide. Even with all the hype about NetWare 4, UNIX, and Windows NT, NetWare 3 is run on more PC-based servers than any other NOS. This means that hundreds of thousands of people have experience administering and supporting NetWare 3. Software developers know how to write and support software that will run on NetWare 3. Hardware manufacturers are well aware of what it takes to design a product that will run with NetWare 3. In short, with NetWare 3 you are far less likely to have either software or hardware compatibility problems and are far more likely to be able to get support if you do.

NetWare 3 Has Extremely Low Hardware Requirements. Novell predicts that NetWare 3 will continue to sell worldwide into the year 2000. One of the primary reasons NetWare 3 continues to sell is its low hardware requirements. The following machine is the minimum platform for a NetWare 3 server:

- An Intel 386
- Four megabytes of RAM
- A 40-megabyte hard disk
- A floppy drive
- A monochrome monitor
- A network card

Not everyone has the luxury of dismissing this machine as obsolete or without value. There are many small offices and schools that continue to run NetWare 3 on an Intel 386 computer (not a 386-*based* computer, but an actual 386 machine). These offices are mostly nonprofit organizations with little or no budget for their network. With NetWare 3's low hardware requirements they don't need to spend a lot of money on their server. They can continue to leverage the investment they have already made. Also consider that a number of nonindustrialized countries have difficulty obtaining new equipment and rely on NetWare 3's low hardware requirements. For hardware reasons alone, NetWare 3 is an ideal solution for anyone on a shoestring budget.

NetWare 3 Has a Huge Number of Support Mechanisms. There are over 60,000 CNEs (Certified NetWare Engineers) and over 100,000 CNAs (Certified NetWare Administrators). All of these professionals are available to support your NetWare 3 system. Likewise, consider the following: there are literally hundreds of forums about NetWare 3 on the Internet, CompuServe, America On-Line, and other private on-line services or electronic bulletin boards (BBSs); there are hundreds of magazines, newsletters, and books about NetWare 3; and there are hundreds of user groups for NetWare 3 administrators, end users, and support professionals all over the world. Last, because of NetWare 3's huge installed base and the length of time it has been in the marketplace, knowledge about supporting and maintaining NetWare 3 servers is common among noncertified professionals as well. All of these factors combine to make support for NetWare 3 easy to get and, therefore, inexpensive, especially when compared to acquiring support for UNIX, NetWare 4, or Windows NT.

Summary

There are two types of approaches to investing in computers, software, and networks. The first approach is to set aside a predetermined amount of revolving capital and continually upgrade your equipment and software. Fortune 1000 companies are forced to use this approach for planning and budgeting reasons. Companies with this approach can afford the increased hardware and support costs of other network solutions. The second approach is to scrutinize every computer-related expenditure your company makes. This approach is common among companies with small budgets, very little cash flow, or other more pressing needs for their money. NetWare 3 is an ideal solution for companies using the second approach. It offers savings in hardware and support that are unmatched by other NOSs. These are the items the conventional wisdom forgets when it declares NetWare 3 irrelevant or obsolete.

THE ARCHITECTURE OF NETWARE 3

Facts and Figures

A NetWare 3 server has the following features and restrictions:

- It can access up to 4 gigabytes of RAM.
- It can access up to 32 terabytes of hard disk space.
- It can attach up to 250 simultaneous users.
- It can mount both CD-ROMs and hard disks for use by clients.
- It can attach OS/2, Macintosh, DOS, Windows, Windows for Workgroups, Windows 95, Windows NT, and UNIX machines as clients.

Dedicated Server

NetWare 3 is a dedicated server solution. A machine must be set aside specifically to run the server software. It can perform no workstation functions. The server can perform other functions, such as providing print services, backup services, database services, and remote dial-in services. These other functions invariably require additional software. In many cases the software necessary is not included with NetWare 3 and will have to be purchased. These pieces of additional software are called NLMs (NetWare Loadable Modules) and are loaded on the server by the server's administrator.

NetWare 3 Requires DOS in Order to Start

An operating system kernel is the foundation for the operating system and all its applications. NetWare 3 loads its kernel as a DOS executable. Therefore, DOS has to be started so that NetWare 3 can launch. This is like a car. A car needs a battery in order to start, but it doesn't need the battery in order to keep running — hence, the practice of roll-starting a car with a dead battery. In some cases DOS is kept on a floppy and the server is booted off the floppy. Then NetWare 3 is loaded. Most servers, however, allocate a small portion of the hard disk for DOS. DOS is booted off of that small portion and then NetWare 3 is loaded. Once NetWare 3 has been loaded DOS can be unloaded. There are ramifications to unloading DOS. For example, without DOS only NetWare partitions can be accessed. For some, removing DOS will cause more problems than it solves. The situation will dictate what is appropriate.

NetWare 3 Is Its Own Operating System

Because NetWare 3 is a dedicated server solution that runs its own software, it is important to separate software that is loaded on the server's hard disk for use by the server itself from software that is loaded on the server's hard disk for use by clients. NLMs are stored on the server's hard disk but are only for use by the server itself. Other software that is stored on the server's hard disk may include word processing, spreadsheet, and other application software that is for use only by the clients that attach to the server. The clients access those applications on the server's hard disk, but execute them entirely with their own hardware. In other words, the client does all the processing. The NetWare 3 server simply holds the executable and data files.

NetWare 3 Runs Software Unprotected

The NLMs that run on a NetWare 3 server run in unprotected mode, in what Intel calls Ring 0. When software runs unprotected, it can access the hardware resources directly. This allows superior performance. However, when software can access the hardware directly, it is possible for two separate programs to access the same resource. When this happens, both programs crash and the server running the programs usually crashes as well. While a properly written program will not crash itself, other programs, or the server, the sad truth is that not every piece of software is properly written. NetWare's unprotected execution of software makes it vulnerable to these poorly written programs. This feature of NetWare is the primary reason some people have a low opinion of NetWare as an application server.

NetWare 3 Uses Its Own Disk Access Method

With only minor exceptions, NetWare 3 accesses only data stored on NetWare partitions. Furthermore, those partitions must be subdivided into volumes. There is a minimum of one volume per partition. These volumes must be mounted before they can be accessed. Volumes are referenced by their name, and the name is usually followed by a colon to indicate that it is the name of a volume. There can be a great deal of variety to the names of the volumes on a NetWare 3 server, but every NetWare 3 server must have a SYS: volume. System directories and files are stored on this volume. In order to access any volume, NetWare 3 needs a disk driver NLM. This special driver and the structure of the volumes allows NetWare 3 to implement security, to maintain extended file attributes, and to support special file formats.

There Are Certain Mandatory Directories on a NetWare 3 Server

When the SYS: volume mentioned above is created, there are a number of directories that are automatically created and used by NetWare 3 extensively. These directories are as follows: SYSTEM, PUBLIC, MAIL, and LOGIN.

The SYSTEM directory is where all the NLMs that the server uses are kept. The server automatically looks in this directory for its NLMs. If NLMs are stored elsewhere, the server must be told explicitly where to find them.

The PUBLIC directory is where most of the utilities used to administer the server are kept. These are all the utilities that are executed by clients, not by the server itself. When a client executes a file from the server, the first place it looks, by default, is in the PUBLIC directory.

The MAIL directory contains one subdirectory for every user defined on the server. Configuration files and scripts that are specific to that user are kept in that user's subdirectory under MAIL. The user is frequently unaware that this directory even exists. It is maintained by the server and requires no intervention by the user or administrator.

The LOGIN directory is a unique directory. It is the only directory available to users before they log in. The LOGIN directory of any server can be accessed by any user. The directory is almost always read-only and contains only basic connectivity utilities.

Summary

NetWare 3 is its own operating system. This fact has a number of consequences. It means that NetWare 3 runs its own type of programs, which it runs unprotected. NetWare 3 accesses its own type of partition, which can be subdivided into volumes. Every NetWare server has a SYS: volume, which always has four important directories created for it. These directories are SYSTEM, PUBLIC, MAIL, and LOGIN.

MANDATORY SERVER SOFTWARE

There are certain pieces of software that must be run at the server or the server will be useless. Once these pieces of software are loaded, the administrator may decide to load additional software in order to enhance or extend functionality. But with the following software properly loaded, the server in question will be fully capable of providing file services to a network. The following pieces of software are listed in the order they are executed when bringing the server up.

The Kernel

This is the file that is written as a DOS executable and requires DOS to be loaded before it can execute. The name of the file is SERVER.EXE. This file provides the foundation for all the NLMs that will eventually be loaded. It also provides the licensing for the server. Upgrading a 50-user version of NetWare 3 to a 100-user version of NetWare 3 is done by simply replacing this one file.

When SERVER.EXE is executed, it must be provided a Server Name and IPX Internal Network Number. The name can be a number or letter sequence as long as it is between 2 and 20 characters in length and contains no spaces or other illegal formatting characters. The IPX internal network number is a number associated with the name and must be a hexadecimal value between 1 and FFFFFFFE. This can be done manually or via a script. A script is a text file. Each line of the text file contains a command that would normally have to be typed in manually. When the script is executed, all the commands listed in the script execute sequentially. This saves a great deal of time when dealing with repetitive or complicated tasks.

SERVER.EXE looks for two scripts when it is launched. The first script it looks for is called STARTUP.NCF. STARTUP.NCF is analogous to DOS's CONFIG.SYS. It loads certain critical device driver NLMs (like the disk driver) and sets certain system parameters. The second script is called AUTOEXEC.NCF. It is equivalent to DOS's AUTOEXEC.BAT file. It is simply a list of commands that the administrator would like executed every time the server is brought up.

The kernel uses a protocol to perform all file, print, and server management related tasks. By default, this protocol is called NCP (NetWare Core Protocol). Any clients who wish to connect to a NetWare 3 server must be running client software that is using this protocol.

The Disk Driver

The disk driver is usually the first NLM loaded at the server after SERVER.EXE is executed. One disk driver NLM is required per disk controller board in your system. For instance, if your server has both an IDE (Integrated Drive Electronics) controller, and a SCSI (Small Computer Systems Interface) controller you would load two disk drivers (e.g., ISADISK.DSK for the IDE controller and AHA1540.DSK for the SCSI controller). The disk driver NLM will require certain information about the controller board, such as IRQ number, DMA number, and other hardware settings. Once a driver for a controller has been loaded, then all NetWare partitions on drives attached to that controller can be mounted and accessed.

The LAN Board Driver

Before any client machine can communicate with the server, the network cards in the server must be initialized. This is done with LAN board drivers. As with the disk driver, one LAN board driver must be loaded for each network card installed in the server and the driver must be provided with the hardware settings for that card. For instance, if you had two Ethernet network cards in your server, then you would load two LAN board drivers (e.g., SMC8000.LAN for one and NE2000.LAN for the other). One other piece of information that should be provided when loading the LAN board driver is the frame type, especially when loading an Ethernet LAN board driver. Frame type is analogous to channels on a CB radio. In order for two CB radios to communicate, they must both be on, have good reception, and be tuned to the same channel. The same is true for two machines on an Ethernet network. The two machines must be attached to the same physical network, must have their Ethernet cards initialized, and must be using the same frame type. If no frame type is specified, then NetWare will use a default frame type. NetWare 2 and NetWare 3 versions 3.11 and older will default to the 802.3 frame type. NetWare 4 and NetWare 3 version 3.12 will default to the 802.2 frame type. It is always best to specify which frame type you wish to use in order to avoid accidentally defaulting to the wrong one.

The Network Protocol

It is imperative to bind a network protocol to the LAN board drivers that are loaded. Network protocols control network and machine addressing, data formatting, and other issues involved in the transfer of information between two machines. Machines running different network protocols cannot communicate. When binding the network protocol, three pieces of information must be specified.

The first is the name of the protocol being bound. IPX (Internet Packet eXchange) is the only protocol available on a NetWare 3 server by default. Other protocols are available, but support NLMs must be loaded before they can be used.

The second piece of information that must be provided is the name of the LAN board driver to which you are binding the network protocol. At least one network protocol must be bound to each LAN board driver.

The third item is a network number that will be associated with that LAN board driver. Network numbers identify the physical cable segment and frame type being used by the LAN board. All machines on the same physical cable segment and using the same frame type must all agree on the same network number. In the case of IPX, the network number is chosen by the network administrator. It must be a hexadecimal number between 1 and FFFFFFFE and must be unique to that physical cable segment.

Summary

There are four pieces of software that must be loaded on a NetWare 3 server before the server is functional. These are the kernel, disk driver, LAN board driver, and network protocol. The network protocol is not loaded but is bound to a LAN board driver. Once these four items are attended to, the server will be available on the network for file services. Additional services and protocols other than IPX require additional software.

MANDATORY CLIENT SOFTWARE

Once the NetWare 3 server is up, software must be loaded on the workstations in order to allow them to attach to the server and access its file services. While the specific files required to attach a workstation to a NetWare 3 server vary from platform to platform and OS to OS, every machine is going to need the following client software.

The Network Card Driver

Obviously, a network board of the proper type must be installed and cabled properly to the same network as the server. In addition, a proper driver must be loaded on the client machine. There are many types of drivers, and some of them are not compatible with the other necessary client software. So, it is imperative to select driver software that is compatible with the other software that is needed. For example, Novell has three different drivers for machines running DOS. These are IPX.COM (officially retired but still used), Multiple Link Interface Drivers (MLIDs, the most common driver found today), and the newer 32-bit LAN drivers (which started shipping in late 1996).

Link Support Layer (LSL)

This software is not always required. It allows multiple protocols to share a single network card. If there is no Link Support Layer (LSL) software, then an equivalent solution must be used, or only one protocol can be used per network card. An example of software that did not use LSL and allowed only one protocol per network card is the retired IPX.COM that Novell provided with earlier versions of NetWare. The MLID referenced above supports and requires LSL.

The Transport Protocol

This software is required. The same protocol must be run at the workstation as is run at the server. If you are running IPX at the workstation, then that workstation can only talk to servers that are running IPX.

The Requester

This software is required. The requester is the "language" that is used to perform network functions. NetWare 3 uses a language called NetWare Core Protocol (NCP). NetWare 3, however, can load additional languages. Example requesters include NETX, VLM, Client Services for NetWare in Windows 95 and Windows NT Workstation, and Gateway Services for NetWare in Windows NT Server.

Authentication

Authentication is not software, but without authenticating, a client cannot access resources on a NetWare server. Without authenticating, a machine is simply attached to a server. Being attached means the client machine has formed a connection to the server but cannot access any resources on the server. Authenticating is the process of acquiring an identity that a server recognizes. Once a user has this identity, called a "username" or "login ID," the server will allow him/her to access those resources that his/her identity has permissions to access. The process of authenticating, or logging in, varies from OS to OS. A DOS workstation, for example, needs to load each of the above pieces of software and then execute LOGIN.EXE. The user will be prompted for a username and a password. If the username and password match, the user will have access to the system. Regardless of what OS the user is running, the type of access he/she has will be determined by restrictions put on their identity by the administrator of the server.

Summary

There are three or four pieces of software that must be run on a client machine before that machine can attach to a NetWare server. These are the network card driver, the transport protocol, and the requester. Some machines may require a link support layer. Once these items are run on a client machine it can attach to a server and access any resources. In order to access resources on a server a user must go through an authentication process to acquire an identity. Once users have acquired an identity, they can access resources on the server. What resources the user can access depends on what permissions and restrictions have been assigned to his/her identity.

MANAGEMENT UTILITIES RUN AT THE CLIENT

Surprisingly, with NetWare most management tasks are not performed at the server but rather are performed at the client. In fact, there are a number of tasks that cannot be performed at the server. Novell intended that this should be the case. The goal was to allow the server to be locked in a room or wiring closet and not be physically accessed except for hardware maintenance. This would allow for maximum physical security of the server hardware.

User Account Management

Management of user accounts on a NetWare 3 server includes the following:

- Creating, deleting, and disabling accounts
- Setting expiration dates for the accounts and passwords
- Deciding whether to require passwords and, if so, how long they should be
- Determining whether or not users should be able to simultaneously log into more than one machine at a time, and what machines they are allowed to log into at all
- Determining what time of day and what days of the week the user should be able to log in
- Deciding how to arrange the users into groups for convenient administration
- Creating and assigning scripts that will execute when the user logs in. These scripts set up the user's environment so that they will be better able to use the server's resources

Novell provides a number of ways to perform the above tasks, but the most commonly used by far is SYSCON.EXE. SYSCON gives the administrator a menu-based utility from which all user administration tasks can be performed (see Exhibit 19.1).

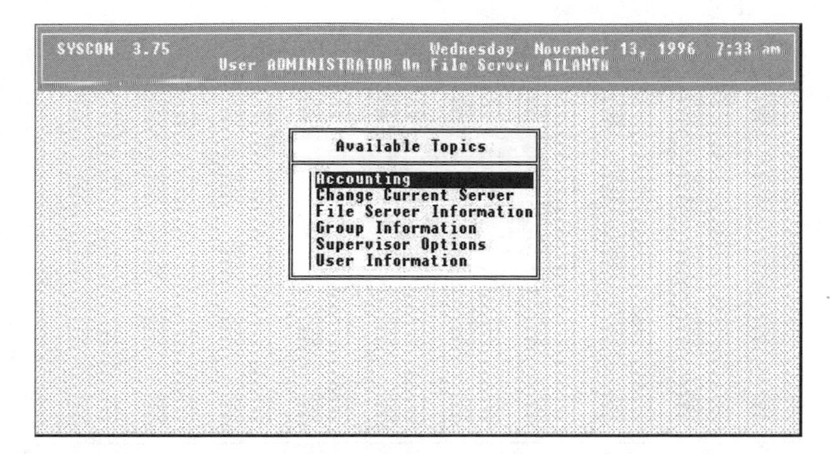

Exhibit 19.1 Syscon.

File System Management

File system management includes the following:

- Creating and deleting files and directories
- Setting and modifying security information
- Moving and copying files and directories from one location to another
- Viewing current file and directory information such as owner, size, creation and modification dates, and security information
- Manipulating the volumes of the server, including setting space usage restrictions on users and directories

There are a number of ways to perform these tasks, including standard DOS, Windows, Macintosh, and OS/2 file management utilities. Non-NetWare utilities, however, rarely allow manipulation of NetWare-specific parameters, such as security and extended attributes. While there are dozens of command utilities provided by NetWare 3, FILER.EXE (Exhibit 19.2) is a commonly used menu-based utility because it combines the capabilities of almost all of the command line utilities.

Remote Server Management

A great many of the management functions described above could be considered remote management given that they are not performed at the server. However, the remote server management tool described here allows an administrator to perform tasks at his/her workstation that would otherwise only be possible if performed at the server console itself. These tasks include:

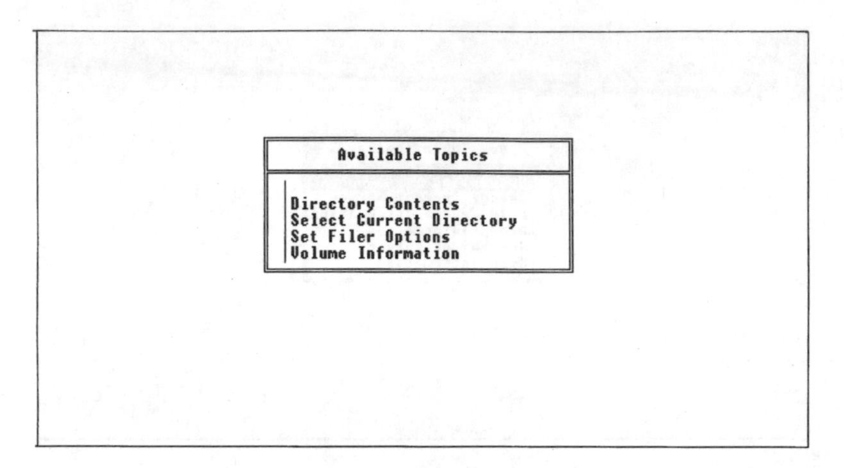

Exhibit 19.2 Filer.

- Bringing down and rebooting the server
- Loading and unloading NLMs
- Mounting and dismounting volumes
- Monitoring server resource utilization
- Monitoring network utilization

The name of the utility that allows this to happen is called RCONSOLE.EXE. RCONSOLE allows a workstation to display the console of a server. If someone was looking, the commands that are typed at the RCONSOLE screen (see Exhibit 19.3) would actually appear on the monitor of the server being remotely managed. This means that anything that can be done while sitting at the server can be done from RCONSOLE, with the exception of physically powering off or powering on the machine (see Exhibit 19.4).

Summary

Almost all administration of a NetWare 3 server is done from the client. In fact, most tasks cannot be completed at the server itself. While there are many administrative utilities, there are three critical server management utilities. These are SYSCON, FILER, and RCONSOLE. By mastering these three utilities, the vast majority of administrative tasks can be accomplished.

IMPORTANT SERVER SOFTWARE

The following is a list of a few key pieces of software that supplement the capabilities of a NetWare 3 server. All of the following are included with NetWare 3 and are executed or loaded at the server console.

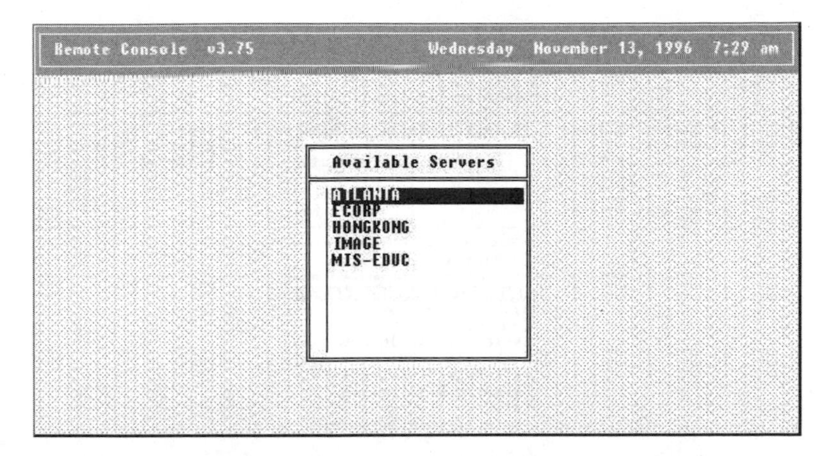

Exhibit 19.3 The Rconsole main screen.

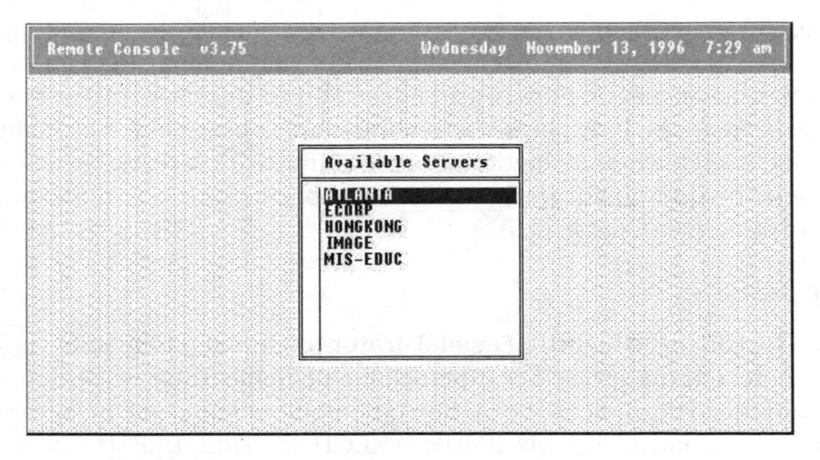

Exhibit 19.4 A server screen taken from Rconsole.

INSTALL.NLM

As its name suggests, INSTALL.NLM is used to install NetWare 3. However, it does much more than that. It is used to create and delete partitions, to create and delete volumes, to edit the STARTUP.NCF and AUTOEXEC.NCF files mentioned earlier, and to install some special server software packages.

MONITOR.NLM

MONITOR.NLM is used for a number of purposes. It checks the processor utilization of the server, how the server is using its memory, and which

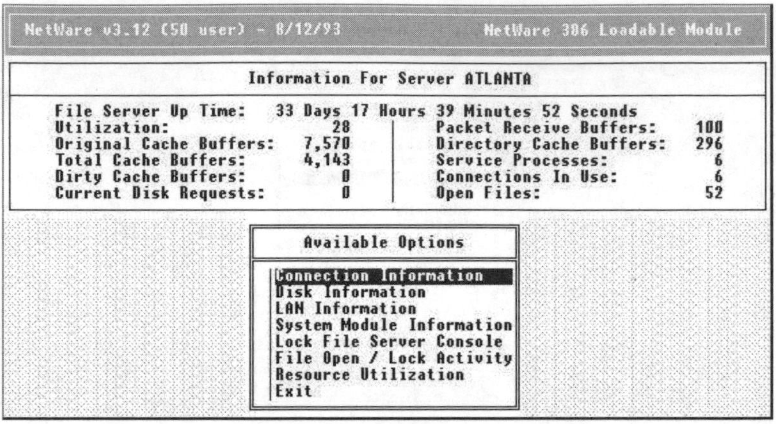

```
NetWare v3.12 (50 user) - 8/12/93                NetWare 386 Loadable Module

                      Information For Server ATLANTA

  File Server Up Time:    33 Days 17 Hours 39 Minutes 52 Seconds
  Utilization:                   28    Packet Receive Buffers:      100
  Original Cache Buffers:     7,570    Directory Cache Buffers:     296
  Total Cache Buffers:        4,143    Service Processes:             6
  Dirty Cache Buffers:            0    Connections In Use:            6
  Current Disk Requests:          0    Open Files:                   52

                         Available Options

                     Connection Information
                     Disk Information
                     LAN Information
                     System Module Information
                     Lock File Server Console
                     File Open / Lock Activity
                     Resource Utilization
                     Exit
```

Exhibit 19.5 Main monitor display.

programs are running at the server and what demand they are putting on the server's resources. It displays the users who are logged in and what files they have open. It allows an administrator to kick users off the system if the need arises. It displays information on disk usage and disk damage. Finally, it displays information about the network: errors, utilization, inbound and outbound traffic levels, and specific LAN board diagnostic information (see Exhibit 19.5).

SET

The SET command allows an administrator to view and to modify various parameters of the server. For example, if an administrator wanted to control how deep the directory structure on the server could get, he/she could issue the command SET MAXIMUM DIRECTORY TREE DEPTH = 20 at the server console. This command is not an NLM. It is simply issued at the server. Changes made with the SET command only last until the server is brought down. Because of this, it is common to see SET commands in a server's AUTOEXEC.NCF file so they are executed every time the server is brought up.

NAME SPACES

In today's world of Apple Macintoshes, IBM OS/2, Microsoft Windows 95, and Microsoft Windows NT, non-DOS file names are common. The problem is that NetWare 3 volumes are capable of storing only DOS files by default. This is easily rectified by the use of name spaces. Each particular type of file format has its own name space NLM. These NLMs are loaded and then the volumes modified to accept the specific type of non-DOS filename in question.

VREPAIR.NLM

Since the NetWare volumes are specially designed to allow for mounting and dismounting, for security, and for extended attributes, no DOS disk repair utility will work on them. If a volume is damaged, it must be repaired by a NetWare disk repair utility. This utility is called VREPAIR.NLM. It can only repair volumes when they are dismounted. It should be run against a damaged volume until it generates no errors. When the volume is clean, it can be mounted and put back into use.

CONCLUSION

NetWare 3 continues to be a popular NOS. It is especially popular with small- and medium-sized companies that want to continue to use the hardware they already own. It offers a tremendous range of functionality even by today's standards. It is an affordable and serviceable network server solution. After comparing the costs of migrating to or installing another NOS against the actual bottom-line benefits of the migration/install, many businesses still choose NetWare 3. For these businesses NetWare 3 does everything they need, and the alternatives offer few tangible benefits to justify their cost.

Dennis Dillman is a senior consultant and instructor for USConnect Milwaukee. He is a Novell Master Certified NetWare Engineer, Novell Certified NetWare Instructor, Microsoft Certified Systems Engineer, and Microsoft Certified Trainer. He is also a member of the faculty at Milwaukee School of Engineering, where he teaches senior and graduate level courses in network management, network design, and telecommunication theory for the Center for Business Management.

Chapter 20
intraNetWare Overview

James E. Gaskin

INTRODUCTION

NetWare has been one of the most stable, recognizable, and trusted brand names in the computer business for the last decade. Why change a good thing by renaming it intraNetWare? In defense of the Novell marketing executives who morphed NetWare into intraNetWare, this is only one in a long string of name changes. In the beginning, Novell went by the name Novell Data Systems, and went bankrupt. It was not until NetWare took off in the late 1980s that many people knew what NetWare was, and the name had already changed twice.

Novell dropped the "Advanced" designation for NetWare 286, renamed to celebrate the hottest new computer platform: the IBM AT — a blazing machine powered by an Intel 286 running at 12 MHz, a full megabyte of RAM, and 40 spacious megabytes of hard disk space. Floppy drives zoomed up to 1.2 MB each, and the keyboard sprouted 12 function keys. So much computing was packed into this package that more keys were needed to control it all.

NetWare 286 mutated into NetWare 3.0 or 386, the first to take advantage of that super-blazing 386 processor Intel released. The name NetWare was still prominent, but the modifiers constantly changed.

Most critics say that Novell NetWare finally became a workhorse operating system with NetWare 3.11. By that time, all the promises made for NetWare 3.0 were actually delivered and working. If Novell executives had been smart, they would have made a bigger noise about 3.11 than just a one hundredth point upgrade. Being somewhat embarrassed about the delay in providing promised features, however, the executives soft-pedaled the upgrade as if all these details were always in the box, and the critics and customers just missed them.

0-8493-9823-1/00/$0.00+$.50

A direct correlation between NetWare 3.11 and intraNetWare should be made here. With 3.11, Novell embraced TCP/IP at the server. Novell servers could sit on the Internet, communicate with remote servers across the Internet through IP Tunneling, and even be controlled through telnet connections with XConsole. This was unheard of in the PC LAN world.

IntraNetWare finishes the integration of NetWare with Internet protocols, applications, and technologies. IntraNetWare is the current pinnacle of LAN technology, and the beginning of Novell's assimilation of all things Internet.

IPX/SPX VS. TCP/IP

IPX lost the protocol war to TCP/IP. This victory says nothing about the relative merits of each protocol. Both IPX and TCP/IP do many of the same things, and each has its strengths and weaknesses. The victory for TCP/IP boiled down to the choice of partners. IPX has Novell backing it, and TCP/IP has the Internet — the Internet won by megabytes (25 each per minute of daily traffic).

TCP/IP is the "protocol of choice" for many companies today. The victory is so complete that companies are adding TCP/IP where it makes no sense, such as within stand-alone groups of PCs and servers.

This does not mean that TCP/IP is a better protocol than IPX. TCP/IP was developed as the protocol for the Internet in the early 1980s and, in fact, has been the only protocol allowed on the Internet since January 1, 1983. The popularity of the Internet overseas has swept TCP/IP along. Companies needing a way to connect from any machine in their building to any other machine often settled on TCP/IP, because a version of TCP/IP is available for every computing platform imaginable. If one can hook it to a network, TCP/IP will run on it.

Much of the entire "open systems" movement has been from the TCP/IP community, despite the fact that neither UNIX nor TCP/IP was an important part of corporate networks. When the web took hold of the world in early 1994, following the release of Mosaic in late 1993, the masses had a reason to use TCP/IP because HTTP (HyperText Transfer Protocol) demanded a TCP/IP platform to run on.

Jumping into the open systems fray when buying Excelan in 1989, Novell has long been one of the major providers of TCP/IP client software. At one time, nearly half the PC-based TCP/IP packages purchased had the Novell name on the box. This success at TCP/IP did not immediately transfer to NetWare, however.

Novell has long been criticized for not following the lead of the customer. Microsoft started berating Novell in the press for not supplying a client

TCP/IP option. Therefore, Novell needed TCP/IP available quickly. Thus was born NetWare/IP.

Besides this, Novell decided that giving the customers what they ask for would speed the acceptance of Novell's own web server software. Since the web server software was originally licensed from an outside company, the web server did not have an IPX option built in for communications.

Novell could have included its own LAN WorkPlace TCP/IP software in the client package with NetWare, but that would have been a clumsy option. LAN WorkPlace was as good as any of the other PC TCP/IP package, but it was still difficult to install, configure, and administer. Novell managers had long been spoiled by the networking ease of NetWare clients, one of the primary advantages of IPX over TCP/IP. Forcing every network manager, no matter how small the network, to install a client TCP/IP package would start a buyer revolt.

Novell executives finally chose the right path through the marketing minefield and made the correct decision. If the customer does not want to get involved with TCP/IP in any form, does not need internal or external web servers, and does not particularly care about the Internet, then NetWare for Small Businesses is the choice for them, and they will get all the new advancements in NDS and network management without touching TCP/IP in the least. TCP/IP support for the server ships in the box, as does the web server software, but each requires an extra installation step. The customer will not run into them by accident; they will have to find them on purpose.

On the other hand, if one needs (or wants) TCP/IP services for a server, a web server, and a client, intraNetWare is the ticket. You get all of NetWare, with the added bonus of excellent TCP/IP support in a variety of ways.

INTERNET TECHNOLOGY INVADES THE NETWORK

The previous section covered the world of TCP/IP, but quickly skipped over the internet technologies that forced Novell management to come up with intraNetWare.

First, there is TCP/IP support from the client to the server. Why is this necessary if IPX works perfectly well in intraNetWare? Because many companies demanded that Novell support them in their goal of supporting a single protocol for all networking. That protocol was TCP/IP, because of the open systems adherents and widespread support of TCP/IP in network building blocks such as routers, gateways, and management tools.

IBM suffered even more culture shock than Novell. For decades, a corporate network meant SNA (Systems Network Architecture), the proprietary enterprise network used by IBM mainframes. Today, companies are giving IBM and their mainframes a choice — attach to the information superhighway or hit the highway. Mainframes are being attached to the Internet to display corporate data to Web-browsing clients every day. The largest growth area for mainframe products today is Internet tools.

Novell had to create intraNetWare, primarily for the sake of HTTP, but also because "The Web" is what people see and demand to reach. Reaching the Web, or more technically, sending HTML (HyperText Markup Language) files over HTTP, requires a Web browser. All Web browsers assume they are running on a TCP/IP protocol stack.

People looking for their "killer application" using Internet technology already have it right under their noses. E-mail is actually the killer application for many companies and for many networks.

Early advances in graphical interfaces and reliable transports meant that LAN-based e-mail clients and servers took a strong lead as computing headed into the early 1990s. NetWare versions included a rudimentary DOS e-mail application up until NetWare 386. Microsoft, cc:Mail (then bought by Lotus and then IBM), WordPerfect Office (bought by Novell, who shed everything but what became GroupWise), and several more systems had great success across the LAN.

These systems' success has waned in favor of Internet e-mail because the LAN-based systems were great within the LAN, but terrible when communicating outside the building. Creating a dial-up network for multiple offices to exchange mail over analog modems and public phone lines, with random quality connections, was difficult and prone to failure. Calling long distances to exchange mail messages is expensive, meaning either message timeliness or budgets were sacrificed.

Second, none of the programs ever made a truly reliable gateway between competing systems. For example, if one has cc:Mail and wants to talk with another department that has chosen Microsoft Mail, more than good luck is needed!

Third, proprietary e-mail vendors were forced to develop the client software for each and every client platform available. If one added a department that used systems (such as Macintosh or UNIX client systems) that did not have a client developed by one's LAN e-mail vendor, then that department would be at the mercy of its technical support department until it could create a new one.

In contrast, the Internet e-mail system was behind in graphical applications but far ahead on standards guiding development. All Internet e-mail servers and clients following the proper standards could, and still can, communicate with each other. One often has no idea whether the message received from an Internet e-mail user comes from a PC, Macintosh, UNIX,

Windows NT, or even a Commodore Amiga system. The message comes, and the user reads it and never worries about it.

Those are the reasons why many companies today are replacing their LAN e-mail systems with Internet-developed e-mail clients and servers. All clients are supported, users can reach anyone on the Internet without worrying about which type of system they have, and long-distance connections happen over the Internet rather than telephone lines. There are no long-distance charges, no analog modems, no hung lines requiring someone to crawl under the desk to find the misbehaving modem, and no delays waiting for the next connection time.

If a company has a newsgroup server or allows network users to read external news servers, then that company needs NNTP (Network News Transfer Protocol) to support a newsreader application. The news client can be within a web browser such as Netscape, or separate such as the ever-popular WinVn shareware reader. Again, NNTP expects to find a TCP/IP protocol suite on the system running the reader.

All of these functions are critical when building an intranet, so NetWare needed to become intraNetWare. This is easy to see if one agrees that the future of networking will center more on the Internet than a "traditional" file and print services network. In other words, what will the network look like in two years? Novell is betting it will look more like the Internet than an old NetWare 386 network.

Information sharing now revolves around the Web server rather than the file server. Novell makes a great file server, meaning it is also a great Web server. The goal of a file server is to quickly provide files to many intelligent clients over a shared network.

Notice how neatly this definition fits with that of a web server. Add in the required capabilities of the near future for corporate networks, such as large graphic files, audio streams, and real-time video, and one realizes that "file" server becomes a misnomer.

Servers must add multimedia extensions in the near future. Novell could go its own way, trying to retrofit NDS to support audio and video streams, but why? The Internet, in the guise of the Web, is doing all that for us. All the "cool" stuff for corporate networks is coming from the Web, so Novell is smart to take the supporting technology from the Web as well.

IntraNetWare

The first item Novell touts on its Web page regarding intraNetWare is the Web server. Following is a breakdown of the Internet-ready features.

- *NetWare Web Server.* The Web server will certainly be a difference between intraNetWare and the next iteration of NetWare for Small Business. This is a good, fast, reliable Web server. It can be put on the Internet or made the core of one's corporate intranet. Either way, it will work well. Having a Web server on the same system holding one's files offers better performance automatically.
- *Multi-Protocol Router.* Novell has always supported routing within the server operating system, but the MPR is a major step above that function. Able to support local and remote connections, the MPR software runs along with the regular server, or on a server operating system dedicated just to routing. Included software makes connecting to one's Internet service provider a snap. Leased lines, ISDN, and frame relay boards are supported within the router, saving money and communication room space.
- *NetWare Internet Access Server.* Often just called the IPX-IP Gateway, this software performs two vital jobs. First, it allows the network client to run Internet software without adding TCP/IP to each workstation. Special client software fools the Internet software into believing the client is running TCP/IP, when really only the gateway server has the actual TCP/IP protocol suite loaded. Second, when using the IPX-IP gateway for an Internet connection, one has an automatic firewall. Since the only TCP/IP software in the network is at the gateway, outside hackers can only see the gateway. They cannot see clients, server, printers, or other network resources running on IPX. In today's world, the ability to quickly and safely provide Internet access for clients is a welcome feature.
- *Novell FTP Services.* FTP (file transfer protocol) provides the communications parameters for file transfers across the Internet from any client to any server. Running FTP Services on a NetWare file server allows any client running FTP client software to send and receive files from that server. Platform independence for file transfer clients was never so quick and easy. IntraNetWare is the first step using Web technology to improve company performance. One can think of NetWare as the larvae stage, and intraNetWare as the butterfly, adding capabilities never dreamed of by the caterpillar.

BENEFITS OF INTRANETWARE FEATURES

Novell initially priced intraNetWare at the same levels as the previous version of NetWare to encourage everyone to get the product with the new features. Of course, if one does not want or need the new features, are they worth the price, even if the price is free? In other words, what does one really gain with intraNetWare?

Any company can create a corporate Web site. A company's morale internally and the image externally are now partially defined by Internet technologies. Purchasing agents worldwide now use search engines to find suppliers. If Company A does not show up in a search, but 50 other companies do, Company A significantly decreases its chances of getting any of that business.

Second, intraNetWare allows a company to provide Internet access to its NetWare clients with the minimum pain and aggravation. If one does not look forward to adding TCP/IP to each and every workstation, use the IPX/IP gateway instead. Not only do users have access to the Internet, but one can control that access. The traditional NetWare security allows control over who sees which application on the server, but a user may load a browser on his/her workstation, even if that is against the rules. No problem, because one can also control each user's access to the gateway itself. Not only that, one can also control what level of Internet access each user has and when he/she can use the gateway. Management time is minimized, access is absolutely controlled, and employees are kept from getting themselves (and the company) in trouble by using "inappropriate" Internet resources.

Third, any type of support a company provides to customers can now be handled over the Internet. Does the company make any products including software? Put patches and updates on the Web server or FTP server, just like Novell and Microsoft do. Is information faxed to customers? E-mail that information, saving time and money. Then use the Web server to make archived copies of each information release available to the world, via the Internet once again. Internal memoranda starting to clog the corporate paper recycling bins? Distribute internal information through e-mail and Web server software, rather than on paper. It may be that paper and distribution cost savings alone will pay for new software and hardware.

Here is an angle that should be emphasized — security. Running intraNetWare with the IPX/IP gateway, network clients continue to use IPX. Outside hackers cannot see past the IPX/IP gateway server because there are no TCP/IP devices or addresses to see.

Some benefits are a bit less immediate than others. Seeing how the NetWare server can become the Web server is simple, and one can appreciate the economies of using existing hardware for multiple functions. Seeing the savings with the IPX/IP gateway is simple, and even the security advantages are apparent to anyone who realizes that hackers require TCP/IP for their skullduggery.

The Internet has always been proud that client and server hardware and operating system platforms matter less than adherence to standards. In the PC LAN arena, NetWare has always been the operating system supporting more clients and connections to other operating systems than any of its competitors. The move to intraNetWare encompasses even more independence of clients and servers, and mimics the highly heterogeneous mix of the Internet itself.

With intraNetWare, Novell makes the leap to the Internet-inspired world of all clients to all servers. All the noise Microsoft is making about Internet Explorer, the new “browser” view of the operating system, as well as the Internet are all part of the same movement. Does that mean Novell and Microsoft are tied in this area?

No, Novell is ahead because of NDS, the IPX/IP gateway, and the power of the NetWare server engine. NetWare clients will have the greatest new features, such as snazzy browser interfaces hiding the complexities of applications, because the client will leverage the advancement Microsoft makes. Where Novell and intraNetWare will take the lead is in providing these advances on top of the same stable platform and secure network operating system they have been selling for nearly two decades. Even Microsoft’s marketing prowess cannot make its limited domain security system stack up against NDS.

FUTURE DIRECTIONS OF INTRANETWARE — BORDER SERVICES AND BEYOND

Novell has enhanced the intraNetWare operating system package with Border Services, and will continue to advance toward better Internet integration while maintaining security.

Among the advantages offered by Border Services, security ranks the highest. Proxy servers both protect the network and improve performance for network clients. A company will migrate to more TCP/IP devices within its internal network, even if the advantage of an IPX/IP Gateway is taken. Protecting those TCP/IP nodes inside the network from outside hackers will be one of the primary features of the Proxy Server software.

Once the network can communicate securely to the Internet, one starts to dream of connecting to other networks. Some of these networks may be within the company, just geographically dispersed. Today, a company’s budget must include routers and leased lines to communicate with these networks. Your boss will wonder why you need a separate set of routers and lines to connect to the Internet and your own network.

One can connect to remote offices via the Internet, using a concept called VPN (virtual private network). IntraNetWare allows remote sites to communicate across the Internet yet remain private. Encryption built into the router connections guarantees that packets to remote offices remain for company eyes only.

Can one company connect its largest customer to a set of internal Web servers? This would allow both companies to see the same information, whether speaking of delivery dates for paper clips or designs of the next Mars landing module. Groups from both companies can participate in discussions, conferences, Web site updates, and even share calendar programs.

This technology is called an extranet. With intraNetWare and Border Services, one can control who at the other company can see your network, when they can see it, and whether they can read and write or just read the information. This means that one can allow outsiders in while maintaining security.

The wall between the Internet and the internal network is crumbling. The wall between a company and its customers will begin crumbling as well. However, well-secured and monitored guests make good customers, and that is the advantage of intraNetWare.

Chapter 21
Migrating to intraNetWare 4.x and Novell Directory Services

Eric Stral

INTRODUCTION

IntraNetWare 4.x is the most powerful release in the Novell NetWare family of network operating systems. It incorporates the power and reliability of its market-dominating predecessors, while incorporating advancements and features that make upgrading to NetWare 4 logical for both small offices and large networking environments. The biggest change is the addition of NetWare Directory Services, which provides network object-oriented administration instead of server-centric administration as in previous LAN operating systems. Every user, group, printer, or other network object can be managed from a single interface regardless of its primary server connection. The efficiencies this introduces means that LAN network operating systems are at last a viable platform for the corporate enterprise.

NOVELL DIRECTORY SERVICES

The Novell Directory Services (NDS) system is a relational database that is distributed across the entire network. NDS software provides users global access to all network resources for which they have been given rights, regardless of where they are physically located. Users log in to a multiserver network and view the entire network as a single information system. This single view is the basis for increased productivity and reduced administrative costs.

0-8493-9823-1/00/$0.00+$.50

NetWare Directory Services, NDS, provides a single point of administration for the entire network. Although multiple administrators may make concurrent changes or deal with individual portions or aspects of the information, they are all accessing the same data. Each network object (user, printer, volume, etc.) has a single global identity. For example, users may have access to multiple servers, but each user needs only one user name and one login into the network.

NDS treats all network resources as objects in a distributed database known as the NetWare Directory. Users can access any network service without having to know the physical location of the server that stores the service; the Directory links Directory Objects to physical resources.

NetWare now offers two installation options, Simple or Custom. The Simple installation option makes a number of assumptions and therefore requires the installer to make fewer decisions. This makes a NetWare 4.1 server easier and quicker to install. The Custom installation option makes no assumptions and requires the installer to provide all installation information. Additional options, such as spanning volumes across multiple drives, and loading and binding TCP/IP and AppleTalk protocols, are available only during a Custom installation.

Administration tasks in the NDS database are done using either the NetWare Administrator or NETADMIN. NetWare Administrator is a graphic utility that allows the administration of all network resources and Novell Directory Services from an MS Windows or OS/2 workstation. NETADMIN is a menu utility that allows you to administer network resources and Novell Directory Services from a DOS workstation.

Novell has added extensive authentication and encryption routines to secure communications between a client and a server. NetWare 4 includes auditing features to track access to the network and (or) specific files.

Enhanced Memory Usage

NetWare server memory pools are managed more efficiently, maximizing the server's use of available memory. Memory, when requested from allocation pools, is returned to the same pool; this minimizes fragmentation and maximizes efficiency. Garbage collection is used to collect unused segments of memory and return them to a common memory pool. Ring 3 memory protection is implemented, for testing of NLMs (NetWare Loadable Modules) prior to production implementation. Programs installed in Ring 3 can crash or be terminated without affecting the operation of the file server.

Disk Storage Enhancements

Block Suballocation better utilizes server hard disk space by allowing disk blocks to be allocated in smaller segments. The beginning of every file is stored at the beginning of an allocated block, with or without suballocation. The balance of a file, beyond the size of the disk block, is allocated in 512K increments to other blocks on the hard disk drive. Previous versions of NetWare used blocks of 4K. So even a 2K file took up 4K of disk space. NetWare 4 changes all that.

File Compression helps optimize server hard disk space. This is a background process that affects NetWare server performance minimally. File compression is managed internally by NetWare. If a disk error or power failure occurs during compression, the original uncompressed file is retained.

Data Migration extends the storage capacity of a NetWare server by integrating an optical disk library into the NetWare file system. Data migrates between faster, low-capacity storage devices (the server's hard disk) and slower, high-capacity storage devices (optical disks in a jukebox).

NOVELL DIRECTORY SERVICES (NDS)

Novell Directory Services is a network-wide database which functions as a directory, or naming service, for all resources on the network. NDS is the network service that controls access to the physical resources of the network.

All users, file servers, and applications on the network can get information on all other network resources because they all use the same Directory. This means that users and programs do not have to log into, or know about, other resources on the network. Each network resource has an entry in the Directory with a unique name. You just request each resource by its unique name; NDS does the rest regardless of the server the user is attached to. Managing NDS, consequently, is an administrator's responsibility, not a network user activity.

Benefits of NDS

- Global database providing central access to and management of network information, resources, and services
- Standard method of managing, viewing, and accessing network information, resources, and services
- Logical organization of network resources independent of the physical characteristics or layout of the network
- Dynamic mapping between an object and the physical resource to which it refers

Composition of the Directory

Similar to the "bindery" in NetWare 3.x, the Directory database consists of objects, properties, and values. The big difference is that the Directory is

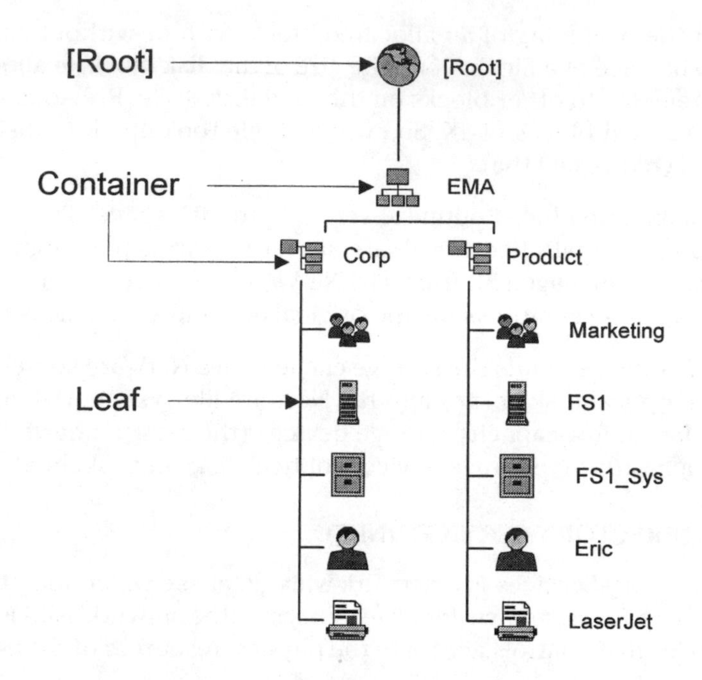

Exhibit 21.1 Three classes of NDS objects.

one large database, rather than the three separate files that NctWare 3.x uses to maintain the same information on a server-by-server basis.

An object represents a network resource contained in the Directory. An object is similar to each record or row of information in a database table. Properties are categories of information that can be recorded about the network resource. It is similar to a field in a database record. A value is the actual information about the network object, contained in the network object property field. Values in an object property are similar to data in a field of a database record.

Types of Objects

NDS objects can be separated into three types or classes: [Root], Container, or Leaf objects as seen in Exhibit 21.1. These exhibits are screen shots from the central NDS management utility NWADMIN.

[Root]. Every Directory must have one, and only one, [Root] object. The [Root] object defines the top of the Directory tree and is created by the installation program when the Directory tree is created. Note that when referring to the [Root] object, brackets ([ ]) must always be used.

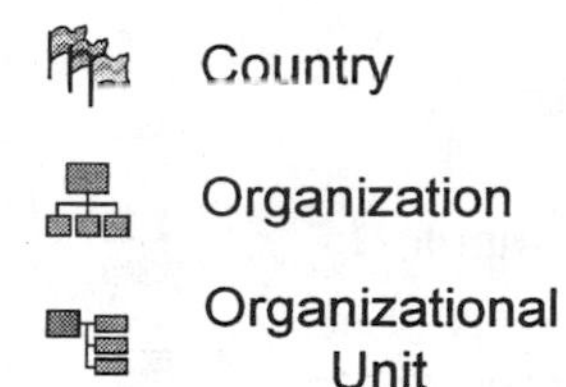

Exhibit 21.2 Three classes of container objects.

Container. Container objects are used to hold or contain other objects. They are used to logically organize and group the objects of the Directory. There are three classes of container objects (see Exhibit 21.2).

A *Country* object designates the countries where your network resides and organizes other Directory objects within the country. This is based upon x.500 standards which generated the framework for NDS structure. The Country object is optional; it is rarely, if ever, used.

An *Organization* object represents a company, university, or a department. It is the first level that can contain leaf objects. Each directory must have one Organization.

An *Organizational Unit* object represents a division, business unit, project team, department, or any unit of organization within a Directory. Organizational Units are used to organize leaf objects below the Organization level.

Leaf. Leaf objects represent the meat of the network, the actual network resources such as users, groups, printers, file servers, and file systems. Exhibit 21.3 displays the various leaf object types.

Directory Tree Structure

The Directory tree is a hierarchical structure that stores and organizes objects in the directory. It will contain the [Root] container and leaf objects. The NDS Directory tree structure is similar to the DOS file structure, except that different container objects have restrictions on where they can be placed and what can be placed in them (see Exhibit 21.4).

Bindery Services

Some applications and services which run in the NetWare 4 environment do not currently take advantage of NDS technology. To enable users of these services to access them from the NetWare 4 environment, Novell created bindery services — which is not the same thing as a bindery.

Exhibit 21.3 Leaf objects.

Container Object	Can Exist In	Can Contain	Example
Country	[Root]	Organization, Alias	US, FR
Organization	[Root] Country	Organizational Unit All Leaf Objects	EKC UCLA
Organization Unit	Organization Organizational Unit	Organizational Unit All Leaf Objects	Marketing Engineering

Exhibit 21.4 Relationship of NDS containers to each other.

With bindery services, NDS imitates a flat structure for leaf objects within a set of Organization (O) and Organizational Unit (OU) objects. Thus, when bindery services is enabled, all objects within the specified container's bindery context can be accessed both by NDS objects and by bindery-based servers and client workstations. Bindery services applies only to leaf objects in the specified container objects. The container object where bindery services is set is called the bindery context. This is a transitional service. Applications and clients should be updated to NDS-compatible clients and applications as soon as they become available. See Exhibit 21.5 for a visual summary of the role NDS plays in the enterprise network.

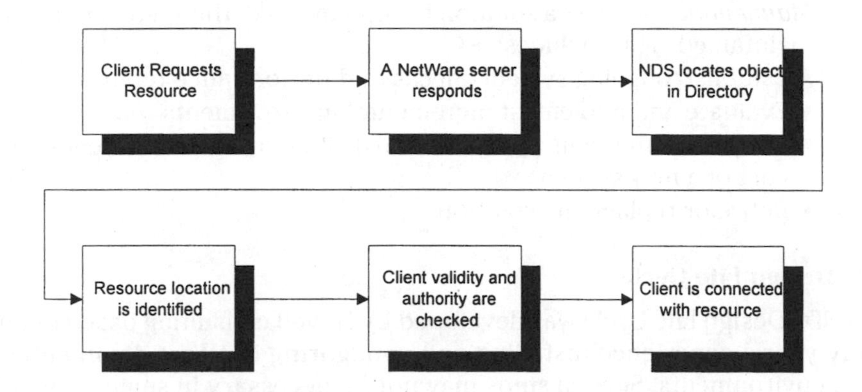

Exhibit 21.5 Implementation process for NetWare 4 and NDS.

Linear System Design Life Cycle

The Linear SDLC is a planning methodology and consists of four phases:

1. *Analysis and specification* — During this phase, project requirements and scope are determined. This includes, but is not limited to, the following tasks:
   - Recognize and state the underlying business need
   - Describe high-level goals, constraints, and resource requirements
   - Gather general information about the project
   - Prepare preliminary schedules
   - Agree on a project charter
   - List team member responsibilities and assignments
2. *Design* — This phase begins the design based upon identified needs. This includes, but is not limited to:
   - Identify generic solutions
   - Evaluate alternatives
   - Design a specific solution
3. *Implementation* — After a specific solution is designed, this phase implements the solution and refines it through feedback from lab testing and pilot implementations. This includes, but is not limited to:
   - Define specific milestone events and set schedule
   - Establish a test lab to verify the validity of a working solution
   - Implement a pilot run of the solution using the feedback to refine the solution

- Implement the final solution
- Document the system or solution and train users

4. *Maintenance* — After a solution is implemented, the system must be maintained. This includes:
   - Establish ongoing system checks and performance reports
   - Evaluate and implement incremental improvements
   - Perform maintenance until the cost of maintenance exceeds the cost of a new system
   - Retire or replace the solution

NDS Design Life Cycle

The NDS Design Life Cycle was developed by Novell consulting based upon many years' experience installing and configuring NetWare 4.x in enterprise environments. Several steps may not be necessary in smaller implementations; however, the overall process/methodology is effective in any size environment. Additional information regarding the NDS Design Life Cycle (Exhibit 21.6) are available from Novell's Design and Implementation Class, *Novell's CNE Study Guide for NetWare 4.1,* and *Novell's QuickPath to NetWare 4.1 Networks.*

Project Approach Phase

- *Determine the Project Approach.* This step helps the project team set realistic expectations, so that the design and implementation proceeds without surprises and all team members know what is expected of them.

Design Phase

- *Design the Directory Tree.* This not only includes Directory tree design, but also includes setting naming standards for NDS objects and accounting for network traffic patterns as a criterion for design.
- *Determine a Partition and Replica Strategy.* This includes planning for multiple NDS partitions, if needed, and the location of the partition replicas on the network. This procedure, if done correctly, will provide stability, fault tolerance, and accessibility to the NDS structure. This procedure is conditional based upon the size of the network.
- *Plan a Time Synchronization Strategy.* This includes the guidelines for determining if the internetwork needs default or configured time synchronization. This procedure is conditional based upon the size of the network.
- *Create an Accessibility Plan.* This is the "who has rights to what" phase of the planning process. This is where login script standards and NDS security standards are set, as well as guidelines for the use of Alias, Directory Map, and Profile objects.

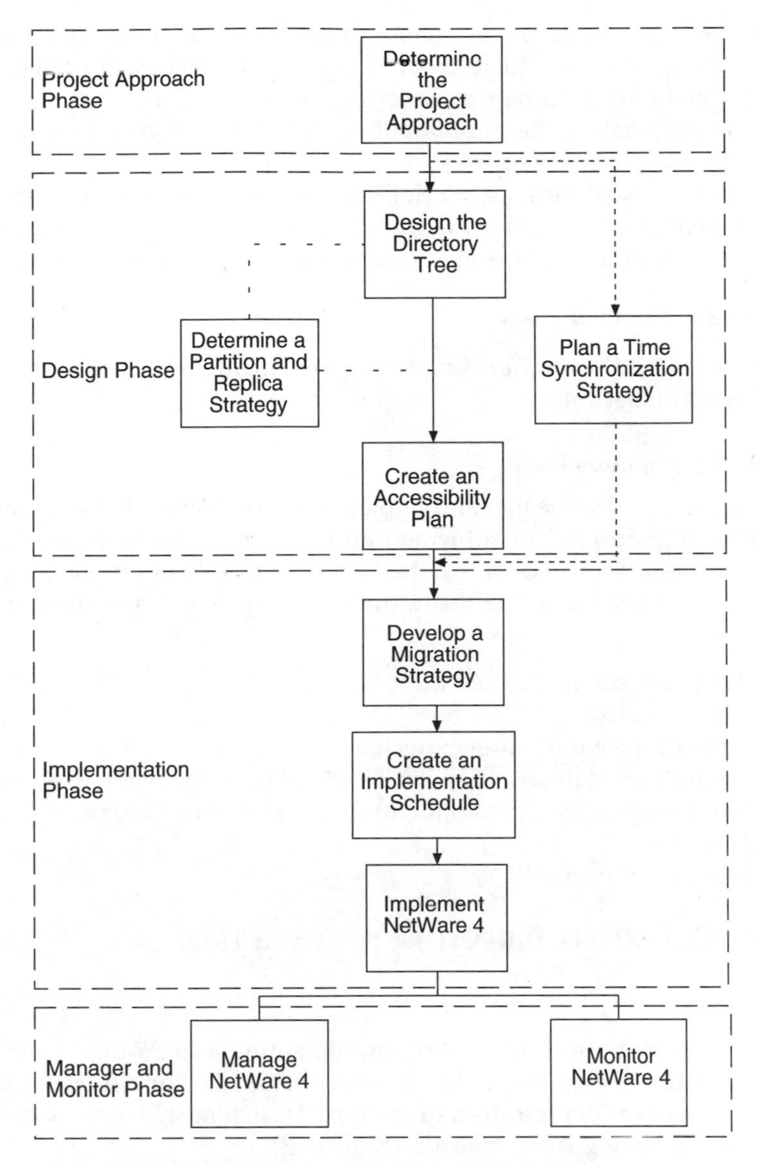

Exhibit 21.6 NDS design life cycle.

Implementation Phase

- *Develop a Migration Strategy.* This includes setting up a test laboratory to develop the procedures for updating workstations and file servers, testing software compatibility, and implementing a pilot installation.

- *Create an Implementation Schedule.* Based upon all the work during the migration strategy phase create a roll-out schedule with milestones. This includes defining tasks, setting timelines, making assignments, and coordinating the implementation with other groups within and without the enterprise.
- *Implement NetWare 4.* Install NetWare on the file servers. This includes installation of new servers and techniques for migrating earlier versions of NetWare and other network operating systems to NetWare 4.1.

Manage and Monitor Phase

- *Manage and Monitor NetWare 4.* Ongoing maintenance and administration of the network.

Required Competencies

Setting up the project design and implementation team is the most critical part of the NDS design and implementation process. The larger the project the more personnel will be needed to implement a solution. Regardless of the number of people on the team, the following expertise will be needed on the team:

1. Team project management
2. NDS expertise
3. Server and workstation expertise
4. Knowledge of the existing network and how it is used
5. Knowledge of existing applications and how they are used
6. Printing
7. Connectivity and WANs

DESIGNING A NOVELL DIRECTORY SERVICES TREE

The Process

This is the most important part of implementing a NetWare 4.1 network. Getting things correct here makes the network easier for network users to use and network administrators to manage. Designing a Directory tree consists of the following steps (see also Exhibit 21.7):

- Creating a naming standards document
- Designing the upper and lower layers of the Directory tree
- Creating the Directory tree structure

Creating the Naming Standards Document

A naming standards document should include the conventions for naming Directory objects including users, printers, printing, queues, servers, and all other NDS objects. In addition, conventions should be established for

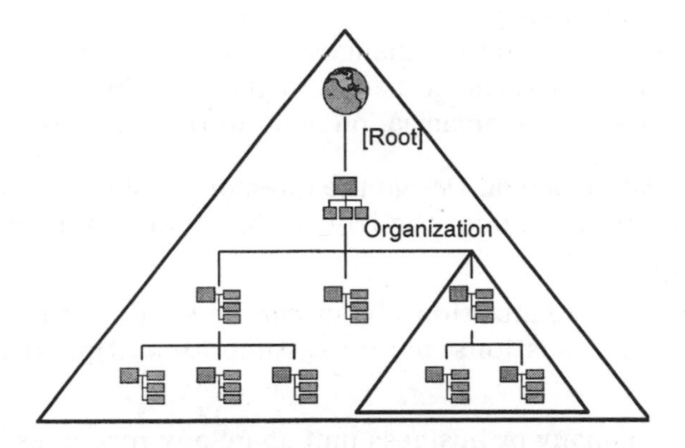

Exhibit 21.7 A directory tree.

certain properties of objects, such as telephone numbers and addresses. Naming schemes provide for consistency in naming all Directory objects. Good naming provides a guideline for network administrators who will manage objects in the Directory tree, eliminate redundant planning, and allow users to identify resources quickly, regardless of who implemented the tree.

The most important recommendation for Directory object names is that they be short and descriptive. Use the following guidelines when creating a naming standards document:

- Use short but descriptive names.
- Use initial capitalization to increase readability.
- Avoid spaces — use hyphens or underscores.
- Limit the number and types of characters used for naming.
- Limit user names to 8 characters when using DOS file systems; this will facilitate the automatic creation of user directories.
- If mixing 3.x and 4.x networks, be careful to use standards that are compatible with both operating systems.

Design the Upper Layer of the Directory Tree

Laying out the upper layers of the Directory tree requires special considerations for networks that span WAN links. The NDS database is a loosely consistent database. Updates and changes are communicated between replica partitions, which may be located across multiple WAN links, on a real-time basis. This can create undesirable WAN traffic. Reasons for designing the upper layers of the Directory tree according to location are as follows:

- Reflects the WAN infrastructure
- Minimizes WAN traffic and related bandwidth costs

- Facilitates partitioning
- Reduces significant future changes to the Directory tree (locations tend to be more permanent than department or divisional groups)
- Allows for the logical organization of network resources.

Use a Pyramid-Like Design. Design the directory as a logical pyramid, as shown in Exhibit 21.7. Under the [root] object, only one Organization object should be used.

More than one Organization object *can* be used, but is not recommended. The key questions, before setting up multiple organization objects, are

- Does the company or business unit share any resources other than email with the other units of the enterprise?
- Does the company share the same WAN or LAN infrastructure?

If the answer to either of these questions is yes, the tree design should still contain one Organization object with divisions defined by Organizational Unit objects.

Don't Span the WAN. Designing the upper layers of the NDS tree according to WAN links provides the following benefits:

- All objects in a partition exist at the same location.
- All users log in and are authenticated locally.
- All replicas can be stored on local servers, reducing synchronization cost of updating replicas over a WAN.

Two design examples can be seen in Exhibit 21.8 and 21.9.

Exceptions to the Rule. *High-Speed WAN Links.* When WAN links are fast enough, as fast as a local LAN connection, the benefits of designing according to location are reduced. Obviously, if a network exists at a single location, WAN considerations are not valid. Time synchronization, partitioning, and replica placement issues across dial-up links may cause more problems than they are worth. If all you have is dial-up service, it may be wise to consider separate Directory trees. Design for the future. When higher-speed lines become available, you'll want to merge trees with the fewest changes possible.

Design the Lower Layers of the Directory Tree

The basic rule for the lower layers of the tree is to create containers for objects that have common access needs to other NDS objects. Generally, people who work in the same department, or physical area, tend to share the same network resources. Consequently, the lower layers of the Directory usually resemble an organizational chart, or workgroups, with departments

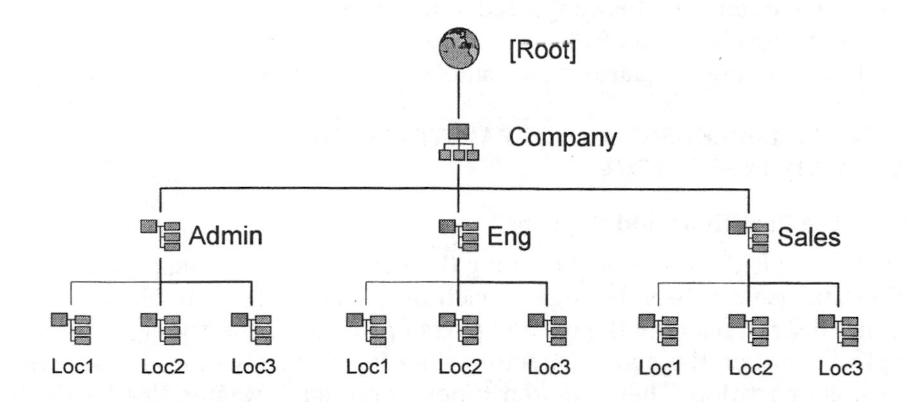

Exhibit 21.8 Function-based directory tree design.

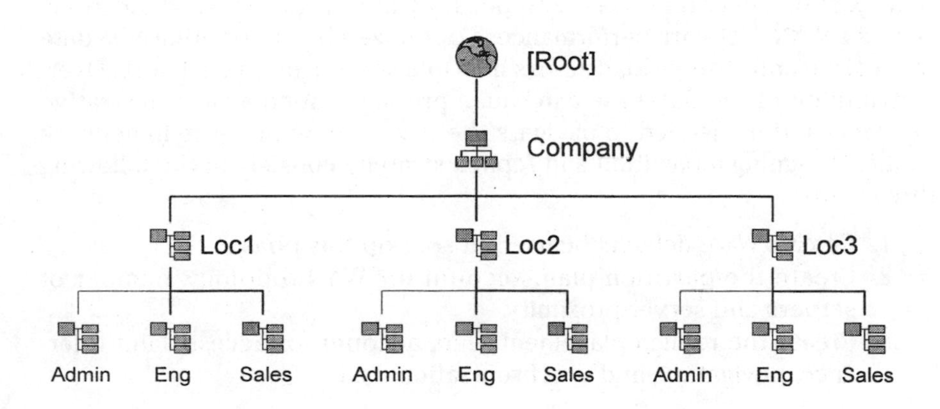

Exhibit 21.9 Location-based directory tree design.

defined as organizational units. Factors that must be accounted for include:

- Does the organization chart, or other resource, accurately represent logical workgroups?
- Are all workgroups included in the Directory?
- Do containers represent groups in multiple locations? Is this unavoidable?

- How are remote sites supported in the tree?
- Are all levels necessary?
- Have naming standards been established and used?

PLAN AND IMPLEMENT AN NDS PARTITION AND REPLICATION STRATEGY

What Are Partitions and Replicas?

Partitioning is the process of dividing the Directory database. Because an NDS database can be very large — each object requires 3K to 5K — partitioning distributes only the needed parts of the database among servers. Replicas contain the actual Directory data for objects within the boundaries of a partition. There are four types of replicas: Master, Read/Write, Read Only, and Subordinate Reference. A *replica ring* is the set of servers that contain a specific partition.

The Process

When partitioning the Directory, you are splitting the database into parts that exist on several servers, and possibly at several physical locations across a WAN. Network performance is optimized by distributing NDS data processing and storage load across multiple servers on the network. Over-distribution of the database can cause problems such as administrative headaches, time needed to navigate the tree, and an increase in network traffic. Designing a partition and replica strategy consists of the following three steps:

1. Can NetWare defaults be used? If so, skip this phase.
2. Create the partition plan, account for WAN topology, number of servers and server proximity.
3. Create the replica placement plan, account for access, fault tolerance, navigation, and synchronization cost.

Novell's installation program provides small and (or) simple networks with a default partition and replica strategy. The defaults are as follows:

- One partition for the entire tree.
- Three partition replicas, placed on the first three servers installed into the Directory tree (one master replica is installed on the first server and read/write replicas are installed on the next two servers). Subsequent file servers will receive no replicas.
- All servers migrated from bindery-based NetWare systems will contain a partition replica of the Bindery Services Context.

No special administrator intervention is needed to implement this strategy, the NetWare installation does it automatically. Novell recommends the acceptance of installation defaults if all the following conditions exist:

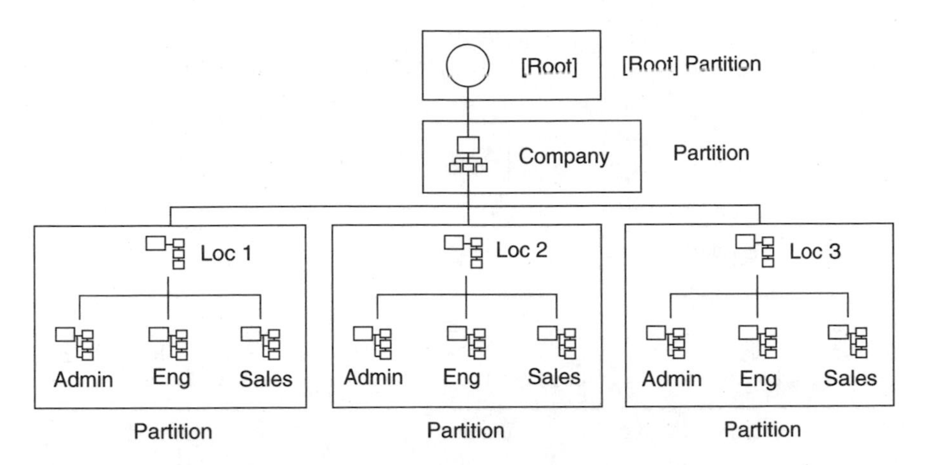

Exhibit 21.10 Partition design by location.

1. No WAN links exist.
2. Fewer servers are holding replicas.
3. Fewer than 5,000 objects are in the Directory tree.

Novell provides the following guidelines when drawing partition boundaries:

1. Determine if the default plan applies to the network structure.
2. Partition the Directory so that all objects, in each partition, are at a single location. Don't span the WAN.
3. If more than 15 servers are holding replicas of the same partition, synchronization traffic can be quite heavy. Consider splitting the partition.
4. Try to get all servers on a replica ring on the same network segment; this will speed replica synchronization.
5. Minimize Subordinate Reference replicas. Subordinate references increase synchronization traffic.

Design partition boundaries around the physical layout of the network infrastructure. This tracks with the approach used in designing the upper and lower layers of the Directory tree. See Exhibits 21.10 and 21.11.

Consider the following rules-of-thumb when partitioning the tree:

- Design upper layer partitions around Organizational Unit container boundaries that represent location or organizational structures.
- Plan partitions that contain objects from a single campus. DO NOT SPAN THE WAN.

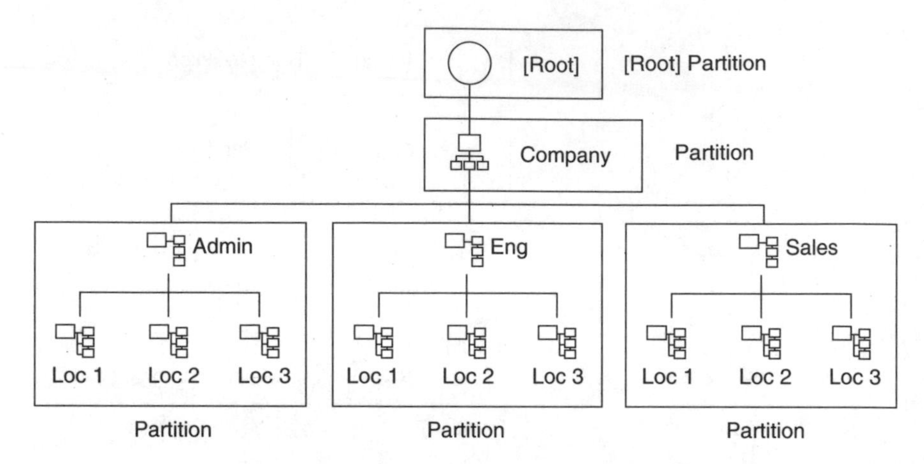

Exhibit 21.11 Partition design across a WAN.

- After installing the first server on the network, install all containers to be located in the Directory tree, then create partitions prior to installing additional servers. The NetWare installation program will create the necessary partition replicas (one master and two read/write).
- Maintain small partitions.
- Place partitions near the NDS objects that use the resources contained within the partition most.
- In a WAN environment, upper level partitions should reflect the physical locations within the enterprise, the number of objects, and WAN costs and performance.
- Lower level partitions should reflect work groups, how the data are used on the network, and the number of objects.

As partitions are planned, the following factors must be considered and balanced. Partitions that contain more than 5,000 objects can cause long delays in synchronizations and high levels of network traffic. Partitions that contain fewer than 100 objects can cause more management overhead than they are worth. A Directory tree design that contains many slow WAN links might require additional partitions. The amount of network traffic the network can support affects the number of partitions of a tree. Regardless of the plan used, splitting and combining partitions is a relatively easy task.

NetWare 4.1 provides a strict set of rules for creating partitions. A partition contains only NDS objects and related data. It cannot include information about the file system directories and files. An NDS object can exist in only one partition. Partitions are stored only on NetWare 4 server. Partitions cannot overlap. Partitions must contain a connected subtree. A single container object cannot exist as a partition without another container or leaf objects contained within it. It would be pointless to have a partition with nothing in it.

Creating the Replica Placement Plan

Replicas can provide three functions: Directory fault tolerance, access performance, and navigation. If a server crashes, another server with a partition replica can provide Directory information until the server is reestablished. Since NDS information is retrieved from the server that is physically nearest to the service requester, proper placement can impact NDS performance. NDS name resolution requires access to NDS databases. To the extent that the information is near the requester, the access of information will be fast or slow.

Types of Replicas

- *Master* — All partition operations (merging, moving, creating, deleting, repairing) occur from the Master replica of a specific partition. There can be only one Master replica.
- *Read/Write* — Contains the same object information as the Master replica, multiple Read/Write replicas can exist for a specific partition. Read/Write partitions cannot be used for partition operations.
- *Read Only* — Contains the same object information as the Master replica, but the information can only be read. Read Only replicas cannot be used for authentication, bindery services, or partition operations. Read Only replicas are almost never used.
- *Subordinate Reference* — Subordinate Reference replicas are created and managed by NetWare. They provide a pointer for NDS for navigating the Directory tree. A Subordinate Reference replica tells NDS where any child partitions exist. Since Subordinate Reference replicas maintain links between partitions, they can generate network traffic.

Replica Placement Design Considerations. Novell provides the following guidelines when determining replica placement:

1. Place a replica close to the users whose objects are in that partition. This allows NDS to respond quickly to user requests for login and access to network resources.
2. Determine a fault tolerance plan and apply it to the network structure.
3. Partition operations require that a Master replica be available to the network administrator; avoid placing Master replicas on the opposite side of a WAN link from the responsible administrator.

4. Replicas are used for fault tolerance. If only one server exists at a location, consider purchasing an additional server to hold a second replica or establish strict backup procedures.
5. If a server has bindery services enabled, list the containers in the server's bindery context, determine the partitions that hold those containers, and plan a Read/Write replica on those partitions on the server.
6. Avoid placing too many replicas on a server. NDS is an application NLM that requires processor cycles. Too slow a server, with too many replicas may make replica synchronization slow for all servers in the replica ring. In addition, it may slow down other server processes.

Design replica placement around the principles of fault tolerance, accessibility, and navigation.

- Create a minimum of three replicas of each partition. Try to place one replica off-site.
- Be sure to replicate the [Root] partition. This is the link that ties the Directory together. Without the [Root], a Directory cannot be reconstructed.
- Create replicas, as needed, for bindery services.
- Consider the Directory Services access performance goals. Place replicas in the location of highest access by users, groups, and services.
- Eliminate any single point of failure.

PLAN AND IMPLEMENT TIME SYNCHRONIZATION

The Process

Time synchronization is the service that maintains a consistent time standard across the network servers. File servers use time to apply time stamps to file and directory operations. Messaging applications use time stamps for messages. NDS uses server time to help properly collate changes to the Directory database by attaching a time stamp to Directory requests and changes. Time synchronization is established among file servers independent of their location in a Directory tree. In addition, time servers need not be in the same Directory tree.

In NDS, changes are made to the nearest replica, and a time stamp is attached to the change. The changes are then communicated to other replicas in the replica ring. Time stamps are used to determine the priority of changes to the database; conflicts are resolved through the use of time stamps. Designing time synchronization consists of the following steps:

1. Determine if NetWare defaults are applicable.
2. Determine time providers and time receivers.
3. Determine the time synchronization communications method.

Time Server Types

A single reference server is a time source that defines network time. It is the king of network time and only used with NetWare's default time synchronization configuration. Single reference servers cannot coexist with other time providers, such as primary and reference servers. They provide time only to secondary servers.

A primary time server is a time source that obtains its time from a reference server. When used, at least two primary servers must be used with a reference server. No more than seven primary time servers can exist in a time provider group. They provide time only to secondary servers. Primary time servers communicate with each other to determine network time. Secondary time servers get their time from any time provider. They are used as part of a default or custom time configuration.

Reference servers act the same way as primary servers except that they do not adjust their internal clock. The reference time server provides a central point of time control for a time provider group. Multiple reference servers and time provider groups can exist on a network; however, they must be synchronized to the same external time source, such as an atomic clock or GPS satellite.

Default Configuration

The first server installed on the network is a single reference server. All other servers installed on the network become secondary servers getting their time from the single reference server. Communication between the time provider and time receiver is through Novell's Service Advertising Protocol (SAP). The default configuration was provided because it is simple and efficient. It does not require any installer intervention to implement and works well for smaller networks (see Exhibit 21.12).

Single Reference Advantages

- It is easy to understand and requires no planning.
- It does not require a configuration file.
- It requires no ongoing maintenance.

Single Reference Disadvantages

- The single reference server must be connected and contacted to every server on the network.
- Using SAP means that a misconfigured server can disrupt the network.
- One time source means a single point of failure.

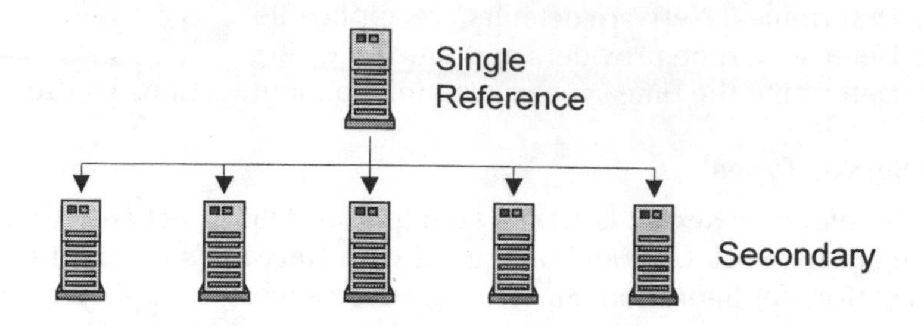

Exhibit 21.12 Single reference server with secondary servers.

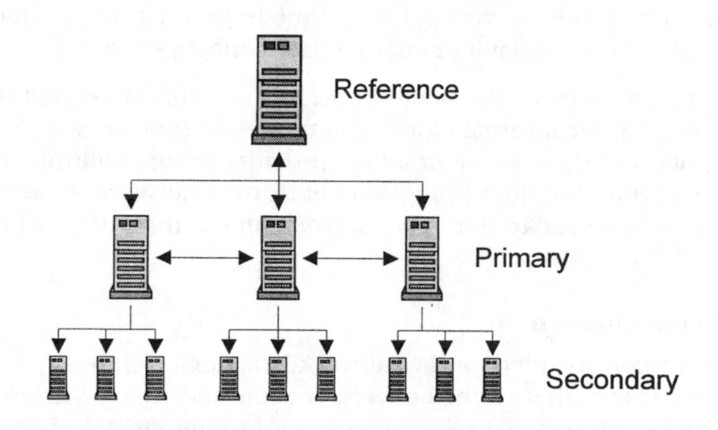

Exhibit 21.13 Custom configuration.

Custom Configuration

A custom configuration uses time provider groups to provide time synchronization for the network. Each group has a reference, and at least two primary, and multiple secondary time servers. Servers can communicate through configured lists or via SAP. Each server is given a configuration file (timesync.cfg), defining the server type, listing authorized time sources for the server (the configured list), and defining the communications protocol (SAP or configured list) (see Exhibit 21.13).

The custom configuration option is especially relevant to complex networks by eliminating single points of failure and reducing network traffic.

Advantages

- The network administrator has complete control of the time synchronization hierarchy.
- Network traffic can be optimized by distributing time sources around the network.
- Single points of failure can be eliminated.

Disadvantages

- Customization requires careful planning.
- Adding new time sources can require that configuration files on several servers be updated.

Design Guidelines

- When using multiple reference time servers, have them obtain their time from a common external time source.
- Have secondary time servers synchronize to time sources; have secondary servers synchronize to other Secondaries as a backup.
- Keep the number of time providers as small as possible to reduce network traffic.
- Use time sources to provide local access throughout the network. Do not allow secondaries to obtain time synchronization across a WAN.

MIGRATING SERVERS TO NETWARE 4

Migration Options

NetWare provides four upgrade and migration methods for version 4.1:

- The NetWare 4.1 Installation Program
- Across The Wire Migration
- Same Server Migration
- In Place Upgrade

The method used depends upon the operating system on the original server, the available hardware, and the risk involved with the technique.

NetWare 4 Hardware Requirements

- IBM PC or compatible with an 80386 processor or faster.
- A minimum of 8 MB ram if installing from a local CD-ROM, 10 MB if installed across the network. Much more is needed for efficient operation. Rule-of-thumb: Minimum required RAM = 6 MB + 1 MB for each 100 MB of disk space + RAM for any installed nlms.
- A minimum of 90 MB of free space on the hard disk: 15 MB for the DOS partition plus 75 MB for a NetWare partition for the SYS: volume.
- Another 60 MB of disk space for NetWare documentation.

The NetWare 4.1 Installation Program

Use this option to upgrade an existing NetWare 3.1x or 4.0x server to NetWare 4.1. This is a menu option when running the INSTALL.NLM on the file server. The NetWare installation program performs the following operations:

1. Copies the server boot files to the DOS partition of the NetWare Server
2. Installs NDS and upgrades the NetWare 3.1x bindery to an NDS database
3. Copies the NetWare 4.1 system, public and NLM files to the volume SYS

Advantages

- The upgrade requires no additional hardware.
- This is the most convenient way to upgrade.

Disadvantages

- A slight chance of data loss exists if the server crashes during the upgrade process.
- NetWare 2 servers and non-NetWare servers cannot be upgraded to NetWare 4.1 using this process.

Across the Wire Migration

This method uses the Migration utility to upgrade an existing NetWare 2.1x, NetWare 2.2, or NetWare 3.1x bindery to a new NetWare 4.1 server. Across the wire migration also can migrate IBM PCLP 1.3 Extended Services; IBM LAN Server 1.0, 1.1, 1.2, 1.3, and Microsoft LAN Manager 2.0 operating systems.

Across the wire migration requires three computers to perform the upgrade: the original (source) server, the new NetWare 4.1 (destination) server, and a DOS workstation with a local hard disk, running NetWare Client software (see Exhibit 21.14).

The Migration upgrade performs the following operations:

1. Migrates a copy of the bindery from the source server to the DOS client workstation and converts it to an NDS database
2. Migrates the translated bindery to the destination server from the DOS client workstation using bindery services
3. Migrates the data files from the source server to the destination server

Advantages

- It is the safest upgrade because the source server is left intact.
- The administrator can migrate all or selected information.
- Multiple servers can be migrated to a single destination server.

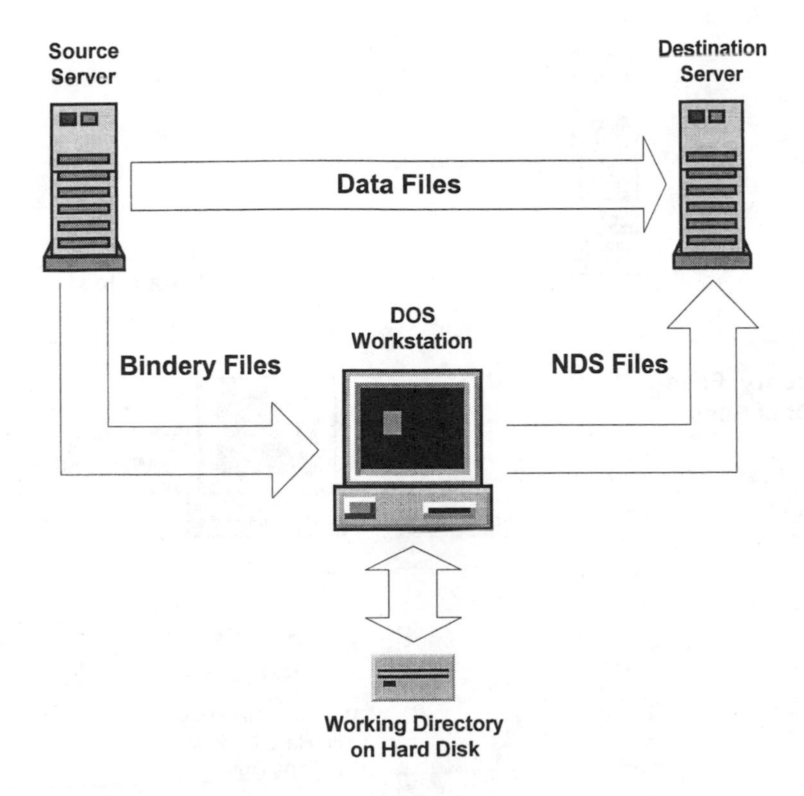

Exhibit 21.14 Across the wire migration.

- Data can be moved from a 286 machine running NetWare 2.x to a machine running NetWare 4.1.
- Network data on non-NetWare servers can be moved to a NetWare 4.1 server.

Disadvantages

- Extra hardware is required, a new file server, and a DOS workstation.
- The resource requirements for the DOS workstation may be significant.

Same Server Migration

This method uses the Migration utility to convert an existing NetWare 2.1x, NetWare 2.2, or NetWare 3.1x bindery to a new NetWare 4.1 server. Same server migration can also migrate IBM PCLP 1.3 Extended Services, IBM LAN Server 1.0, 1.1, 1.2, 1.3, and Microsoft LAN Manager 2.0 operating systems. Same server migration requires two computers to perform the upgrade: the original (source) server and a DOS workstation with a local hard disk running NetWare Client software (see Exhibit 21.15).

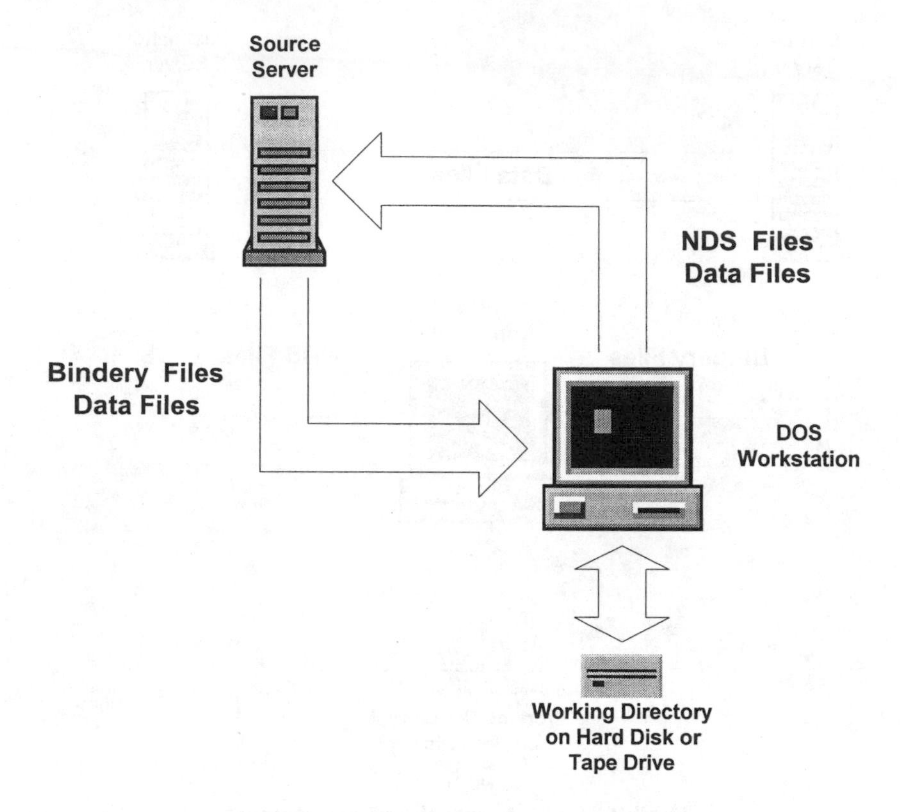

Exhibit 21.15 Same server migration.

To perform the upgrade the following steps must be completed:

1. Back up the server's data files.
2. Migrate the bindery from the existing server to the DOS client workstation and translate it to a NetWare 4.1 format.
3. Install NetWare 4.1 using the NetWare installation program on the same hardware as the old server.
4. Restore the data files from the backup device to the new NetWare 4.1 server.
5. Use the migration utility to migrate the translated bindery to the new NetWare 4.1 server from the DOS client workstation.

Advantages

- No additional server hardware is required if the old server is capable of running NetWare 4.1.
- NetWare 2.x and non-NetWare servers can be upgraded to NetWare 4.1.

Disadvantages

- There is risk to the Bindery data during the conversion since the original Bindery is destroyed during the upgrade.
- All file attributes may not migrate if the backup software does not support NetWare file attributes.
- The DOS workstation may require a large hard disk, depending on the number of users and data files on the file server.
- Not all tape devices work. The tape device must map to a DOS drive letter.

In Place Upgrade

Use the In Place Upgrade to upgrade an existing NetWare 2.1x or 2.2 server to NetWare 4.1. The In Place Upgrade uses the 2x upgrade program to reformat existing server hard disk partitions into NetWare 3.1x partitions. Then use the NetWare 4.1 installation program to upgrade the server from NetWare 3.1x to NetWare 4.1x. Only one computer is required, the server being upgraded. To perform the upgrade the following steps must be completed:

1. Verify that the server can run NetWare 4.1.
2. Back up the file server.
3. Upgrade the file system to NetWare 3.1x using 3xupgrde.nlm.
4. Use FDISK to create a DOS primary partition; set it as active and format it.
5. Create a boot directory and copy the boot file to the directory.
6. Upgrade the operation system to NetWare 4.1 using the NetWare 4.1 installation program.

Advantages

- No additional server hardware is required.

Disadvantages

- If the server is running NetWare 2.x using an IDE hard disk and a DOS partition does not exist, one cannot be added when the upgrade to NetWare 3.x is performed.
- If the upgrade fails, the server must be restored to its original state and the migration restarted.

The Migrate Utility

The Migrate Utility converts NetWare 3.x binderies and data to NetWare 4.1 format. Exhibits 21.16 through 21.20 includes screen shots for the major components of the migration utility.

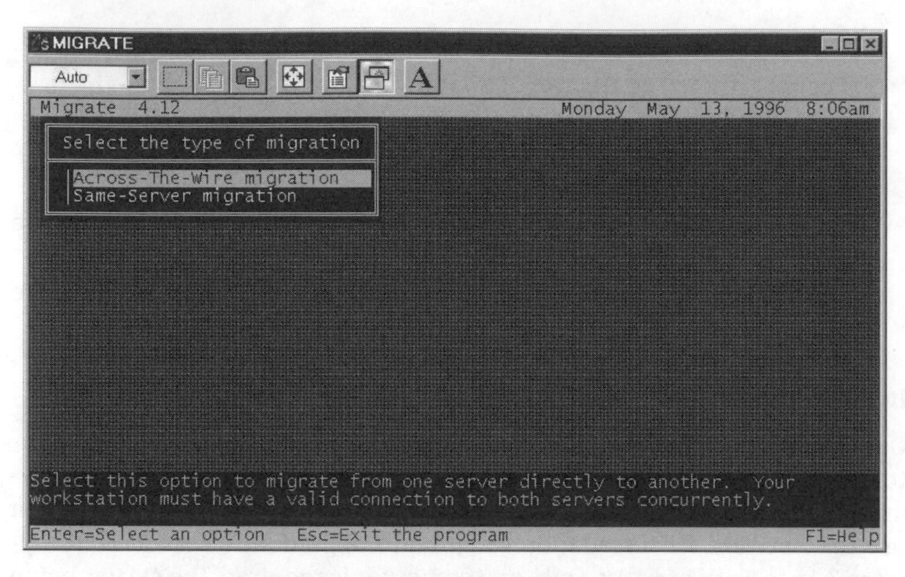

Exhibit 21.16 Selecting the type of migration.

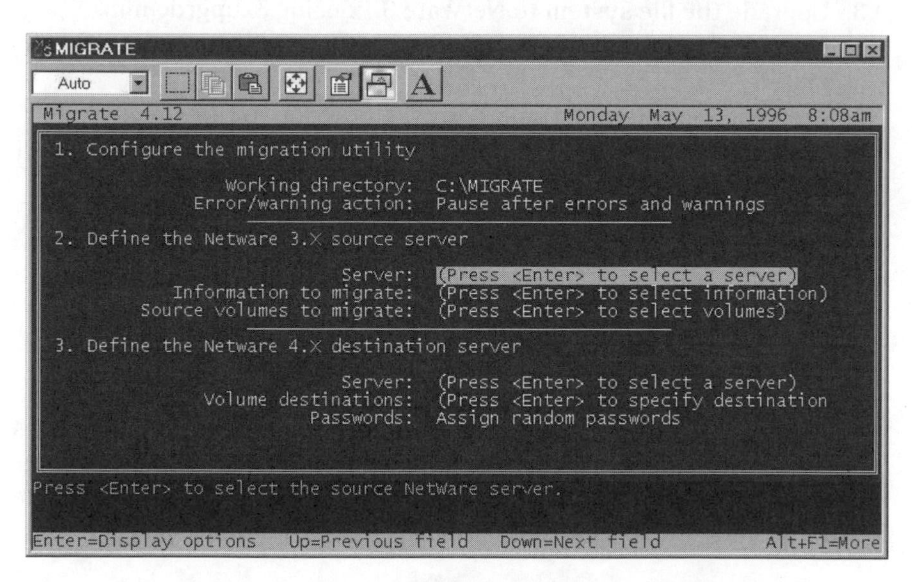

Exhibit 21.17 Configuring the migration utility.

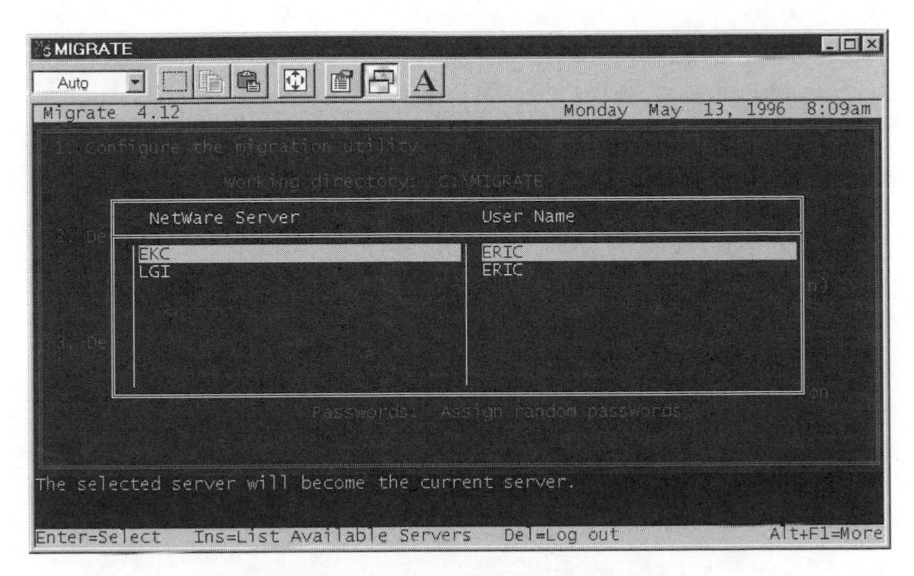

Exhibit 21.18 Selecting the NetWare server.

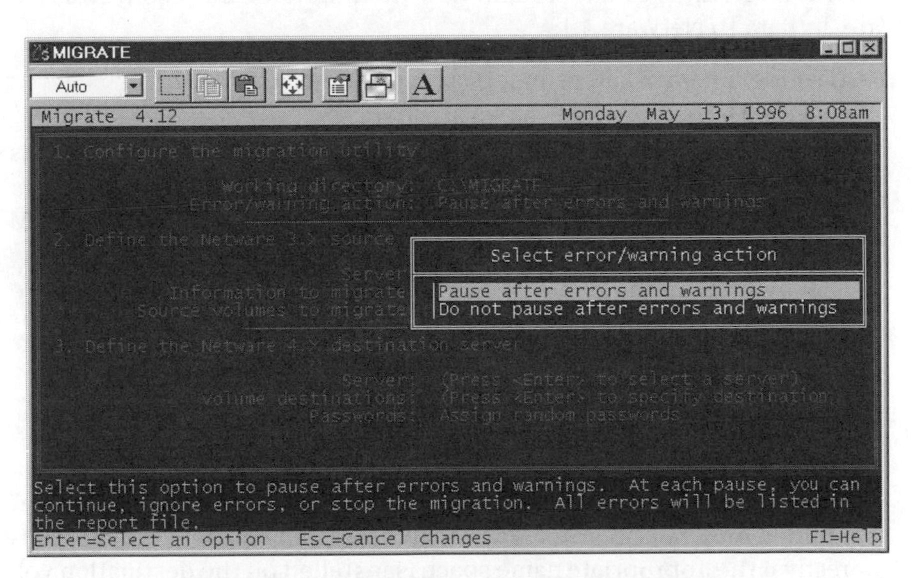

Exhibit 21.19 Selecting the error/warning action.

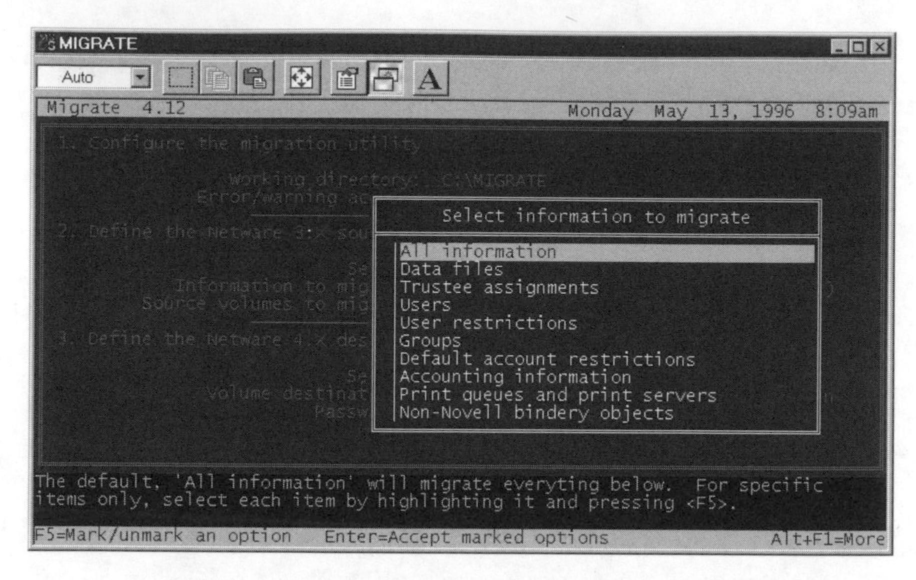

Exhibit 21.20 Selecting the information to migrate.

Migrate Utility Restrictions. The Migrate Utility, while effective, is not perfect. It is imperative to understand how the Migrate Utility converts NetWare 3.x data to NetWare 4.1:

- User login names and print job configurations are copied.
- Users with the same user name are merged.
- Passwords are not migrated. The administrator can assign passwords or the migrate program will assign passwords. The passwords assigned by the utility will be saved in "new.pwd" on the SYS:SYSTEM directory.
- The system login script is not migrated; user login scripts are migrated.
- The DOS and NetWare attributes of files are migrated.
- User and group trustee rights for directories and files are migrated.
- User account restrictions are copied.
- Only DOS directories and files that conform to DOS naming conventions (8.3 format) are migrated. Files saved with special "long name" utilities may not migrate. Macintosh and OS/2 files will migrate correctly if the appropriate name space is installed on the destination volume.
- Subdirectories deeper than 25 levels are not migrated.
- The printing environment is not migrated. Use the migprint.exe utility to migrate printing objects. It is recommended that new printing

objects be created, rather than migrating NetWare 3.x printing objects.

After the Upgrade...

After upgrading the operating system, migrating binderies, and data files the following tasks must be performed:

- Change user passwords or have users input new passwords.
- Update workstations to VLM clients. This should be performed before converting 3.x servers to 4.1.
- Make sure the compatible frame types are installed on the file server and the workstations. Ethernet 802.3 should be replaced with Ethernet 802.2 on the entire network, both frame types can be installed on servers until the conversion is completed.
- Assign Directory object and property rights to Directory objects that were upgraded from bindery objects.
- Modify the system and user login scripts to reflect changes in the server name or directory paths that might have been altered.
- Copy the NetWare 4.1 login.exe file to SYS:LOGIN of all non-NetWare 3.x servers on the network. Login scripts from the NetWare 4.1 server do not execute properly until this is done.
- Immediately back up the server after making the above changes.

Do not attempt an upgrade to NDS without spending plenty of time planning your tree as well as imaging the future of your enterprise. Make sure you have an accurate view of the current network configuration, including WAN links, printer configuration, geographical location of organizational work groups, and any current naming standards. Only with a complete picture of today's network can you plan how to get to tomorrow's network.

Reading List

Configuring NetWare 4 for the Mobile User, Marcus Williamson, Novell Research, July 1994.

Novell's Information Services Standard Guidelines, NetWare 4.1 Design and Implementation Student Guide, Novell Education, 1995.

Novell Consulting Services Tool Kit, CD-ROM, Subscription Available from Novell. Call 1-801-429-5387 for subscription information.

The Novell Consulting Services™ Tool Kit (NCS Tool Kit) is an electronic information base containing information especially geared for members of Novell's Consulting Partners Program. This includes design methodology, and planning information for a whole range of Novell products and services..

Novell Application Notes, Call 800-377-4136 for subscription information.

Novell's CNE Study Guide for NetWare 4.1, David James Clarke, IV, Novell Press, 1995.

Novell's Guide to NetWare 4.1 Networks, Jeffrey F. Hughes and Blair W. Thomas, Novell Press, 1996.

Novell's QuickPath to NetWare 4.1 Networks, Jeffrey F. Hughes and Blair W. Thomas, Novell Press, 1995.

Inside NetWare 4.1, Doug Bierer, New Riders Publishing, 1995.

Chapter 22
Windows NT Architecture

Gilbert Held

Windows NT is a 32-bit, preemptive multitasking operating system that includes comprehensive networking capabilities and several levels of security. Microsoft markets two version of Windows NT: one for workstations — appropriately named Windows NT Workstation — and a second for servers — Windows NT Server. This article, which describes the workings of the NT architecture, references both versions as Windows NT when information is applicable to both versions of the operating system. Similarly, it references a specific version of the operating system when the information presented is specific to either Windows NT Workstation or Windows NT Server.

Readers should note that the basic architecture described in this chapter is retained by what was originally referred to as Windows NT Version 5 and was renamed Windows 2000. Windows 2000 may be officially released by the time you read this handbook. A key difference between Windows NT and Windows 2000 is the support of the latter for 64-bit operations. This support will allow Windows 2000 to operate on a new generation of Intel processors expected to become available during the first year of the new millenium.

ARCHITECTURE

Windows NT consists of nine basic modules. The relationship of those modules to one another, as well as to the hardware platform on which the operating system runs, is illustrated in Exhibit 22.1.

Hardware Abstraction Layer

The hardware abstraction layer (HAL) is located directly above the hardware on which Windows NT operates. HAL actually represents a software module developed by hardware manufacturers that is bundled into Windows NT to

0-8493-9823-1/00/$0.00+$.50

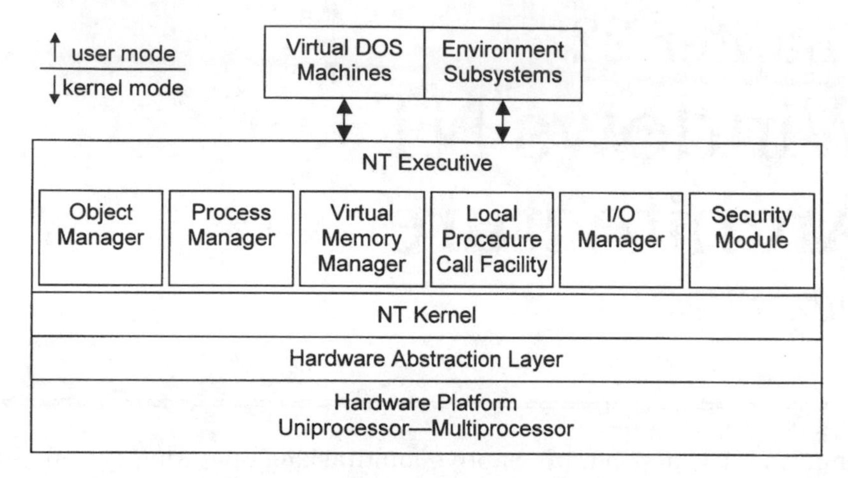

Exhibit 22.1 Windows NT core modules.

allow it to operate on a specific hardware platform, such as Intel X86, DEC Alpha, or IBM PowerPC.

HAL hides the specifics of the hardware platform from the rest of the operating system and represents the lowest level of Windows NT. Thus, HAL provides true hardware platform independence for the operating system. Using HAL, software developers can create new software without a lot of knowledge about the hardware platform. This allows software developers to provide enhanced performance capabilities, such as additional device drives. Hardware vendors can provide the interface between the operating system and the specific hardware.

Kernel

The kernel represents the core of the Windows NT operating system. All operating systems have a kernel. The key difference between the Windows NT kernel and those found in other operating systems is the tasks managed.

The Windows NT kernel manages thread dispatching. (A thread is a basic item that can be scheduled by the kernel.) The kernel is also responsible for scheduling and processor synchronization when the hardware platform has multiple processors.

To perform scheduling, the Windows NT kernel attempts to dispatch threads for execution in a way that promotes the most efficient use of the processors in the hardware platform. The actual dispatching of threads is based on their priority, with Windows NT supporting 32 priority levels to maximize processor use.

The kernel always resides in real memory within the hardware platform's RAM and is nonpayable to disk. When NT controls a multiprocessor platform, the kernel will run on all processors at the same time and communicate with each other to govern the distribution of threads.

The NT Executive

The NT Executive can be considered a common service provider because it is responsible for providing a set of services to all other operating system components. The Windows NT Executive is the highest level within the kernel mode of the operating system.

As indicated in Exhibit 22.1, the Executive consists of six core modules that provide an interface between users and computers (represented by Virtual DOS Machines and Environment Subsystems) and the kernel. Virtual DOS Machines support DOS or 16-bit Windows 3.X applications. Windows NT provides support by creating virtual machines and then implementing the required environment within such a machine, resulting in the term *virtual DOS machines.*

In comparison, *environment subsystems* are environments that may be required to operate on top of Windows NT. Examples of currently supported environment subsystems include OS/2, POSIX, and Win32 (the Windows NT subsystem).

Object Manager

The object manager names, retains, and provides security for objects used by the operating system. In a Windows NT environment, an object represents physical items as well as the occurrence of defined situations. Thus, an object can represent directories, files, physical hardware ports, semaphores, events, and threads. An object-oriented approach is used to manage objects. If network managers are using Windows NT, they can view the status of event objects through the NT Event Viewer, which is provided in the operating system as an administrative tool.

Process Manager

In a Windows NT environment, a process represents an address space, a group of objects defined as a resource, or a set of threads. Thus, each of these entities is managed by the process manager. In doing so, the process manager combines those entities into a virtual machine, on which a program executes. Here the term *virtual machine* represents a set of resources required to provide support for the execution of a program. Windows NT permits multiple virtual machines to be established, allowing multiprocessing capability.

Virtual Memory Manager

Windows NT uses a special file on the hardware platform's hard disk for additional memory beyond available RAM. That file is referred to as a virtual memory paging or swap file and is automatically created when the operating system is installed.

The Virtual Memory Manager manages the use of virtual memory as a supplement to physical RAM. For example, when one program cannot completely fit into RAM because of its size or the current occupancy by other executing programs, the Virtual Memory Manager might swap one program currently in memory to disk to enable another program to execute, or it could swap portions of the program requesting execution between RAM and the hard disk to execute portions of the program in a predefined sequence.

Although the operation of the Virtual Memory Manager is transparent to programs using it, network managers can change the paging file size. To do so, they would first select the system icon in the Control Panel and then select the Virtual Memory entry from the resulting display. This action results in the display of a dialog box labeled Virtual Memory. Exhibit 22.2 illustrates the Virtual Memory dialog box with its default settings shown for a Pentium processor.

Although Windows NT automatically creates a virtual memory paging file and assigns an initial file size based on the capacity of the system's hard disk, the operating system does not know what applications the network manager intends to run or the size of those applications. Thus, if network managers frequently work with applications that require a large amount of memory, they should consider raising the default setting.

In Exhibit 22.2, Windows NT provides a pseudoconstraint on the sizes of the paging file. That constraint is in the form of a range of values defined for the size of the paging file; however, that range is a recommendation and is not actually enforced by the operating system. For example, to set the initial size of the paging file to two megabytes, the user would type 20 into the box labeled Initial Size and then click on the Set button. Similarly, if users want to raise the maximum size of the paging file to 100 MB, they would enter that value in the appropriate location in the dialog box and click on the Set button.

Local Procedure Call Facility

Programs that execute under Windows NT have a client/server relationship with the operating system. The Local Procedure Call Facility is responsible for the passing of messages between programs.

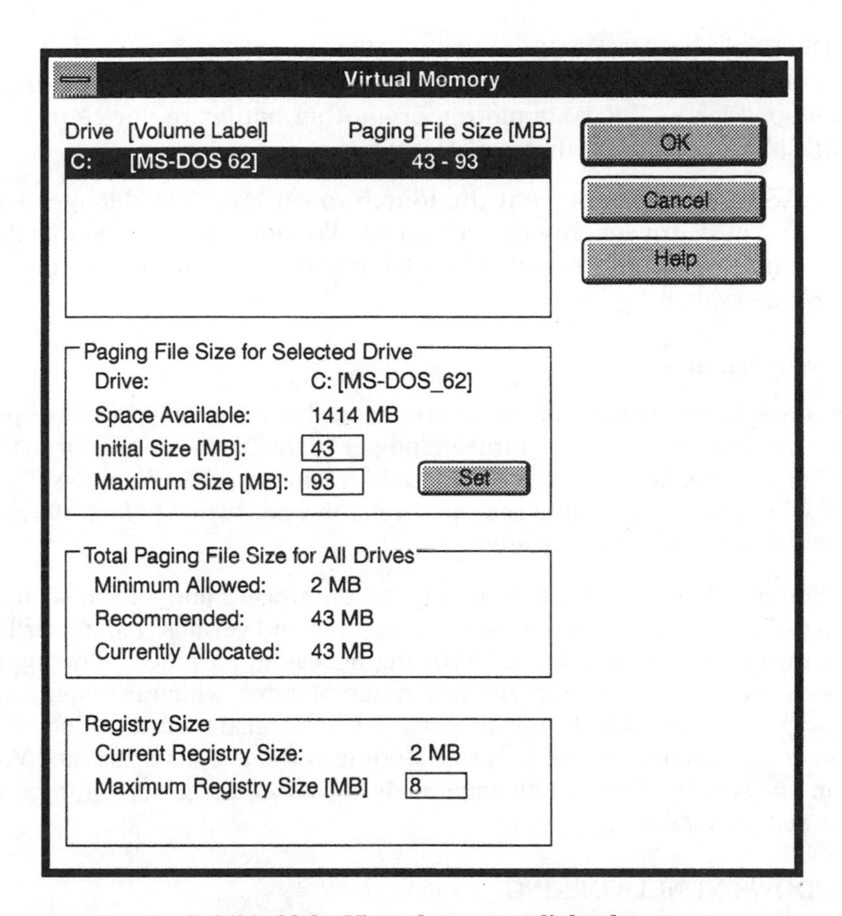

Exhibit 22.2 Virtual memory dialog box.

I/O MANAGER

The Input/Output (I/O) Manager is responsible for managing all input and output to and from storage and the network. To perform its required functions, the I/O Manager uses four other lower-level subsystems — the Cache Manager, file system drivers, hardware device drivers, and network drivers.

The Cache Manager provides a dynamic cache space in RAM that increases and decreases based on available memory. File system drivers provide support for two file systems, the file allocation table (FAT) and the high performance file system (HPFS). The FAT file system provides backward support for DOS and 16-bit Windows 3.X-based programs, whereas the HPFS enables support of the new file system for Windows NT 32-bit applications.

The hardware device drivers used in Windows NT are written in C++ to provide portability between hardware platforms. This allows a driver developed for a CD-ROM, a plotter, or another hardware device to work with all Windows NT hardware platforms.

Network drivers represent the fourth lower-level I/O Manager subsystem. These drivers provide access from Windows NT to network interface cards, enabling transmission to and from the network and the operating system.

Security Module

Windows NT includes a comprehensive security facility built into the operating system. Once the user turns on power to the hardware platform, this facility is immediately recognizable. Unlike Windows 3.X, Windows 95, or DOS, Windows NT prompts the operator for a password before allowing access to the computer's resources.

Windows NT security works by the log on process and a local security subsystem that monitors access to all objects and verifies that a user has appropriate permission before allowing access to an object. The log on process is linked to the Security Reference Monitor, which is responsible for access validation and audit generation for the local security subsystem. Another component of the Security Module is the Security Account Manager. The Security Account Manager maintains user and group information on a secure database.

WINDOWS NT NETWORKING

One of the biggest advantages associated with the use of Windows NT is its built-in support of many transport protocols. The Windows NT networking architecture was established in a layered design that follows the seven-layer ISO Open System Interconnection (OSI) Reference Model. Exhibit 22.3 illustrates the general correspondence between Windows NT layers and OSI Reference Model layers.

The environment subsystems represent virtual DOS machines as well as 32-bit applications operating on top of NT. At the presentation layer, the Network Provider module is required for each network supported through a redirector. At the session layer, the Windows NT Executive uses a server and redirector to provide capability for a server and workstation, respectively. Both components are implemented as file system drivers and multiple redirectors can be loaded at the same time, so a Windows NT computer can be connected to several networks. For example, NT includes redirectors for NetWare and VINES, enabling an NT workstation or server to be connected to Novell and Banyan networks.

Exhibit 22.3 Correspondence between Windows NT and OSI reference model layers.

OSI reference model layers	Windows NT layers			
Application	Environment subsystems			
Presentation	Network provider			
Session	Executive services			
	Server		Redirector	
Transport	Transport driver interface			
Network	NetBEUI	DLC	TCP/IP	NWLink (SPX/IPX)
Data link	NDIS			
	NIC drivers			
Physical	NIC			

At the transport layer, the transport driver interface (TDI) provides a higher-layer interface to multiple transport protocols. Those protocols, which represent operations at the network layer, include built-in NT protocol stacks for NetBEUI, used by the LAN Manager and LAN Server operating systems; Data Link Control (DLC), which provides access to IBM mainframes; TCP/IP for Internet and intranet applications; and NWLink, which represents a version of Novell's SPX/IPX protocols. Through the use of TCP/IP, a Windows NT computer can function as a TCP/IP client, whereas the use of NWLink enables a Windows NT computer to operate as NetWare client.

At the data link layer, Windows NT includes a built-in Network Device Interface Specification (NDIS). NDIS enables support for multiple protocol stacks through network interface card drivers. Thus, NDIS allows a network interface card to simultaneously communicate with multiple supported protocol stacks. This means that a Windows NT computer could, for example, simultaneously operate as both a TCP/IP and a NetWare SPX/IPX client.

UPGRADE ISSUES

The key differences between NT 3.5 and 4.0 are speed and user interface. Windows 4.0 added the Windows 95 user interface to NT. In addition, a recoding of the operating system makes it slightly faster than 3.51. However, because the difference in cost between a Pentium and Pentium Pro microprocessor is a few hundred dollars, it may be more economical to purchase the more powerful processor and retain the familiar Windows 3.51 interface. This could eliminate the costs associated with retraining employees.

Conversely, if an organization has already migrated to Windows 95 or is planning to migrate to that operating system, the network manager may want to consider Windows NT Version 4.0. Its use of the Windows 95 interface may be well known to some or most of the organization's employees who will be using NT, which should minimize training costs while providing a slightly improved level of performance.

CONCLUSION

The modular design of the Windows NT architecture makes it both portable and scalable. Windows NT's hardware abstraction layer allows the operating system to run on different hardware platforms. Currently, Windows NT runs on Intel X86, Digital Equipment Corp. (DEC) Alpha, MIPS RISC (reduced instruction set computing), and the PowerPC series of microprocessors jointly manufactured by IBM Corp. and Motorola.

Besides being highly portable, Windows NT supports scalability, which allows the operating system to effectively use multiple processors. Thus, when network managers evaluate Windows NT Server as a platform for different applications, it is important for them to note that they have several options for retaining their investment as applications grow.

For example, because of its scalability, network managers could replace a uniprocessor Intel Pentium motherboard with a dual- or quad-processor motherboard. If this replacement does not provide the necessary level of processing power, network managers might consider migrating hardware to a high-level PowerPC or a DEC Alpha-based computer. If that migration is required and the applications continue to grow, network managers could use multiple processors to ensure scalability.

Chapter 23
Windows NT Workstation vs. Server 4.0

Stewart Miller

Windows NT workstation and server have always shared the same user interface, application programming interface, API, and kernel architecture throughout both products. However, price, licensing, and optimization were another category entirely. These categories were divided into the networked desktop operating system and the high-performance server. Windows NT Server and Workstation 4.0 also share this same philosophy and are separated accordingly.

PERFORMANCE

In terms of Windows NT Workstation performance, its goal is to yield a high level of performance for one desktop user. However, the Windows NT Server is designed to yield a high level of performance when implemented as a server operating system that has several users concurrently connected to the server. Users can have a very responsive desktop which offers high-speed graphics and the power to switch rapidly between multiple tasks. The server, though, finds sharing files, printers, and Web pages a higher priority than tasks responding to user input or providing better performance in graphics. Windows NT 4.0 has a more optimized performance in the server and workstation applications due to its ability to reduce scalability bottlenecks in the Windows NT Server, tuning network caches to meet current usage and configuring network components to utilize as little memory as possible on the Windows NT Workstation 4.0.

In order for Windows NT 4.0 to be tuned for the best possible performance as a workstation or server, the concept of task scheduling comes into play. Task scheduling yields a very responsive atmosphere for the user. Windows NT Workstation task scheduler splits its time into small time slices to permit several tasks to be loaded and unloaded at a very fast pace,

0-8493-9823-1/00/$0.00+$.50

without the user encountering any delays. The benefit is that users can easily and rapidly switch from one task to the next while the system is still responsive. Windows NT server task scheduler splits its time into bigger time slices which permit the server to cope with network requests without stopping. This can play a critical role with symmetric multiprocessing (SMP) applications where thread and cache synchronization are crucial in each and every processor.

Enhanced performance is also achieved with memory allocation. The applications within Windows NT Workstation 4.0 are allocated with the smallest amount of memory at the time the application is loaded, since Windows NT Workstation users will often load and unload many applications during the work day. The Windows NT Server applications, however, are usually given all of the memory they request at the time they load. In most cases, servers are configured with a greater amount of memory than a workstation; server applications are not loaded and unloaded as often on a server as they are on a workstation.

CRITICAL DIFFERENCE

One item which is a major separation in the Windows NT Server is that access to important system resources is fairly and dynamically singled out. Only a few worker tasks (threads) service the queue of incoming user requests with dynamic load balancing throughout all computers. In addition, increased access performance is provided to protect virtual memory space and network I/O. However, in contrast, the Windows NT Workstation does not have the same amount of dedicated resources to handle a significant file server load. Instead, any incoming network requests are sent through one queue.

Windows NT Server's file cache is the number-one memory priority, as it increases network performance. When compared with Windows NT Workstation, the "any" processes in the users' top level is given the number-one memory priority to achieve the best responsiveness from the system. It should be noted that optimization can affect the performance of each operating system. When there is an increase in the number of clients connecting to each system, the Windows NT Server can reserve a certain amount of the computer processing time (as much as the whole processing power) for network file sharing, because Windows NT Workstation keeps network processor time at a minimum to allow system responsiveness to remain at an acceptable level for user input and local operations.

PROCESSING POWER UTILIZATION

There are several advantages of keeping Windows NT Workstation processing power at a minimum. In terms of local interactive performance, as the number of clients increases, the Windows NT Workstation performance can remain high for local applications. However, as processing power is exhausted by networking operations, the performance can decrease as several users log on to the network. Windows NT Server 4.0 can provide a high level of network performance as the number of clients increases when you allow processor utilization to scale with the number of connecting clients. As the number of clients connecting to Windows NT Server increases, the client delay remains low until more Workstation clients connect to the Windows NT Server.

In terms of performance optimization, Windows NT Workstation should yield high user responsiveness and consume low memory. However, Windows NT Server 4.0 is designed to provide fast network throughput and a faster I/O. User responsiveness always takes priority over network sharing performance in the Windows NT Workstation. However, Windows NT Server will always give priority over user input for file server performance graphic redraw capabilities.

ce that both the Workstation and Server components of onsistently provide a consistent kernel architecture, user API. Both tuning and optimization are tailored appropriately ry user. Windows NT Workstation performance advantages ly powered networked operating system. In terms of multi- abilities, Windows NT Server 4.0 is the best answer for your networking needs.

ieve the best intranet/Internet capabilities for your organization when you combine the Windows NT Server, Windows NT Workstation, and (or) the Windows 95 client running Microsoft's Internet Explorer. These solutions provide you with an excellent means for storing and retrieving corporate information throughout your company infrastructure for a variety of projects. Certain information can be carefully tucked away on your private network and kept for internal use only, while you can use the Internet portion of your Windows NT Server product to provide your customers with full information 24 hours a day on any of your products or services. In fact, the trend has been turning for many companies. Some organizations are turning their Internet sites into an order entry facility to maintain a low-cost solution to achieve sales and earn revenue 24 hours a day without having to pay for a 24-hour staff order line.

FUNCTIONAL DIFFERENCES

There are several functional differences. Windows NT Server and Windows NT Workstation 4.0 are two very different products designed to complement each other and meet some very specific needs of your company. In terms of licensing, Windows NT Server supports 5, 10, or more concurrent users with more users requiring additional licensing. In comparison, Windows NT Workstation can support a maximum of 10 computers connected at any given time. This includes sharing such as file and print, and peer Web services.

Each system was designed with very separate goals in mind. Windows NT Workstation was primarily set up to be responsive to the local user yet consume very little memory. In contrast, Windows NT Server 4.0 was designed to make network performance its number-one priority. It is set up to utilize all of its available memory and processing power to provide the fastest network file access possible.

In terms of similarities, both the Workstation and Server components of Windows NT have the same Windows 95 interface. This is a very important feature as users can either migrate from the Windows 95 environment or switch from the Workstation to the Server without having to relearn a whole new set of commands. Furthermore, the Windows 95 interface allows you to execute operations in a simple and straightforward GUI. In addition, both Windows NT Workstation and Server have the Win32 API as well.

In terms of memory considerations, the Workstation and Server components are different. Windows NT Workstation requires a minimum of 12 MB RAM, but recommends at least 16 MB of RAM or more. In comparison, the Windows NT Server requires a minimum of 16 MB RAM but recommends at least 32 MB of RAM or more. Hard disk space requirements also differ by at least 50 MB. Windows NT Workstation requires at least 110 MB and Windows NT Server requires at least 160 MB. The number of processors each supports also differs greatly. Windows NT Workstation can support 2 processors, while Windows NT Server can support up to 32 processors.

While Windows NT Workstation does not support fault tolerance, Windows NT Server supports Mirroring, Duplexing, and RAID 5. The number of inbound dial-in connections which the Workstation can support is only one, whereas the Server components can support up to 256 connections. Windows NT Workstation can support peer file and printer servers, while Windows NT Server can also support them, but requires client access licenses.

In terms of servers there is a big discrepancy between each component. Windows NT Workstation cannot support DHCP, DNS, WINS, and Index servers. However, Windows NT Server can support all of them. The Windows NT Workstation also cannot support Web page authoring. However, the server component can by utilizing the included application Microsoft FrontPage.

There are also several services which Windows NT Server can run, but Windows NT Workstation cannot. These include: services for the Macintosh, file and print services for NetWare, Directory Services Manager for NetWare, as well as the ability to run Microsoft BackOffice and BackOffice Server Logo Applications.

Administrative control also differs significantly for each Windows NT component. Windows NT Workstation offers local administrative control and restricted remote administration. However, Windows NT Server offers centralized administrative control across all desktops and servers. In terms of performance, Windows NT Workstation gains the highest advantage through applications in the user's foreground, and gives a minimum amount of memory at startup. In addition, the scheduler uses small time slices to achieve the best user response. In contrast, the Windows NT Server gains performance through network services, which is given the highest system priority. The file cache is preserved higher than all of the other services, while applications are given the most amount of requested memory at startup. Furthermore, the scheduler uses bigger time slices to respond to network requests.

FOCUS: WINDOWS NT SERVER 4.0

Network managers and system administrators will find that Windows NT Server offers much more than a new user interface and end user features. The server component itself functions on a 486 or better and can exist quite well in the memory and disk settings described as its minimum requirements in the preceding section of this chapter. However, both memory and disk considerations will generally increase as you support more users on your overall network.

In terms of functionality, Windows NT Server has all of the main features that Windows NT Server 3.51 had. This includes native support for the NetBEUI protocol, NetWare compatible IPX, and TCP/IP Internet standard as well. In addition to the file sharing, printer sharing, and NetWare gateway support and functions being standard on NT Server, you will obtain remote access support for up to 256 concurrent users. However, the number of users on your system is dependent on having enough hardware and licenses available to accommodate all of the concurrent users accessing your network resources at any given time. Functions also extend to multidomain administration and directory replication. In addition, Windows NT Server 4.0 can also support IP address allocation utilizing DHCP Internet name assignment which uses Windows 95, 3.x, Windows NT Workstation, and Windows Internet Name Service (WINS), DOS, OS/2, and the Macintosh Operating System.

NEW INTERFACE

At first glance, you may think you are staring at Windows 95, but you are not. You gain several new features with this new user interface — not the least of which is better Internet support, improved administration tools, and several performance enhancements outlined above. In comparison, both Windows NT Server and Workstation 4.0 do have the same look and feel as the Windows 95 operating system. The advent of one unified user interface across several operating system platforms will provide a simple and direct means to working on multiple platforms. Administrators gain an advantage as they must also work in the same type of operating system environment as the user.

However, this new user interface does not necessarily change the way an administrator might work on NT Server or the way a user works with NT Workstation. The server itself still retains several key administration tools, including the User Manager for Domains, Server Manager, Disk Administrator, Event Viewer, Performance Monitor, DHCP Manager, WINS Manager, Network Client Administrator, License Manager, and Migration Tool for NetWare. Even the Remote Access Administrator is still an integral part of the Windows NT. However, it now resides in the Administrative Tools menu.

The system policy editor is a new component that is compatible with both Windows NT and Windows 95. It actually replaces former User Profile Editors from previous Windows NT Server versions. Specifically, Windows NT Server 4.0 has new tools, which include an Administrative Wizard, System Policy Editor, a new set of Windows NT Diagnostic Tools, and a Network Monitor.

The biggest change is represented in the Administrative Wizards, which offers simple guide procedures from the time to you start up your Windows NT Server. It also allows you to add user accounts, manage administrative groups, control filer and folder access, add print drivers, add/remove programs, install modems, establish network client installation disk sets, and manage license compliance.

INTERNET

One of the best enhancements in the Windows NT Server 4.0 includes its Internet features and the Internet Information Server (IIS), which is a Web, ftp, and gopher server. In addition, the Windows NT Server also has a point-to-point tunneling protocol (PPTP) as well as security settings for TCP/UP. This NT Server is set apart as a server which can adequately satisfy the Internet or intranet demands of your organization.

Administrators will also find Internet features such as a UNIX domain name service (DNS) integrated into the operating system. When a working

DNS host is available, an administrator is allowed to simply type in a DNS name or IP address to be recognized as a valid name. In order to deal with DNS support of only static addressing, Windows NT has a combination of DNS support with its own WINS, which yields a true dynamic DNS. Whenever the WINS client tries to resolve a name, it primarily looks at the WINS database and only then looks at DNS. Therefore, either a DNS or WINS name will work. In a DNS client, either one will actually show up as a DNS name.

Windows NT Server 4.0 also has a Web administration tool as part of the functionality of the Windows NT administrative tools. In terms of security, you can use a Web browser which supports the ability to log in directly to Windows NT or one which supports secure sockets layer (SSL).

LICENSING

Windows NT Server 4.0 also supports two very distinct forms of licensing. The first is called per-user licensing, which requires that an access license be issued for each user who is going to connect to the network. However, that license allows connection to several systems. The second is called concurrent connection licensing. This option licenses a server for a specified number of users.

CONCLUSION

This chapter has pointed out differences in the Windows NT Workstation and Server components. However, one function which is meant for both the Windows NT Workstation and Server will allow network managers and administrators to utilize setup tools to permit a Windows NT system to be totally preconfigured. In terms of performance, Windows NT will have an improved network transport interface which will increase performance on a fast Ethernet. In addition, a new print spooler will provide a better client response by off-loading page rendering to the server.

When looking at the integration of the Windows NT Server and Workstation 4.0 components, you will gain a combination of performance improvements, new features, and a more enhanced user interface. As a previous Windows NT user, you will no doubt find the functional improvements to be well worth the time and expense in upgrading. This new version is a superior upgrade in many ways. Both of these components are separated into distinct parts to provide your network administrator and users with the ability to have your network grow to be a mainstream system for high-end users. This operating system will no doubt take your organization and carry you into a global status.

Chapter 24

Evaluating Client/Server Operating Systems: Focus on Windows NT

Gilbert Held

Although the market penetration of Microsoft Corporation's Windows Microsoft New Technology operating system at the client/server level only recently exceeded 10 percent, actual development of this client/server operating system began in 1988. Windows NT was primarily developed by a team of former Digital Equipment Corp. employees originally responsible for the development of DEC's VAX operating system, which explains why people familiar with VAX note many similarities between the two operating systems.

Windows NT version 4.0 represents the current version of the system and sports the Windows 95 interface. Both the client, known as Windows NT Workstation, and the server, known as Windows NT Server, are marketed by Microsoft. The key difference between the two versions of Windows NT concerns scalability and network and file support.

Windows NT Server, as its name implies, was developed as a network-based server. The operating system is optimized to support network requests to the server, which include the access and transfer of information in the form of files.

In addition, because the server is the focal point of a client/server network, Windows NT Server is scalable and able to operate on a computer containing up to eight processors, with an increase to 12 planned for the new Windows 2000 version. An organization can therefore purchase a computer with a single processor that has a motherboard designed to support

0-8493-9823-1/00/$0.00+$.50

up to four processors and add processors as computing requirements increase.

In comparison, Windows NT Workstation was developed to favor workstation processing. This means that the workstation operating system does not have to be scalable to support additional network-related processing nor concerned with providing network services to users, which is a key function of Windows NT Server.

SYSTEM REQUIREMENTS

Both Windows NT Server and Windows NT Workstation currently operate on three hardware platforms: Intel X86-based processors, DEC AXP, and MIPS Reduced Instruction Set (Reduced Instruction Set Computer) computers.

Windows NT Workstation requires 12 MB of RAM when operating on Intel platforms and 16 MB for use on Digital Equipment Corporation and MIPS RISC-based computers. Although Windows NT Server also operates with a similar amount of RAM, the amount is sufficient only when the server supports less than 50 workstations. As workstation support requirements increase, RAM requirements can increase; the applications supported by the server as well as the number of workstations supported govern the amount of RAM required. Although Microsoft publishes a guide to the minimum amount of memory that should be installed for use with Windows NT Server, networks with 50 or more workstations should use servers with a minimum of 32 MB of RAM.

The introduction of Windows 2000 will considerably boost system requirements. When Windows 2000 was in early beta testing as Windows NT Version 5, suggested memory requirements were 64 MB for the workstation version and 128 MB for the server version. Although this author was able to operate the workstation beta on a 133-MHz Pentium and the server version on a 200-MHz Pentium, both ran sluggishly. As a minimum, a 300-MHz Pentium should be considered for operating this new version of the NT operating system.

SYSTEM FEATURES

Both Windows NT Workstation and Windows NT Server include a core set of features that differentiate them from other client/server operating systems, such as Novell's NetWare and UNIX-based systems. Those features include the scalability of the server operating system as well as security, performance monitoring, and network support enhancements. Examining these features provides IS and network managers with a basis for comparing and selecting a client/server operating system.

Scalability

As discussed previously, Windows NT Server is scalable, capable of running on a multiprocessor computer. In addition, because Windows NT operates on Intel, DEC Alpha, and MIPS machines, users can migrate their server from a relatively low-cost Intel single-processor platform to a multiprocessor higher performance DEC Alpha or MIPS platform while continuing to use NT.

NetWare is currently restricted to operating on Intel platforms and cannot provide the migration path associated with the use of Windows NT. Although UNIX operates on a variety of hardware platforms ranging from PCs to Sun Microsystems, Hewlett-Packard, and IBM high-performance workstations that function as servers, differences between the version of UNIX supported by each platform may inhibit organizations from migrating applications from one UNIX-based system to another. NT thus provides the only server-based operating system that is truly portable for applications migrated from one vendor platform to another.

Performance Monitoring

Both Windows NT Workstation and Windows NT Server include a performance-monitoring capability that is extremely valuable for determining when or if the current hardware platform should be changed. In addition to monitoring, for example, processor metrics, NT's Performance Monitor program also generates alarms and plots the performance of a large number of network-related metrics, such as session timeouts, retries, and number of open connections.

Under NetWare, performance monitoring is accomplished through the use of the MONITOR utility program, a passive tool that cannot be used to set alarms. Although several third-party products that perform in a manner similar to Windows NT in a NetWare environment are available, these products are not bundled with NetWare and require the expenditure of additional funds. Some versions of UNIX include a built-in performance monitoring capability, but most versions of it also depend on the use of third-party products. Windows NT thus provides a built-in performance monitoring capability that remains in place if the need to migrate to a different hardware platform arises.

Security

Users of Windows NT notice immediately that once the system is installed, unauthorized individuals cannot power the hardware off and back on and take control of the computer by inserting a disk into the A drive. Instead, NT has a User Manager that limits access to the computer to predefined

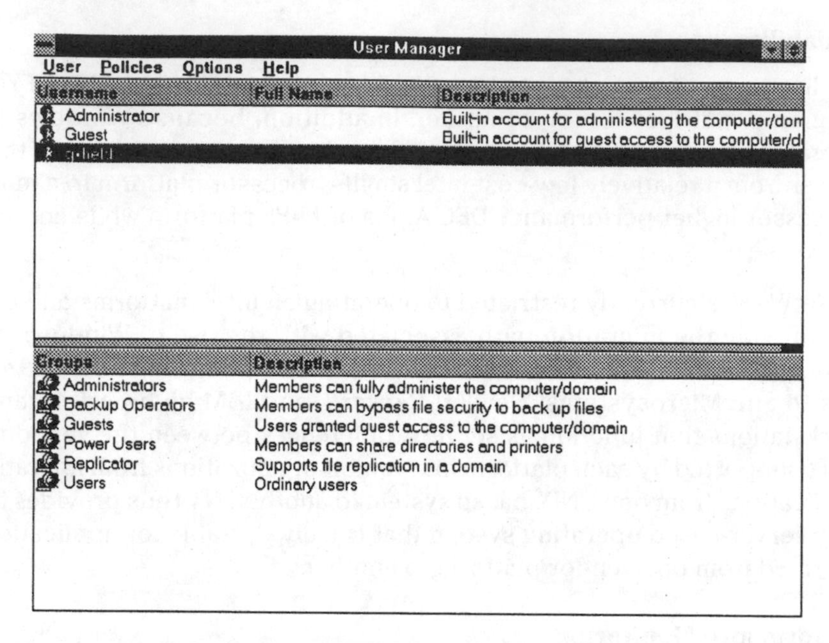

Exhibit 24.1 Windows NT user manager screen.

users, such as the administrator and any guest or individual accounts that permit employees to use a workstation or a server.

Exhibit 24.1 illustrates the Windows NT User Manager screen common to both server and workstation versions of the operating system. To illustrate the utility of the User Manager, Exhibit 24.2 shows the user properties it supports. Note that an authorized person can assign and change user passwords as well as control the password's expiration and, if so desired, disable an account.

NetWare is similar to Windows NT in its support of predefined groups, such as guests and administrators (called supervisor under NetWare). The key difference between NetWare and Windows NT lies in the ability of client workstations in the latter system to control access to the computer and in ease of use.

In the Group Memberships dialog box, for example, simply clicking on a button after highlighting an entry changes a user's group membership. Although NetWare has a similar feature, it requires slightly more manual intervention, which becomes significant if a server supports a large number of users. In the case of UNIX, each version of the system differs in its security and network administration, so the administration training effort increases if computer platforms are switched.

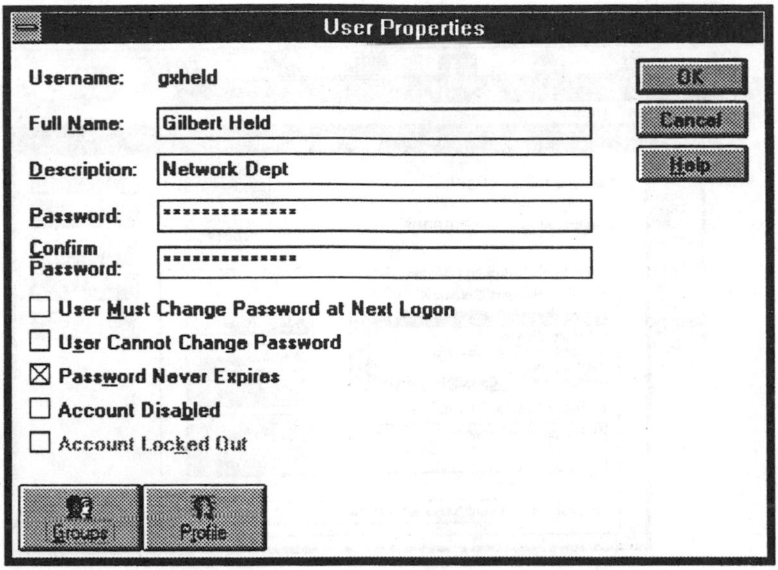

Exhibit 24.2 User properties supported by the Windows NT user manager.

Network Support

One of the most important features of any server is its ability to support organizational networking requirements. Windows NT truly excels in this area, because it supports a diverse and comprehensive range of network protocols ranging from Network Basic I/O System to AppleTalk, IBM's Data Link Control (Data Link Connections), NetWare's NWLink Internetwork Packet eXchange/Sequenced Packet eXchange, and TCP/IP (Transmission Control Protocol/Internet Protocol).

In addition to supporting a large variety of networking protocols, Windows NT includes direct support for more than 20 network adapter cards and indirect support for more than 100 additional adapters. Direct adapter support is in the form of drivers included in the operating system, and indirect support is in the form of vendor-supplied drivers that typically accompany the hardware on a disk with files that must be loaded by the operating system.

Exhibit 24.3 illustrates the Windows NT Network Settings dialog box. This dialog box is invoked from the icon labeled Network in the Windows NT control panel, a method similar to the one employed by users of Windows and Windows 95 to control network settings. Thus, in addition to supporting a wide range of network protocols, use of Windows NT may

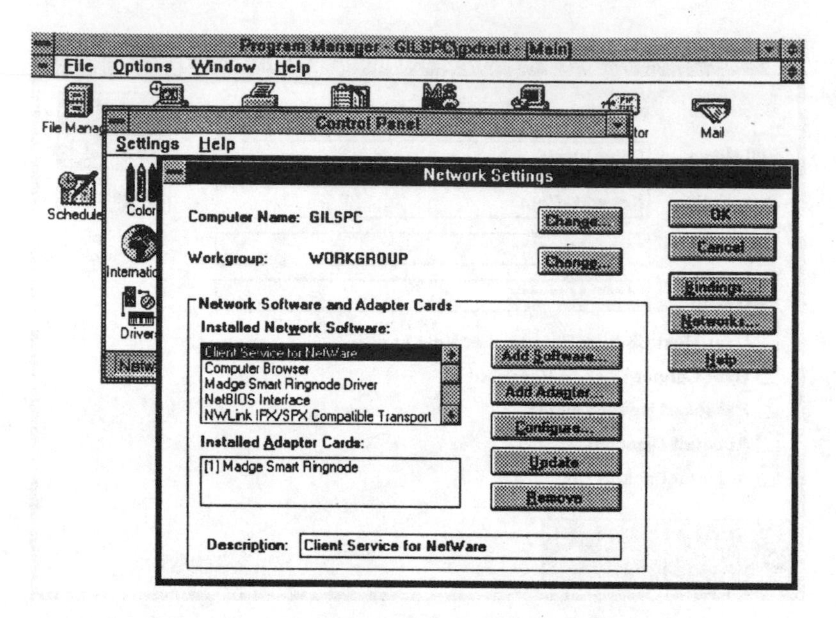

Exhibit 24.3 Windows NT network settings dialog box.

significantly reduce training and administrative costs for many organizations.

Although NetWare supports several network protocols, its main built-in support is limited to IPX/SPX, the native method of communications used in a NetWare environment. The use of other network protocols commonly requires the acquisition of drivers from the Novell bulletin board or user forum and the loading of those drivers on workstations and servers. Similarly, most UNIX systems are limited to a built-in support of TCP/IP, requiring other network protocols to be acquired and loaded onto workstations and servers.

Another advantage of Windows NT is its capability to support multiple network bindings, or the association of a network protocol to a network adapter or a network service to a network protocol. It is important to note that Windows NT supports more than ten simultaneous bindings, which makes a server most suitable for supporting NetWare, UNIX, AppleTalk, and other network protocols transported from different client or server platforms to a Windows NT server.

In addition, because Windows NT Workstation also supports the same bindings capability as Windows NT Server, a Windows NT Workstation can be used to obtain simultaneous access to NetWare servers and Windows NT servers. This is accomplished through use of the Frye Utilities for NetWare

Management program, a NetWare-based utility program that provides NetWare server performance metrics, operating under Windows NT. In fact, owing to Windows NT's extensive support of different network protocols and network adapter cards, it is quite common for users of Windows NT Workstation to employ that operating system to connect to multiple server platforms.

Advantages of Other Operating Systems

Although Windows NT Server includes features that make it the server of choice for many organizations, Novell's NetWare continues to excel in two key areas:

- Ability to support thousands of users
- Suitability of the directory structure for an enterprise network

Unlike Windows NT, which can only keep pace with NetWare 4.1 when supporting up to 100 to 200 users, NetWare supports several times that number of users without significant performance degradation. This means that large organizations may require multiple Windows NT servers in place of one large NetWare 4.1 server.

As for directory services, under NetWare 4.X, a hierarchical enterprise-wide structure can be set up that lets users on one network easily access resources on other segments. In comparison, Windows NT's Domain Services, although efficient in terms of set up and resource management, uses a replication scheme that becomes inefficient for large groups. For this reason, NetWare is probably the network of choice for large organizations with thousands of workstations.

The previously described situation will change with the introduction of Windows 2000. Windows 2000 includes a hierarchical directory service, known as Active Directory, which in many ways resembles Novell's Directory Service (NDS).

IMPLEMENTATION CONSIDERATIONS

Although Windows NT is relatively easy to install, careful planning is required to ensure that its rich set of options are appropriately configured and implemented.

Before installing the operating system, it is prudent to review the bundled products distributed with Windows NT, such as the built-in FTP (file transfer protocol) server capability included in Windows NT Server. Before simply clicking on the appropriate button to install the FTP software module, IS staff must determine whether to set up the module to permit anonymous access or access only to persons with accounts on the NT server. Similarly, with the introduction of Windows NT 4.0 Server, Microsoft added a bundled Web site software module that turns an NT server into a World Wide Web server. Using the software, however, requires the correct configuration of several software settings, such as IP address and address mask, so that the server can recognize requests directed to it.

Because of the number of variables that may require settings, an appropriate testing procedure is key to any implementation plan. A well-thought-out test plan should ensure that each service works according to the desired configuration of the operating system. For example, installation of FTP as an anonymous FTP server on a Windows NT server should be tested by accessing the server with an FTP client and verifying the configuration of the FTP server module.

In addition, if the server will be connected to the Internet as well as the organization's private network, the connection should be verified through both the internal (i.e., private) network connection and the Internet. This dual verification ensures that access through the Internet, resulting in data flowing through a router with filters set up as a firewall barrier or as a firewall, allows access to the FTP server. If it does not, IS staff then have time to request the necessary modifications to the router or firewall before moving to a production environment.

Another important consideration in an implementation plan involves global applications, such as calendaring or e-mail. Although available products vary greatly in terms of requirements, they share many common elements, including the need for a predefined naming structure. Instead of simply naming e-mail post offices as PO1, PO2, and PO3, for example, it is more effective and efficient for users and staff both in terms of meaningfulness and diagnostic testing to use site names such as ATLANTA, BOSTON, or MACON.

CONCLUSION

Windows NT Server is a robust, scalable operating system with substantial built-in security, performance-monitoring, and networking capabilities. IS managers in organizations whose future networking requirements are anticipated to grow should consider using the Windows NT Server operating system because it runs on different platforms and supports the use of multiple processors. This allows the organization to retain the use of the operating system as its networking requirements grow.

In addition, because Windows NT Server's security, performance-monitoring, and networking capabilities now exceed the features offered by NetWare and most UNIX systems, NT is a more robust operating system. Perhaps this explains why several trade journals predict that Windows NT will encompass more than 50 percent of the server market by 2000.

Although Windows NT offers several advantages over competitive network operating systems, it is important to remember that technology does not remain static. Novell's planned new version of NetWare may provide some of the key features now associated with Windows NT. In addition, NetWare continues to retain several advantages over Windows NT that make it more suitable for large networks; however, those advantages will probably disappear when Windows 2000 is released.

Because most IS managers and network administrators must make decisions based on released technology, it is important to note that Window's NT Workstation provides access to NetWare servers, and Windows NT Server can be used in a NetWare IPX/SPX networking environment. This means that the use of both Windows NT platforms coexists with NetWare as well as with TCP/IP UNIX-based systems. Thus, another key reason to consider the use of Windows NT is to be able to use one or more of the new applications being developed to run on an NT server without having to change an existing network.

Chapter 25
A Look At Linux

Daniel Carrere

Computer operating systems, since their inception, have suffered greatly from lack of portability (the ability to run an operating system on another platform/hardware architecture for which it was not originally designed) and closed operating system source code. Luckily, that suffering seems to be almost over, thanks to the development of an operating system based on open standards and open source code. Linux, an operating system conceived in the mind of a Finnish college student by the name of Linus Torvalds and contributed to by scores of programmers across the Internet, has effectively positioned itself to combat the once thought implacable foe of closed operating system standards and closed operating system source code. It does so by allowing anyone who has the need for the source code to be able to obtain it and modify it to their necessary specifications without restriction.

Linux, unlike many other operating systems today, is freely available. A person can install the operating system itself on an infinite number of computers without having to pay for per-server or per-client licenses. Projects which demand the deployment of multiple systems can realize significant savings. But the savings on licensing fees is somewhat minor in light of even more significant benefit of using Linux as the operating system of choice for systems deployment. As is well-known, server and client licenses are only one component in the Web of total cost of ownership of systems deployment. A much greater benefit of Linux is that your business will not be left out in the cold in the years to come due to the lack of source code availability or the stringent licensing restrictions imposed by some of the popular closed/proprietary systems in the market today. With a port of Linux available to virtually every computing platform known, a company should be able to deploy Linux on almost any system imaginable. With all the features of Linux and the support that it is receiving in the marketplace, Linux has a bright future in store.

Linux can be administered at the system console, over a telnet session, using the X Windowing system graphical user interface (GUI), or through a Web browser.

0-8493-9823-1/00/$0.00+$.50

Linux is used at companies such as Hewlett Packard who used Linux to develop their international, enterprise-wide distributed printing system. NASA has effectively used Linux in monitoring their experiments in space. The USPS uses Linux for bar-code reader recognition for purposes of routing mail. Linux aided the graphics design group Future Domain in rendering the imagery for the movie *Titanic*. These are just a few examples of Linux use. There are countless other examples and it is those other examples that will comprise the rest of this document.

A CASE FOR LINUX

In today's world of heterogeneous networking schemas, mixed protocol offerings, and the demand for high availability, many businesses are looking for systems that provide a "one size fits all" solution. Linux can effectively provide such a solution because of its ability to be host to a wide array of protocols and services. Linux can be used for Web, ftp, and e-mail, firewalling (ipfwadmin or ipchains), network file systems (NFS), compute servers, file (NFS and SMB) servers, and print servers (lpd and SMB).

Linux aids business in replacing some of its aging systems with new Linux systems without incurring significant or strenuous costs.

The standard Web server that ships with Linux is Apache, the predominant Web server according to NetCraft (http://www.netcraft.com) which studies Internet usage and the operating systems and Web server software used to provide such solutions.

LINUX TERMINOLOGY

The term Linux, in and of itself, refers to the kernel, or the core, of the computer operating system. Effectively, one could think of the kernel as being similar in function to the engine within a car. Since without the engine, the car effectively is rendered helpless to transport itself uphill or on flat land. Operating systems are not unlike motor vehicles in that they have a central engine called the kernel. A complete operating system involves a tool set that allows you to create, read, update, and delete files. Furthermore, a complete operating system also involves a user interface which can present itself in many a form. The two primary interfaces under Linux are the console's command line and the X Windowing System's graphical user interface (GUI). It is the kernel, the tool set, and some form of user interface that effectively form an operating system.

When one refers to a Linux *distribution,* that person is referring to the kernel and the tool set along with a user interface assembled by a particular organization or business. Differences among distributions include installation features and package (program archive) management. Some of the most popular Linux distributions include, but are not limited to, RedHat Linux, Caldera OpenLinux, Slackware Linux, Debian/GNU Linux, Linux PPC, Linux Pro, MKLinux, and S.u.S.E Linux. Most distributions are essentially the same except that some include value-added packages. RedHat is known for providing excellent systems administration and management tools. Additionally, RedHat appears to have the most straightforward and flawless system installation routine. It is not the purpose of this manuscript to be distribution specific. However, due to the differences between distributions (and their associated file system standard), the problems addressed and the locations of programs throughout this document specifically target Red Hat Linux (http://www.redhat.com), specifically version 5.2.

BRIEF LISTING OF MAJOR BUSINESS APPLICATIONS ON LINUX

As Linux gained popularity the number of business applications developed to run under the operating system increased. By late 1998 major industry software vendors such as Oracle, Informix, Sybase, DB2, Inter Base, and Pick relational database management systems became available as native Linux binaries. Others to join the Linux camp were Netscape (with its servers — FastTrack — and clients — Communicator, Navigator) and Corel (with its productivity applications, including WordPerfect).

ECONOMIZING WITHOUT SACRIFICE

Linux allows firms to economize without sacrificing by its license which is protected by the GNU Public License (GPL, see http://www.gnu.org/ for details). Linux costs nothing other than the time and facilities if the consumer has a fast connection to the Internet (i.e., T1) and is willing to spend time online to download the desired packages. Alternatively, one can purchase a distribution on CD-ROM from one of many distribution companies or CD-ROM vendors.

PROBLEMS ADDRESSED

The first step in designing your Internet server's service offerings is to determine what services you wish to offer your customers. Do they need Web service, e-mail, and file transfer protocol (FTP) capabilities? And what about the server messaging block (SMB) — the protocol that allows a computer to *appear* to be an NT server/Microsoft workgroup hosting a network share and network printers. Also, what about NFS, Network File System, which affords UNIX systems the ability to make file offerings to other hosts. If you aim to have wide dissemination of disparate information types through a heterogeneous mix of networking protocols, then Linux might just be the umbrella that you have been looking for.

Network File System (NFS)

> NFS is a file transfer protocol for heterogeneous networks, implemented in 1984 by Sun Microsystems Inc. Similar in function to the Defense Advanced Research Projects Agency File Transfer Protocol, which allows files to be transferred between systems for use. NFS allows computer systems to share files with other systems; a file remains on the remote system but appears to be activated, or mounted on, and to actually belong to the local system.
>
> *UNIX: An Open Systems Dictionary,* William H. Holt, Rockie J. Morgan

Creating an NFS share under Linux involves modifying the /etc/exports file, creating the mount point, and issuing the mount command on the NFS client (with appropriate options when using the command line), or just a few clicks (if using the RedHat Linux control panel or linuxconf, which will be discussed later, under the X Windowing System). The /etc/exports file defines which filesystem the remote system is offering (exporting) to its clients.

Three Steps to an NFS Mount

1. Login as the root user to accomplish this task.
2. Create the point at which you wish to locally mount the NFS volume via mkdir:

   /mnt/**directory_name**

   where **directory_name** stands for a filename of your choice.
3. Modify /etc/exports on the server (remote machine) as shown in Exhibit 25.1.

Samba: Bridging the Gap between UNIX and Windows

Samba is a suite of programs that allow a UNIX system to share its files with computers using the Microsoft networking protocol. By using the Microsoft networking protocol, Linux and other UNIX systems can service requests from clients of the DOS, Microsoft Windows, or IBM OS/2 persuasion.

Using Samba can allow users of a Microsoft workgroup to drag-and-drop or cut-and-paste their files into their home directory /home/username/public_html so that their files can be served by the Apache Web server. This means that users of a LAN are no longer required to use an FTP program to transfer their files to their home directory to provide their files to the web. This also means that Linux can share its printers with Microsoft clients and vice versa.

Below is a listing of documentation that you may find useful when working with Samba to connect Microsoft clients. The paths indicate the location of

```
# This is the first line of the /etc/exports file.

/mnt/cdrom me.sys.com(ro)

# The ro option in () stands for read only and is a good option to use
# with read only media as it avoids the generation of error messages
# from attempting to write to the media. The line above creates a
# publicly accessible volume that can be accessed by the client with
# the domain name of me.sys.com
# Note: "/mnt/cdrom" defines the point within the file system where a
# CD-ROM is mounted to share with its clients (as well as locally).

# The following command could mount a CD-ROM in the file system
# at /mnt/cdrom
# mount -t iso9660 /dev/cdrom /mnt/cdrom"
# The above line assumes that the mount point (file system directory)
exists.
# If you receive an error stating that the mount point doesn't exist,
you can make
# the directory with the following command:
# "mkdir /mnt/cdrom" given at the command line.

# This is the last line of the /etc/exports file.
```

Note: A # defines a comment (which means that programs ignore the text to the right). Comments serve as documentation. What better way to document something than within?

Exhibit 25.1 /etc/exports program.

---

the documentation within the Linux filesystem provided that the Samba package was installed.

- /usr/doc/samba-1.9.18p10/docs/NT4_PlainPassword.reg
- /usr/doc/samba-1.9.18p10/docs/Passwords.txt
- /usr/doc/samba-1.9.18p10/docs/Win95.txt
- /usr/doc/samba-1.9.18p10/docs/Win95_PlainPassword.reg
- /usr/doc/samba-1.9.18p10/docs/WinNT.txt

If you do not have these files, visit http://www.samba.org/to obtain and install Samba.

Exhibit 25.2 illustrates logging into a Microsoft networking session. Here we will be accessing a Windows workgroup (provided by Samba) entitled MYGROUP and a share name of Daniel's Share. Note: Daniel must log into Microsoft networking (on the Windows client machine) with the same username and password as on the Linux machine.

The login form will allow you to log into the Samba server and access the share and printers for which you are authorized by passing your login password from Microsoft networking to Samba.

Note: If, when you log into Microsoft networking and click on the Network Neighborhood icon, you see //**workgroup name**/IPC, the password

Exhibit 25.2 Authenticating yourself to microsoft networking to access Samba services. To access Samba, you must log in to Microsoft networking in much the same manner as you would for a Microsoft workgroup.

---

authentication schemes being used between Samba and Windows are not equivalent. For example, either Samba is using encrypted password authentication scheme and Windows is using plain text password authentication scheme, or Samba is using a plain text password authentication and Windows is using encrypted. By default, Samba uses plain text as does Windows. In order to log in, the two password schemes must be equivalent. In other words, either both client and server must encrypt passwords or both must use plain text passwords.

Enabling Encrypted Passwords

The Samba package to which I will be referring here is samba-1.9.18p10.

The default here is not to use encryption for password authentication. Hence, the line

```
encrypt passwords = no in/etc/smb.conf
```

To enable encrypted passwords, two things must be done:

1. Enable encrypted password support within the Samba configuration file so that when Samba reads it, Samba acts accordingly.
2. Compile Samba so that it can support encrypted passwords.

See /usr/doc/samba-1.9.18p10/docs/ENCRYPTION.txt for more information on encrypted password usage.

Enabling Plain Text Authentication on Windows 95/98

The updated file VRDRUPD.EXE in Windows 95 disables the use of a plain text authentication to the server. Plain text authentication can be enabled by using the registry editor (REGEDIT.EXE).

The registry key to edit with REGEDIT is

HKEY_LOCAL_MACHINE\System\Current\ControlSet\Services\VxD\VNETSUP\

To edit this key, select Edit -> New-> DWORD from the menu. Change the name of the DWORD value from New Value #1 to EnablePlainTextPassword. Set the value to 1.

Alternatively, you may obtain

/usr/doc/samba-1.9.18p10/docs/Win95_PlainPassword.reg

and run this file under Windows 95 to update the registry without going through the above-mentioned steps.

ENABLING PLAIN TEXT AUTHENTICATION ON WINDOWS NT

To update the Windows NT registry, you may edit it manually using REGEDIT32.EXE. The key that needs to be edited to enable plain text password authentication is

HKEY_LOCAL_MACHINE\System\Current\ControlSet\Services\rdr\parameters\

Steps:

1. Add the key EnablePlainTextPassword.
2. Set the value to 1.

Alternatively, you may obtain

/usr/doc/samba-1.9.18p10/docs/NT4_PlainPassword.reg

and run this file under Windows NT to update the registry without going through the above-mentioned steps. Note: To update the registry, you must be logged in as the administrator.

/etc/passwd Format

username:password:userid:groupid:home_directory:shell

Example:

daniel:x:500:500::/home/daniel:/bin/false

/bin/false is a program that literally does nothing. By having a shell that does nothing you block a given user from accessing their account via telnet or FTP utilites and keep some of the doors closed that may aid the unruly from accessing your system. A user with an entry such as in the example above will be allowed to use Samba, but not telnet and FTP utilites which require shell processing. Users will be able to cut and paste to their user-owned share and map the drive using the standard Microsoft Network Neighborhood options.

Mapping A Drive to Samba Under Windows 95/98

Double-click on the Network Neighborhood icon and find the computer and its associated shared volume and perform a single right-click on that share. Next, choose Map Network Drive.

Note: The client computer (a Windows 95/98/NT) machine must have the necessary protocols loaded, the client program for Microsoft Windows file and print sharing. See also Exhibit 25.3.

Additionally, the password given to Microsoft Windows must be identical to the user's password in /etc/passwd or /etc/shadow

LOOKING AT LINUX: A PICTORIAL PERSPECTIVE

In the following images, I am graphically showcasing some of the tools available under Linux. With each image I will provide commentary as to its significance and describe the provided functionality of the application(s) contained within the screenshot.

THE X WINDOWING SYSTEM: PROVIDING SUPPORT FOR GRAPHICAL APPLICATIONS

In Exhibit 25.4, you see a relatively busy desktop. To show you what a desktop under the X Windowing system can look like, I started X (with the startx command from the system console and then from X started a few applications to provide a point of reference.

APACHE: SERVING THE WORLD

What is Apache?

> Apache was originally based on code and ideas found in the most popular HTTP server of the time. NCSA httpd 1.3 (early 1995). It has since evolved into a far superior system which can rival (and probably surpass) almost any other UNIX based HTTP server in terms of functionality, efficiency and speed. Since it began, it has been completely rewritten, and includes many new features. Apache is, as of January 1997, the most popular WWW server on the Internet, according to the Netcraft (http://www.netcraft.com) Survey.
>
> — Apache Group (http://www.apache.org/docs/misc/FAQ.html#what)

```
# This is the first line of the /etc/smb.conf file.
# This is the main Samba configuration file. You should read the
# smb.conf(5) manual page in order to understand the options listed
# here. Samba has a huge number of configurable options (perhaps too
# many!) most of which are not shown in this example
#
# Any line which starts with a ; (semi-colon) or a # (hash)
# is a comment and is ignored. In this example we will use a #
# for commentary and a ; for parts of the config file that you
# may wish to enable
#
# NOTE: Whenever you modify this file you should run the command "testparm"
# to check that you have not many any basic syntactic errors.
#
#======================= Global Settings ==========================
[global]

# workgroup = NT-Domain-Name or Workgroup-Name
workgroup = MYGROUP

# server string is the equivalent of the NT Description field
server string = Samba Server

# This option is important for security. It allows you to restrict
# connections to machines which are on your local network. The
# following example restricts access to two C class networks and
# the "loopback" interface. For more examples of the syntax see
# the smb.conf man page
; hosts allow = 192.168.1. 192.168.2. 127.

# if you want to automatically load your printer list rather
# than setting them up individually then you'll need this
printcap name = /etc/printcap
load printers = yes

# this tells Samba to use a separate log file for each machine
# that connects
log file = /var/log/samba/log.%m

# Put a capping on the size of the log files (in Kb).
max log size = 50

# Security mode. Most people will want user level security. See
# security_level.txt for details.
security = user

# Most people will find that this option gives better performance.
# See speed.txt and the manual pages for details
socket options = TCP_NODELAY

# Browser Control Options:
# set local master to no if you don't want Samba to become a master
# browser on your network. Otherwise the normal election rules apply
; local master = no
# Domain Master specifies Samba to be the Domain Master Browser. This
# allows Samba to collate browse lists between subnets. Don't use this
# if you already have a Windows NT domain controller doing this job
; domain master = yes
```

Exhibit 25.3 Sharing a volume: the samba way.

```
# DNS Proxy - tells Samba whether or not to try to resolve NetBIOS
# names via DNS nslookups. The built-in default for versions 1.9.17 is
# yes, this has been changed in version 1.9.18 to no.
dns proxy = no

#============================ Share Definitions =====================
[homes]
comment = Home Directories
browseable = no
writable = yes

# Un-comment the following and create the netlogon directory for
# Domain Logons
; [netlogon]
; comment = Network Logon Service
; path = /home/netlogon
; guest ok = yes
; writable = no
; share modes = no


# NOTE: If you have a BSD-style print system there is no need to
# specifically define each individual printer
[printers]
comment = All Printers
path = /var/spool/samba
browseable = no
# Set public = yes to allow user 'guest account' to print
guest ok = no
writable = no
printable = yes

# This one is useful for people to share files
;[tmp]
; comment = Temporary file space
; path = /tmp
; read only = no
; public = yes

# A publicly accessible directory, but read only, except for people in
# the "staff" group
;[public]
; comment = Public Stuff
; path = /home/samba
; public = yes
; writable = yes
; printable = no
; write list = @staff
# Other examples.
#
# A private printer, usable only by fred. Spool data will be placed in
# fred's home directory. Note that fred must have write access to the
# spool directory, wherever it is.
;[fredsprn]
; comment = Fred's Printer
; valid users = fred
; path = /homes/fred
; printer = freds_printer
; public = no
```

Exhibit 25.3 (continued).

```
; writable = no
; printable = yes
# A private directory, usable only by fred. Note that fred requires
# write access to the directory.
;[fredsdir]
; comment = Fred's Service
; path = /usr/somewhere/private
; valid users = fred
; public = no
; writable = yes
; printable = no

# a service which has a different directory for each machine that
# connects this allows you to tailor configurations to incoming
# machines. You could also use the %u option to tailor it by user
# name. The %m gets replaced with the machine name that is connecting.
;[pchome]
; comment = PC Directories
; path = /usr/pc/%m
; public = no
; writable = yes

# A publicly accessible directory, read/write to all users. Note that
# all files created in the directory by users will be owned by the
# default user, so any user with access can delete any other user's
# files. Obviously this directory must be writable by the default
# user. Another user could of course be specified, in which case all
# files would be owned by that user instead.
;[public]
; path = /usr/somewhere/else/public
; public = yes
; only guest = yes
; writable = yes
; printable = no

# The following two entries demonstrate how to share a directory so
# that two users can place files there that will be owned by the
# specific users. In this setup, the directory should be writable by
# both users and should have the sticky bit set on it to prevent
# abuse. Obviously this could be extended to as many users as
# required.
;[myshare]
; comment = Mary's and Fred's stuff
; path = /usr/somewhere/shared
; valid users = mary fred
; public = no
; writable = yes
; printable = no
; create mask = 0765

[Daniel's Share]
comment = This a test
path = /home/daniel
valid users = daniel
public = no
writable = yes
printable = no
```

Exhibit 25.3 (continued).

```
# This is the last line of the /etc/smb.conf file.

Note: That in order to have printing available, you must have your
   printer configured. The file
that stores your printer configuration is /etc/printcap

# This is the first line of the /etc/printcap file.
#
# This printcap is being created with printtool v.3.27
# Any changes made here manually will be lost if printtool
# is run later on.
# The presence of this header means that no printcap
# existed when printtool was run.
#
# The entry below is for my Hewlett Packard LazerJet Printer
#
##PRINTTOOL3## LOCAL ljet4 300x300 letter {} LaserJet4 Default {}
lp:\
:sd=/var/spool/lpd/lp:\
:mx#0:\
:sh:\
:lp=/dev/lp1:\
:if=/var/spool/lpd/lp/filter:

# This is the last line of the /etc/printcap file.
```

Exhibit 25.3 (continued).

Apache has been the most popular Web server on the Internet since April of 1996. The January 1999 WWW server site survey by Netcraft found that over 53 percent of the Web sites on the Internet are using Apache (58 percent if Apache derivatives are included), thus making it more widely used than all other Web servers combined.

The Apache project is an effort to develop and maintain an open-source HTTP server for various modern desktop and server operating systems, such as UNIX and Windows NT. The goal of this project is to provide a secure, efficient, and extensible server which provides HTTP services in sync with the current HTTP standards.

Note: You may reach the graphical Apache configuration tool (as seen in Exhibits 25.5–25.8) while running the X Windowing System by typing comanche into a terminal command line.

In Exhibit 25.6, you see the main menu for Apache configuration. By default, Apache will answer on TCP port 80 for whatever host is in /etc/HOSTNAME . To find out what your hostname is, you can type echo $HOSTNAME at a command prompt.

Apache allows virtual hosting services. To set up a site that you wish to virtual host, you can add the site by clicking on Add. Visit the Apache Web site (http://www.apache.org/docs/vhosts/index.html) or(http://www.apache.org/) for more details on virtual host configuration.

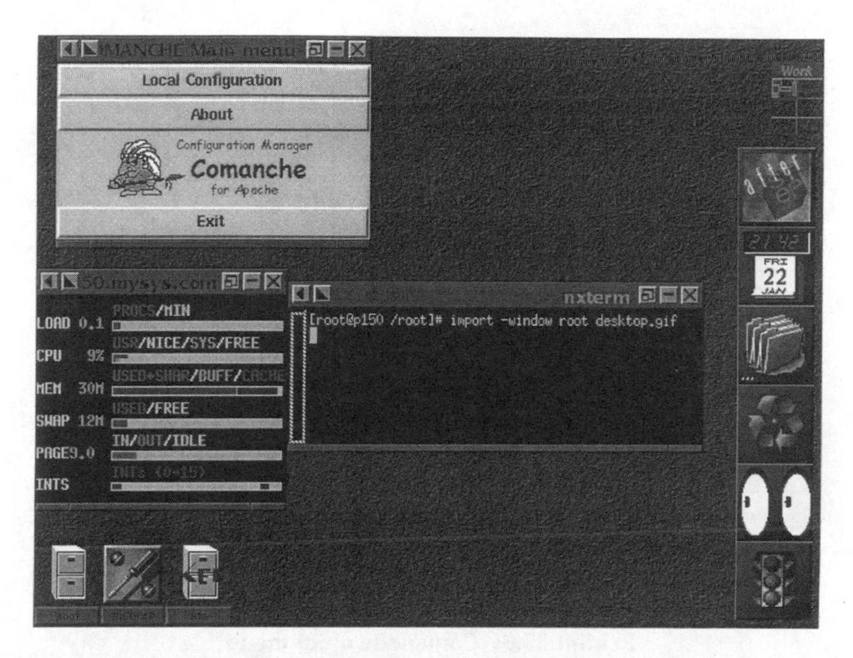

Exhibit 25.4 An example X Windowing system desktop with several applications running. Here we see xosview which displays system information, nxterm a terminal window, comanche — a graphical configuration utility for the Apache webserver, a minimized (iconified) version of the GNU Image Manipulation Program (gimp), and the X File Manager (xfm).

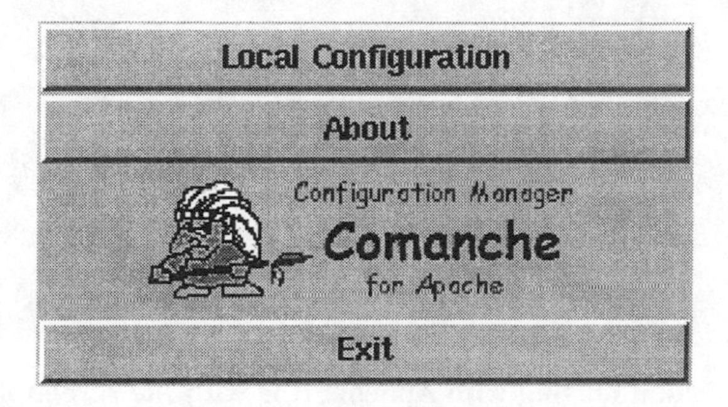

Exhibit 25.5 Comanche: Apache graphical configuration.

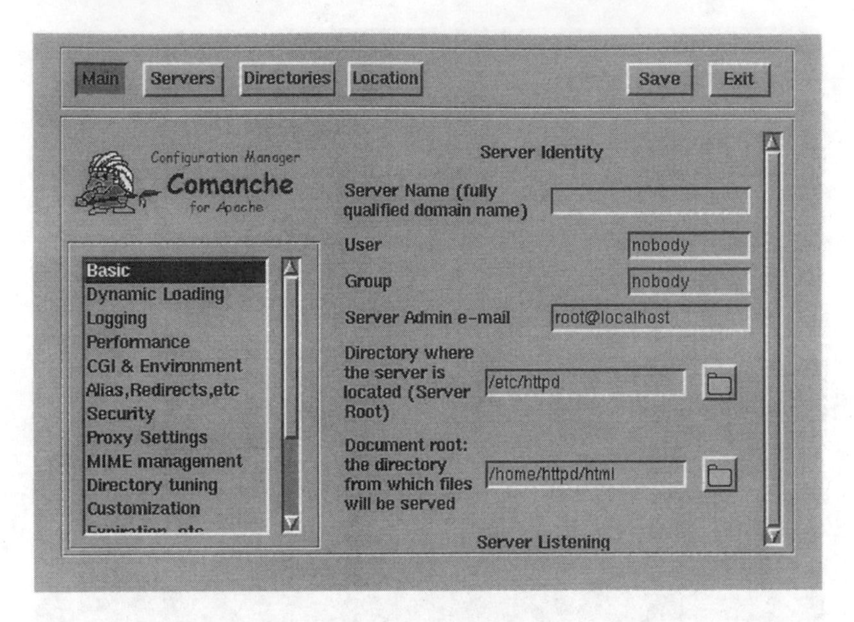

Exhibit 25.6 Comanche main menu.

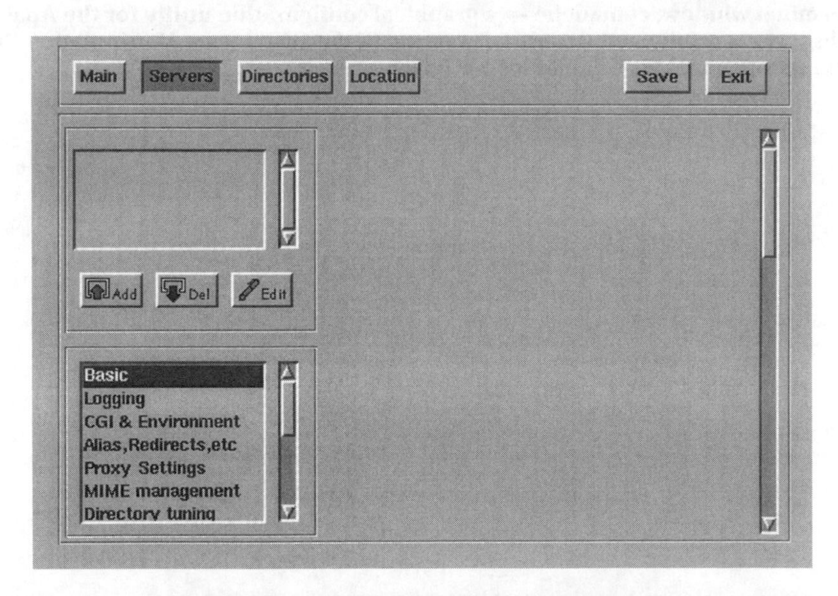

Exhibit 25.7 Virtual hosting with Apache. It is with the screen below that you can provide virtual hosting services. Please see (http://www.apache.org/docs/vhosts/index.html) or (http://www.apache.org/) for configuration as well as conceptual information.

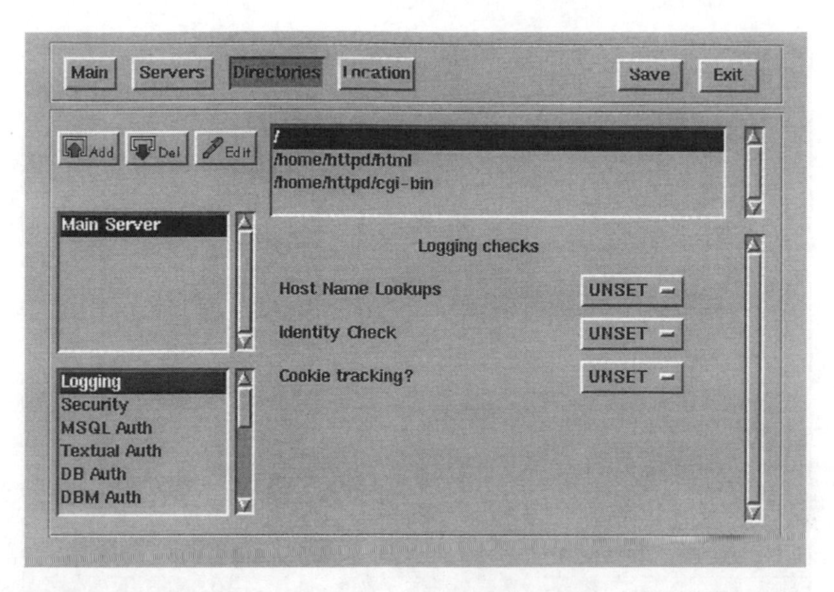

Exhibit 25.8 Apache directory specifications for host and virtual hosting configurations.

> Apache was one of the first servers to support IP-based virtual hosts right out of the box. Versions 1.1 and later of Apache support both, IP-based and name-based virtual hosts (vhosts). As the term IP-based indicates, the server must have a different IP address for each IP-based virtual host. The benefits of using the new name-based virtual host support is a practically unlimited number of servers, ease of configuration and use, and requires no additional hardware or software. The main disadvantage is that the client must support this part of the protocol. The latest versions of most browsers do, but there are still old browsers in use who do not.
>
> — Apache Group

As you will notice in Exhibit 25.8, the directory section allows you to specify the location of your HTML files that will comprise your homepages as well as those for virtual hosts (if you have added any). Unless you have a specific reason to change this, you may leave it as it is by default.

In Exhibit 25.9, you are shown a screenshot of linuxconf. To run linuxconf, you must be running the X Windowing System. To launch linuxconf, simply type linuxconf within a terminal window. Note: The predecessor to linuxconf is control-panel. The control-panel provides similar functionality and aids in the ease of system management tasks. The control-panel can be invoked by typing control-panel in a terminal window while running the X Windowing System.

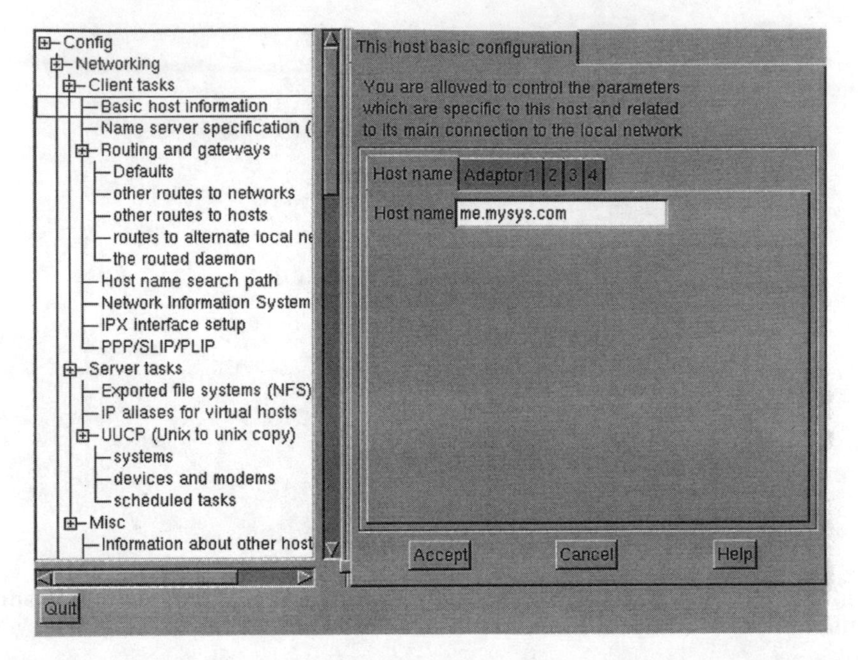

Exhibit 25.9 Linuxconf — systems administration in a singular tool. From within and without leaving linuxconf, you are afforded the ease of configuring network settings (hosts, name servers, routes, gateways, and interfaces), server tasks (NFS, IP aliases for virtual hosts, UUCP), user accounts, policies, file systems, and control panel all from a singular location.

FTP: Anonymous and Real

The standard FTP server that comes with a Linux distribution, specifically RedHat Linux, is Washington University's File Transfer Protocol Daemon (WU-FTPD).

Wuarchive-ftpd, more affectionately known as wu-ftpd, is a replacement ftp daemon for UN*X systems developed at Washington University (*.wustl.edu) by Bryan D. O'Connor. wu-ftpd is the one of the most popular ftp daemons on the Internet, used on many anonymous ftp sites around the world. Now that we know a bit about the freely available FTP service under Linux, let us look at accounts and configuration aspects.

FTP Configuration. The two FTP configuration aspects that I will touch on here deal with limiting access (either to only anonymous or only real account user connections). By default, both anonymous and real accounts (accounts that have an entry in /etc/passwd) are allowed access, provided they supply proper identification.

Anonymous Account FTP. Anonymous is an FTP account that allows an organization to offer its nonsensitive files to all who need them without requiring a specific account to be associated with a given user. For mass distribution of public files, anonymous FTP attempts and does a good job at providing a one-size-fits-all solution. Anonymous connections use the individual's e-mail address as the password. WU-FTPD can use RFC822 to verify an e-mail address's validity in the regard to form (i.e., someone@somehost.com) but cannot verify the address's existence or an individual's regard to ownership of the e-mail address.

Real Account FTP. Real account FTP means that a given user has a valid entry in /etc/passwd that allows login to the machine and unique identification. A *real* user, generally speaking, will have a home directory /home/**username** (where **username** corresponds to the entry for the user's login name in /etc/passwd). (See Exhibit 25.10.)

Making a Directory for Users to Upload Files. To provide a directory to which anonymous users can upload files, type mkdir/home/ftp/incoming at a command line. Then type chmod u=rwx,g = rwx,o=rwx/home/ftp/incoming or chmod a=rwx/home/ftp/incoming to make the directory readable, writable, and executable by all. Additionally, type chown root:ftp/home/ftp/incoming and chmod g+s /home/ftp/incoming. These commands (specifically the chown command) will keep users from deleting an uploaded file; the commands change the ownership of the directory (file) to root and the users are not the root user.

Additionally, you should confirm the owner of the directory, by typing ls –l/home/ftp. If the owner and group pair is anything other than root:ftp you should change the owner of the directory by using chown –R root:ftp/home/ftp/incoming.

CONCLUSION

This chapter was written to introduce a robust, quality, and cost-effective toolkit from which any organization can construct an Internet or network solution that successfully meets its desired objective. It was not my intention to cover every Linux topic as that could fill the shelves of a library. It was, however, my intention to introduce some of the technologies and services that Linux is capable of providing. Furthermore, it was also my goal to do so in a manner that provides enough of an introduction to configure the service and then guide the reader to further reading to extend the modeled configurations.

```
# *** This is the first line of /etc/ftpaccess . ***

classallreal,guest,anonymous *

# The line above specifies what types of users are
# allowed in. By default all users from each class
# are allowed in.

# Note: To limit your service to only anonymous
#       logins, you could change the line above to
#       "class all anonymous *"
#       Alternatively, you could limit your service
#       to only real users (those users with an
#       entry in /etc/passwd) by "class all real *."

email root@localhost

loginfails 5

readme   README*   login
readme   README*   cwd=*

message /welcome.msg  login
message .message      cwd=*

compress   yes  all
tar        yes  all
chmod      no   guest,anonymous
delete     no   guest,anonymous
overwrite  no   guest,anonymous
rename     no   guest,anonymous

log transfers anonymous,real inbound,outbound

shutdown /etc/shutmsg

passwd-check rfc822 warn

# The above line checks if the "password" (E-mail) address
# for an anonymous user conforms to standard for E-mail
# addresses and defined in Request For Comments 822 (RFC822).

# (i.e. someone@something.com).

# *** This is the last line of /etc/ftpaccess . ***
```

Exhibit 25.10 /etc/ftpaccess. * program.**

References

Internet Linux Resources:

- Linux Resources: http://www.linuxresources.com/
- Linux Organization: http://www.linux.org/
- Linux International: http://www.li.org/
- Linux Journal: http://www.linuxjournal.com/
- Linux Online UK: *Is Linux Year 2000 Compliant — Answer: Yes*: http://www.uk.linux.org/mbug.html
- Linux Journal: *Linux is Reading Your Mail USPS*: http://www.ssc.com/lj/issue52/2985.html

Apache Resources:

- Apache Organization: http://www.apache.org/

Linux and UNIX Books:

1. Blair, John D., *Samba: Integrating UNIX and Windows*, published by SSC, ISBN: 1-57831-006-7
2. Bentson, Randolph, *Inside Linux: A Look at Operating Systems Development*, published by SSC, ISBN: 0-916151-89-1
3. O'Reilly and Associates, *Linux in a Nutshell*, ISBN: 1-56592-167-4
4. Raymond, Eric S., *Linux Undercover: Linux Secrets as Revealed by the Linux Documentation Project,* published by Red Hat Software, ISBN: 188817205-3
5. Sery, Paul G., *LINUX: Network Toolkit,* published by IDG Books, ISBN: 0-7645-3146-8

Chapter 26

With Linux and Big Brother Watching over Your Network, You Don't Have to Look over Your Shoulder (or Your Budget)

Daniel Carrere

Many systems administrators in today's networking world find themselves attending to multiple systems. As such, these administrators want to automate monitoring and problem detection. To effectively monitor a system, one must consider many aspects, each of which is vital to system availability. The aspects that require a system monitor's attention are the states of the services being provided (DNS, NNTP, FTP, SMTP, HTTP, and POP3), the states of the server's hardware (disk space usage, CPU usage/utilization), and the states of core operating system aspects (essential system processes). In addition, administrators also wish to be able to determine the system uptime, as well as any warning or status messages. Most importantly they want to be able to accomplish all of these tasks without repetition.

One of the most effective ways to accomplish all these aims without repetitition and also to provide the results within a singular interface for ease of analysis is with the synergy created by using Linux and the Big Brother UNIX Network monitor. Furthermore, by using a combination of Linux and Big Brother administrators are afforded the ability to view the status of systems and their associated processes without having to use specialized software to view the results since the results are formatted into

0-8493-9823-1/00/$0.00+$.50

an HTML document which can be served by the Web server. As a benefit of having a TCP/IP connection available to the monitoring server, administrators can monitor their systems from anywhere in the world provided the machine serving the results is using a routable network layer address and is not blocked via a firewall. Last but not least, using a monitoring system comprised of Linux and Big Brother involves no additional costs for software (neither system nor monitoring).

COMPONENTS OF A LINUX AND BIG BROTHER UNIX NETWORK MONITOR

There are two core components of a Linux and Big Brother network monitoring solution: (1) the Linux operating system and (2) the Big Brother Networking monitor. Each of these two core components are comprised of numerous components, but this article will discuss only the monitoring components and how they relate to, integrate into, and operate within the Linux operating system.

Big Brother: The Five Core Parts

Big Brother is composed of five core components. Those five core components are the central monitoring station (also called the display server), the network monitor, the local system monitor, pager-programs, and intramachine communications programs. Additionally, these five components involve two key programs: bb, the client that runs on the machines being monitored, and bbd, the server program (daemon) running on the central monitor/display server.

Component One: The Central Monitoring Station (Display Server). The central monitoring system/display server accepts the system status reports from the systems being monitored. Through the generation of HTML results, the reports generated can be viewed on virtually any computing platform available today. The format of the results has the ability to be customized by simply modifying one of the Bourne shell scripts.

Component Two: The Network Monitor. The network monitor of Big Brother operates using ICMP echo requests (pings). The network monitor contacts each host system listed in its host file. It runs on any UNIX machine and periodically contacts every element listed in the **directory_path_chosen_for_installation**/bb/etc/bb-hosts file via ping. The results are sent to the central monitor to update the system status.

Component Three: Local System Monitor. It is the local system monitor's duty to keep a check on the disk utilization, CPU utilization, and system processes. After determining the status of these system aspects, the central monitor is updated. In the event of problems, the local system monitor has the ability to contact the system administrator.

Component Four: Pager Programs. The Big Brother client program resides on the system being monitored. It sends the monitoring information to the display server which forwards the information, using the Kermit modem protocol, to the administrator's pager.

Component Five: Intra-Machine Communications Programs. The Big Brother client program sends its status information to the specified display and pager servers to TCP port 1984 (this port number was chosen by Big Brother's creator, Sean MacGuire, in reference to George Orwell's book, *1984*.

UNDERSTANDING LINUX AND BIG BROTHER UNIX NETWORK MONITORING

The workings of the Big Brother network monitor are such that there are two core aspects that fit well with the traditional understanding of client/server computing. By this, I mean that the monitored stations function as servers (serving information to the display server after obtaining system information using their location client applications), serving their system and processes status to a central monitor which functions as a client. The central monitor polls the servers to obtain status information in a manner similar to any client/server interactive query session. The central monitor, the machine collecting the information about the other hosts, acts as a client. Once the central monitor obtains the information, the Web server running on the central monitor serves the statuses, represented by colored spheres, to requesting Web clients in the form of an HTML document so administrators can determine the status of their network at a glance.

ECONOMIZING WITHOUT SACRIFICE

When you combine Linux and Big Brother, you gain in several areas. The most important benefit is that you have stability afforded by the Linux operating system. When you are monitoring machines, the last thing that you want is to have your monitor machine fail. When monitoring using the Linux operating system, you gain stability at least in light of the operating system employed. When combined with high quality hardware, you have a winning solution. To further ensure that you have reliable monitoring, you can have redundant display servers. This can easily be accomplished through the simplicity of Big Brother's structure as a group of Bourne shell scripts that can be modified by a text editor (vi, emacs, pico, etc.).

DETAILS OF THE SERVICES THAT CAN BE MONITORED

Big Brother can monitor your connection, CPU utilization, disk utilization, DNS availability, HTTP (HyperText Transfer Protocol) service availability, IMAP (Internet Message Access Protocol) service availability, MRTG (Multirouter Traffic Grapher) service availability, msgs, POP3 (Post Office

Protocol 3) service availability, processes (specified system processes), SMTP (Simple Mail Transport Protocol) service availability, SSH (secure shell) service availability, and Telnet service availability.

Many of these services are monitored via an ICMP echo request ping command to determine if the system is reachable. Although the nonreturn of a ping to a given host is not indicative of a host failure, it does alert the system administrator that the matter needs to be investigated in order to determine if there are problems along the transmission line that provides connectivity to the machine. Additionally, in the event of an unsuccessful ping and verification that the transmission line is functioning properly the system administrator should then check the cabling to the machine as well as the network interface card(s) into the machine to determine the source of the problem.

Services Offered by Linux Monitored by Big Brother

- DNS
- HTTP (HyperText Transfer Protocol)
- IMAP (Internet Message Access Protocol)
- MRTG (Multi Router Traffic Grapher)
- msgs
- POP3 (Post Office Protocol 3)
- procs (specified system processes)
- SMTP (Simple Mail Transfer Protocol)
- SSH (Secure shell)
- FTP
- NNTP (Network News Transport Protocol)
- Telnet

DNS availability is accessed via the use of a nameserver lookup.

HTTP server process/daemon availability is accomplished by using a session of lynx (a text-based Web browser) to check for a valid HTTP response as well as output.

TCP port 80 is the port queried on the host unless an alternate port is specified in the /etc/services file.

IMAP availability verification is accomplished by querying the server on TCP port 143 unless an alternate port is specified in the /etc/services file.

POP3 availability verification is accomplished by querying the server on TCP port 110 on the host unless an alternate port is specified in the /etc/services file.

Procs (processes)

The means by which Big Brother determines whether critical system processes are running is through a Bourne shell script located in the directory in which you installed Big Brother. The specific script location is **/directory_path_chosen_for_installation**/bb/etc/bbdef.sh. In the event that a system process defined as critical within this script is no longer running on the system (as determined by issuing the ps command) Big Brother can use its paging facilities to contact the administrator.

SMTP (Simple Mail Transfer Protocol)

The standard TCP port (port 25) that sendmail runs on is examined. Note: Sendmail is a mail transfer agent (MTA) that is used to route messages from one system to another. Sendmail was developed at the University of California at Berkley and is the dominant mail transfer agent on UNIX systems.

The connection. The connection is verified via the ping command. The script that inquires about this aspect of system availability is

/directory_path_chosen_for_installation/bb/etc/bb-network.sh

NNTP (Network News Transport Protocol). By default, the TCP port examined for NNTP is TCP port 119.

FTP Service. The standard FTP port (TCP port 21, unless an alternate port is specified in the /etc/services file) is contacted to see if a connection can be established.

Telnet Service. For telnet availability status, Big Brother contacts TCP port 23 (unless an alternate port is specified in the /etc/services file).

HOW BIG BROTHER MONITORS LINUX'S HARDWARE UTILIZATION

Big Brother monitors Linux's hardware utilization via the standard UNIX commands available on a Linux system. These commands are df (to determine the percentage of disk space used) and uptime (which displays the time of day, duration of uptime, number of users, and the process load average — which is indicative of CPU utilization). The Big Brother scripts process the output of these commands and warn the system administrator of problems based on thresholds. For example, the threshold for disk utilization is 95 percent capacity.

As we can see within the output of the command displayed in Exhibit 26.1, the utilization on the/usr partition is at 74 percent. The Big Brother client program, running on the station being monitored, would send a message to the Big Brother server process, running on the display

Exhibit 26.1 Output of df command.

Filesystem	1024-blocks	Used	Available	Capacity	Mounted on
/dev/hda2	249855	24153	212800	10%	/
/dev/hda5	398124	2551	375012	1%	/home
/dev/hda1	1435168	793760	641408	55%	/mnt/win
/dev/hda9	50717	1883	46215	4%	/tmp
/dev/hda6	1193895	835034	297173	74%	/usr
/dev/hda10	117087	32	111009	0%	/usr/local
/dev/hda8	23391	13	22170	0%	/usr/misc
/dev/hda7	101471	23550	72681	24%	/var
/dev/fd0	1423	299	1124	21%	/mnt/floppy

server, that would indicate that the disks are in good condition from a standpoint of utilization.

Another Big Brother script would run the uptime command and use its output to assess the processor's utilization. CPU utilization is assessed by the load average returned from the uptime command:

7:55pm up 52 min, 2 users, load average: 0.22, 0.06, 0.02

BIG BROTHER: THE USER INTERFACE

The interface to any system is crucial. Big Brother's interface allows the many aspects under the monitoring software's watchful eye to be collected and displayed from a central location that requires no operating system-specific specialized software. All that one needs to effectively interact with the Big Brother monitor display is a Web browser. The aspect of the system generating reports in HTML thus allows one to monitor the performance of their systems from anywhere in the world via a consistent user interface across all computing platforms.

In Exhibit 26.2, you will see the top level interface to Big Brother's monitoring results. When the monitor is running on a group of specified hosts, the URLs or IP addresses for each host will be displayed horizontally and have an entry for each row. We see the various system aspects that are under Big Brother's supervision. In Exhibit 26.2, we see the titles: conn, cpu, disk, dns, ftp, http, imap, mrtg, msgs, pop3, procs, smtp, ssh, and telnet, which represent the various services being monitored by Big Brother. For example,

	conn	cpu	disk	dns	ftp	http	imap	mrtg	msgs	pop3	procs	smtp	ssh	telnet
www.someurl.com	[]	[]	[]	[]	[]	[]	[]	[]	[]	[]	[]	[]	[]	[]
or														
10.10.10.1	[]	[]	[]	[]	[]	[]	[]	[]	[]	[]	[]	[]	[]	[]

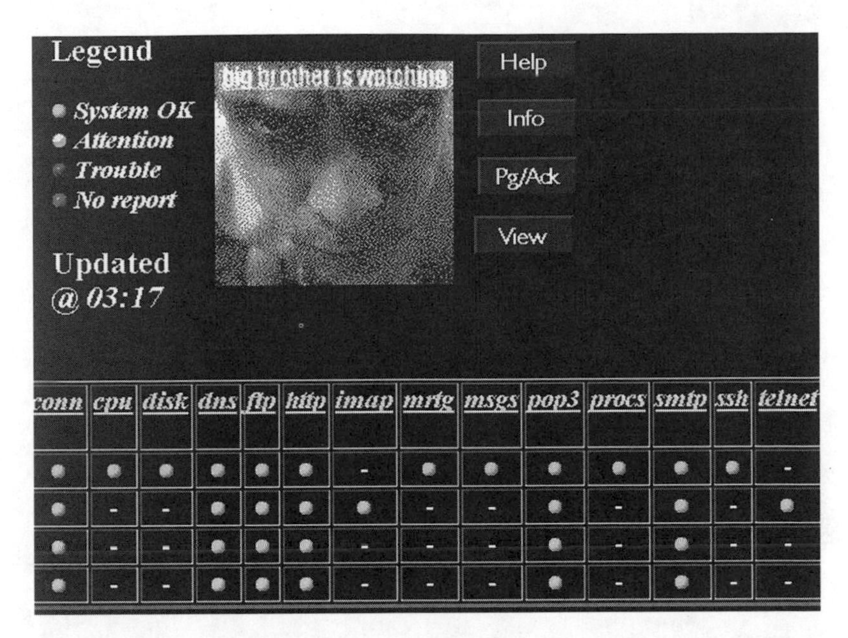

Exhibit 26.2 Top level interface to Big Brother's monitoring results.

Drilling Down For Detail

The Big Brother systems and network monitor provides drill-down facilities in that it allows one to click on one of the colored spheres in a column to display details about the system service or system aspect in question.

Below, in Exhibit 26.3, you see the detail portion of the http service by clicking on one of the green spheres above under the http column in Exhibit 26.2. Shown below is the status of one of our Web servers as well as details about its uptime.

AUTOMATICALLY NOTIFYING THE SYSTEM ADMINISTRATOR IN THE EVENT OF A PROBLEM VIA PAGER, E-MAIL, OR BOTH

In the event of a system problem, Big Brother has the ability to notify the system administrator via pager by utilizing the Kermit protocol. Below is a screen that displays a paging function such that system users can manually page the system administrator in the event of a problem. Alternatively, one is able to send the administrator an acknowledgment of system status.

Below, in Exhibit 26.4, we see the form that users can complete in order to manually page the administrator or send the administrator and acknowledgment.

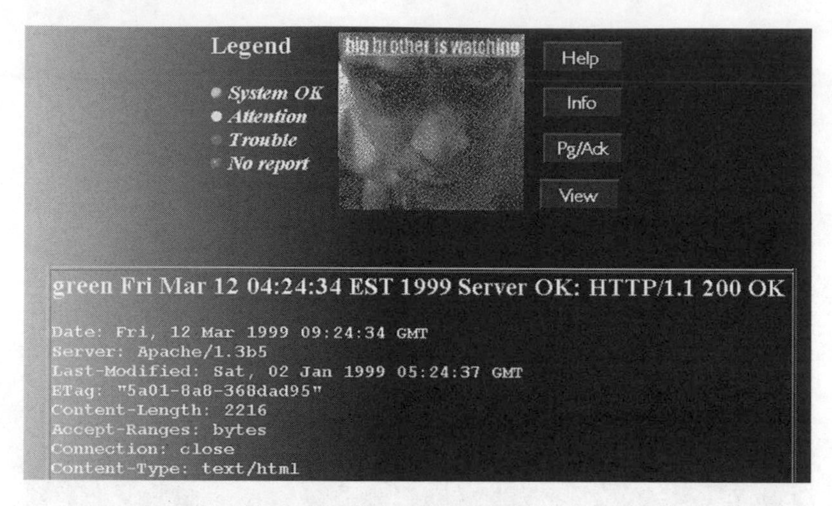

Exhibit 26.3 The details of the HTTP service and its status.

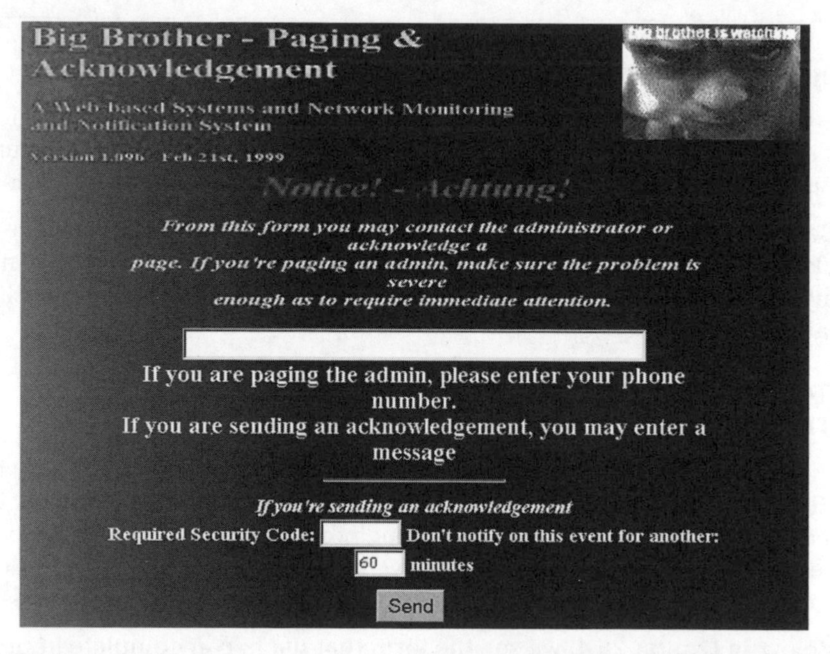

Exhibit 26.4 The interface to the manual paging facilities of Big Brother. Users can submit a page to alert the administrator or send acknowledgment to the system administrator.

The Paging Facilities: Codifying System Statuses

The pager facilities of Big Brother provide a means for the administrator to be able to determine system status from a three-digit code. The three-digit code comes before the IP address of the host in sequence (in the page) so the administrator first notices the system status and then if the status demands immediate attention based on its relation to system availability; the system administrator then knows the network address/IP address of the machine requiring attention.

Pager Codes. In the event of a problem with one of the systems being monitored, the administrator can either be paged automatically by the system or manually by a user. The format used to send the administrator a page is [3-digit code] [IP address]:

[3 DIGIT CODE]

100	Disk Error. Disk is over 95% full...
200	CPU Error. CPU load average is unacceptably high.
300	Process Error. An important process has died.
400	Message file contains a serious error.
500	Network error, can't connect to that IP address.
600	Web server HTTP error — server is down.
7–	Generic server error — 7 + server port number i.e., 721 = ftp down
800	DNS server on that machine is down
911	User Page. Message is phone number to call back.
999	The host reporting an error could not be found in the etc/bb-hosts file.

CONCLUSION

The intended purpose of this article was to introduce the reader to the Big Brother system and network monitor and how it can effectively integrate with Linux to offer an effective, efficient, and stable system and network monitoring solution. It was my intention to provide an overview of the services and how they are monitored as opposed to providing an installation manual because the accompanying online documentation does an excellent job guiding one through the installation with its various configuration aspects. I would like to thank Sean MacGuire for writing such a useful program that stresses simplicity, modularity, and extensibility in design and for making such a package available in source code. It is my belief that Big Brother can effectively provide a monitoring solution for a Linux network or any other UNIX network in which system monitoring is desired.

References

Big Brother UNIX Network Monitor (Main Page):
For downloading and information: http://maclawran.ca/bb-dnld/index.html
Multi Router Traffic Grapher (MRTG):
http://ee-staff.ethz.ch/~oetiker/webtools/mrtg/mrtg.html
http://www.ee.ethz.ch/stats/mrtg/

Chapter 27

Changing Server Operating Systems

Howard Marks

Most organizations today are on their second or third generation of servers. Starting with beefed-up PCs scattered throughout the organization for workgroup automation, servers have evolved into an integral part of the organization's overall computing infrastructure.

Few of us can, however, rest on our laurels. Change is inevitable. Whether it is a simple upgrade to the current version of the server operating system you're using or a wholesale conversion to a new one — the process is similar.

WHY CHANGE?

The first question you need to ask yourself, your management, or your client is: Why are we changing server operating systems?

In our consulting practice, we've been involved in operating system conversion projects for a myriad of reasons, some good (like adding new functionality, improving management through the addition of a global directory service, or replacing a system that is no longer well supported by its manufacturer), others not as good (like replacing a system that has falling market share or is less fashionable than it once was).

WHY NOT CHANGE?

Any major operating system change is going to have associated with it some degree of disruption. Users will need IDs in the new operating system's directory, rights or permissions to resources on the new operating system, logon scripts, and assorted other administrative details that you haven't done on more than a one-by-one basis in a long time. In addition, it is rare that workstations are able to access the new server operating systems with the same protocols, network components, etc. as they used for the old. So while we'll talk about changing server operating systems, there's always some impact on user workstations as well. The more workstations

0-8493-9823-1/00/$0.00+$.50

you have, the more user disruption there will be. And then there are the routers, which may need a complete reconfiguration since your new server operating system uses TCP/IP as opposed to the protocols used by your old one.

Think It Through Carefully

Just as the three keys to real estate are location, location, and location, the three keys to a successful server operating system rollout are planning, planning, and planning. We've been involved in projects where the project manager has assumed that old server hardware would be available on which to install the new OS. But halfway through the conversion only half of the user update plan had been completed, leaving users who needed to continue accessing the old operating system and therefore no server hardware free for server operating conversion. The only other choice was to leave half the corporation's users without access to any system — definitely an unacceptable amount of downtime.

Look at Apps and Services. Step one in the planning process should be to review the applications and services, like Internet gateways and FAX servers currently in use. This will help you identify any operating system or protocol dependencies. Make sure to include workstation dependencies like specific versions of network software components. Several network managers were confounded to discover when they upgraded from NetWare 3.x to 4.x. that the client software was not completely compatible with all their applications. And that was the same vendor's software.

Workstations or Servers First. Step two is to decide whether you want to update servers or workstations first. Updating servers first, especially if you can use tools like File and Print Services for NetWare (see below) that allow workstations to access new servers without new software, allows you to visit each workstation only once. Since the most time-consuming portion of the project is reconfiguring the workstations — not building new servers — anytime you can avoid visiting the workstations more than once is a good thing.

Updating workstations first, on the other hand, allows you to incorporate workstation operating system upgrades in the same project. For example, from Windows 3.x to Windows 95 or NT.

Gradual or Blitz Conversion. There are two schools of thought about how to make the transition. In general, we fall into the blitz conversion school. Get the pain over with as fast as possible. Where the scope of the project doesn't allow a wholesale conversion in a single weekend, pick departments or other user groups to update in batches. This is also called a cutover approach because you're cutting all the users over in one fell swoop.

Train First. There will be some changes from the users' viewpoint. If you're making the change in server operating systems to add new services, there may be significant changes. The key to a successful project is to train the users before you update, not after. If you wait, your help desk will have to answer an extremely high volume of calls stemming from user ignorance. Training also gives you an opportunity to get the users to "buy in" to the changes. You have the chance to stress the benefits the new software will provide instead of letting users run headlong into all the headaches and disruptions. We've found users are very tolerant if they know it will be better later and very intolerant if they don't have a clue.

Create New Servers

The single most important recommendation we can make is never, repeat never, attempt an operating system migration, or even a significant upgrade, in place. It may appear attractive to take the perfectly good server that you bought just last year and put the new operating system on it. The truth is that the process is at best complicated. Changing operating systems on the same hardware typically means following the multiple step process below:

1. Make the server unavailable to the users to prevent data changes.
2. Backup the server (typically, a several-hour process).
3. Reformat the disks on the server.
4. Install the new operating system.
5. Restore the backup (typically, another long process).

Trouble lurks at almost every step. We had one client who didn't discover that the backup tapes made on his NetWare server using Novell's SMS (Storage Management Service) couldn't be restored to a non-NetWare server until after he reformatted the disks on his server. Moreover, he was using the same backup program on both the NetWare and Windows NT servers.

The biggest advantage of using a new server is that you can pull the plug on the conversion at the last minute. If you start an in place conversion early Saturday morning and things go wrong, you have to make the decision to restore the old operating system early Sunday to get the system back to the state it was in when you started before the users come in on Monday morning. If you built a new server, you could just turn the old server back on Monday morning, giving you several more hours, even weeks, to work out the problems. Even better, you can build the new server during the week and only have to transfer the data over the weekend.

If you have multiple servers, you can buy just one new server and upgrade your servers one at a time, using the server released as you upgrade each to upgrade the next. The cost of the hardware is nothing compared to the cost of user downtime if you hit unexpected problems. And you *will* hit unexpected problems.

Is Peaceful Coexistence a Better Idea?

If you, like many IT managers, have come to the conclusion that you need to add new services like intranet Web services, client server applications, and databases that aren't available for your current server operating system or that your current system isn't well suited to your needs, you may be tempted to make a wholesale conversion of all of your servers.

Think twice before taking such a precipitous step. In today's environment, it may be easier to add UNIX or Windows NT servers to host these new services to your existing infrastructure than to replace all your existing servers with new ones. This is especially true if the new services will only be accessed by some of your users. Servers running a new operating system will probably require new requester clients on all workstations (as discussed above), which further complicates a wholesale conversion. If only some users require the new services, then you also have a smaller training cost. You can choose to train only these users instead of everyone in your company.

Managing a Mixed NetWare Windows NT Network. Perhaps the most common coexistence scenario is the addition of Windows NT servers to run Exchange, Internet Information Server, SQL Server, or other applications to a primarily NetWare network. In this scenario, you need to pay particular attention to two points: providing a single user database for both systems through some common directory service, and minimizing the change required on user workstations.

We've found that when installing applications on a user's workstation without making changes to the underlying operating system, network components and protocol stacks wherever possible is much easier, more reliable, and faster than the alternative. We will therefore install Windows NT application servers with the IPX protocol — the default protocol of older NetWare networks — and install the clients for SQL Server, Exchange, or other application without more significant changes.

Directory Coexistence Tools. Several tools are available to allow Windows NT and NetWare servers to share a common directory database, eliminating the need to create and maintain user IDs and passwords for each user in separate NetWare and Windows NT databases. These products can eliminate huge amounts of network administration and should be considered for any environment that has NetWare and Windows NT servers for more than a few months.

Novell's NDS for Windows NT. Novell's NDS for NetWare allows the administrator of a NetWare centric network to replace the Microsoft domain architecture with Novell's NetWare Directory Services. Using NDS for NT, you replace the central domain services DLL on your NT servers with a Novell provided replacement. This redirects all domain directory access requests to a NetWare server running the NDS for NT NLMs.

Domains become simply another type of NDS container object. Users, servers, and other resources simply are contained in, or have rights to, the domain container. This also allows you to eliminate trust relationships between domains, as a user in the NDS structure can have rights to multiple NT domains.

Novell is promising a new version of NDS for NT in the near future that will maintain the NDS database on Windows NT servers, providing the global directory services advantages of NDS on a pure NT network. You manage the resulting directory database through Novell's Nwadmin program but can make changes to the users and groups in a domain through the usual Windows NT management tools. NDS for NT makes this happen by redirecting the domain management APIs to the NDS database.

On the downside, Microsoft has made several confusing announcements about their support of users running NDS for NT. As of this writing, their current position is that they will not support such users that are having security problems.

NetVision's Synchronicity. Even before Novell released NDS for NT, Orem, UT-based NetVision released its similar Synchronicity for NT. Rather than replace the domain database in Windows NT, Synchronicity adds a service to a domain controller in each domain and an NLM on each server that has a read/write replica of the NDS database to synchronize the NDS and domain database. An event-driven system, it automatically passes updates from NDT to NT and vice versa as the changes are made.

This approach avoids some possible compatibility problems with Windows NT Server applications. NetVision has also expanded Synchronicity, adding versions to automatically create and maintain Exchange mailboxes and to synchronize with the Lotus Domino and Notes databases.

Microsoft's Directory Services for NetWare. If you have primarily NetWare 3.x or even NetWare 2.x servers and are adding Windows NT servers or replacing them with Windows NT servers, Microsoft's Directory Services Manager (DSMN) for NetWare allows you to manage the NetWare binderies on multiple NetWare 3.x servers from a Windows NT domain. DSMN allows you to copy the user information, except passwords, from one or more NetWare binderies into a Windows NT domain.

Once you've uploaded this user information into the domain database, it is automatically written back to all the NetWare servers' binderies. Not only does this mean that you now have a single point of administration for both your NetWare and Windows NT servers but also that your NetWare bindery-based servers have a synchronized directory for the first time.

Once you implement DSMN, you have to be careful that no one uses NetWare administration tools like Syscon, or even SETPASS, to modify the bindery on a NetWare server as the synchronization is one way — from the domain database to the NetWare bindery, not the other way around. One trick would be to rename the NetWare administration files to nonexecutable names and then create text files that display whenever Syscon etc. were run that remind you of the correct procedure. Do this on ALL the NetWare servers. By the way, Microsoft provides a replacement utility for SETPASS.

Moving from NetWare to Windows NT. This is today's most common conversion. Therefore, several vendors have written tools to make the conversion easier. There are, however, a few potential problems that you'll have to be aware of.

1. Windows NT servers default to an internal IPX network number of 0. This confuses some IPX routers including NetWare servers.
2. Novell's Client 32 for Windows 95 has no uninstall. If you install it as part of a workstation upgrade, then later remove your NetWare servers, you won't want Client 32 anymore. But you'll need to rebuild the workstations in order to remove it.
3. If you use multiple IPX packet types, make sure to set them explicitly on NT servers.

NetWare Migration Tool. Microsoft's NetWare Migration tool, included in the standard Windows NT Server package, will copy NetWare users, groups, and data from a NetWare server to a Windows NT server. It will even convert user restrictions to Windows NT rights and NetWare trustee rights to Windows NT file and (or) directory permissions. It will allow you to set users' passwords, add all users who are security equivalent to supervisor to the administrators group, and even run a trial migration to check for potential problems like duplicate user IDs. Duplicate user IDs might crop up when consolidating multiple bindery-based NetWare servers into a single NT domain.

The migration tool transfers data at up to 100 MB/min. This is a big improvement over the backup/restore method and about twice as fast as XCOPY.

Note that the migration tool is not NDS aware so you may need to run multiple migrations for different bindery emulation contexts if moving from NetWare 4.x to Windows NT. It also doesn't move some information, including printers, print queues, and user passwords.

File and Print Services for NetWare. Perhaps the most powerful tool for easing transitions between NetWare and Windows NT is Microsoft's File and Print Services for NetWare (FPNW), which is sold as part of the Services for NetWare product which also includes Directory Services for NetWare. A Windows NT Server running FPNW emulates a NetWare 3.x file server so that workstations running NetWare clients can access file and printer resources on Windows NT servers without modification.

FPNW runs as a service on a Windows NT server. Once it is installed, the administrator of that Windows NT server can create emulated NetWare volumes out of directories on the server's disks in much the same manner as creating data share points. Through User Manager, he or she can grant users in the Windows NT domain NetWare login access and maintain their login scripts.

When moving from NetWare to Windows NT, FPNW is a key tool in a "server first" transition. We frequently create a new Windows NT server, use the migration tool to move all the data, and install FPNW to allow users to access the new server without requiring any changes to the workstations. This allows us to take our time updating workstations at our leisure. We change the login script on the old server to map drives to the new server.

Gateway (and Client) Services for NetWare. If you want to update your workstations first, you'll typically have to visit each workstation twice: once to add the network software and requester to access the new Windows NT server and a second time to remove the NetWare requester, client, or shell. Microsoft's Gateway Services for NetWare (GSNW) allows you to visit each workstation just once by giving workstations running Microsoft clients for Microsoft networking access files on NetWare servers.

Once the gateway service is running on a Windows NT server, it logs into your NetWare server and allows you to create Windows NT shares that access the volumes on the NetWare server. The NetWare server always sees the machine running the gateway as a single user. That user must have all the trustee rights you want to grant to any users on Windows workstations. You can then control the access permissions of the users via Windows NT security.

The gateway works by translating Microsoft-style Server Message Block (SMB) requests into Novell-style NetWare Core Protocol (NCP) requests and forwarding them to the NetWare server. It then sends the resulting data to the workstation via SMB.

Moving from LAN Server/LAN Manager/Warp Server to Windows NT

IBM's LAN Server and Warp Server share a common heritage with LAN Manager though Microsoft's MS-NET. This means that workstations use the same SMB protocol to communicate with LAN Server servers that they use to talk to Windows NT servers.

The SMB protocol uses NETBIOS to identify workstations and servers. Unfortunately, each of these products uses a different implementation of NETBIOS over TCP/IP, so an OS/2 Warp workstation cannot access Windows NT servers data via TCP/IP.

We recommend installing the NetBEUI protocol on your Windows NT server and performing a "server first" migration from these systems. Existing workstations using NETBIOS protocol will be able to access the server, and you can update the workstations to your preferred protocol later.

CONCLUSION

Changing operating systems can be a traumatic experience for a network manager. Careful planning and the appropriate tools can help you keep both your users and your family happy. Avoid installing the new server operating system on the same hardware as your current server. It leaves you no fallback position in case of unexpected problems. Training your users in advance will make the transition even smoother as they won't be harassing your help desk technicians through lack of understanding. Also, glitches in the transition are more likely to be tolerated by users eagerly anticipating increased services. And finally, if you have no clear business-driven need to change operating systems — don't.

Chapter 28
UNIX as an Application Server

Scott Koegler

INTRODUCTION

Networks have evolved to become the mainstay of corporate America. Millions of personal computers are connected to network servers running applications ranging from personal productivity to corporate accounting, yet the demands on the network continue to increase. These demands typically take two forms. The first is the availability of application software specifically tailored to an organization's operations. The second is the need to run processor-intensive applications in a personal computing environment. Both of these situations require increased horsepower from individual computing components, and in some cases the required horsepower just isn't available from networked components.

Today's network architectures have been maximized to provide superb performance for file and print sharing tasks, but still may not always offer the level of performance required. While the network operating system can provide network services, it is not typically optimized for application processing. Even in those cases where an application can be run on a network server, doing so may degrade the performance of the entire network, making the proposition counterproductive.

Even with the stunning advances in hardware systems and programming techniques, many companies find that the specific applications that fit their operation were created to run in multiuser environments rather than in network environments. In many cases this is the result of migrating an application from a mainframe system, and resulting programs remain optimized for the multiuser rather than networked systems.

IS UNIX STILL GOOD FOR ANYTHING?

UNIX has a long history in the computing world. The simple fact that it's still around must mean more than that it's hanging on. In fact, UNIX has enjoyed a resurgence of interest over the last couple of years, largely

0-8493-9823-1/00/$0.00+$.50

because of the growth of the Internet. But UNIX has other strengths beyond the Internet.

UNIX's original advantages were with its file system, multitasking kernel, and easy configurability. None of those early considerations has deteriorated, and they are still not fully duplicated in the newer operating systems. To be sure, both NetWare and NT have advantages with regard to market availability, commercial support, and a wide range of software options, but the basics still can be filled easily and inexpensively with UNIX services. UNIX is a mature operating system and has been fairly well debugged and enjoys a wide, mostly independent, support group via the Internet. Its scalability, as of this writing, puts NetWare and NT Server in the little leagues.

INTEGRATING WITH THE NETWORKING ENVIRONMENTS

Networks have become commonplace, and the architecture has embraced Ethernet as the predominant standard. Novell NetWare continues to maintain a 60 percent plus share of network server installations, and Microsoft NT server continues to build an increasing market share. In the past, these operating systems imposed their own network protocols on their environment. Customer demand forced simplified integration and both vendors supplied solutions and the third-party solutions, but the connectivity was generally less than seamless. The stunning rise in popularity of the Internet and its associated TCP/IP protocol has had the effect of forcing these two major network operating system suppliers to support this protocol as a ground-leveling standard. The net result to the user is that both Novell and Microsoft now supply the TCP/IP protocol as a part of their base product offering. In fact, Novell now offers TCP/IP as a selectable replacement for its longstanding IPX protocol.

What this means in a practical sense is that it's easier than ever for an organization to implement a variety of operating systems based on business needs and application requirements rather than on the networking vendor's dictates. It is now commonplace to find a corporate IT shop with a mixture of Novell and Microsoft servers on the same network.

WHY BOTHER WITH ANOTHER OS?

While NetWare is a highly efficient file and print server, with Internet services included as part of the file server function, it doesn't offer the application services available in Windows NT. NT can provide a combination of file and print services along with robust application service support and Internet services as well.

UNIX can supply nearly all the same facilities and, with TCP/IP now a networking mainstay, integrates into the LAN environment almost without being noticed. But, since there are few practical differences between the available solutions, there needs to be some compelling reason to increase the complexity of the corporate LAN, even if only marginally.

The most often cited reasons for installing a UNIX server in a LAN environment fall into two categories: application support and price. In short, if an application is only available as a UNIX application, the choice is reduced to finding an application that runs on the existing network or installing the necessary UNIX resources. Organizations have somewhat more flexibility when installing a new application than when trying to integrate a legacy system into a LAN environment. The good news in either of the situations is that connecting an existing UNIX application to a network is likely easier than either converting the function to a different application that is LAN based, or redeveloping the application to run on the LAN. The combination of Windows and a good terminal emulator can deliver a very powerful combination that leverages the strengths of all the available applications and operating systems.

The second powerful consideration in favor of UNIX is the wide range of vendors offering similar versions of the same operating system. A byproduct of this competition is the availability of highly competent operating systems at very low prices. This is not to say that all UNIX OS are the same. Some proprietary versions, like IBM's AIX and DG's version weigh in at the same or greater costs than the popular NOS (Network Operating Systems). But at the other end of the range are systems like SCO's UNIX and at the extreme low end, Calera's Linux. To be sure, there are differences in the variations in both capabilities and support, and these differences need to be factored into any decision; but depending on the level of in-house expertise, nearly all can be viable solutions.

One of the other things that can provide a benefit in the UNIX environment is the existence of a mature and widespread user support community. The UNIX community has been the premier example of noncommercial systems development for many years. Programs ranging from small utilities to full applications and even the operating system itself have been available through the efforts of programmers around the world. Possibly the best example is the Linux version of UNIX that was developed by a loosely coordinated group of programmers who each contributed their efforts to developing a state of the art OS that is available, complete with source code, simply by downloading it. The commercial versions of Linux add several features that make this user-supported system viable in a corporate environment. But at under $500 for a full license, it's a tough competitor.

OPERATIONAL CONSIDERATIONS

The introduction of multiple hardware and software platforms into a computing environment is a normal, if unwelcome, reality. It increases the range of knowledge required for internal support and adds to the list of vendors involved in the organization, and because of these factors should not be entered into simply for the sake of adding a new environment. However, given a valid business reason for doing so, there are tools and systems that minimize the impact.

Modern IT systems provide increased management capabilities through a variety of facilities. The availability of the SNMP (Simple Network Management Protocol) has made the management of heterogeneous networks possible by providing a common protocol that nearly all system software and hardware vendors have implemented. With this kind of consistency it is now possible to view a range of network services through a common console. This means that the differences in management of various platforms are minimized through the common interface. Day-to-day activities can be incorporated into a centralized system that decreases the load on the IT staff and reduces the integration of yet another operating system to a routine function.

NETWORK SERVER VS. APPLICATION SERVER

The role of the server needs to be taken into consideration when contemplating adding computing resources. In fact, one of the main reasons for adding a UNIX server to an existing networked environment is the different function the server will fill.

In a typical networked environment the processing load is distributed, not always evenly, across the client PCs, the server, and the network cable. Client/server systems attempt to do a better job at this load distribution, but these systems have never really delivered on their promise of low cost, quick development, and high performance. In fact, because they distribute the load, they put increased demands on all the network components, leading to across-the-board upgrades for the entire network.

While certain applications like personal productivity, word processing, and spreadsheets certainly work efficiently as client-side applications, accounting, client information, and other database-type applications depend on centralized processing of some type. This may amount to file and print services, but more often involves some data retrieval that depends on a server process to maintain the indexes or process SQL queries. It is these more demanding services that impose processing loads on the server and end up degrading network performance as a whole. When the processing load increases and users begin to complain about the quality of service on the network, the upgrade spiral begins again. If the bottleneck is diagnosed as a server problem, the options for upgrading server hardware come into focus.

There is an increasing variety of server hardware available with increasingly faster CPUs and disk drives. The market for multiprocessor servers is even approaching the commodity level with systems of up to six CPUs in a single cabinet. So, while it is now possible to increase the horsepower available to the network server, running the server and an application in the same device concentrates the load. This is where the addition of an independent server can be justified.

While there are numerous application types that can be run on a UNIX server, the most obvious include databases, mail systems, and Inter/intranet servers. Relegating these processes to a separate device takes them out of the normal processing environment and segments the load. Any network administrator who contends with increasing loads on server processes knows that users notice when network performance degrades. It may be possible to avoid the eventual performance hits that result from increasing application support by strategically distributing particular application services to independent servers.

WHICH OS FOR WHICH SERVICE?

Operating systems are designed to provide certain specific services to multiple connected computers and devices. Basic services such as file and print services were the initial reasons for installing NetWare and NT servers. All else being equal, there is little reason to add a new operating system to a network simply to deliver either of these services. However, if they already exist, or if they are being added for other reasons, there may be solid reasons to take advantage of one or both of these facilities.

File Service

UNIX can provide the same traditional services as part of its original range of processes as a multiuser processor. The file access optimization technique of directory caching and indexing pioneered by UNIX is now incorporated in nearly all server OS. This means that UNIX can often provide excellent performance as a file server. Because of this, it may be advantageous to make use of otherwise unused disk space on the UNIX server. The issues involved in doing so will revolve around providing access to the files within the existing applications and directory structures already in place within the network.

Novell's NDS (NetWare Directory Service) provides possibly the best unified directory structure available by combining all available computing resources into a tree-structured directory. Adding components to one part of the tree makes the new component available to any other component based on assigned access rights. A UNIX server (or any other server) that can participate in the NDS structure can be integrated seamlessly into the network and becomes an additional storage resource. Users and application software may never be aware of the actual server software running in the new device. Selecting a version of UNIX that supports NDS would be important in a NetWare environment because it would ease the integration of the new device and lessen the requirement for users to cope with another set of functions and commands.

Microsoft's ADSI (Active Directory Service Interfaces) does some of the same directory maintenance as NDS, but is not as widely deployed. This means that it is still unlikely that there are UNIX versions available that support ADSI. ADSI will support NDS, so there will be some long-term viability and cross-system access, but the control of directory access is likely to continue to be a major issue for debate for some time, as both Microsoft and Novell see it as strategic to their business.

Generally speaking there is nearly always a method for making a viable connection to a UNIX server across a LAN. All UNIX servers support TCP/IP, and most support Novell's IPX/SPX. There is less support for Microsoft's NetBUI protocol, but it is increasingly likely that TCP/IP will be the common denominator in an integrated environment. Depending on the applications and access requirements, simple direct attachment to the UNIX system may suffice, which decreases the necessity for a sophisticated directory system like NDS or ADSI. In fact, for small installations, it has been said that NDS can be more burdensome to set up and administer than the advantages it provides.

Printing across the Network

The second mainstay of the NOS is print service. Once again, UNIX can provide excellent print services. When integrated into an NDS structure, UNIX printers become additional print resources to the network. But when these remote printers are used as fully competent print servers rather than as simply remote printers, there may be some productivity gains to be had. This possibility exists but depends on several variables that include the structure of the network, topology, and where the print jobs are serviced.

For simplicity, most UNIX versions provide print capability along with their overall file and connectivity services. As with file service, it is doubtful that adding a UNIX server for the purpose of providing print services could be justified without an underlying plan for the use of the UNIX system for other reasons. Nonetheless, if UNIX servers already exist within a network, their use as print servers may provide another resource for distributing the load.

INTERNET/INTRANET SERVICES

The demand for Internet and (or) intranet capabilities is increasing. Companies who don't already have some kind of Internet capabilities are likely candidates for the near future. The Internet was born on UNIX systems, meaning only that they are natural hosts for most Internet services. While both Novell and Microsoft offer Internet server capabilities within their respective server OS offerings, using a company's production server as an Internet server can present problems. For these reasons it may be advantageous to consider a UNIX platform for a company's primary Internet server.

Most organizations provide Internet services so those users outside the company's walls can have access to various information that the company wants to make available. In some cases this amounts to advertising and promotional material, or data files. It may include direct access to corporate resources and data if the system is used as an intranet for employees and partners. In all cases, security and performance are issues to consider.

The issue of security is beyond the scope of this article; however, whatever system is deployed, it must provide some kind of user validation and firewall capabilities to provide the right information to the right users, and deny access to all others. The most secure way to separate access is to separate the systems used to provide access. In considering the installation of Internet capabilities, consider both the risks and advantages of providing these services from within your main network server.

Performance, or degraded performance, is an obvious consequence of adding Internet services to an existing server. It is not necessarily true that setting up Internet services on either a NetWare server or an NT server will degrade the system's performance. For many companies, simply enabling the service will provide both the services and performance needed. In other circumstances, doing so can have significant impact on the entire network. The difference lies entirely on the services provided and the load produced by the combination of the applications and number of users accessing them.

It isn't possible to cover the range of applications and environments here. Particular attention must be devoted to the mid- and long-term use of the installation, and the relative capacities of the servers that will be hosting the Internet services. This analysis is similar to that required for any other application installation.

Nearly every UNIX server software vendor offers Internet server software. It is either a part of their base offering, or offered as an option. The cost of operating systems software is always a consideration, but not always an indication of a product's worth. And the fact that server software is bundled as a part of the system doesn't always mean that the included services will meet your organization's needs. One application worth mentioning here is the Apache HTTP Internet server (www.apache.org). This highly competent, user-supported application is available as a free download, and currently is estimated to support nearly 50 percent of today's Internet sites. Apache is developed and supported by a user/developer community around the globe. For some organizations this may raise concerns about vendor responsiveness since there is no commercial vendor. But the success of this application is possibly directly attributable to that fact, and that the system is supported by its developers who all share a common interest in the system.

On the other hand, there are still compelling reasons to select systems like Windows NT for installations that require their proprietary services. There is an increasing variety of applications and development kits that depend on server-side facilities like Microsoft's Active Server extensions, and LiveWire from Novell. Just like the other applications discussed earlier, if the application you've chosen requires these facilities, you may have few practical choices.

COPING WITH A MIXED ENVIRONMENT

Today's distributed systems can become highly complex even when the network is homogeneous. Maintaining a consistent OS environment lets network administrators develop a depth of understanding that often leads to better efficiencies in both day-to-day operations and in troubleshooting. If the option exists to maintain a single-OS network rather than to introduce multiple OS, the choice is clearly to keep things simple. In reality, there are numerous forces and considerations that lead to the addition of multiple operating systems, and specifically UNIX. But this doesn't mean that the network will immediately become compromised because support time is diluted.

The addition of proper network management systems that consolidate and even automate routine maintenance functions can make the difference between stability and panic. The popularity of the SNMP (simple network management protocol) for managing distributed systems can make this kind of support possible. There are numerous supporting application providers for the various operating systems. One foundation application is HP's OpenView that supports plug-in modules for monitoring different SNMP-enabled systems. Be sure to check with your OS vendor to determine their support for SNMP. There are also different levels of support within the SNMP framework. Systems that provide interactive support and control in addition to simple monitoring facilities will pay off when the time comes to actually perform maintenance on a remote unit. Active maintenance allows such capabilities as remote reset and configuration.

A quick search of the Internet for "SNMP UNIX" using your favorite search engine will provide a good sampling of providers of SNMP management systems.

ALL TOGETHER NOW

If you don't have to, don't do it. But if your business requires computing services that go beyond the basic offerings of your current NOS, or if your users are already screaming about slow response times, adding a UNIX server to your LAN can be a good thing if done properly.

Scott Koegler is VP of I.T. for a national printing franchise and specializes in networking subjects. He can be reached at skoegler@usa.net.

Section V
Application Servers

During the initial developmental stage of client/server processing, the role of the server was primarily restricted to functioning as a repository of shared files. Thus, the server was often referred to as a file server. As client/server technology progressed in tandem with the development of LAN technology, servers were developed to satisfy different applications. Because printing and faxing were commonly performed applications suitable for sharing over a network, they were a natural target for software developers looking to develop products that could be shared by employees within an organization. Thus, fax and print servers were introduced as logical extensions to file servers, operating either as modules on the same hardware platform or as stand-alone network-connected devices.

As the utilization of LANs progressed, a wide range of applications were developed to support the client/server processing model. Each of these applications resulted in the development of modular software packages that either operated on an existing server or required the use of a dedicated server. They had the ability to operate on a multifunction server or the requirement to operate as a stand-alone server, based upon the processing performed by the software and the end-user population accessing its resources.

In this section we will focus our attention upon a number of applications that operate in a client/server environment. Such applications run the gamut from print and fax servers to voice and communications servers. Although applications may not appear to have anything in common with each other, they are all server-based applications. Thus, we will examine them as an entity in this section.

In Chapter 29, the first chapter in this section, Allen Waugerman, with assistance from employees at Lexmark, introduces us to the popular print server. In his chapter "Print Servers and Network Printing," Waugerman provides us with detailed information concerning the role of the print server in a network and its selection parameters, and describes a subject gaining the attention of both individuals and corporations — printing through the Internet.

Max Schroeder introduces us to the fax server in Chapter 30, "Fax Servers." Schroeder provides us with a basic overview of the operation and utilization of fax servers as well as describes some of the management and administration considerations that should be reviewed prior to implementing a fax server solution.

Chapter 31 turns our attention to data communications. Gerald L. Bahr introduces us to several methods that can be employed to support communication to and between networks. In his chapter "Remote Access Concepts," Bahr first distinguishes the difference between a remote node on a network and remote control. Once this is established, he turns his attention to the advantages and disadvantages of the use of modem pools, ISDN, Frame Relay, and other techniques for accessing a server.

In a second chapter focused on remote access, Denise M. Jones examines the development of remote access communication policies and rules. In her chapter entitled "Communication Servers," Jones first reviews the various methods that can be used to provide connectivity between distant devices and then turns her attention to access policy issues that system administrators must consider.

To enhance our knowledge and understanding of electronic mail, two chapters are included in this section covering this topic. The first, Chapter 33, provides us with an introduction to the Simple Mail Transport Protocol (SMTP) and the installation and configuration of an electronic post office. In this chapter entitled "Internet Mail Server Overview," Lawrence E. Hughes defines the components of a mail server and takes us on a tour of the life of a typical Internet electronic mail message. Using this as a base, Hughes discusses the environment required to operate an Internet mail server as well as methods to test and troubleshoot the server. This chapter is followed by Lee Benjamin's "An Introduction to Microsoft Exchange Server," which provides an overview of how this popular commercial product can be used to improve an organization's business process. In this chapter Benjamin discusses trends in messaging and collaboration and also the infrastructure required to implement these technologies, Internet connectivity, and the administration process.

Judith M. M Myerson turns our attention to two very specialized types of servers. In Chapter 35 "Virtual Server Overview," Myerson introduces us to a lightweight server that you can literally carry in a briefcase. In Chapter 36, "Text-to-Speech Server Overview," Myerson examines how you can obtain the ability to have colleagues around the world edit a speech or paper without having to know the language it was written in. She acquaints us with this emerging technology and discusses its utilization.

We continue our examination of specialized servers in Chapter 37 with Robert Brainard's chapter, "CD-ROM Servers." Brainard examines the architecture involved in providing network users access to CD-ROMs and the different types of servers that can be used to satisfy requirements ranging from a few persons in a department needing to share a common CD-ROM to satisfying the data retrieval requirements of an enterprise requiring a large CD-ROM shareable jukebox.

In concluding this section Nathan J. Muller in Chapter 38, "Choosing and Equipping an Internet Server," examines the platform architecture, operating system, and features necessary to consider to construct an Internet server.

Chapter 29
Print Servers and Network Printing

Allen Waugerman

With help from Don Wright, Richard Russell, Randy Sparks, Mike Sublette, Jeff Watrous, and Vic Mollett, of Lexmark

WHAT IS A PRINT SERVER?

The term *print server* has changed a great deal in the past few years. Not long ago when you said print server you meant a host such as a personal computer or a UNIX machine that would have a printer attached to its parallel or serial port that could be shared by many network clients. The print server would provide spooling and scheduling functions. This print server would be dedicated to printing or used for multiple functions such as file or Web serving. Although this was the primary method of sharing a printer a few years ago, things have changed.

The printing industry has now moved to what is called the *network print server.* The term refers to two similar but different network devices: internal and external network adapters. The internal adapter is designed to be installed under the covers of a specific family of printer and allows the flexibility of placing a printer on any node of the network.

The external network adapter is designed to connect virtually any printer, serial or parallel, of different technologies to the network. These adapters come in one-port (one parallel port) to three-port (two parallels and one serial, or three parallels) models. This allows up to three printers to be attached to the network with only one network connection or node.

As more and more businesses, companies, and schools grow their networks, or create new networks, the network print server has found its way into the standard setup. The network print server has many advantages over the old-style host-based print server. As with most good ideas, the network print server was created out of necessity. Printers need to be out in the open where the users can pick up their printouts — not locked inside the server room. The network print server takes up little to no room and

0-8493-9823-1/00/$0.00+$.50

allows the printer to be located anywhere on the network. The external network print server allows a total printing solution to be located anywhere on the network with only one network node attachment. For example: an advertising department may need to print many types of documents to meet their needs. By locating an external network print server in the advertising work-group area, they could have a high-speed laser on parallel port 1 printing everything from concepts to final camera-ready copies, a color laser or inkjet on parallel port 2 to print transparencies for customer presentations, and a vinyl cutter on the serial port cutting out the letters for a sign.

Managing network print servers is generally accomplished through management utilities such as Lexmark's MarkVision and HP's JetAdmin. These management tools give you everything from "paper out" alerts to job statistics, and all of this can be done remotely. These management tools are growing very sophisticated very quickly and are becoming a standard tool in IS departments around the world.

Price is another advantage the network print server has over the old print server. Network-attached print servers can be found for under $500, while a PC or UNIX machine to act as print server host may cost thousands. All of these advantages help control the total cost of printing. The network printers become a lower unit cost per network node. The users now print jobs to the appropriate printer to save on supplies. The IS team is more efficient now that they have great management tools. And the whole team saves time and money by having the printers accessible and always ready to use.

WHAT'S IMPORTANT IN A NETWORK PRINT SERVER AND HOW TO CHOOSE ONE

Network topologies vary greatly from network to network, but most installations have one thing in common. They are running either Token Ring or Ethernet to the desktop workstations and printers. These two environments can be split into two more categories according to speed. It is important to find a network print server that supports and auto-senses both speeds and connectors. The auto sensing makes installation and moving print servers to different places in the network a breeze. Token Ring networks in today's setups are running either 4 or 16 MB. Connection types could be a UTP (unshielded twisted pair) RJ45 connector or an STP (shielded twisted pair) DB9 connector.

As with Token Ring, Ethernet also has different speeds and connector types. Today's Ethernet environments are primarily running 10baseT or 100baseTX and use an RJ45-style connector. There are many other Ethernet topologies out there, but they are not widely used in connecting printers to the network. Getting a network print server that can support both 10 and 100 MB is very important in today's fast growing networks. More and more companies are running out of bandwidth with their 10-MB networks and are quickly moving to 100 MB. If you have a network print server that will auto-sense between the two speeds, the network print server will not even need to be touched to work in this new environment.

Determining what print server functions are needed is next on the list of considerations. If you recall, there are two types of network print servers, internal and external. If one printer can meet the entire needs of a workgroup, installing an internal network print server in that printer may be the best solution. If you need to install multiple printers, or printers that do not support internal network print servers, then an external network print server is the answer. There may be an old workhorse printer that has been serving the needs of its users for years and the only thing it lacks is the ability to be networked. A low-cost, single-port external-network print server works great in this case. Somc external network print servers even allow an external modem to be serial attached to deliver inbound fax capability. What a great way to eliminate those expensive fax supplies and that annoying curled up paper.

PROTOCOLS

What protocol is used in the world of printing? It would be nice to answer that question with a generic answer like "the printer protocol," but in today's complicated network environments there is no such thing. It sometimes takes multiple protocols to support the entire needs of a company. There are four main protocols that you will find in very small to very large network environments.

TCP/IP is emerging as the mainstream protocol. With the explosion of the Internet, more and more applications, including printing, are using TCP/IP as their default protocol. With this is mind, picking a network print server that supports the primary TCP/IP applications is critical. It is important to look for support of FTP, TFTP, and LPR/LPD for total flexibility. But some network print servers do not support all of these TCP/IP printing methods. Some vendors also support a proprietary application that allows for more printer-specific functions in certain environments such as Windows 95 and NT. These applications are often designed around the vendor's management tools.

While TCP/IP is the emerging protocol in today's world, Novell is the networking leader, and supporting this environment continues to be very important. With that said, there is one thing to keep in mind. It is Novell's objective to become a pure Internet/intranet network software leader by the end of 1998. When you hear the words Internet and intranet, TCP/IP should leap to mind. But for some time to come the current NetWare solutions will still need to be supported. The network print server should support Novell NetWare 3.11 or later and Novell NetWare 4.x or later including NDS (Novell Directory Services).

Although AppleTalk is not talked about much these days, it continues to be the workhorse protocol in schools and graphic design houses. As far as support, there is not much to say. The adapter either supports it or it doesn't.

The last protocol for discussion is DLC (Data Link Control) — a hardware address-based protocol that is primarily from the days of IBM and OS/2, but you sometimes find it in some NT environments. In most cases, you need a vendor-specific proprietary application running on the host to print.

As networks get busier, print server performance is becoming more of an issue. A network print server can be measured just like everything else in the PC market — faster is better. The first concern of users who find out that they have to share a network printer is that their prints jobs will take longer to get printed. In some cases that may be true, but if you pick the right network print server for your printing needs, you will never get a call about slow printing. There are a number of factors that determine how fast a print server will be: processor speed, available memory, code efficiency, data delivery capability (e.g., Ethernet 10/100), etc. In general, however, you won't know if you have an efficient mix of the contributing factors without testing the throughput of the device. To do this, you must focus your performance testing on the print server. Outside factors like data rendering, printer speed, and network traffic should be eliminated. Test your print server on a quiet network using a prerendered large print job that the printer can consume easily. This allows the print server to be the gate, and the data you gather will be a true measure of your print server's throughput.

Unlike their workstation-based predecessors, network print servers do not have keyboards and displays by which they can be configured. Similarly, the limited "display" capabilities that network print servers include — usually at most a few LEDs — don't provide much of a means to indicate the device's status. This is where management tools such as MarkVision and JetAdmin come into play. They run on workstations or the file server and provide a means to configure the network print servers on the network, as well as to monitor their status.

Management utilities are generally included with the network print servers, so there is no explicit cost involved. But since management utilities are generally customized for the brand of network print server with which they are sold, there is no opportunity to mix and match between the print servers and the corresponding management utility. The capabilities of the management utility can have an important impact on the total cost of ownership of the printing solution, so it's important to analyze the management utility that is provided with the network print server. Make sure you can learn and use it easily and that it gives you the control information you want.

The management utility performs the initial configuration of the network print server. The extent of the configuration required depends on the network environment in which the print server will operate. Some environments, such as a simple TCP/IP network, might require nothing more than the basic IP parameters to be set. In a more complex environment, like NetWare, the print server may have to be configured more extensively with server names, print modes, advertising options, and more.

The configuration work required does not stop with the network print server itself. Workstations and file servers in the network environment often need to be configured to "push" jobs to the network print servers. Again, the extent of this configuration depends on the nature of the network environment in which the adapter is to operate, and can be as simple as creating a simple LPR port or as extensive as the creating queues and print server software on a Novell NetWare server.

Finally, the management utility plays an important role in monitoring the status of the network print servers and the printers they are servicing. Management utilities send data to and from the network print servers in order to establish the status of the devices, and in many cases, to maintain an ongoing representation of that status. The various utilities use different mechanisms of communicating this information, and some can involve a substantial amount of network traffic. The most efficient mechanism involves the ability of the network print server to report problems or changes to its own state or the state of the printer to which it is connected without requiring the management utility to periodically poll the devices.

Small and medium businesses should look for support of Microsoft's SBS (Small Business Server) solution. This support provides the functionality, from one click of the mouse, to automatically discover printers, automatically create printer objects, and install drivers, all with no need for dedicated IT (information technology) support.

Price also drives network print server selection decisions. This market is becoming very competitive, which helps the consumer. This is not to say that there is still not a diamond in the rough. If you weigh price against performance, features, functionality, management tools, and total cost of management, the field narrows considerably.

INSTALLING AND SETTING UP A NETWORK PRINT SERVER

This information is presented in a generic format due to the fact that different vendors use different techniques. We'll cover the basics and highlight some issues to watch.

Getting the network print server on the network is not very difficult. When installing an internal network print server, you only need to install the adapter into the connector/slot of the printer and connect an active network node cable. Since this is an integrated solution, usually no setup is needed for the adapter to talk with the printer. In the case of the external network print server, you need to supply power, a network connection, and attach the serial or parallel printer. Some vendors require special cables for these connections; check with each vendor on the requirements. Depending on what type of printer is attached to the external network print server, port setting changes may need to be made. An example may be to turn off ECP (Extended Capabilities Port) on the parallel port or change the baud rate of the serial port. Each vendor has different techniques for setting such parameters.

Getting the adapter up on a TCP/IP network is as simple as setting a TCP/IP address, netmask, and gateway. If the network is running DHCP, BOOTP, or RARP, the adapter will automatically obtain these necessities. If not, these parameters will need to be set manually. This can be done by many different methods. If you are running a different protocol on the network you may be able to use the management tools supplied to that protocol to make the TCP/IP settings. If the adapter is an internal model, you can use the printer's operator panel. After the address is set, simply send the jobs via FTP, LPR, or another supported method.

Setting up the adapter for NetWare is not as simple as TCP/IP unless the management tools are really good. Below is an overview of general NetWare configurations.

NetWare Configuration

In general there are two steps to configuring an embedded print server.

1. Generating the objects on the file server(s).
2. Configuring the embedded print server to use those objects.

Step 1 can always be accomplished using NetWare's pconsole or nwadmin utilities. Step 2 has always involved using the company's proprietary utilities to configure their print server. The best utilities also allow object creation (Step 1) in both a bindery and an NDS environment, making this task significantly easier for the supervisor/administrator.

Print Server Bindery Configuration. For the single port adapters, the object creation is the same — one print server object and one queue, which points back to the print server object. The adapter will always need the print server object name and generally will need the file server name. Some adapters in the past tried to search ALL file servers on the network. This method generates much network traffic and generally can't be done in a reasonable amount of time on larger networks.

There seem to be two schools of thought for the multiport adapters. The first is to assign a print server name to each port, allowing the network adapter to login in multiple times getting the appropriate queue's print jobs to the appropriate port of the adapter. This has the disadvantage of using extra "licensed" connections on the file servers (costing money). The second approach involves using the objects themselves to hint the adapter into getting the jobs from the queue to the proper port. In one company's case, this involves generating an extra print server object per port that has the name of the original print server objects plus a port extension. The queue then adds this extra print server object to its allowed service list.

Print Server NDS Configuration. Like the bindery configuration above, there are multiple methods for the adapters to be NDS configured. Minimally, the adapter needs the print server name (possibly per port), the NDS tree name, and the context that the print server object(s) resides in (to minimize search time). Due to the underlying nature of NDS, that generally gets hidden from the users, some companies also require file server names whose volumes host the print queues.

Again, some companies assign a print server name to each port of the adapter, requiring multiple licensed connections to the NDS tree. Some companies have opted to use the printer objects to give hints as to which queue's jobs are supposed to go to which adapter port. This might involve a naming scheme for the printer objects or possibly utilize the print object number to map it to a port.

Not all network print servers require so much configuration by the administrator. Look for simple graphical interfaces that clearly identify problems and alerts. Avoid configuration restrictions that violate your network naming standards.

EMERGING STANDARDS FOR THE INTERNET AND PRINTING

Overview

The growth of the Internet has far exceeded the expectations of the entire information technology industry. The Internet is a global internetworking infrastructure that connects industry, commercial, and personal computer users needing to communicate. This infrastructure is used by most people for electronic mail (e-mail) and for retrieving a wide variety of information using the World Wide Web. While much work has been done by various consortiums and companies to improve the Web, and especially the appearance of information when accessed using Web browsers, little effort has been spent to address the use of this infrastructure to send print documents to remote printers.

What Is Printing through the Internet Used For?

Today's fast-paced corporate environment has moved communications from first-class mail to overnight mail to fax and, most recently, e-mail. Even the most mundane of communications is now sent via fax or at least overnight mail. While fax and overnight mail are efficient in what they do, each has a number of problems:

- Fax:
  - Typically, low resolution and therefore low quality
  - Often expensive when involving long-distance phone charges
- Overnight Mail:
  - Not as fast as fax
  - $10 to $20 for a typical document

In addition to the need to communicate and send documents from one company or individual to another, the application of Internet technologies within a corporation (often called an "intranet") has grown even faster than the public Internet. As such, tools and programs to allow users to select, install, and use a wide range of printers and print servers connected to an intranet are also needed.

Here are several examples of using the Internet for printing:

- A business analyst wants to print another company's financial report, stored on a public Web server, on a shared departmental printer. The analyst locates a suitable printer using a Web browser and then submits the print request to the printing system by providing the URL of the document. The document is retrieved and printed by the printing system, which then notifies the researcher.
- An independent insurance agent wants to print a copy of a report on a public printer at the home office of one of the insurance companies that she represents. She then chooses Print from her application's file menu, and enters the URL of the home office's public printer. The request is transmitted to the printing system in the home office and printed.
- While more and more computer users have access to printers at home and in the office, both color and black-and-white, many of these are

just too small or too slow to be used when we have a really big job. The Internet Printing Protocol could be used to access the print shop down the road — Kinko's, SirSpeedy — for school work or business. Work can be done at home or even on the road in the hotel, and then the report, the proposal, or the set of transparencies can be printed on a high-quality printer at the print shop.

- There is now a vast and growing amount of information available to each of us every day from many sources, including magazines and other periodicals covering everything from worms to words, from gambling to gardens, from astronomy to antiques. IPP could be used by a "clipping service" to scan the periodicals covering the subjects you are interested in and create a custom newspaper or newsletter just for you — send it to you through the Internet and print it on paper on the printer for your office reading in the morning.

Due to the very nature of the Internet, any solution for printing must accommodate the heterogeneous nature of the computers and printers connected to it. One way for the computer printer industry to accomplish this and be able to deliver solutions to its customers is to develop open, public standards addressing these needs.

INTERNET PRINTING STANDARDS EFFORTS

In an environment like the Internet and even an intranet, a wide variety of computers and printers will be interoperating. Without open, platform-neutral, international standards, this interoperability would not be possible. Recognizing this, the printer industry through the Printer Working Group and the Internet Engineering Task Force (IETF) has developed a standard for Internet printing (that also works for intranets.) This effort, known as the Internet Printing Protocol, or IPP, is now emerging as one of the best solutions for distributing printing within and between enterprises.

Using the IETF's standard development model, a series of IPP proposals called *Internet Drafts* have been created which document all the components of an Internet printing solution. Over time, many of these documents will become *Requests for Comment* (RFCs), which are the proposed, draft, and final versions of these standards. These documents are available from the PWG's website (http://www.pwg.org/ipp) as well as from the IETF's site (http://www.ietf.org).

Several points about the design of IPP are important to consider:

1. IPP is an object-oriented protocol. It has a defined a set of objects:
   - An IPP printer
   - A print job

These objects have attributes and operations that are defined to act upon them. Operations like:

- Print Job
- Cancel Job
- Get Jobs

and Attributes such as:

- Job Name
- Printer Name
- Copies, priority, media size, and many others

2. The protocol is designed to be scalable — it must be implementable on servers (like NT or Sun Solaris) but also down in the printer itself. In many environments, no traditional print server will be available. The printer will connect to the network, accept the print job, and print the document.
3. It must be relatively simple — easy to implement and use. To that end, there are few mandatory operations or attributes defined by IPP that must be supported; there are only six mandatory operations.
4. IPP operates over HTTP, the underlying protocol of the Web today. An IPP operation is simply an HTTP post and the response back from that post. While IPP operates over HTTP, a Web browser is not required to use it. A complete Internet Printing System might use a browser to help locate a printer and view its status, but IPP itself will not require the use of that browser in order to print. Because of the simplicity of its design, the modification or enhancement of an existing Web server to support IPP is expected to be straightforward.
5. Because it operates over HTTP, all the HTTP security methods such as basic and digest authentication and transaction layer security (or TLS) are supported. The proper use of these security techniques can be applied to prevent "SPAMMING" and other unauthorized usage of an Internet printer, as well as to provide security for the content of the document being printed, including personal, financial, and other confidential information.

While the documents describing IPP are, in many cases, too technical for the general reader, they are important for understanding and planning for this emerging print technology. Administrative issues, including modifying or configuring firewalls to pass IPP, must be addressed early in the planning process. It is expected that embedded print servers (both internal and external) as well as other servers that offer print services (such as Windows NT Server and Novell NetWare Servers) will office IPP-based printing in the very near future.

SUMMARY

Printing for work groups has changed dramatically over the last few years. Print server technology now allows multiple printers to be placed where it's most convenient for users and provides many management utilities to track usage and troubleshoot problems — even simple ones like running out of paper. Decision factors in determining which network print server solution is right for you include cost, performance, protocol support, ease of configuration, and ease of learning the management utilities.

The growth of the Internet and the World Wide Web have vastly exceeded the expectations of the original creators of both. Many people believe that they are a fad, but the reality in today's world is that a worldwide communications network has become a requirement for many businesses and individuals. IPP brings this worldwide communications network to printing.

Chapter 30
Fax Servers

Max Schroeder

This chapter will provide a basic overview of fax servers and focus on some of the management and administration considerations that should be considered before implementing a fax server solution on your network.

As with any other server, a fax server provides the network with the benefits typically associated with the sharing of devices — in particular the convenience of both sending and receiving faxes at the desktop. A fax server can also deliver many other even more significant business benefits when implemented properly. In fact, a fully integrated network fax system can open up entirely new application and communications areas.

Since fax servers are based on existing fax standards, a brief review of those standards is in order. Current fax technology is based on the Group 3 fax standard, which was approved by the International Telecommunications Union (formerly the CCITT) in 1980. There is also a Group 4 digital standard, but since so many of its features have been included in Group 3 — including digitizing — Group 4 is not of immediate relevance here. Group 3 is an analog standard that, when introduced, used a compression technique called Modified Huffman (MH). Maximum transmission speed was defined as 9600 bps. Today, the defined transmission speed has been increased to 14,400 bps, and even more powerful compression techniques have been added to the standard. The ITU is soon expected to approve a maximum speed of 28,800 bps.

However, the transmission speed of a modem in a fax server is dictated by the speed of the receiving machine. So even with an upgrade to 28,800 bps, the majority of the installed base of fax machines will still be running at 9600 to 14,400 bps. Also, "low" resolution is defined as 98 vertical pixels by 203 horizontal pixels. "High" resolution offers an increase in the vertical pixels to 196. This is both the good news and bad news for fax servers. The bad news is that you still cannot send at the basic resolution of laser printers (300 dpi) although that may change soon. The good news is generating faxes electronically dramatically increases the quality of the image, since the distortion caused by the physical mechanics of the scanning process is eliminated.

0-8493-9823-1/00/$0.00+$.50

With these basic functional parameters in mind, it is important to look at some of the distinguishing characteristics of fax servers. In managing and administering a fax server, the first question to ask is how it is to function in the context of the business. Fax servers can offer much broader benefits than the mere sending and receiving of individual faxes at user desktops. Fax servers can be used to extend the functionality of workflow applications, transaction processing systems, document management platforms, customer service solutions, and other strategic IT assets. A fax server that is not designed to interface effectively with these other networked systems will be difficult to integrate and maintain as it is asked to support an increasingly diverse set of business processes. So a basic caution for the ongoing management and administration of a fax server is to select one that has an open architecture. This typically includes APIs and other programming support, as well as pre-engineered links to popular applications such as Lotus Notes, SAP R/3, and the like.

Integration with the full range of corporate IT assets should therefore be strongly investigated before implementing any fax server solution. For example, if an organization is a heavy user of the Internet and a messaging platform such as Microsoft Exchange Server, then a browser-based interface for the fax server and an Exchange "Connector" are probably mandatory. Without them, the administration of user access via the Web or Exchange's management facilities will become problematic. If hard-copy documents are going to be scanned for faxing or e-mailing as attachments, then support for the associated hardware, such as HP's ScanJet 5, should also be fully supported by the fax server software. Other technologies and applications such as Interactive Voice Response (IVR), remote access, telecom accounting, customer and enhanced fax services, such as those from Xpedite and MCI, are also key considerations.

The various ways in which the fax server will be applied to business requirements are also important to consider because of another key consideration associated with the long-term manageability of any enterprise fax solution — capacity. Typically, one fax port is required per 8 to 18 people, depending on the nature of the business and the urgency of the fax traffic involved. The fax server can also be configured to dedicated specific fax ports to priority usages. Additional information on this subject is available on the Web at www.eema.org/committees/fax/faxplatform.doc.

Other factors that will influence the number of lines required will be the type of modems used. The subject of "Class" has caused much confusion over the years. Class 1 is a series of Hayes AT commands that can be used by fax software to control fax boards. Class 1 modems are very limited in capacity. For example, with Class 1 modems both the ECM/BFT (error-correction mode/binary file transfer) and the T.30 (data packet creation and decision-making necessary for the call set-up) are done by the host computer, which places heavy demands on its processor.

Class 2 was published as a "ballot standard" by the EIA/TIA in 1991. The "ballot" failed due to dissension over some technical issues. However, a number of manufacturers jumped in and began manufacturing modems based on the "ballot" specifications, resulting in a large base of Class 2 devices and thereby creating an unofficial *de facto* standard. The TIA continued with the specification development, eventually producing the official Class 2 standard. To distinguish it from the former "ballot standard" it is called Class 2.0. Class 2/2.0 modems have the ability to buffer small amounts of data, which reduces some of the load from the PC. However, fax servers still have to deal with the burden of managing multiple interrupt-driven serial ports.

Manufacturers of high-end fax modems designed for network faxing determined not to use the Class 1 or 2 approach. For reliability, scalability, speed, and the ability to provide intelligent communications features, they decided to add robust processing power to their boards. Thus, the CPU in the host fax server communicates with the CPU on the fax board. These high-end modems are the optimum choice in any network environment for many reasons, including speed, port availability, and reduced cost-of-ownership. For additional information, please refer to Davidson Consulting at www.pdavidson.com (which includes white papers on cost comparison) or to Network Subscription Service: Enterprise Faxing (published by Network Subscription Service Ltd. in the U.K.: Network House, PO Box 297, Bedford, MK4411YR, Fax: 44 (0) 1933 59021).

High-end fax modems also provide support for automatic inbound routing, which is obviously another key communications management issue. Recently, the ITU approved a routing method entitled T.33 that allows for additional digits to be added to a fax number so that the receiving device can route it automatically. However, until this is more widely implemented it will be of little value. Fortunately, there are currently several useful ways to route messages over the enterprise network:

- *Manual Routing* — Quite simply, this requires manual intervention by an administrator who scans cover pages to determine who the fax should be routed to. While this technique may seem clumsy, it is still far superior to physical distribution of paper faxes. It is also more secure, since the routing administrator can only view the cover page.
- *OCR Routing* — By using Optical Character Recognition, a cover page can be scanned for recipient information, allowing the document to be routed to the proper person with a fair degree of accuracy. However, until handwriting recognition is further improved, some manual intervention will be required.

- *DTMF (Dual-Tone Multifrequency Routing)* — This simply means that the sender supplies additional numbers ("tones") when sending a fax. The primary limitation with this technique is that the person sending the fax must know the recipient's DTMF code.
- *CSID Routing* — The line you see at the top of a fax page is called the Call Sender Identification (CSID). In the U.S., for example, it is a legal requirement to include the fax number of the sending machine or device. This phone number can then be inserted into a look-up table to be used to route the incoming document based on its origin. So, for example, salespeople who handle specific customer accounts could receive all faxes coming from those customers.
- *DID Routing* — DID stands for Direct Inward Dialing (in the U.K., Direct Dial Inward [DDI]) and is the best routing choice available today. A similar technology is also available with digital lines such as ISDN, T1, and E1. Basically, this consists of one or more trunks between the corporate office and the Central Office (CO) of the local carrier. This is a one-way trunk that only allows for incoming calls. The carrier leases a set of phone numbers that can be assigned to individual employees. When one of these numbers is dialed, the call is routed through one of the trunks and the phone number is passed on to the company's Private Branch Exchange (PBX). This technique allows a large set of phone numbers to be routed through a limited number of trunks, thus reducing costs. In the U.S., some local carriers are leasing blocks of numbers for as little as $1 per number per month.

The other consideration in configuring fax routing is how to deliver it to the user. In the past, users were often given a desktop fax application and viewed their faxes there. However, the growing emphasis on a more unified approach to messaging has motivated many organizations to route faxes as e-mail attachments, thus giving users a single communications management interface.

Other organizations use a shared work group printer to deliver faxes. This is typically done to give several users access to certain types of incoming faxes, such as customer orders. Other organizations send faxes directly into their workflow/document management applications, so that they can automatically trigger a network-based business process, such as the resolution of a customer complaint.

In today's high-pressure business environment, fax servers can also deliver significant value by enabling knowledge workers to get access to faxes when they are out of the office. Thus, in addition to providing internal routing, the fax server should provide some means of remote retrieval or remotely triggered fax forwarding. This can be done through dial-up access, or by using a browser-based agent for situations where the Internet is used for remote access. Users may want to view the faxes they've received back at the office on their laptops, or may instead choose to have them sent to a conventional fax machine at the hotel, office, or airport lounge where they are at the moment.

Fax servers can also reduce transmission costs in a variety of ways. The most popular method is known as Least Cost Routing (LCR). With LCR, faxes are routed over the corporate WAN or VPNs to the fax server closest to the ultimate destination of the fax. Thus, for example, a company with headquarters in New York and a sales office in London would route any faxes bound for Great Britain from New York to the London office's fax server first, and then on to its in-country destination.

Another way to achieve lower transmission costs is to employ an outside fax service which has its own global network and multiple points-of-presence (POPs) in other countries. This technique can be especially useful for large-scale broadcast faxing, where the service provider's economies-of-scale become particularly applicable. Service providers who use the Internet for this purpose are now emerging, offering even lower costs for international fax transmissions. By delivering documents to these providers electronically, fax servers are the ideal conduit for interfacing with these services.

There are a variety of features that make the daily administration and maintenance of the fax server much simpler. First of all, access to administrative facilities from any network or remote PC is an obvious requirement. Automated self-diagnostics for potential problems such as failed ports are also essential. To avoid duplicating administration, synchronization with existing directory services is a real time-saver. There are a variety of parameters associated with fax server use, and each of these must be controlled by the server administrator. These include rights to such operations as modifying shared phone books, the ability to set up temporary fax forwarding, choice of supported file formats, etc. The ability to set and modify these parameters both individually and by logical groups can greatly streamline operations.

Because the fax server is a telecom asset, in addition to being part of the data network, server administrators should be aware of corporate cost-recovery requirements. Fax servers can supply complete data on fax activity, including the customer and (or) project associated with each call. These activity reports, combined with third-party tools such as Crystal Reports, are important for controlling corporate telecommunications costs.

The growing trend toward integration of telecom and data technologies makes the choice of a fax server particularly critical today. For many organizations, it may be best to work with an outside organization with experience and expertise in CTI (computer telephony integration). The rapid changes taking place in the public switched network, the Internet, and corporate intranets will all have an impact on how fax is applied to business. Fax usage continues to grow rapidly even as e-mail and other types of communications flourish. So the real question is how to integrate fax into the ever-expanding communications landscape, in order to provide reliable, automated, image-based communications between internal users and the outside world.

That integration requires robust links to corporate messaging platforms such as Microsoft Exchange and IBM/Lotus Notes, as well as next-generation platforms such as Octel's Unified Messenger. By integrating tightly with these platforms, fax servers can be administered and configured as part of the overall IS environment, reducing operational costs and delivering additional functionality to every IT asset across the enterprise.

Max Schroeder is the Chief Operating Officer of Optus Software, Inc., a long-time leader in the fax server industry. He is also the former President of the International Computer Fax Association and the current chairman of the Computer Fax Committee of the Electronic Messaging Association.

Chapter 31
Remote Access Concepts

Gerald L. Bahr

INTRODUCTION

As we near the end not only of this decade but of the century, my crystal ball tells me that we are somewhere in the middle of the information age. With not only the ability, but the need to have ready access to all sorts of information from anywhere on the globe, we have come to expect that we will have ready availability to any information that we need, day or night. Whether we are in the office, telecommuting, or temporarily mobile, we naturally expect to have access to the resources necessary for us to do our job efficiently and accurately.

This chapter will discuss two different approaches — remote node and remote control — with variations on remote control for accessing data, whether we are temporarily mobile or hardly ever seen in the office, building on the resources of Microsoft's Windows NT 4.0.

DRIVING FORCES

With the advent of the Internet, intranet, telecommuting, e-mail, and voice mail, we have come to expect that we can have ready access to any data warehouse that will make users more productive, entertain, or make our lives more productive.

It is estimated that many large corporations are spending between $500,000 to over $1,000,000 per year to support remote access for their employees. This does not include the costs for the laptops that are used in the office as well as on the road. It is directed only toward server hardware and software and supporting equipment and dial-up access lines.

TWO APPROACHES — REMOTE NODE VS. REMOTE CONTROL

The remote node approach lets the remote client act just like another computer on the network. All files that are accessed by the remote client are

0-8493-9823-1/00/$0.00+$.50

transferred across the line. The responsiveness that the remote client experiences depends on the size of the file and speed of the line.

The remote control approach has two variations. These two variations use a computer as the "server" located at the central site to provide all the computational power and ability to communicate with all local resources. Remote control sends only key strokes from the remote client to the server, and screen updates from the server come across the line to the remote client. This can easily increase the responsiveness to the remote client by two to four times or more than that of the remote node approach.

Both the remote node and remote control hardware configuration and topology are essentially the same. What makes the difference in operation is the software on the Windows NT Access Server. Each of them has advantages and disadvantages.

Remote Control — Two Variations

Variation 1. Software is loaded on a dedicated "server" computer running Windows NT. This software supports someone calling in and taking control of that computer. Literally, anything that can be done from the local keyboard can be done from the remote client. This method requires a dedicated computer for each simultaneous caller. If remote clients are using different operating systems, you will most likely need to have different telephone lines directed to a specific server supporting that operating system.

Variation 2. This variation of remote control runs software that allows many simultaneous users (primarily determined by the applications being accessed and the configuration of the computer) to call into the same computer. Each remote client runs in its own environment within Windows NT; thus each remote user looks like a separate entity on the operating system and the network.

Since neither one of these methods requires serious computing power on the remote client side (all computational work is done on the server), the user might get by with a 286 running DOS or Windows 3.1 or a 386 or higher class machine running Windows 3.1, Windows 3.11, Windows 95, or Windows NT 4.0 Workstation. This can allow some companies to take advantage of the capital investment that has not been depreciated yet.

Advantages

- Client workstations generally do not require upgrades very often.
- Deploying a new application is simplified because it is handled at the central office.
- Upgrades are centrally administered.

- Even though the application is running on the central server, users can normally still load or save files to local drives, print to locally attached printers, cut and paste information between a remote and a local application, or drag and drop to copy files in the background.
- The look and feel of the applications is the same on the road as in the office.
- The remote application performs as fast as it would when running directly on the LAN.
- Users have access to the same database, along with all other productivity applications they are familiar with.
- Large databases are left intact; they do not have to be segmented.
- Sensitive information is not replicated to laptops; it can remain within the secured boundaries of the LAN.
- Remote workstations, which lack sufficient processing power, are able to run work group applications on-line from almost anywhere.

Disadvantages

- Normally there is a large up-front cost.
- It does not fit well into small systems.
- If a server fails, all of the users on that server are not serviceable.
- It is potentially a large network traffic generator.
- It may need to be placed on its own network segment.
- It requires a powerful computer for the server.
- It requires the purchase of additional software for the server.
- Some applications, like e-mail, are more difficult to use in the off-line mode, because remote control does not automatically download the files needed to work off-line.
- If the single session option is selected, it might require different telephone numbers for each remote client operating system.
- Off-line work is more difficult, because files are not available on the local workstation.
- This may require more support lines because on-line time may be longer (some off-line functions not supported).

Remote Node

With the introduction of LANs (Local Area Networks) the philosophy of shared wiring, high data transmission rates, and local computing (smart devices) came into being. This approach uses the LAN wiring system to download executable files as well as large image and database files to the local desktop computer.

Remote node access mirrors this same approach. Since you must transfer all files across the modem line (those which you are using to gain access), you must have all of your applications loaded on the remote client. Even with the applications loaded locally, the user may get discouraged with the responsiveness of the system if the file accessed is more than 50 to 100 KB.

Advantages

- Smaller up-front cost.
- Look and feel (visual) of being locally connected to the network.
- Might be able to use an older, less costly computer for server.
- Packaged with Windows NT.
- All commands, either command line or Windows, work just the same as if you were in the office.

Disadvantages

- Replication with several large databases takes too long and consumes a lot of disk space on the client system.
- To make replication more effective, large databases may have to be segmented into smaller databases. This can dramatically complicate the management of these databases.
- Because of the replication architecture, laptops carry sensitive corporate information. If your laptop is lost or stolen, security breaches can (and often do) occur.
- The client component requires a workstation with a minimum of a 486 processor, 8 MB of RAM, Windows 3.1 or Windows 95, and a minimum of 500 MB of disk space.
- Could generate some serious network traffic, depending on the number of users per server.
- Remote users often get discouraged waiting for large files to load.

Both remote control and remote nodes have been developed to support remote users. The one which is best for you depends on your computing environment and the method you use to provide the most features, functionality, ease of use, and security. Luckily, products are available that will support either approach on Windows NT 4.0 Server operating system.

NT REMOTE ACCESS SERVER

Dialing In

Determining the Number of Lines Required for Support. When considering the number of hunt lines you will need to support the number of remote users you have in your organization, keep in mind both the number of users and the type of applications which your users will be accessing. The telephone companies typically use a 1:10 ratio for the number of lines required vs. the number of subscribers. Keep in mind, however, that statistically the phone is only used for about 3 to 5 minutes at a time and usually with long periods of time in-between.

This ratio may be reasonable for remote users who only access their e-mail; however, if users will be using the network for on-line file updates or any heavy office work, the line usage rate will go up considerably. Local virtual office users may tie up the line for 8 hours or more at a time, while other users who are traveling may very well tie up a line for an hour or more at a time.

Local users will probably use the lines more in the daytime, while those who are traveling will most likely use the lines more during the evening hours. If you are looking for nonblocking access, you will need to consider incorporating a high number of lines for users — conceivably even a 1:1 ratio. Most modem pool systems range from between 1:3 to 1:8 lines to accommodate the number of users needing remote support.

Be sure that the number of lines that you finally determine to incorporate are placed in hunt groups (rotary) into common servers so that one number will serve many common users simultaneously.

Methods of Access

Modem Pools. Modem pools allow you to share the expense of modem hardware as well as the number of dial-in lines necessary to support the number of users who will be calling into your system. A modem usually supports analog voice dial-up lines with speeds normally at 33.6 Kbps. Some modems support 56.6 Kbps today, but are still having problems in trying to establish a real link at that speed. Keep in mind that many times a 33.6 Kbps modem won't establish reliable communications even at 33.6 Kbps speed because of noisy lines. This problem gets worse as you try to establish an analog link across the country. Aren't we glad that ISDN (Integrated Services Digital Network) is coming?

A modem pool can be configured from stand-alone modems that are already available or can be rack mounted with modems that are built to fill specific slots within the rack cabinet. If you anticipate the need for more than 12 modems, it would be wise to look at a rack-mounted unit that will provide for better support and cleaner installation.

A Windows NT Server with RAS (Remote Access Service) can support up to 256 simultaneous users and provide you with a stable, secure environment for dial-in clients. Choosing hardware to support that many users is often not easy. Most interface devices provide access to multiple modems or ISDN terminal adapters; however, they are only available in 4-, 8-, or possibly 16-port versions. This may be sufficient to get you started, but large systems will probably grow out of this configuration very quickly.

ECCI has a selection of devices that will allow you to provide from four-port ISA adapters to a rack-mounted modem pool unit that will allow you to reach or exceed Windows NT's limit of 256 sessions or beyond with port speeds up to 115-kbs transfer rates. If you anticipate serving more than 64 dial-in devices from a single server, you need to consider using a SCSI interface to external support boxes. This will allow you to chain more than seven rack-mount-type boxes that will support multiple modems. With this approach, you can physically support more than the 256 simultaneous sessions.

Be careful, however. Don't exceed the load demand that the server itself can handle. This will vary depending on the manufacturer, the number of CPUs, and the amount of memory and I/O bus that your server can handle. If you anticipate numerous lines being serviced by a single server, you will want to use I/O boards that support their own CPU on board. The boards will cost more, but they will help to offload the server CPU and will allow you to provide high performance with less expensive server hardware.

ISDN Access. With the advent of telecommuting and Internet access, ISDN for wide area access has grown considerably in the last two years. The advantages of ISDN are many.

Advantages

- With ISDN you have dial-up access just the same as with the POTS (Plain Old Telephone System) that we are all used to. The only caution is that it is not available in all parts of the country.
- Instead of being limited to the high end rate of 33.6 Kbps with the modems mostly in use today, you can get access at 128 Kbps. With compression, a realized throughput can be as high a 500 Kbps.
- You get two telephone lines that can be used for either voice or data or a combination of both. If you are using the two B (bearer channels) for data, you will have a data rate of 128 Kbps.
- Without band signaling on the D channel you can see (using caller ID) who is calling, even when both B channels are being used. If you answer the phone, then the data automatically rolls back to 64 Kbps. When the call is completed, it will return to the 128-Kbps data rate.
- ISDN will connect within 3 to 5 seconds vs. the 20 to 40 seconds for the POTS network we are used to.

Cautions

- Not all areas of the country support ISDN at this time, so be sure to check with your local telephone company to determine availability.
- Because it is fairly new, however, not all personnel are familiar with it so you need to be specific with the carrier as to what you want.

- Standards are not fully developed yet, so stay with vendors who are reputable and well established in the market. As in the early stages of modem usage, not all manufacturers support the same features, particularly compression technology. This is probably still the biggest problem between ISDN equipment manufacturers, so if you plan to use compression, it would be wise to use the same vendors and equipment on both ends. You may wish to verify that the manufacturers you are considering will support the same compression and will be able to communicate with each other.

FRAME RELAY — PUBLIC OR PRIVATE NETWORK

Frame relay is a technology that you will need to support if you have your own private network which you wish to use to connect remote offices together. It works well for bursty traffic and is typically found in computer WANs (Wide Area Networking). If your company is planning to connect to the public networks that are primarily dedicated to computer networking, it is almost certain that you will be using frame relay technology.

If you have your own WAN implementation, you will only need to know the technology being implemented when you purchase and configure equipment. In this case, you will want to use a router to interface between the WAN and the LAN side of the networks. If you connect to a public network such as the Internet, you will want to place a firewall between the WAN and LAN interfaces.

Routing

Windows NT Routing vs. an External Router. Windows NT 4.0 will support routing of various protocols. The most popular and the one I recommend is TCP/IP (Transmission Control Protocol/Internet Protocol). This protocol is extremely robust and has been designed and tested for the worst conditions in WAN applications. All WAN external routers support TCP/IP, so you are virtually guaranteed support. Windows NT 4.0 will support it in native mode and, unless you have a good reason to do otherwise, I would recommend that this be your protocol of choice.

Windows NT Server 4.0 supports MultiProtocol Routing as part of the operating system. This is an integral part of the operating system that lets Windows NT route a number of protocols simultaneously. This is fine for small applications, since the routing that Windows NT supports is not what you would call fast or fully functional.

If you are anticipating the server to support a small number of users connected to a small LAN (less than 25 nodes) and not connected to any other network, then you might consider using the routing services provided by Windows NT 4.0. It supports Novell's IPX/SPX transport, so you don't have to deploy TCP/IP on your internal network to give your enterprise Internet access. This provides some degree of security since IPX/SPX does not support client-to-client communications, thus preventing someone from the outside gaining access to any devices other than servers running IP. In this environment, this would make for a good solution since it also negates the need for a firewall.

If, on the other hand, you have a lot of users that will be calling in and (or) the remote access server is connected to a large LAN, then you will want to strongly consider an external, dedicated router as the interface between the LAN and WAN that you interface with.

An external router typically will support more protocols and be able to process them faster and in greater detail (i.e., filtering, routing, etc.) than would the routing features of Windows NT 4.0. Because an external router is dedicated to only one service, it can also handle the number of packets that are required of it much faster than the Windows NT server that you are asking to do double duty.

Keep in mind that if you want a speedy response and wish to reduce your possibility of failure, you may want to consider using more units to provide the functions that are critical to your organization's needs.

Firewall for Internet Connection. A router was originally used to provide firewall activities. Many people still use a router to serve that purpose as well. There is, however, a philosophical difference between a firewall and a router. In essence, a router is configured to pass all traffic unless explicitly told not to. A firewall is configured not to pass any traffic unless explicitly told to.

This makes a lot of sense since both can be a bit tedious to configure. You might not think of all the rules or ways in which you want to deny others access to your network; therefore, if they must go through a firewall you have had to take action to allow them in. Just make sure to be careful to whom you authorize access in your company.

The reason for this discussion will become clear when we discuss later another alternative for remote users to access your network via the Internet using PPTP (Point-to-Point Tunneling Protocol).

IP ADDRESSING (PPP FOR DIAL UP)

Dynamic IP Addressing

TCP/IP is the leading protocol for connecting dissimilar platforms and operating systems. It has been tested under virtually all adverse conditions, and because it has been around for a long time, it has developed into a very robust and routable protocol. It is based on a four-octet address, however, which has limited its growth to the number of "hosts" (the name given to any computer that uses TCP/IP). TCP/IP has been divided into four classes of addresses. All the classes are assigned, with the exception of the class "C" address, and it is being "subnetted" to help give everyone reasonable access until the new numbering scheme called IPv6 can be effectively implemented.

Because all companies are faced with the same dilemma — running out of IP addresses — and because it provides easier maintenance of an IP network, many companies are implementing DHCP (Dynamic Host Configuration Protocol) servers.

DHCP Servers. If you have chosen to use TCP/IP as your primary protocol, then the protocol that you will use between the remote client and the server for dial-up access will be PPP (point-to-point protocol). This is the IP protocol that supports dial-up access. Since the number of IP addresses available is truly limited, you will want to use DHCP, which dynamically assigns IP addresses to users as they call in.

The main advantage of using DHCP is that it assigns addresses from a predefined pool of addresses, thus allowing users to share the same scarce set of addresses. It will also allow users to dial in via remote offices and use the company WAN (thus saving long distance charges) without worrying about the subnet class of address that the remote office might be using. In essence, it will give you and the user far greater flexibility.

Use of the DHCP relay agent will allow the server to relay DHCP messages to other DHCP servers. This allows your network to communicate with full-fledged DHCP servers on one network even if your network only has a DHCP relay agent.

DNS/WINS (Windows Internetworking Name Server). Because IP addresses are much harder for humans to remember than names, a system has been put in place that allows us to use names when we want to talk to other computing devices, while still allowing machines to talk to each other by numbered addresses.

This technology is referred to as DNS (Domain Name Service) for the Internet. Microsoft has developed another approach called WINS (Windows Internetworking Name Server) for networks that use Windows NT-type network operating systems.

The good thing is that Windows NT 4.0 supports both of them, either separately or simultaneously. Thus you can decide which is best for you and your particular environment.

USING THE INTERNET FOR REMOTE ACCESS

The Internet will soon provide us the ability to log on to the Internet from a local ISP (Internet Service Provider) in our local area and then gain access to our company data via the Internet. This obviously saves us long distance access charges, but can expose our company proprietary information to the world if we are not careful. Internet access adds an alternate method to gain access.

PPTP

Enter Microsoft's sponsored PPTP protocol. This specification is being jointly developed by 3Com Corp., Santa Clara, CA; Ascend Communications Inc., Alameda, CA; U.K.-based ECI Telematics; Microsoft Corp., Redmond, WA; and U.S. Robotics Inc., Skokie, IL. This new specification will allow companies to establish virtual private networks by allowing remote users to access corporate networks securely across the Internet.

Once PPTP is widely supported, the remote user would simply dial into a local ISP and then his work station wraps the data packets into an encrypted "tunnel" using PPTP. The data are then sent across the Internet into the corporate Windows NT 4.0 remote-access server. Once the PPTP packets reach the Windows NT server, they are routed, decrypted, and filtered (if internal routing is used) before being forwarded to the corporate LAN.

The advantage of this approach is not only the reduction of line charges, but the reduction in modem pools, and supporting equipment that must be maintained and upgraded.

Another nice feature is that PPTP requires few changes in client configuration. The ISP will need to make minor upgrades to their sites. Microsoft has also promised to make the specifications available to anyone. At this writing they are supposed to have submitted it to the Internet Task Force (a body that helps determine standards used on the Internet).

Scalability

Since Window NT 4.0 supports SMP (symmetrical multiple processors) and over 4 GB of memory with a wide variety of network interface adapters with over 256 modems, ISDN equipment, WAN adapters as well as LAN adapters — it is possible to start out small and allow your servers to grow with you as your needs grow.

If you use the server for remote node service, it will be able to handle more users with a given configuration than if it were providing remote control services. This is because it is much easier to route information than it is to run applications and then provide screen updates.

Although you could provide all the services on one server, there is a more important question. Should you? Keep in mind that the more you have on one piece of hardware, the more you have at risk when the hardware or software fails. Generally good engineering practice dictates that you spread your risk by having more than one way of routing or gaining access to the information that your users need access to.

Disaster Prevention

Backup/Restoration. When considering disaster recovery, disaster prevention is always the best policy. It has been proven many times that the cost of replacing the data (whether it be customer data or configuration data like that on a communications server) is almost always more than it is to replace the hardware and operating system software.

It is therefore imperative that a backup plan be developed and implemented. Next to enforcing security policies, this is probably the most important function that the network administrator can perform. The frequency with which the data will change will determine the frequency of backup. At the very least, two full backups should be made every month, with one copy maintained off site. This ensures that if one copy is destroyed by fire, flood, or theft you have another one that is no more than a month old. Obviously, if the data changes very frequently or involves a large number of changes, then weekly or even daily backup should be considered.

Also, make sure that you can restore the data that was backed up. Many people try to restore the data that was backed up, but don't have the proper drivers and (or) procedures necessary to restore the data from a total system failure.

Power Protection/UPS

The other major area for prevention consideration is power protection. This means providing clean, isolated power running on a true UPS system. You should not employ the switched-type UPS systems that is typically used in the small LAN servers or workstations. The true UPS units provide better line regulation and lightning protection than the switched units. Also make sure the UPS will support the server for a minimum of 15 minutes under full load, or longer if the rest of the servers/hosts are also on a UPS system. Remote users should have the same access time or notice to complete and log off as the local users do.

Security

The best security you can have for your system is a well thought out security policy procedure which is consistently enforced. The proper use of passwords should make them expire after a period of time, and certain secure logoff procedures should be enforced. These are all part of providing security to business-critical information.

For users at virtual offices, dial-back procedures should be implemented. With the right equipment and caller ID, this can be made rather automatic. This, however, may require them to have two different logon names and passwords to support the times when they travel.

Fortunately, Windows NT 4.0 supports multilevels of C2 security. Encryption has been added to this multilevel security with multilevel passwords and privileges, roving callback, encrypted logon and data, and file level security to protect data privacy and network resource integrity. Dubbed CAPI (for Cryptography API), it supports all of the most common encryption methods, including Data Encryption Standards and public-key encryption. It can handle digital signatures and transactions — where one party is validated by a licensed third party, commonly referred to as certification authority.

EQUIPMENT REQUIRED — CLIENT

Remote Control

Manipulation of files and calculations are made on the server; the computer that is carried on the road does not need the power or application support that is needed in the remote node configuration. In many situations a remote user might very well get by with a 286 processor with 2 MB RAM and a 40-MB hard drive running DOS or Windows 3.1. The biggest problem you might have will be finding a laptop of this vintage with a UART fast enough to handle the speed of a 28.8-Kbps modem.

If Windows 95 is used, then you will need to support a 386-class machine with 16 to 24-MB of RAM and at least a 500-MB hard drive in order to provide you with adequate service.

Remote Node

When you are using remote node operation, the client must have all applications loaded locally in addition to all of the files/databases that the user needs while away from the office. You need to have a minimum of a 486-class laptop with a large hard drive (today that means greater than 1 GB), a minimum of 16 MB of RAM (preferably 32 MB), as well as the other peripherals that you have determined the user needs.

Application Server Requirements

Remote Control Server

1. A minimum of 200 MHz Pentium or higher microprocessor with 64 MB of base memory (add an additional 8 MB for each concurrent user) with an EISA or PCI bus architecture
2. 32-bit disk controller recommended

3. 1GB hard disk drive recommended, not including application and data storage
4. A high-density 3.5 diskette drive
5. SVGA (800 × 600 × 16) video adapter
6. Intelligent multiport adapter or SCSI adapter supporting differential SCSI (depending on your modem pool selection)
7. 32-bit network interface card

NOTE: If your are going to use SMP in the server, depending on the application, a single four-processor SMP Pentium server could support 40 to 60 concurrent users — depending upon applications that are being accessed by the remote users.

SUMMARY

Windows NT 4.0 is an operating system that supports many applications. In this chapter we have looked at how it can support multiple forms of remote access. It can support small businesses with just a few remote users up to large corporations with hundreds of remote users.

The most cost-effective approach is to use the RAS (Remote Access Service) that comes packaged with the operating system. This approach will provide all the security supported by Windows NT and is fairly easy to set up. RAS will only support remote node access, which will provide many users access to their data warehouse. The negative side is that the files which are accessed will be downloaded across a rather slow 28.8-Kbps line (unless you are using ISDN) when compared to the interoffice LAN. The advantage is that you can work off-line with your laptop since you have all of your files on your local computer. When you get back to the office, you can use Windows NT's "My Briefcase" to synchronize the files that have been updated on your machine as well as those that were changed in the office while you were gone.

Remote control access provides the fastest access to your data back in the office. It will support multiple users simultaneously on the same hardware platform. It can provide all of the services you are familiar with in the office. Depending on the configuration, however, it may be slightly more difficult to work off-line on some applications, such as your e-mail. It is more expensive to implement initially because of the additional software that you will need to purchase as well as a more powerful computer which you will want to use for the server. It might be offset, however, because you can very possibly get by with remote clients that are less powerful. Depending on the applications that your company uses, you might even be able to use some older 286 processor machines running DOS applications.

If you prefer the remote control approach, but have a limited budget to get started, you may want to look at the less expensive remote control system that still provides most of the advantages of the multisession system. This requires a dedicated computer for each simultaneous user, but if you have only a few users who need access for a reasonably short period of time, this is a good choice.

Microsoft Windows NT has many options, but one of its strongest suits is its built-in multilevel security. When it comes to providing access to remote users, you are providing an opportunity for unauthorized users to gain access to your company's confidential information. This is an area that cannot be underestimated. Windows NT permits you to have a good start on providing the level of security that your company expects.

Although Windows NT will support up to 256 simultaneous connections, for reliability reasons (unless you have a very powerful computer with excellent power and line protection), it is not recommended that you use it in this mode. You should spread the risk of failure among multiple platforms if you begin serving between 60 and 100 simultaneous users in your organization.

Chapter 32 Communication Servers

Denise M. Jones

The benefits communication servers offer to the networking world, including branch office integration, deployment of enterprisewide and Internet e-mail, and the acceptance of mobile forces, have been leaked to the masses. IS managers around the world are being barraged with requests for connections, connections, connections. This chapter will provide an overview of the latest technologies communication servers are utilizing to satisfy this demand for enterprisewide connectivity and provide guidelines for matching the most cost-effective access method with each dial-up scenario.

WHO'S CLAMORING FOR CONNECTIONS?

Communication servers enable people to change the way they do business, giving them the flexibility to review e-mail, place orders, send faxes, edit files, run reports, and get real-time information from anywhere, at any time of day. There are three basic types of network users who require the functions performed by a communication server.

Telecommuters

Telecommuters are at-home employees who generally connect to the network from the same location each session. Their most significant problems are updating work documents and sharing those documents with others, accessing e-mail, and producing correspondence and presentations. They can often work without the network, but make occasional connections to access network files and resources. Telecommuters also need access to printers, fax machines, and e-mail systems.

Traveling Sales Force

Traveling sales people need many of the telecommuters' capabilities, including access to e-mail and network files, but instead of always connecting from the same location, they often move around from hotels to airports to regional offices. Salespeople connect to check inventories, enter or

0-8493-9823-1/00/$0.00+$.50

check on the status of an order, update customer accounts, and perform other functions that require access to large databases that can't be copied to a notebook hard drive.

Roving Executives

Roving executives want to stay in touch with their organizations and be informed about ongoing project status and general operations. They often perform many of the same tasks listed for the telecommuter or traveling salesperson, such as updating work documents or changing presentation, but their main priority during a dial-up session is sending or answering e-mail or scheduling information with other employees.

THE METHODS OF CONNECTING

When you consider the growing numbers of users requiring network communication services, the goal of maintaining any sort of coherent computing environment is clearly a challenge. Keeping workers up-to-date requires more than just a laptop and a modem. Those who frequently work from home or on the road know that you need a digital lifeline to the office to stay current. There are numerous approaches to dial-in remote communications, but all techniques fall under two fundamental classes: remote control or remote node.

Remote Control

With the remote-control approach, a modem-equipped PC on a network runs remote-control software, such as Symantec's PC Anywhere. This PC is referred to as the host. Exhibit 32.1 provides an illustration of a typical remote-control scenario.

At the remote site — which could be a hotel room, your home, or a client's office — you use a modem-equipped laptop or desktop PC running the client portion of the remote-control software to dial up the remote-control host. The software lets you utilize the host PC as if you were sitting at its keyboard by transmitting your keystrokes across the modem link and putting them into the keyboard buffer of the host PC.

A remote-control setup is a typical solution for users of large databases, accounting applications, and other shared files on the network. For these situations, the realistic approach is to leave the applications where they are and provide access to them for remote clients. Because the actual volume of data in a remote-control session is relatively low (only keyboard, video, and mouse movement updates), response time is good.

In addition to above-average performance, security remains intact, data stays on the host or network and can be backed up centrally, and multi-user applications work as they should. However, you do pay for this sometimes

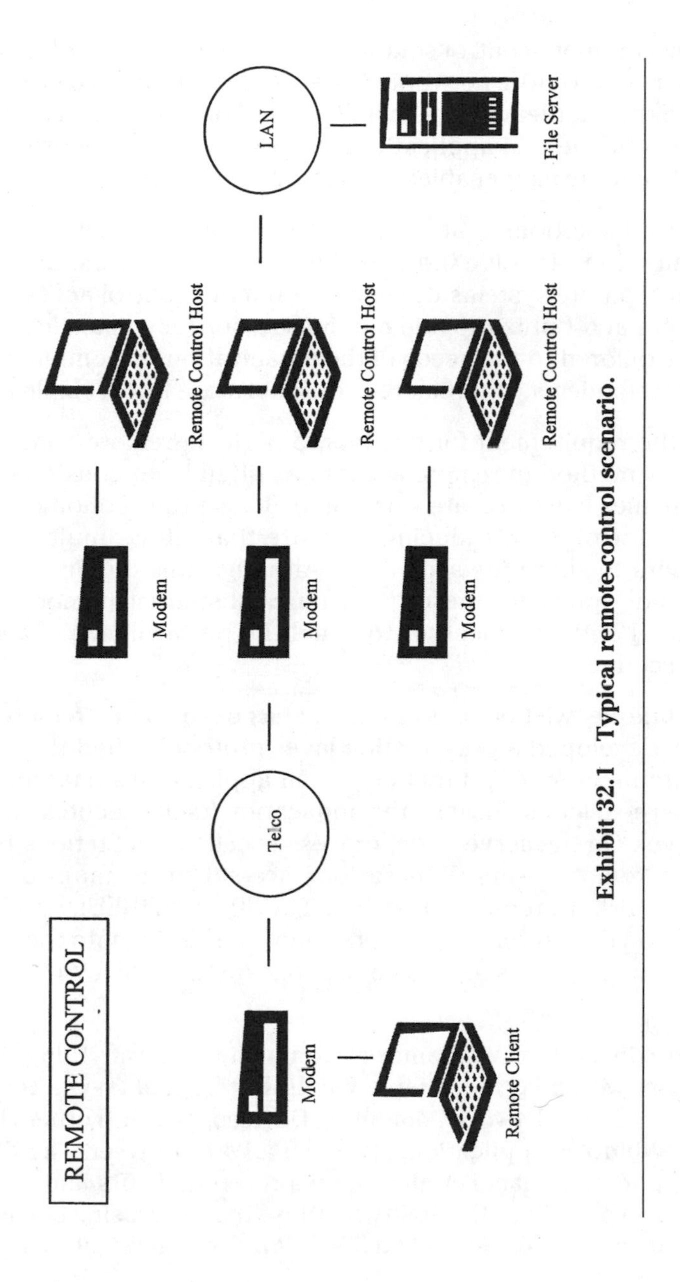

Exhibit 32.1 Typical remote-control scenario.

spectacular connection. Two machines are required for a remote-control process, and a modem link must be maintained during the entire dial-up session.

Installing a remote-control solution is simple if setup involves only one host PC and not an entire network. However, this approach causes security problems if the user leaves the host PC turned on and logged onto the network. Most commercial applications implement some security, but it is only effective if the user enables the security function.

Centralized solutions that put security and access under tighter controls usually don't involve individual users' desktop PCs. Instead, they depend on separate systems designed for remote-control access. Vendors like Chatcom and Cubix provide multiprocessor (asymmetrical) systems that can be tailored to the needs of the organization by combining several dial-in hosts, modems, and remote-access software into a single chassis.

Due to the requirement for two PCs to perform remote control, implementing this method of remote access has often been considered expensive. Companies like Citrix are working to change the economic feasibility of remote control by introducing software that allows multiple remote-control users to share the same host. Although this divide-and-conquer theory sounds appealing, these PCs require substantial memory (64 MB or more) and a Pentium processor to equal the performance of a standard one-to-one connection.

Citrix's unique twist on remote control has been coined remote application. Citrix developed a presentation layer protocol called the Intelligent Console Architecture (ICA) that allows an application's user interface to execute on the client side while the application itself executes on the WinFrame server. On the server side, processes called WinStations run applications for remote users. WinStations are set up to monitor either a specific network protocol such as IPX, TCP/IP, or NetBEUI, or to monitor serial ports on the server. When users connect, they log into the WinFrame server as WindowsNT users and control one instance of a WinStation on a processor.

Client PCs in a Citrix WinFrame environment can be anything from a relatively low-powered PC to a simple Windows terminal (Wyse Technology, Inc., Tektronix, Inc., Network Computing Devices, Inc., and others build terminals for Windows applications) or even a Web browser. The WinFrame Remote Application Manager allows low-powered clients and clients running over low-speed links the ability to utilize the processing power of a Citrix WinFrame server to run 16- and 32-bit Windows applications.

Despite the fact that Citrix can run 16-bit Windows applications, it is not recommended. Win16 applications run in WOW mode (Win16 on Win32) under WinFrame. This causes 16-bit applications to consume more system resources. This advice is also applicable to some DOS applications.

Remote Node

Although remote-control technology has made significant improvements in performance and management techniques, the advent of client/server computing has brought remote node to the forefront of communication server technology.

Remote node is comprised of a remote bridge or router, referred to as a remote access server, which allows the remote PC to become a full-fledged node on the network. Data packets issuing from the remote PC travel across the modem link, where a remote access server using remote-node software forwards the packets on to the LAN. Any packets from the LAN bound for the remote PC are also forwarded across the dial-up modem link by the remote access server. Exhibit 32.2 provides an illustration of a remote-node scenario.

Remote node establishes a link to a remote disk drive or printer much in the same way a LAN rerouter adds LAN-based devices to a PC. The drives are connected to the remote PC only by a temporary connection. In the case of a LAN, the connection is the LAN cabling and interface cards. In a remote-node scenario, the connection is via modem, T1 line, or ISDN connection, to name a few options.

It is convenient to have access to another disk drive containing data that are not available on the local drive of the remote PC. But the amount of information you can access is limited by the bandwidth of the remote connection. On a LAN operating at 10 MB per second, the access speed is acceptable. However, a 28.8-Kbps modem connection will be, at best, 100 times slower than a LAN. It is easy to see that remote-node access is useful for retrieving small files, but not for running applications from the host machine.

Remote node is appropriate when applications can be executed on the remote PC and relatively small amounts of data are transferred between the network and the client. Sending mail messages, accessing client/server databases such as Lotus Notes, or retrieving a document from the LAN can be handled easily with a remote node connection. For example, because word processing documents are relatively small, users can expect decent performance running Microsoft Word on the remote PC and retrieving files from the network.

If, however, the user did not have Microsoft Word loaded on the laptop or remote desktop PC, all the executable and supporting files would need to be transferred across the modem line as the application is running.

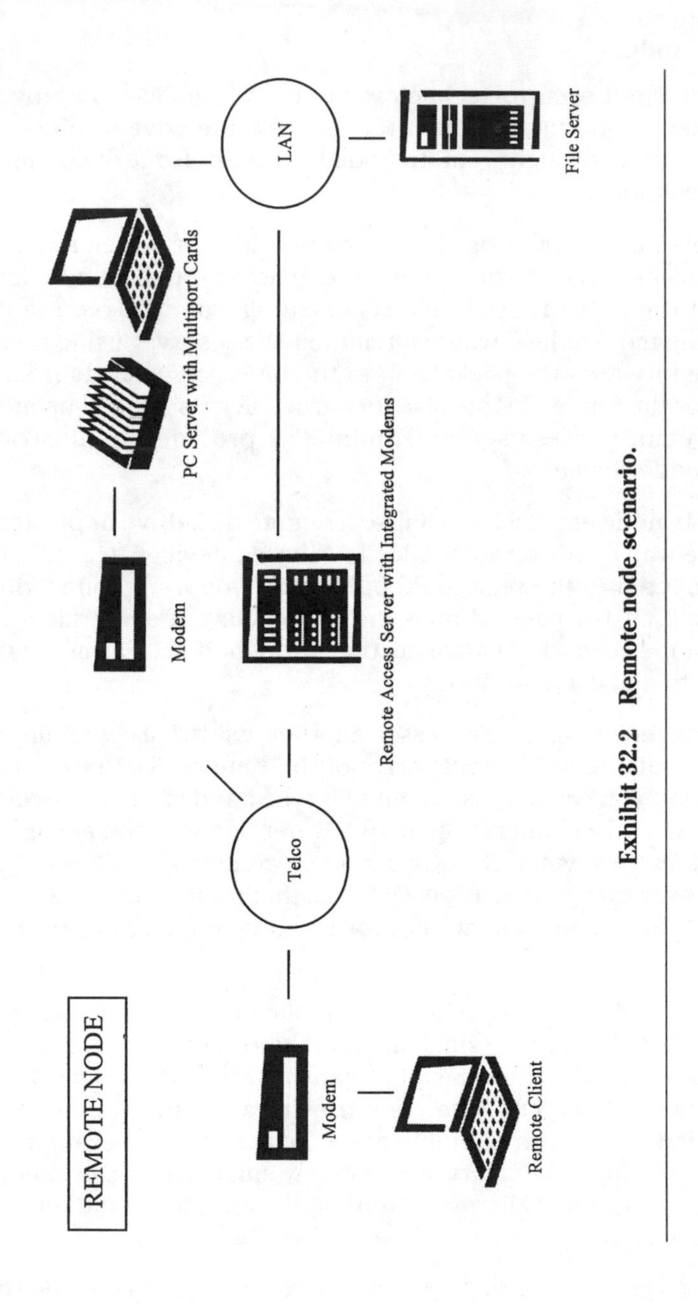

Exhibit 32.2 Remote node scenario.

Performance will be terrible and users may be tempted to contact members of the IS staff or help desk to find out what went wrong.

A variety of remote-access servers exists to deliver remote-node connections to a corporate network. Some products are software only, such as Microsoft RAS, and require only the addition of a third-party mulitport device to provide a LAN port on the host side of the connection. Other servers come in the shape of small appliances, like the Shiva LANRover or Cubix's WorldDesk Commuter. These devices come with integrated remote-node server software, unlimited number of client licenses, 6- or 12-port configurations, and integrated modems.

COMBINING CONNECTIONS

Both remote-control and remote-node technologies provide advantages and disadvantages in each dial-up scenario. Corporations have begun to deploy combinations of remote-node and remote-control technologies as they grapple with the escalating needs of their burgeoning remote-client community. This mix creates confusion, cost, and technical-support administrative overhead.

Doling out and managing communication resources has become an arduous task.

- Network administrators must create and enforce remote-access usage policies.
- Users are expected to adhere to policies they often do not understand.
- Optimal usage of communication server resources is never realized.
- Unnecessary costs take place because of a disregard for or lack of understanding of the technologies involved to provide enterprise-wide network connectivity.

CREATING AND ENFORCING REMOTE-ACCESS COMMUNICATION POLICIES

System administrators are tasked with creating remote-access policy. The simplest type of policy allows remote-node access for one set of users and remote-control privileges for another. More sophisticated situations include granting remote-node and remote-control access based on the application type, time of day, and communication speed required.

Manually enforcing such a policy is a time-consuming and often impossible task. The users of communication server resources are often non-technical, widely dispersed, and cannot be forced to comply with corporate policies.

Network administrators' plates are full without having the task of policing their remote-client community. Remote clients often do not understand

Exhibit 32.3 Time required for tasks after installation for 100 remote clients.

2 or more hours training each remote client	200 hours
4 or more hours in help-desk support for each remote client	400 hours
An equal number of hours of general network administration time	600 hours
	1200 hours or more

that violation, of what may seem to be an arbitrary policy, can have serious consequences for all users. The breaking of any remote communication policy can produce very unpredictable, but usually negative results. Long delays in accessing information, needless phone charges, and increased labor costs are just a few examples.

Administration Costs Can Be Staggering

The hidden costs of remote-access communication policy administration will far exceed initial product costs. For example, network administrators at a medium-sized company trying to support 100 remote clients might spend considerable time within the first three months of a communication server installation, as indicated in Exhibit 32.3.

Plus incorrect technology choices made by the remote clients will create dissatisfied employees, waste many hours of labor, incur unnecessary phone charges, and add a lot more cost to the projects these users are working on. Add another 100 users and costs would double.

Adherence to Remote-Access Communication Policy

The remote client who adheres to a typical remote-access communication policy fumbles with different technologies under various circumstances. For example, he may access e-mail by remote node and use a remote-control session for a custom sales application. The decision as to which connection method is most appropriate is often unclear to nontechnical users.

Without a set of automated guidelines, nontechnical users are left to ponder:

- "If I want to enter an order, should I use remote node or remote control for the best performance?"
- "If the file I need is 300 KB in size, which remote-access method should I use? Remote node or remote control?"
- "If I want to look at my e-mail, which company phone number should I connect through?"
- "How do I use an application that I have at work but don't have loaded on my laptop?"
- "If I want to do a quick look-up of a large spreadsheet file, should I use remote node?"

The questions associated with remote-access communication methods go on and on. Each question may generate a help desk call, or the remote client may find out the answer to his question the hard way. Either technique to problem solving is burdensome and costly.

RULES-BASED REMOTE-ACCESS COMMUNICATION

One company is battling the confusion of remote communications. Cubix Corporation has introduced a new technology called WorldDesk Applink to bring transparent remote access to the dial-up community. WorldDesk Applink provides a seamless combination of remote-node and remote-control connection methods to the end user through a single interface. Unlike other remote communication products, Applink maintains a desktop image for the remote user that hides the details of the access method used for each session. The primary advantage of WorldDesk is that the remote user conducts computing sessions from a remote location in the same manner that a LAN-based desktop PC operates applications.

This transparent access is made possible through a simple rules database maintained by the LAN administrator. Each rule is set through a combination of parameters like the ones illustrated below:

- Drive Type: Local or Network
- File Path, File Name, and File Size
- Time of Day
- Communication Speed
- User, Group, Cluster

The following examples show how these and other decision parameters can be combined with operators to create and manage effective user policy, which is transparent to users and which optimizes use of available communication server equipment.

"If %filesize>100K and %drivetype = network run application at server"

"If %user = "boss" run application at server"

Applink shuffles the decision-making procedures of each remote-access session back to the hands of the network administrator. With the introduction of the Internet into the remote-access equation, it is imperative that other remote-access communication companies shift to providing this type of transparent connectivity.

COMMUNICATION CONCLUSIONS

After installing any communication service, network administrators should create an internal beta group for testing. These behind-the-scenes guinea pigs will help weed out the majority of user problems before the system goes completely on-line. Use the comments that are gathered during these beta tests to outline useful tips for future user training sessions. It is vital that everyone be aware of and capable of using these new network communication services. If user education is ignored, chances are that a communication server will become a cost-intensive, nonproductive application on the network.

Denise M. Jones is Marketing Communications Manager at Cubix Corporation, located in Carson City, NV, where her principal responsibilities are in public and press relations. Cubix Corporation designs, manufactures, markets, and supports clustered remote-access servers, clustered application servers, and remote-network-user client products for dial-up access to branch offices and corporate LANs.

Chapter 33
Internet Mail Server Overview

Lawrence E. Hughes

HISTORICAL PERSPECTIVE

Historically, e-mail on PC-based systems has been done using proprietary LAN-based systems (e.g., cc:Mail or MS Mail). Most of these systems have some scheme for linking multiple LANs (connected by WAN or dial-up links) together into an "enterprise" mail system (e.g., Microsoft's Message Transfer Agent for MS Mail). They also typically have "Internet gateways" (e.g., Microsoft's SMTP Gateway for MS Mail) available to allow their users to exchange mail with direct users of Internet mail. Furthermore, if two such proprietary LAN-based e-mail systems both have Internet mail gateways (whether the two LAN-based systems are from the same or from different vendors), it is possible for users of one LAN-based mail system to exchange mail with users of the other LAN-based mail system, using the Internet as a "backbone."

However, there are numerous problems associated with using Internet mail gateways, such as reliability, performance, inability to pass through binary attachments, and features such as delivery notification not working (for mail that crosses a gateway). Even Microsoft's new e-mail system (exchange server) is clearly a proprietary system that uses a gateway to allow exchange of mail with users of, or via, the Internet (although they now call it an "Internet connector" — which is akin to calling a trailer a "mobile home" to make it sound more attractive). Exchange with the Internet mail connector can support standard Internet POP3/SMTP clients and, with Internet information server, allow you to read e-mail with a Web browser.

Recently, there has been a major increase in the number of people who have direct access to the Internet (increasing by double-digit percentages every month). There has also been a major improvement in the quality and features of direct Internet mail clients (also called SMTP/POP mail clients), such as Qualcomm's new Eudora 3.0 Professional. Therefore, there has

0-8493-9823-1/00/$0.00+$.50

been a corresponding increase in the demand for Internet-style (SMTP/POP3) mail servers, both from Internet service providers and as the primary e-mail system on an organization's "intranet." In the UNIX market, these have been available for some time as unsupported freeware (the dreaded SendMail, which requires an 8-th level UNIX guru to install and configure). A few brave souls ported the readily available SendMail source code to Windows NT (but even these have been sold for roughly $500 a copy).

Coincident with the growth of the Internet has been a major shift in market share of server platforms from Novell Netware (and to a lesser extent from UNIX) toward Microsoft Windows NT. Currently, about 30 percent of all server platforms are running Windows NT Server. Unlike earlier Microsoft operating systems (MSDOS, 16-bit Windows), Windows NT is a viable, robust platform with all the features necessary to support "real" Internet server products, including a large, flat address space, preemptive multitasking, process spawning, daemons (NT services), TCP/IP socket API (WinSock), and synchronization primitives. In fact, Windows NT has much better support for "threads" and Symmetric Multi Processor (SMP) support than most UNIX variants, and hence should be able (with a good server design optimized for Windows NT) to outperform a UNIX Internet server on equivalent hardware platforms.

However, Microsoft has been very late to market with native Internet server products for Windows NT, especially in the area of e-mail. For example, their Internet Information Server included only WWW, FTP, and Gopher).

A few companies tried to fill this vacuum by creating commercial-grade, supported Internet mail systems. I myself created a native Windows NT e-mail server (called KIMS), but the financial backing never came together on that product, and I have since left the company where that work was done. My current employer, Software.com (based in Santa Barbara), was one of the first, and so far one of the most successful, of these vendors. They sell a product called Post.Office for several UNIX variants and for all four flavors of Windows NT (Intel, Alpha, MIPS, and PowerPC). An early version (1.x) of this product was the starting point for Netscape's mail server product (although the two products have since diverged significantly). Since then, Software.com has merged with two other Internet infrastructure software companies, both based in the Boston area. One of these was Accordance, which had created a very high-end message store (still used by AT&T in its WorldNet service). The other was the division of Banyan Systems that created "Universal StreetTalk" (a generalized network directory product). The resulting organization is still known by the name Software.com, Inc. (see www.software.com for information).

Note that you can obtain a free license to install, evaluate, and even use a copy of Post.Office for up to ten users (see www.software.com). Unfortunately, only very limited technical support is available for the freeware version. For full technical support or for more than ten users, you should contact Software.com in Santa Barbara at (805) 898-8917 for details on a commercial license.

WHAT EXACTLY IS AN INTERNET MAIL SERVER?

A mail server is the e-mail equivalent of a phone company's "central office switch." Using the same analogy, a mail client is the e-mail equivalent of a telephone desk set. Few users of the lowly telephone realize just how little of the overall telephone system is in their desk set, and what complexity lies hidden behind the little hole in the wall that it plugs into, that makes it work. Similarly, most people think of the mail client program as the "e-mail software," without any real knowledge of what goes on behind the scenes to make that client program useful.

In the case of Internet Mail, there is a worldwide system of mail servers, DNS servers, and routers that allow you to easily exchange mail with any other Internet users, anywhere in the world, whether they are at their computer, or if it is even turned on at the time you send your message.

A mail server consists of four parts:

- An incoming SMTP process, that accepts incoming SMTP connections (from mail clients or other mail servers) and puts the messages received into the message store (and/or forwards them onward to the outgoing SMTP.
- An outgoing SMTP process that relays received mail addressed to accounts on other mail servers by making connections to the incoming SMTP process on other mail servers (typically to the final destination on which the mail recipient has an account, although it is possible for mail to go through several mail servers before reaching the final destination).
- A POP process that allows a mail client program to retrieve messages from the message store.
- A message store that is used to temporarily store messages received from the incoming SMTP process, before they are retrieved (at the request of a mail client) by the POP process.

There may also be a component that maintains the e-mail account database, or the mail server could use the native account database of the host operating system.

THE LIFE CYCLE OF A TYPICAL INTERNET MAIL MESSAGE

A typical message life cycle for an Internet Mail message between two users with accounts on the same Internet post office would be as follows:

- Andy (whose address is andy@boys.com) starts his mail client program, composes a new message to Bob (bob@boys.com), and sends it.
- Andy's mail client program makes an SMTP connection to his smtp host computer (as configured in his mail client, say "smtpmail.boys.com").
- The incoming SMTP process (on smtpmail.boys.com) accepts the connection from the mail client program, receives the message, sees that it is addressed to an account in this mail domain (boys.com), so it stores it in Bob's area of the message store.
- At some later time, Bob brings up his mail client program and tells it to check for new mail. His mail client makes a POP3 connection to the POP process on his pop host computer (as configured in his mail client, say "popmail.boys.com" — which might actually be an alias for the computer smtpmail.boys.com, or possibly a different computer altogether, but in either case is the computer where the POP process runs that uses the message store written by the above incoming SMTP process).
- The POP process checks and sees there is an unread message waiting in the message store for Bob, so it reports this to Bob's mail client program. His mail client program retrieves the message and then tells the POP process to delete the message from the message store.
- Bob sees a new message in his "inbox" (recently received messages), which he can then read, file away, or print.

The life cycle for a message between two users with accounts on two different post offices is somewhat more complicated:

- Andy (whose address is andy@boys.com) starts his mail client program, composes a new message to Alice (alice@girls.com), and sends it.
- Andy's mail client makes an SMTP connection to his smtp host computer (as configured in his mail client, say, "smtpmail.boys.com").
- The incoming SMTP process on smtpmail.boys.com accepts the connection, receives the message, sees that it is addressed to an account not managed by this post office (it's for girls.com), so it forwards it to the outgoing SMTP process.
- The outgoing SMTP process sees that the message is for the post office for mail domain girls.com, so it queries the local DNS server (which in turn may query other DNS servers elsewhere on the Internet if it doesn't already know the answer) to find out what computer the mail domain girls.com prefers to use to handle incoming SMTP mail.

Say it gets the answer that node "smtpmail.girls.com" is the preferred server. The outgoing SMTP process then makes an SMTP connection to smtpmail.girls.com (which could be halfway around the world) and sends the message to the incoming SMTP process on that computer.

- The incoming SMTP process on computer smtpmail.girls.com accepts the connection from the outgoing SMTP process on computer smtp-mail.boys.com, receives the message, sees that it is addressed to an account in the mail domain it manages (girls.com), so it stores the message in its message store in Alice's area.
- At some later time, Alice brings up her mail client software and tells it to check for new mail. Her mail client makes a POP3 connection to the POP process on her pop host computer (as configured in her mail client program, say, "popmail.girls.com").
- The POP process on popmail.girls.com checks and sees there is an unread message waiting in the message store for Alice, so it reports this to Alice's mail client program. Her mail client program retrieves the message and then tells the POP process to delete the message from the message store.
- Alice sees a new message in her "inbox" (recently received messages), which she can then read, file away, or print.

CREATING THE ENVIRONMENT REQUIRED TO RUN AN INTERNET MAIL SERVER

From the above scenarios, it should be obvious that the following things need to be in place for a given post office to participate in the worldwide Internet e-mail system:

1. The various post office processes must be installed and running on a computer that is always running and connected to the Internet.
2. The local DNS server must have been configured to be able to answer queries (from local mail client programs or remote mail servers) concerning the IP address for all relevant nodenames, as well as "MX" records that indicate the preferred mail server(s) for this domain.

Ideally, for a fault-tolerant system, there should be a backup DNS server and post office at another site (geographically isolated from your site) that can receive and temporarily hold mail until your site is back on-line should something happen to your site (such as a power or network outage). Often Internet sites work out reciprocal arrangements to serve as backups for each other.

If you don't already have an Internet service provider (a company that has a high-speed link into the main Internet and sells access to smaller organizations), you must contract with one to provide you with service. All levels of service (from a 56 kbit/second line to ISDN to T1) are available from various ISPs, depending on what kind of total bandwidth your organization is going to need.

As for a DNS, some ISPs will provide DNS service to you, but typically it is better to manage your own local DNS. A DNS server is included in NT Server 4.1.

You should seriously consider installing a "firewall," such as Checkpoint's Firewall 1 (available for Windows NT — see www.checkpoint.com), to control access to your local nodes from nodes outside your domain. If you do this, you will need to ensure that the DNS, SMTP, and POP3 protocols can pass correctly through it in both directions.

To verify correct operation of your DNS, you should try to do the following:

- Ping your local nodes using just nodename (e.g., masternode) and fully qualified domain names (e.g., masternode.mydomain.com), from nodes in your IP domain.
- Ping nodes outside your IP domain (e.g., ftp.microsoft.com).
- Ping your local nodes that will provide DNS and Internet Mail services using fully qualified domain names from one or more nodes outside your IP domain.

The next step is to add "MX records" to your local DNS to inform remote Internet post offices of the preferred computer to receive mail for your IP domain. It is possible to add multiple MX records with different priorities (the lower the priority number the more likely that node is to be used). If you have multiple "gateway" computers that should share the load of incoming mail, you would enter them with the same priority (in theory other post offices should randomly choose among these — to ensure this, it is possible to modify your DNS to present the next nodename in a group each time it is queried, but this is not for the faint of heart). If you have backup post office(s) at remote sites, these should be entered with higher-priority numbers than your local post office(s).

To test that this has been done correctly, you should obtain an "nslookup" utility (usually supplied with a DNS) to manually retrieve MX records for your domain, from a node outside your IP domain. The details of how to do this are beyond the scope of this chapter, but instructions should be included with your DNS server package.

Finally, you should select a computer on which to install your post office software that meets the following criteria:

- Has sufficient CPU power, memory resources, and disk space to support a post office. (The product documentation should include recommended guidelines, but in general, a small post office will require a single 166-MHz Pentium CPU, 48 to 64 MB of RAM, and several hundred megabytes of disk space exclusively for the use of the post office.

- Will be running at all times (i.e., not someone's personal workstation that is turned off at night or on weekends). Ideally, it should also be on a very good UPS, and be able to reboot automatically upon restoration of power.
- No other really heavy-duty applications should be on the same computer, such as Exchange Server or SQL Server. It is reasonable to have file and print sharing on the same computer unless these services are very heavily used. It should also be reasonable to run the DNS server and even a firewall on the same computer. A lightly used Web server and ftp server should also be able to coexist with a mail server. The mail server computer should also be able to handle a reasonable number of RAS dial-in ports. If you do put a lot of services on a single computer, you should use Performance Monitor to ensure that there is ample memory to prevent excessive paging, and that CPU utilization rarely exceeds 90 percent or so.
- Only one product that acts as a server for the SMTP or POP3 ports (25 and 110, respectively) can be installed on a given machine. For example, you cannot install both an Internet post office and exchange Internet connector on the same computer.

INSTALLING AND CONFIGURING POST.OFFICE

Once all this underlying structure has been set up, installation of Software.com's Post.Office product is not much more difficult than installing an application program, such as Microsoft Word. It is much simpler than installing Windows NT.

There are three basic steps involved:

1. Create an NT service account. This is just an NT account that has its rights limited to exactly those needed to run Post.Office. The minimum right required is "Logon as a service." If you also wish to use the "integrated accounts" feature (use the password of an associated NT account for an e-mail account), the following additional rights are required: "Act as Part of the Operating System," "Increase Quotas," and "Replace a Process Level Token." This account can be a domain account, or a local account on the computer that will run Post.Office, but rights must be granted on the local computer (i.e., log in as Administrator of your local computer, except on a Domain Controller, where this is not possible).
2. Run the InstallShield "setup" program to install the Post.Office executables, install the service, and start it running. You will also specify your IP domain name and a password for your postmaster account (for initial validation in the next step). You can verify that

this has been done correctly by checking for the service "Post.Office MTA" in the Control Panel/Services applet.

3. Use a Web browser to complete the installation, which involves specifying the e-mail account information for the postmaster account (e.g., full name, e-mail address, etc.). Additional user e-mail accounts can be created at this time as well.

TESTING AND DEBUGGING YOUR NEW INTERNET MAIL SERVER

At this point, it should be possible to use any Internet-compliant "SMTP/POP3" mail client program (e.g., Eudora) to test your new Internet mail server. If this does not work for some reason, you can use the NT Telnet client to debug the installation:

To test the incoming SMTP process, connect to the nodename that Post.Office is running on, using port 25 (you can also specify the port as "SMTP"). You may wish to select the "Local Echo" option in Terminal/Preferences so you can see what you are typing. You should see a welcome message on your screen:

220 <nodename> ESMTP server (post.office vx.y ¼) ready <date/time>

You can try a few commands to see if it is working:

HELP
HELO test.com
QUIT

To test the POP process, connect to the nodename that Post.Office is running on, using port 110 (you can also specify the port as "POP3"). You should see a welcome message on your screen:

+OK <nodename> POP3 server (Post.Office vx.y ¼) ready <date/time>

You can try a few commands to see if it is working:

USER <some valid email account name>
PASS <valid password for above email account>
LIST
QUIT

FUTURE DEVELOPMENTS

A newer protocol (IMAP v4) has been defined and is currently being implemented by several vendors as an alternative (both can run in parallel) to POP3. This new protocol will allow selective retrieval of parts of the message (say just "headers"), to allow more control over exactly what is downloaded (among other enhancements). This will make it better suited for access to e-mail over dial-up modem links. Not that this new protocol is only between your mail client and your local mail server, so it does not matter whether anyone you exchange mail with supports it or not. However, both your local mail server and your mail client must support it (IMAP v4 clients are just now under development).

SUMMARY

Windows NT is entirely capable of running an Internet-style mail server (SMTP/POP3) comparable in all ways to, and interoperable with, SendMail running on UNIX. This chapter describes such a server and how to install it. Windows NT allows it to be far easier to install and manage than SendMail ever was. (SendMail typically requires a real UNIX guru with years of experience, and several days of hard work, to install).

Chapter 34 An Introduction to Microsoft Exchange Server

Lee Benjamin

Microsoft Exchange Server embraces Internet standards and extends rich messaging and collaboration solutions to businesses of all sizes. As such it is the first client/server messaging system to integrate e-mail, group scheduling, Internet access, discussion groups, rules, electronic forms, and groupware in a single system with centralized management capabilities.

Microsoft Exchange Server provides a complete and scalable messaging infrastructure. It provides a solid foundation for building client/server solutions that gather, organize, share, and deliver information virtually anyway users want it. Microsoft Exchange Server was designed from the ground up to provide users and administrators with unmatched open and secure access to the Internet. Native SMTP support, support for MIME for reliable delivery of Internet mail attachments, and support for Web browsers and HTML ensures seamless Internet connectivity.

Since its introduction in the spring of 1996, customers have been evaluating and deploying Microsoft Exchange Server along with Microsoft Outlook and other clients in record numbers. This chapter offers an overview of how Microsoft Exchange Server can help organizations improve their business processes, work smarter, and increase profits through improved communication. The topics to be covered include:

- Trends in messaging and collaboration
- Infrastructure for messaging and collaboration
- Redefining groupware
- Internet connectivity
- Easy and powerful administration
- Building a business strategy around Microsoft Exchange Server

0-8493-9823-1/00/$0.00+$.50

Microsoft Exchange Server is part of the Microsoft BackOffice integrated family of server products, which are designed to make it easier for organizations to improve decision making and streamline business processes with client/server solutions. The Microsoft BackOffice family includes the Microsoft Windows NT Server network operating system, Microsoft Internet Information Server, Microsoft Exchange Server, Microsoft SQL Server, Microsoft SNA Server, and Microsoft Systems Management Server.

TRENDS IN MESSAGING AND COLLABORATION

Information, both from within organizations and from outside sources, is becoming one of the most valuable commodities in business today. Never before has so much information been so readily available. Nor have there been such high expectations for how much individuals will be able to accomplish with this information. To take advantage of this information, businesses are rethinking every aspect of their operations and reengineering business processes to react more quickly, become more responsive, provide better service, and unify teams separated by thousands of miles and multiple time zones.

Until now, organizations looking for a messaging system had two choices: either a host-based system that provided beneficial administrative capabilities but was costly and did not integrate well with PC-based desktop applications, or a LAN-based system that integrated well with PC-based desktop applications but was not scalable and was less reliable than host systems.

Unifying LAN- and Host-Based E-Mail

Microsoft Exchange is not a response to any one single product. Rather, it is the evolution of messaging products in general. For the past 10 years, Microsoft has been a leader in LAN messaging solutions. In 1987, Microsoft released the first version of Microsoft Mail. This product was significant in two ways.

First, as a client/server implementation of messaging, Microsoft Mail was a test platform of what Microsoft Exchange Server would become. Second, Microsoft added a programming layer (API) to the product. One might say that this was the grandfather of MAPI, the Messaging Application Programming Interface upon which Exchange is built. Microsoft Mail for PC Networks now has an installed base of well over 10 million copies. Over the years, customers have told Microsoft what they wanted in their next-generation messaging system. Microsoft Exchange is that product.

Microsoft Exchange Server delivers the benefits of both LAN-based and host-based e-mail systems and eliminates the shortcomings of each approach. It integrates e-mail, group scheduling, electronic forms, rules, groupware, and built-in support for the Internet on a single platform with centralized management capabilities. Microsoft Exchange Server can provide everyone in the organization, from professional developers to administrators to end-users, with a single point of access to critical business information. It makes messaging easier, more reliable, and more scalable for organizations of all sizes.

Technology is not only changing how businesses process and assimilate information, it is affecting how this information is transferred, viewed, and acted upon. Electronic messaging plays a pivotal role in this process. The annual growth in individual electronic mailboxes is phenomenal. It has been estimated that there was an installed base of more than 100 million mailboxes worldwide at the end of 1996. Five key trends have led to this growth:

- *Growth in PC use.* It has been estimated that an installed base of more than 200 million personal computers exists worldwide. Performance increases and price decreases have expanded the demand for PCs both at home and in the workplace. In addition, the rapid acceptance of the Internet by companies as a marketing vehicle indicates that the PC has reached the mainstream for more than just recreational or business uses.
- *Adoption of the graphical user interface.* The intuitive, icon-based graphical user client interface has made e-mail applications easier to use and has made it possible to create more sophisticated messages than with previous MS-DOS and host-based e-mail applications.
- *Integration of messaging in the operating system.* As messaging functionality has been integrated into the operating system, every application has become "mail-enabled." Users are able to easily distribute documents and data from within applications without having to switch to a dedicated e-mail inbox. Messaging functionality in the operating system also provides a platform for critical new business applications such as forms routing and electronic collaboration.
- *Client/server computing.* Client/server computing combines the flexibility of LAN-based systems — for easier management and extensibility — with the power and security of mainframe host-based systems. In addition, client/server technology has made electronic messaging systems "smarter" so they can anticipate problems that previously required human intervention, thus reducing overall support costs.
- *Growth of the Internet.* The growth of the Internet is perhaps the most important platform shift to hit the computing industry since the introduction of the IBM personal computer in 1981. The most explosive expansion is expected to be in the use of Internet technologies to improve communication within organizations.

Anticipating these trends, Microsoft Exchange Server was developed to unify host-based and LAN-based environments that have historically been separate. Microsoft Exchange Server incorporates both messaging and information sharing in a unified product architecture. By taking advantage of client/server technology, organizations receive the scalability benefits of host-based environments and the flexibility of LAN-based environments.

The Microsoft Exchange Product Family

The Microsoft Exchange product family consists of:

- The Microsoft Outlook family of clients (Outlook, Outlook Express, and Outlook Web Access) for Microsoft Exchange Server, which includes e-mail, rules, public folders, sample applications, native Internet support, rapid and easy electronic forms, and an application design environment that makes it easy for users to create groupware applications without programming. These clients are based on protocols such as MAPI, IMAP4, POP3, and LDAP. Outlook is the premier client for Microsoft Exchange Server and is available for Windows 95 and Windows NT. A version of Outlook is also available for the Windows 3.x and Apple Macintosh System 7.x operating systems. Other clients such as POP3 and Web-based clients can also access Microsoft Exchange Server.
- Microsoft Exchange Server, which consists of core components that provide the main messaging services — message transfer and delivery, message storage, directory services, and a centralized administration program. Optional server components provide seamless connectivity and directory exchange between Microsoft Exchange Server sites linked over the Internet, via X.400 or other messaging systems. Microsoft Exchange Server supports SMTP and X.400 as standards to ensure reliable message transfer for systems backboned over the Internet or other systems. It also provides outstanding NNTP (Network News Transport Protocol) and Web access and integration, enabling customers to easily access all types of Internet information.

Microsoft Exchange Server must also be a platform for an assortment of business solutions, which organizations of all sizes can implement to meet a wide range of key challenges, including:

- Making it easy for salespeople and support technicians to find product information
- Improving an organization's access to market information
- Allowing users to receive timely and accurate information regarding sales and product activity
- Making individual and group scheduling fast and easy
- Improving customer tracking

- Allowing all technicians and engineers to share common customer technical issues
- Continuing to expand the network and add new applications while maintaining complete compatibility with existing systems

INFRASTRUCTURE FOR MESSAGING AND COLLABORATION

Microsoft Exchange Server combines the best features of both host-based and LAN-based E-mail systems with some additional benefits all its own. The result is a messaging system that is easy to use and manage and that moves messages and files through the system quickly, securely, and reliably, regardless of how many users or servers the organization has.

Universal Inbox

The Universal Inbox in the Microsoft Outlook Client lets users keep all messages, forms, faxes, and meeting requests in one location, where they can be easily accessed. Users can search and sort these items using a wide range of criteria — such as addressee, topic, or date of receipt — to quickly locate the information they need.

In addition, server-based rules automatically process incoming messages, including those from the Internet, even when the user is out of the office. These rules can be configured to file incoming messages in appropriate folders or to respond immediately with specified actions, such as forwarding messages to another person, flagging them for special attention, or generating a reply automatically.

Tight Integration with Desktop Applications

Because Microsoft Outlook is tightly integrated with the Microsoft Windows operating system and the Microsoft Office family of products, it is easy for users to learn and use. Microsoft Outlook actually ships with both Exchange Server and Microsoft Office 97. With new features such as "journaling" (which allows a user to find a file based on when that file was used, rather than by file name), Outlook can keep track of what users do every day.

Fast, Secure, and Reliable

Microsoft Exchange Server takes full advantage of the robust client/server architecture in Windows NT Server to get messages to their destinations quickly, whether across the hall or around the world. It also provides tools for easily tracking messages sent to other users of Microsoft Exchange Server and via the Internet to users on other systems, to confirm that they arrived and that they were read. Support for digital encryption allows users to automatically secure messages against unauthorized access, and digital signatures guarantee that messages get to their recipients without modification.

In addition to these security features, Microsoft Exchange Server also takes advantage of the security features built into Windows NT Server to prevent unauthorized users — inside or outside the organization — from accessing corporate data.

Remote Client Access

Local replication is the ability to do two-way synchronization between a server folder and a copy of that folder on a local or portable machine. Local replication is initiated by creating an offline folder — a snapshot or "replica" — of the server-based folder the user wishes to use while disconnected from the server. (The use of offline folder synchronization is discussed further in a subsequent section.)

Scalable

Built on the scalable Windows NT Server architecture that supports the full array of Intel and Digital Alpha-based servers, Microsoft Exchange Server scales to meet a range of requirements — from those of a small, growing office to those of a multinational corporation.

It is easy to add users to existing servers and new servers to an organization as it grows. Routing and directory replication occur automatically between the new and existing servers at each site. Plus, optional connectors are available to connect computers running Microsoft Exchange Server to the Internet and X.400 systems.

REDEFINING GROUPWARE

In addition to e-mail, which allows users to send information to each other, Microsoft Exchange Server and Microsoft Outlook support groupware applications that help users share information by retrieving it wherever it might be — without the traditional complexities of navigating through a maze of network servers or jumping between multiple screens and applications.

A built-in suite of groupware applications in Microsoft Exchange Server gives users a headstart with group scheduling, bulletin boards, task management, and customer tracking. Because these applications are designed to integrate tightly with the Windows operating system and Microsoft Office, Microsoft Exchange Server provides an ideal platform for integrating business solutions with desktop applications.

While some Microsoft Exchange Server applications are ready to go right out of the box, and many more are also available on the Web, you can also easily customize them using Microsoft Outlook and extend them using popular development tools such as the Visual Basic programming system, the Visual C++ development system, Java, and ActiveX components.

The concept of discussion groups and bulletin boards are nothing new to Internet users. With the Microsoft Exchange Internet News Service, the complete set of Internet newsgroups are easily available to users through public folders. Organizations can make public folder information available to internal or external users of the Web without storing information in redundant locations or manually reformatting information into hypertext markup language (HTML) format, via Outlook, Outlook Web Access, or any NNTP newsreader client. Users can also access newsgroups hosted on Exchange Server using the native NNTP protocol.

Microsoft Exchange Server also allows users to communicate with each other and to share information from any time zone or location. This is especially important for mobile users, who need to break through traditional organizational boundaries to communicate with the enterprise. Group scheduling and public folders help users work together more effectively, whether they are across the hall, across the country, or around the globe.

Group Scheduling

A full-featured personal calendar, task manager, and group scheduler in its own right, Microsoft Outlook has been incorporated into Microsoft Exchange Server to provide a fully extensible system that can act as a rich foundation for business-specific, activity-management applications. It takes full advantage of the advanced client/server architecture and centralized management features in Microsoft Exchange Server.

Microsoft Outlook is a tool for scheduling group meetings, rooms, and resources. To schedule a meeting, users can overlay the busy times of all the attendees in a single calendar to automatically schedule a meeting, conference room, and any other resources required. The Microsoft Outlook contact-management features provide users with easy access to the names and phone numbers that are part of their daily work.

Public Folders

Public folders make it easy for users to access information on a related topic all in one place. Documents can be stored in public folders for easy access by users inside and outside an organization. These folders are easy to set up without programming; relevant documents can be dragged into the folder. Microsoft Exchange Server uses these public folders as containers for groupware and custom applications.

Bulletin Boards

Support for bulletin boards enables organizations to easily share information throughout the enterprise. Information is organized so that users can easily find what they need, leave messages, and communicate about the topic.

It is interesting to note that Internet users have been working with bulletin boards for many years using Usenet Newsgroups. Microsoft Exchange Server uses the same Internet Standard, called NNTP. Sample bulletin board folders, which are easily customizable, are included with Microsoft Exchange Server.

Outlook Forms

Electronic forms are easy to create and modify in Outlook so users can send and receive structured information. Traditionally advanced features such as drop-down lists and validation formulas are easy to get at and use. More sophisticated capabilities are accessible from Outlook's rich programming extensibility interfaces. In addition, Outlook forms are automatically rendered to the Web so any user can get to them.

Public Folder Replication

One of the key strengths of Microsoft Exchange Server is its ability to distribute and synchronize shared information through the Microsoft Exchange Server replication system. It is possible to have multiple synchronized copies of folders in different locations regardless of whether users are connected over a LAN or WAN, or the Internet or X.400 backbone.

Replicating information in this way means that synchronized copies of a public folder can reside on multiple servers, distributing the processing load and improving response time for users accessing information within the folder. It also means synchronized copies of a public folder can reside at several geographically separated sites, significantly reducing the amount of long-distance WAN traffic necessary to access information. If a server holding one copy of a public folder becomes unavailable, other servers holding synchronized copies of the same folder can be accessed transparently, greatly increasing the availability of information for users and resulting in a highly reliable system.

Microsoft Exchange Server offers users the unique benefit of location-independent access to shared information. With replication, the physical location of folders is irrelevant to users, and Microsoft Exchange Server hides the sophistication of public folder replication. Users need not be aware of where replicated folders are located, the number of replicated copies, or even that replication occurs at all. They simply find information more easily than ever before.

With the Microsoft Exchange Internet Mail Service users can replicate public folders and groupware applications throughout a distributed organization, even if they do not have a wide area network. Managing public folder replication is very easy. Using the graphical Microsoft Exchange Server Administrator program, system managers need only select the servers that will receive replicas of the public folders.

Offline Folder Synchronization

Microsoft Exchange Server allows users to automatically perform two-way synchronization between a server folder and a copy of that folder on a local PC. For example, a user can create an offline folder — a snapshot or "replica" — of a customer-tracking application to take on a business trip and update it based on interactions with customers during the trip. Then, when the user reconnects to the server — either remotely by modem or by connecting to the LAN upon returning to the office — the folders can be bidirectionally synchronized with the server. Changes, including forms and views, made on the local machine are updated to the server, and changes to the server-based folders automatically show up on the user's PC. Offline folder synchronization lets users maintain up-to-date information without having to be continuously connected to the network.

Creating an offline folder is different from simply copying a server folder to the hard disk, because an offline folder remembers its relationship with the server folder and uses that relationship to perform the bidirectional update. Only changes — not the whole folder — are copied, which helps minimize network traffic.

Microsoft Exchange Server supports multiple simultaneous offline folder synchronization sessions from many different locations. Built-in conflict resolution for public folders ensures that all the changes are added. The owner of the folder is notified if there is a conflict and can choose which version to keep. With the powerful server-to-server replication technology in Microsoft Exchange Server, this information can then be automatically replicated to users of your system around the world.

Easy-to-Create Groupware Applications

Microsoft Exchange Server delivers a scalable set of tools that lets almost anyone — even users who have never programmed — to develop custom groupware applications. It also gives professional programmers all the power they need to build advanced business software systems. Microsoft Exchange Server includes these key development features for both users and programmers:

- *Fast applications development without programming.* Users can build complete groupware applications — such as a customer-tracking system or an electronic discussion forum — without programming.

Assuming they have the appropriate permissions, users can simply copy an existing application (including forms, views, permissions, and rules) and modify it as they wish with the functionality available in the Microsoft Outlook Client. They can easily modify existing forms or create new ones with a menu choice to Design Outlook Form, which requires no programming knowledge.

- *Central application management.* Once users complete an application, they will usually hand it off to the Microsoft Exchange Server administrator for further testing or distribution to others within the organization. The Microsoft Exchange Server replication engine manages the distribution of the application or any new forms that may have been revised or created for existing applications. One can also replicate these applications from one Microsoft Exchange Server site to another over the Internet by using the Microsoft Exchange Internet Mail Service.

Both of these capabilities translate into reduced cycles for creating, modifying, and distributing groupware applications. That means end-users can build applications that are valuable to them without having to wait for a response from their IS departments. Even if an application turns out to be less useful than the creator hoped it would be, the development cost is minor.

The IS department can also benefit because it can customize those applications that do turn out to be worthwhile, since forms created or modified with Microsoft Outlook are extensible with Visual Basic Script programming system. In addition, Outlook Forms can be further extended with other programming tools such as Visual C++, ActiveX Controls, and Java. By using the Microsoft Exchange Server replication engine, revisions and new applications can be deployed inexpensively as well.

The speedy application design and delivery process made possible by Microsoft Exchange Server enables the people who have the best understanding of the functionality needed to respond quickly to market requirements. As a result, an organization can reduce the costs of adapting and rolling out those applications. Whenever an application is rolled out within an organization, it is usually only a matter of time before the applications developer hears from users about how it could be improved. Many applications provide limited functionality — once the barrier is reached, they cannot be customized any further and require redesign from scratch with a more powerful tool.

Thus, forms designed by end-users can be customized by professional developers using the full power of Visual Basic Script. Other work group application design tools either require a high degree of programming skill or quickly run out of steam, as a particularly useful application requires additional functionality. Microsoft Exchange Server opens the door between end-user application design and the full power of the Windows APIs available through more powerful programming languages.

Exchange Server takes these forms even further by automatically rendering them to the Web as HTML forms. By leveraging a technology known as ActiveServer Pages, included with the Microsoft Internet Information Server, forms and the information in them can be seen by any user accessing Exchange from a browser anywhere on the World Wide Web (if they have the appropriate permissions of course).

MAPI: Messaging Application Programming Interface

The MAPI subsystem is the infrastructure on which Microsoft Exchange Server is built. Messaging client applications communicate with service providers running on the server through the MAPI subsystem. Through broad publication of Microsoft messaging APIs, and because of the robust messaging and work group functionality defined in them, MAPI has become a widely used standard throughout the industry for messaging and groupware clients and providers.

MAPI-compliant clients span a variety of messaging- and work group-based applications and support either Windows 32-bit MAPI applications on Windows 95 or Windows NT, and 16-bit MAPI applications running on Windows 3.x. Each of these types of applications can access the service provider functionality needed without requiring a specific interface for each provider. This is similar to the situation where applications that use the Microsoft Windows printing subsystem do not need drivers for every available printer.

Messaging applications that require messaging services can access them through any of five programming interfaces:

- Simple MAPI (sMAPI)
- Common Messaging Calls (CMC)
- ActiveMessaging (formerly known as OLE Messaging and OLE Scheduling)
- MAPI itself
- (In the near future) Internet Mail Access Protocol, or IMAP

Client requests for messaging services are processed by the MAPI subsystem — either as function interface calls (for sMAPI or CMC) or as manipulations of MAPI objects (for OLE Messaging or MAPI itself) — and are passed on to the appropriate MAPI-compliant service provider. The MAPI service providers then perform the requested actions for the client and pass back the action through the MAPI subsystem to the MAPI client.

Third-party programming interfaces that can be built upon MAPI are frequently employed. Because MAPI is an open and well-defined interface, a proprietary third-party API can be implemented on top of MAPI without having to revise the MAPI subsystem itself. Thus, customers and vendors can implement their own MAPI solutions that meet their particular needs without incurring the development costs that would otherwise accrue on other messaging infrastructures.

INTERNET CONNECTIVITY

Extensive built-in support for the Internet in the Microsoft Outlook Clients, as well as the Microsoft Exchange Internet Mail Service, Microsoft Exchange Internet News Service, and Outlook Web Access, makes it easy for organizations to use the Internet as a communications backbone, to make Internet newsgroup data available to their users through public folders, and to make messaging and public folder information available to the ever-growing numbers of Internet Web users.

The Microsoft Exchange Internet Mail Service provides high-performance multithreaded connectivity between Microsoft Exchange Server sites and the Internet. It also supports MIME and UUENCODE (and BINHEX for Macintosh) to ensure that attachments arrive at their destinations intact. Built-in message tracking helps ensure message delivery. Standards-based digital encryption and digital signatures ensure message security.

These capabilities make it possible for organizations to use the Internet as a virtual private network to connect Microsoft Exchange Server sites over the Internet and to route messages using the TCP/IP SMTP or X.400 protocols. You can easily control who sends and receives Internet mail by rejecting or accepting messages on a per-host basis.

Integrated Internet Support

The Microsoft Outlook Clients include built-in Internet mail standards to allow users, connected locally or remotely, to reach other Microsoft Exchange Server sites and virtually anyone else using any Internet service provider. Native MIME support allows files to be transported reliably over the Internet. Support for Post Office Protocol, Version 3 (POP3), PPP, and IMAP4 ensures compatibility with all SMTP e-mail systems.

The Microsoft Exchange Inbox — a version of Microsoft Exchange Client that does not include Microsoft Exchange Server–specific functionality — is built into the Windows 95 operating system. This feature makes Internet mail easy to set up and access. Any user with an Internet mailbox via POP3 can use the Internet Mail Driver for Windows 95 in the Microsoft Exchange Inbox. Similarly, any client that supports POP3 can connect to a Microsoft Exchange Server. Outlook Express is an Internet Mail and News Client that ships with Microsoft Internet Explorer 4.0 and supports the SMTP/POP3, LDAP, NNTP, and IMAP4 protocols.

Direct Connections over the Internet for Mobile Users

The Microsoft Exchange Inbox and Microsoft Outlook clients can also leverage the Internet in another way — as an alternative to dialup connections.

Outlook clients and Microsoft Exchange Server both have built-in support to connect to each other securely over the Internet. Mobile users can use a local Internet service provider (ISP) to connect to the Microsoft Exchange Server site located back in their organizational headquarters. Once this connection is established, users have full access to all server-based functionality, including directory services, digital signature and encryption, group scheduling, free/busy checking, and public-folder applications.

Support for Internet Newsgroups and Discussion Groups

As previously mentioned, the Microsoft Exchange Internet News Service can bring a Usenet news feed to Microsoft Exchange Server, from which administrators can distribute the feed to users through the public folder interface in Microsoft Exchange Server. Items within a newsgroup are assembled by conversation topic — the view preferred by most discussion group users. Users can then read the articles and post replies to be sent back to the Internet newsgroup.

Using the standard Microsoft Exchange Client Post Note feature, users can post a new article or a follow-up to an article or send a reply to the author of an article. Users have all the composition features of the Microsoft Exchange Inbox for composing posts to discussion groups. As with e-mail, however, the extent to which these composition features can be viewed by other users depends on the encoding format used.

The Internet News Connector automatically uses UUENCODE or MIME to encode outgoing and decode incoming post attachments. Thus, when users see an attachment in a post, they need only double-click and watch the attachment pop up. There is no waiting for the decoder to process the file.

Outlook Web Access

This capability provides a different, but equally important, kind of integration with the Internet. Outlook Web Service translates the information stored in Microsoft Exchange Server folders into HTML and makes it available — at the document or item level — as a uniform resource locator (URL) to any user with a Web browser. This capability teams up with the Microsoft Internet Information Server (IIS), which hosts the URL. As a result, organizations with documents or discussions they want to make available to Web users inside or outside their organization can accomplish this without storing the information in two different places, manually changing its format into HTML, or requiring that everyone use the same kind of client.

EASY AND POWERFUL CENTRALIZED ADMINISTRATION

While Microsoft Exchange Server offers the tight integration with desktop applications previously available only with LAN-based e-mail systems, it also offers the centralized administrative capabilities previously available only with host-based systems. Its easy-to-manage, reliable messaging infrastructure gives administrators a single view of the entire enterprise.

Easy-to-Use Graphical Administration Program

Microsoft Exchange Server includes a number of tools that help administrators reduce administration time while keeping the system running at peak performance. The graphical Administrator program lets administrators manage all components of the system, either remotely or locally from a single desktop. Built-in intelligent monitoring tools automatically notify the administrator of a problem with any of the servers and can restart the service or the server if necessary. Microsoft Exchange Server integrates tightly with Windows NT Server monitoring tools as well, so administrators can even create new user accounts and new mailboxes in one simple step for those users.

Information Moves Reliably

To keep the right information flowing to the right people, users need to be able to count on reliable message delivery. Using powerful monitoring and management tools, Microsoft Exchange Server helps ensure that the entire organization enjoys uninterrupted service. It even seeks out and corrects problems based on administrator guidelines. If a connection goes down, Microsoft Exchange Server automatically reroutes messages as well as public folder and directory changes, balancing them over the remaining connections. This greatly simplifies administration and ensures reliable and efficient communication.

Microsoft Exchange Server Components

Let's take a closer look at Microsoft Exchange Server components: private folder, public information store, directory, directory synchronization agent (DXA), and message transfer agent (MTA) objects reside in the server container on Microsoft Exchange Server.

Each server installation of Microsoft Exchange Server automatically contains an instance of the directory, the information store, and the MTA. These Windows NT-based services control directory replication and mail connectivity within a site. Directory and public folder replication between sites, as well as mail connectivity between sites and with other mail systems, are controlled through the Administrator program.

Private Folders: Central Storage for Private User Data. Private folders provide central storage for all the mailboxes that exist on that server. Users have the option to store messages locally, but server-based private folders are recommended for security, management, and backup purposes. Synchronizing server folders to the local machine is the best of both worlds and is the default configuration for people who travel with their computer.

Public Information Store: Centrally Replicating Global Access Store. On each server, the public information store houses data that can be replicated throughout the organization. Using the Administrator program to customize this replication, you can allow some data to be replicated everywhere, while other data is replicated only to key servers in each site. Data replication can be tightly controlled because rich status screens, available at all times, enable the administrator to track the replication of data throughout the enterprise.

Exchange Directory Replication and Synchronization. The Microsoft Exchange Server directory provides a wealth of customizable end-user information and covers all the routing information required by a server. Automatic replication of directory information between servers in a site eliminates the need to configure servers.

Directory synchronization has been perceived as the single biggest weakness in LAN-based messaging. Microsoft Exchange Server changes this perception with a process that keeps directories automatically synchronized on a daily basis. This makes it possible to communicate quickly and easily with users on a wide range of messaging systems such as Microsoft Mail for PC Networks, Microsoft Mail for AppleTalk Networks, Lotus cc:Mail, and optionally other messaging systems.

Message Transfer Agent. The MTA delivers all data between two servers in a site and between two bridgehead servers in different sites. The MTA is standards-based and can use client/server remote procedure calls (RPCs), Internet Mail (SMTP), or X.400 to communicate between sites.

All transport objects that enable connectivity to other sites and other mail systems reside in the Connections Container in Microsoft Exchange Server. These objects can be accessed directly through the Administrator program. The Connections Container on a Microsoft Exchange Server site houses four objects that enable site-to-site connectivity: Microsoft Exchange Site Connector, Microsoft Exchange Internet Mail Connector, Microsoft Exchange X.400 Connector, and the Remote Access Service (RAS) Connector.

Single Interface for Global Management

All objects are created and managed through the Administrator program using the same commands. A Microsoft Exchange Server installation can be implemented using a wide range of connectivity options that are all managed through a single interface. The exchange of all site-to-site information — from user-to-user messaging to data replication to route monitoring — is handled through mail messages. This single administration infrastructure greatly simplifies management of the rich functionality of Microsoft Exchange Server.

Microsoft Mail Connector. The Microsoft Mail Connector, included standard with Microsoft Exchange Server, provides seamless connectivity to Microsoft Mail Server for PC Networks, Microsoft Mail Server for AppleTalk Networks, and Microsoft Mail Server for PC Networks gateways. It uses a "connector" post office that is structured as a Microsoft Mail 3.x post office. Each Microsoft Exchange Server site appears to Microsoft Mail Server as another Microsoft Mail post office. A Microsoft Exchange Server site can connect directly to an existing Microsoft Mail post office, allowing you to replace — not just supplement — an existing Microsoft Mail MTA. No additional software is required.

Lotus cc:Mail Connector. The Microsoft Exchange Connector for Lotus cc:Mail also provides messaging connectivity and directory synchronization. Customers can co-exist and send information easily between these e-mail systems, and then later migrate when they are ready with the Lotus cc:Mail migration tools that are also included.

Microsoft Exchange Internet Mail Service. The Internet has long used several e-mail standards. RFC 821 (also known as Simple Message Transfer Protocol, or SMTP) defines how Internet mail is transferred, while RFC 822 defines the message content for plain-text messages. RFC 1521 (Multipurpose Internet Mail Extensions, or MIME) supports rich attachments such as documents, images, sound, and video. Microsoft Exchange Internet Mail Connector supports all three standards. It also enables backboning between two remote Microsoft Exchange Server sites using the Internet or other SMTP systems, making it an important component for customers who rely on SMTP connectivity to communicate with members of their own organization, as well as other organizations.

Microsoft Exchange X.400 Connector. The X.400 Connector supports three different connectivity options — TCP/IP, TP4, and X.25 — between Microsoft Exchange Server and other X.400-compliant mail systems, and between two different Microsoft Exchange Server sites over an X.400 backbone. The X.400 Connector supports both 1984 and 1988 X.400 communication and includes support for the latest X.400 protocol, File Transfer Body Part (FTBP). The X.400 Connector also enables backboning between two remote sites using a public X.400 services, such as MCI or Sprint.

Dynamic Dialup (RAS) Connector. The RAS Connector object is a special-case site connector. It uses dial-up networking (also known as RAS, or Remote Access Services) instead of a permanent network connection, thereby enabling dial-up connectivity between two Microsoft Exchange Server sites. The administrator configures when the connections should be made, and Microsoft Exchange Server connects to the other site at that time. This connector is also standards-based, using the Internet Point-to-Point Protocol (PPP). The Dynamic Dialup Connector can be automatically invoked by the Internet Mail Service and the Internet News Service so companies can participate in the Internet without the added cost of a permanent connection.

Client Support

Microsoft Exchange Server supports clients running the Windows NT and Windows 95, Windows 3.1, Windows for Workgroups, MS-DOS, and Macintosh System 7 operating systems so that users work within a familiar environment. It uses the built-in network protocol support of Windows NT Server, specifically TCP/IP and NetBEUI. In addition, its network-independent messaging protocol enables Microsoft Exchange Server to work cooperatively with existing network systems such as Novell NetWare.

You can also install and use the Microsoft Outlook Client for Windows 3.x on a Novell NetWare 3.x client running a monolithic IPX/SPX NETx, ODI/NETx, ODI/VLM, or LAN Workplace for DOS (version 4.2 or later) with no modification to the client. Microsoft Exchange clients communicate with the Microsoft Exchange Server computer by using DCE-compatible remote procedure calls, which are forwarded within an IP or SPX packet using the Windows Sockets interface.

Manage all Components from a Single Seat

Because the connectors for Microsoft Mail, Lotus cc:Mail, the Internet, and X.400 systems all function as core parts of Microsoft Exchange Server rather than as add-on applications, they take advantage of the message routing, management, and monitoring features built into Microsoft Exchange Server. They also integrate with the administrative tools provided in Windows NT Server. By using and extending tools found in Windows NT Server, Microsoft Exchange makes use of strong authentication, provides an easy-to-use backup facility that does not require the system to be shut down to save data, and features an extensive dial-in facility that can manage up to 256 connections on a single server.

Monitoring tools include extensions to Windows NT's Performance Monitor, as well as both Server and Link Monitors that inform network administrators when there is a problem or delay in the system. Microsoft Exchange Server makes use of the Windows NT Event Log to store all types of information on the operating status of the system. This monitoring capability lets the administrator set up an automatic escalation process if a service stops. For example, if the MTA service stops, the monitoring system can be configured to automatically restart it or to notify specific individuals who can determine an appropriate action.

Easy Migration

Built-in migration tools make it easy to convert user accounts to Microsoft Exchange Server. These tools work with the existing system and the Administrator program to copy and import addresses, mailboxes, and scheduling information from existing systems. It is also easy to automatically upgrade client software from the server. Migration tools are included for Microsoft Mail for PC Networks, Microsoft Mail for AppleTalk Networks, Lotus cc:Mail, Digital All-in-One, IBM PROFS/OV, Verimation MEMO, Collabra Share, and Novell Groupwise.

BUILDING A BUSINESS STRATEGY AROUND MICROSOFT EXCHANGE SERVER

Businesses of all types and sizes can implement Microsoft Exchange Server as their information infrastructure. It supports all e-mail, information exchange, and line-of-business applications that help organizations use information to greater business advantage. Microsoft has worked closely with customers throughout the development of Microsoft Exchange Server to help ensure that it meets the needs of even the largest and most complex systems. The following are some common examples of how customers are implementing Microsoft Exchange Server.

Downsizing

Many large organizations will migrate their e-mail systems from a host mainframe to a client/server system based on Microsoft Exchange Server. Microsoft Exchange Server provides the security and robust operations capabilities of the mainframe in a more flexible, inexpensive, scalable, and manageable implementation. It also includes migration tools that make it easy to move users from existing LAN-based and host-based e-mail systems.

Customers who are downsizing operations can develop applications for Microsoft Exchange Server using popular languages and development tools not applicable for mainframe computers. These customers require the flexibility that only a family of clients such as Outlook can offer.

Connecting Multisystem Environments

Customers with multiple personal computing and network platforms can use Microsoft Exchange Server to link all their users together. Organizations can benefit from the simplified administration of having just one server and a single client interface that supports all popular computing platforms. In addition, Microsoft Exchange Server allows organizations to use the Internet as a communications backbone to connect to and share information with other geographic locations of their own organization as well as with other companies.

Upgrading Current Microsoft Mail Systems

Many customers have built powerful messaging systems — including electronic forms and mail-enabled applications — with Microsoft Mail. All of their existing messaging investments will seamlessly migrate to Microsoft Exchange Server, allowing them to gain the new capabilities that Microsoft Exchange Server offers without losing access to their mission-critical applications already in place. Customers of other LAN shared-file system based e-mail systems will enjoy the same benefits.

The real test of Microsoft Exchange Server capabilities is in real-life business solutions. The following are just a few of the solutions that can be implemented using the Microsoft Exchange product family.

Customer-Support Systems. Organizations have always struggled with the costly problem of duplicating efforts because individuals do not know that others have already tackled the same issues. A customer-support system can remedy this problem by allowing support technicians to document and share their experiences and acquired knowledge with their colleagues in other support centers. This sharing helps keep organizations from "reinventing the wheel," because all employees can see and use the information and ideas generated by others. It also allows technicians to automatically route product bug reports to the engineering staff at the home office.

Customer Account Tracking. Providing superior customer service with distributed sales teams requires excellent communication among all team members and a shared history of customer contact. Inconsistent communication with customers is one of the main reasons companies lose customers to competitors.

An account-tracking system improves the management of customers by enabling account managers to see at a glance whenever anyone in the company has made contact with a customer account. A customer- tracking system also helps identify solid new sales opportunities and pinpoint customer problems that require immediate attention. Because many account managers travel extensively, this information must be accessible both from the office network and from remote locations such as hotel rooms, airports, or home.

Sales Tracking. Today, every organization that manufactures a product worries about the high cost of carrying large inventories of finished goods and supplies. A sales-tracking application can help businesses make better manufacturing planning decisions by helping sales managers and marketing executives get up-to-the-minute information, including sales volumes by region, product, and customer. This information makes it possible to identify regions or products that require special attention and to make more informed projections of demand for each product.

Product Information Libraries. The key to excellent customer service is providing customers with the right information, right now. A product information library application can help organizations improve customer service by providing salespeople with up-to-date, correct information.

This online library must contain a variety of interrelated information, including word-processing documents, spreadsheets, presentation graphics slide shows, e-mail messages from product managers, and, increasingly, multimedia elements such as images, sound clips, and videos. Sales reps can have read-only access to this library, while product managers at any location can change and modify those items that pertain to their particular products.

Such an electronic library of product information eliminates the need to continually distribute new printed product information to the sales force, which in turn eliminates the problem of disposing of expensive inventories of obsolete brochures and data sheets when products change.

A Market and General Information Newswire. Today's rapid business pace requires that managers stay in constant touch with business trends that will affect their markets and customers. A newswire application provides an easy way for employees to stay in touch with important trends, the needs of customers, and their competitors without a separate specialized application.

SUMMARY

By integrating a powerful e-mail system, group scheduling, groupware applications, Internet connectivity, and centralized administrative tools all on a single platform, Microsoft Exchange Server makes messaging easier, more reliable, and more scalable for organizations of all sizes. Microsoft Exchange Server is also a highly extensible and programmable product that allows organizations to build more advanced information-sharing applications or extend existing applications easily, based on existing knowledge. In addition, it provides the centralized administrative tools to keep the enterprise running securely behind the scenes.

As a result, Microsoft Exchange Server can help organizations save time and improve all forms of business communications, both within and beyond the enterprise. By the time this chapter goes to press, the next version of

Microsoft Exchange Server will already be available. New functionality in its clients, more integration with the Internet and Web, and greater scalability and performance are just a few of the improvements in store for customers.

Microsoft Exchange Server was designed to handle today's messaging and collaboration requirements. It is built on existing Internet standards and is designed to easily adopt new and emerging technologies to provide the best platform to its customers. Messaging is an evolutionary technology and Microsoft Exchange Server provides the foundation for any organization's messaging and collaboration growth.

Chapter 35
Virtual Server Overview

Judith M. Myerson

Have you ever wished for a lightweight server that you can carry in a briefcase? Have you failed to bring up a tower server within a reasonable time? Time lost in getting the information you need to make critical decisions may have cost you a large share of the market.

You may want to consider virtual server technology as a more economical, faster way of getting, storing, and manipulating information while you work on a major network problem. This allows you direct control of the data and even the operating system other than the one you usually work with.

Let's look a bit closer at the virtual server technology's brief history. In the 1960s, the dominant information technology strategy was host computing, better known as a centralized computing resource serving a number of "dumb" terminal workstations (see Exhibit 35.1). We have seen how mainframes evolved over the years to PC network operating systems[1] of the 1980s and then toward multilayer client/server environments[2] of today. The array of servers is vast: from desktop and tower servers (application, file, print sharing, e-mail or database) to lightweight models that you can easily hold in your hands (not your arms). It is not at all uncommon to move files through diverse platforms, such as from an IBM 370 mainframe at point A, to a UNIX server at point B, and then finally to a Honeywell mainframe at point C, via various network operating systems.

Good examples of how the servers can affect the network's performance are the servers on the Novell-based LANs connected to a fiber optic network. If a server housing the Novell operating system, for instance, is down for a long time, you will have the headache of trying to fix the problem while changing production schedules. When this happens, you probably wish for a virtual server that you can use to test applications in a single mode environment.

0-8493-9823-1/00/$0.00+$.50

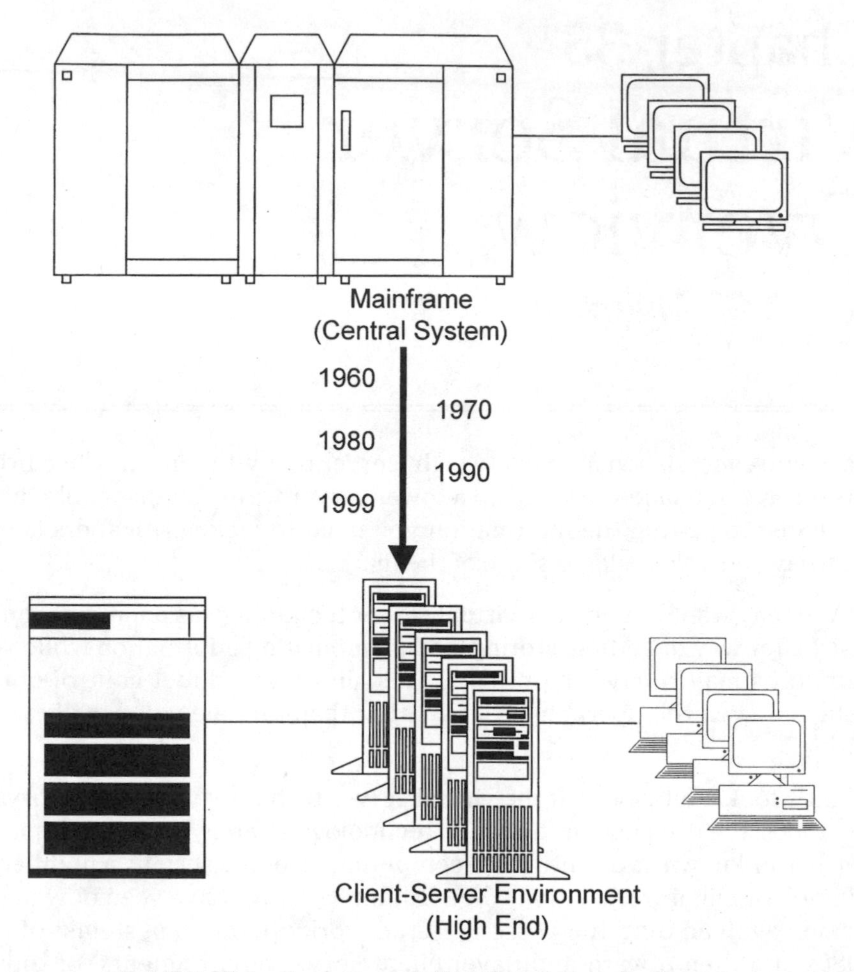

Exhibit 35.1 Evolution of client/server environment.

The smaller, traditional servers are dedicated to specific applications (print, CD-ROM, remote access), while the larger ones are used to store files and databases, run heavy duty applications, share printers, and manage databases and electronic messages. What is relatively new is the idea of using an external removable device (hard or cartridge) as a virtual server to run multimedia and video applications. Lightweight and portable, the server is capable of holding files and applications up to 2 GB.[3] To maximize a cartridge's performance, use a SCSI interface card rather than a parallel port, as it moves data more quickly from the computer to the device and vice versa. As you probably already know, the SCSI interface allows you to have a cascade of up to seven devices.

There are, however, some drawbacks. One, the SCSI-based cartridge may run somewhat slower than an external hard disk drive. Two, you have to be careful in handling the cartridge when you put it in or take it out of the device. Shutting the power off while the cartridge is running will render it useless. Three, the size of a cartridge is limited to 2 GB,[4] in contrast to the internal 16.8 GB (or more) disks as the new standard for the newer computers. Four, the device runs only in a single mode environment. Yet, these drawbacks seem trivial in any of four scenarios:

- You need wider bandwidth to run multimedia and video applications.
- The network is degrading or is completely down due to packet delays, traffic overload, and coordinated denial of service attacks.
- You go from multi-user environment to a single user mode to test special or classified applications.
- You are using a laptop while traveling to a distant office.
- Upon arriving at the distant office, you use a desktop computer to make presentations from the virtual server.

Just make sure that the laptop or desktop computer you are using has the programs to access the cartridge(s) as well as the applications on them.

Despite its size, you should make a virtual server fault-tolerant as much as possible. According to Witherspoon,[5] faults come in two basic categories: data loss and energy. In addition to server disk failures, you can lose data when you erroneously delete or corrupt a file, or unintentionally introduce a virus into the system. You should protect the disk by including, for instance, an uninterruptible power supply (UPS) in your hardware configuration. You should consider additional measures such as mirroring the cartridge(s) and preventing intruder access to them.

It is a good idea that you periodically check with an ADP security officer that you are allowed to carry cartridges from one place to another. You may need instructions on how to protect the cartridges and where, when, and how to report a security incident. It does not matter if that data are unclassified or sensitive unclassified (financial, personnel, and so on). Data declared unclassified at your home office in the United States may become very sensitive if stolen or compromised by foreign nationals. One country, for instance, may consider one type of data sensitive, while another may treat it as not. To be on the safe side, encrypt the data before you depart and decrypt them after you arrive.

WHAT EXACTLY IS A VIRTUAL SERVER?

A virtual server is the equivalent of a human operator. By the same analogy, using a virtual client is equivalent to using a telephone desk set to send a request to the operator for information and to receive a call from the operator who then displays the requested information on the screen. Similarly, most people think of a virtual server as a black box server without any real knowledge of what goes on behind the scenes to make the server useful.

A virtual server may consist of several partitions, but it may not be possible to put all partitions on one cartridge due to its limited storage size. It is more likely to have a cascade of cartridges (up to 7) to accommodate the partitions you want. A Windows NT server system, for instance, may consist of the following partitions:

- *System area,* including operating system software and swap area.
- *User area*, including the areas to which each user is assigned according to the level of file and directory permissions in this and other partitions.
- *Server and related applications*, including Internet Information Server, Visual Basic, Microsoft Office, and C/C++.
- *Utilities*, including disk management partition and network performance monitor.
- *WWW published files*, including all files to run a Web site under Netscape and Microsoft Internet Explorer.

THE LIFE CYCLE OF A TYPICAL VIRTUAL SERVER

Let's suppose that a user just arrives at an office from a distance and sits in front of a client workstation already connected to a virtual server. A typical life cycle would be as follows:

- Bob takes the drive out of his briefcase, places it in a secure area, and makes sure it is not left unattended.
- He very carefully inserts the cartridge (NT server) into the device.
- He then turns on and boots the computer, and waits for it to run its course.
- Bob enters the password to the computer.
- From the menu, he selects the NT server option and presses the ENTER key.
- He is ready to use the virtual server connected to the computer either as a console or in a client workstation mode.

CREATING THE ENVIRONMENT REQUIRED TO RUN A VIRTUAL SERVER

It is important to develop a performance assessment checklist to ensure a virtual server is optimized. Borrowed from Winterspoon, the following components need to be considered for optimization:

- Disk drive seek times
- Number of SCSI devices

- Workstation or server load capacities
- Workstation memory
- Disk fragmentation
- BIOS and VLSI chip sets and their efficiencies
- Software performance
- Client requests and server paging
- Graphic screen refresh rate

The specifications for the first three components may be less for a virtual server than those for a traditional desktop or tower server. They, however, may be compensated for by an increase in workstation memory and graphic screen refresh rate, an improvement in software performance and an optimization of disk space.

While a fault-tolerant system is ideal, the reality is that you need disk management tools as a preventive measure against server crashes. A data file, for instance, can be fragmented into small pieces scattered all over the cartridge. The smaller the piece and the larger the size of the cartridge (1 to 2 GB) disk, the higher the chances are for the files to fragment.

To better control wasted space, use a disk partition tool to divide a cartridge into *slices,* and choose an optimal size for a partition you wish to create. The larger the partition, the more wasted space there will be for short text documents. Given the correct partition size, partitioning improves performance by allowing you to separate different groups of users and data. If one partition, for instance, ran out of space, the other partition would not be affected. To take advantage of this feature, be careful how you take out or put in the cartridge from its case. If you remove it at the wrong time (while it is running) or handle it roughly (while inserting it), you may completely damage it losing the data in all partitions in that cartridge.

With a cascade of SCSI devices, you can have a smaller number of larger partitions on each cartridge. This is especially useful if you want to set up a virtual Windows NT server. In this multiple device arrangement, the cartridge in one device, containing the operating system software and swap space, is called the *system disk*, while the cartridges in other devices may be called *nonsystem* or *secondary disks*. When you have more than one *disk*, you can modify the sizes of partitions on *secondary disks* without having to shut down the system or reload the operating system software.

Now, let's take a look at a good example of how reusing a document (an object) can create a security problem. If you save a sensitive text document, later edit it making it smaller than the original one, and resave it, the chances are high that the residual data containing the sensitive data will be attached to the revised document. While this residual data are transparent to a user, it is not to a system administrator. With a disk management tool, the administrator can look at the residual data (and completely remove them).

INSTALLING AND CONFIGURING A VIRTUAL SERVER

Once all this underlying structure has been set up, the installation of a virtual server is not much more difficult than for any other sever type. Due to its size, it may actually be easier to install and configure it.

There are six basic steps:

1. Install a SCSI interface card.
2. Run "Install" setup program and follow instructions.
3. Plug one end of a SCSI cable to the card and the other end to a SCSI device.
4. If necessary, build a cascade of up to seven devices, and assign a unique ID for each.
5. Plug in a power cord and turn on the device's power.
6. Boot the computer, format the cartridge, and create partitions.

TESTING AND DEBUGGING YOUR NEW VIRTUAL SERVER

It is not possible to use the *ping* and *tracert* (or *trace route*) commands to check your local computer and the virtual server, as IP addresses are not needed to make a connection to the server. There are other ways of testing and debugging. For instance,

1. Go into the DOS mode.
2. Enter the drive letter assigned to the drive.
3. Enter dir to get a list of files and subdirectories.

If you see the list scrolling down the screen, continue with the following tests:

1. Use a disk management utility to check for sensitive residual data attached to the files and remove them.
2. Check the files for unlabeled sensitive data and mark them based on the sensitivity level.

If a message says that the drive is invalid, make sure the power is turned on. Repeat the above process. If you still get the same message, you may have a defective cartridge. For this reason, it is extremely important that you periodically back up or mirror the cartridge to another device. Destroy the defective cartridge containing sensitive data according to federal procedures.

FUTURE DEVELOPMENTS

Current trends suggest virtual servers will have a higher storage capacity with faster transfer speed. We hope to see something better than a SCSI interface that will allow a cascade of more than seven devices with an alarm system for unauthorized removal or improper handling of the cartridges.

SUMMARY

For more than 30 years, we have seen mainframes evolve into multilayer client/server environments. Today, smaller servers are dedicated to specific applications, while the larger ones are used to run heavy-duty applications and share files and printers. What is relatively new is the idea of using an external, removable device as a virtual server to run multimedia and video applications. The current drawbacks of this server will become history as technology improves. Carrying a lightweight virtual server in a briefcase is the wave of the future.

Notes

1. Examples include MS Net (Microsoft), Netware (Novell), and LAN Manager (Microsoft).
2. Multilayer client/server environments aim to be more flexible in distributing load among the servers and proving response time.
3. An external tape drive, for instance, is strictly used for backup and recovery, although its capacity is much higher than 2 GB.
4. A cascade of seven drives gives a total of 14 GB.
5. Witherspoon, Craig & Coletta, *Optimizing Client/Server Networks,* Chicago, Illinois: IDG Books (Compaq Press), 1995.

Chapter 36
Text-to-Speech Server Overview

Judith M. Myerson

Have you ever wished that you could get your colleagues at the banks around the world to edit your speech in English and none of them could read or speak it? Have you brought up a Web page in a language you do not understand?

You may want to consider text-to-speech server (TTS) technology as a more economical and faster way of getting the information to global decision makers. Keep in mind that the TTS technology does not translate; it only converts text to speech. If you have a working knowledge of, say, French language, you may type a French phrase in a document in English, listen to the document as it is displayed on a Web page, and proofread it.

The TTS technology, however, has its hardware limitations, partially due to the way the digital signal processing chips are built and the way today's computers are configured. For instance, it cannot handle more than six or nine languages,[1,2,3] while a translation language program can translate Web pages, documents, and e-mail messages in 25 languages and an OCR program can scan foreign documents in 33 languages. Someday, the TTS technology will have the capability of *translating* and *converting* text to speech in more than 50 languages. This will be possible when today's hardware and software limitations are gone.

Let's take a closer look at the TTS technology's brief history.[4] Twenty years ago, choppy speech was the norm of the early TTS systems on standalone personal computers. These systems were constrained by (1) the small size of memory and disk space, and (2) the concatenation approach[5] in synthesizing speech. As a result, many people had difficulty in understanding the inaccurately pronounced words, and words that were not in the database were not pronounced. There was no way to adjust intonation to emphasize words or show emotions.

0-8493-9823-1/00/$0.00+$.50

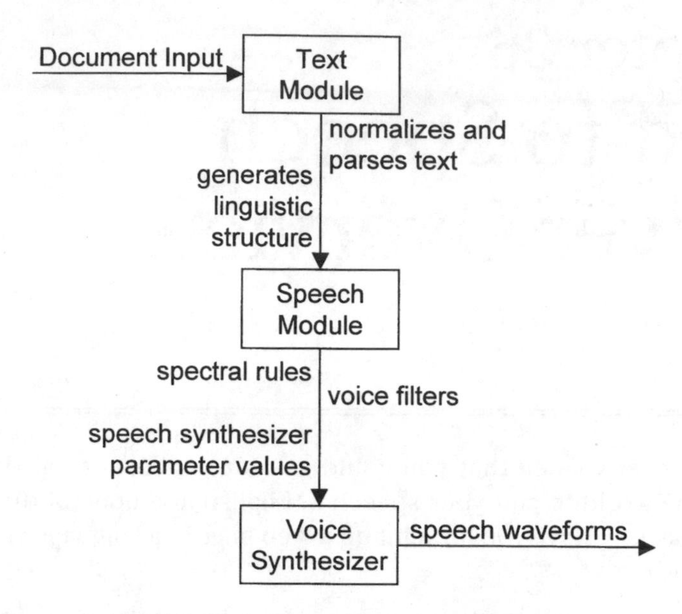

Exhibit 36.1 Three processing modules.

Today, most TTS systems use an approach based on storing recordings of actual human speech units. To store these recordings, an enormous amount of memory is required. To reduce memory requirements, the number of sentences and phrases than can be spoken must be limited. Another way of reducing memory is to record and store smaller speech units, usually phonemes, instead of complete words, phrases, or sentences. With this approach, unlimited vocabulary is possible. The speech produced by this methodology is often human-sounding, but choppy. As a result, applications using this methodology have one or two speaking voices with no tonal variations.

To get a more natural, human-sounding speech, Eloquence Technology, Inc. (ETI),[6] for instance, has taken a completely different approach. Its TTS engine "produces speech entirely by using a set of linguistically oriented rules and models based upon many years of analyzing human speech, rather than the more common concatenated approach." As shown in Exhibit 36.1, it is composed of three separate processing modules:

- Text module that "normalizes and parses the text and generates a linguistic structure."
- Speech module that "applies the rules for spectral values and durations, and applies voice filters to produce the appropriate set of speech synthesizer parameter values."

- Voice synthesizer module (software only) that "produces the speech waveforms that are inputs to a PC's sound card, telephony board or other appropriate device [as well as an audio software product]".

By setting different parametric values for input into the Voice Synthesizer, different voices can be generated.

In 1983, ETI first used the Delta system in synthesizing speech.[7] As this system evolved into the ETI's main development tools over the years, the DECtalk family of products achieved, in 1985, the highest test scores of all TTS systems on the market.[8] These products were very popular with the visually impaired, and speech disabled. In the latter part of 1995, the ETI's first version of Eloquence (TTS) was released.[9]

To speak over the World Wide Web,[10] ETI-Eloquence uses Real Player Plus 5.0 or G2[11] compression and streaming technology. The rationale behind the ETI's choice is that this technology allows quicker downloading of compressed speech files,[12] although these files can consume an enormous amount of disk space. The longer the text, the more space a speech file needs.[13]

In keeping up with the demand for heavy duty TTS applications in distributed client/server environments, the ETI-Eloquence SAPI Server[14,15] was conceived and implemented. It has taken advantage of the Microsoft Speech Application Programmer's Interface (SAPI) to control its TTS engine, and uses the Eloquence Command Interface (ECI) to provide an interface between customized applications and the TTS system. The choice of Windows NT as the platform to run the ETI-Eloquence[16] evidently indicates a continuing shift in the market share of server platforms from Novell Netware (and, to a lesser extent, from UNIX[17]).

WHAT EXACTLY IS A TEXT-TO-SPEECH SERVER?

A TTS server is the equivalent of a human relay operator. Using the same analogy, a TTS client is the equivalent of using a telephone desk set to fax a text document to the operator or to receive a call from the operator who read aloud the faxed document. Similarly, most people think of the TTS system as the TTS software without any real knowledge of what goes on behind the scenes to make that program useful.

A TTS server should have multiple disks as speech files take up much disk space. The server may consist of eight partitions, as follows:

- *System area,* including operating system software and swap area.
- *User area,* including the areas to which each user is assigned according to the level of file and directory permissions in this and other partitions.
- *TTS applications,* including ETI-Eloquence SAPI Server, Elocutor, ECI, Microsoft SAPI.

Exhibit 36.2 Male and female voice characteristics.

	Male	Female
Pitch	110	203
Speed	176	176
Volume	26987	37640
Head size	50	50
Pitch fluctuation[a]	30	30
Roughness[b]	0	0
Breathiness[c]	0	50

[a] *Pitch fluctuation* is how much the pitch fluctuates in a given sentence. The greater the pitch fluctuation, the more excited the speaker sounds.

[b] *Roughness* adds a raspy quality to the voice.

[c] *Breathiness* makes the voice whispery.

- *Related applications,* including translation and OCR software.
- *Utilities,* including disk management, partition, and network performance monitor.
- *Real Player tools,* including RealPlayer Plus 5.0 or G2.
- *Customized WAV files,* including those developed and compressed in other partitions.
- *WWW published files,* including all files to run a Web site.

THE LIFE CYCLE OF A TYPICAL TEXT-TO-SPEECH SYSTEM

A typical TTS life cycle for a text document between two users with accounts on the same IP domain would be as follows:

1. Carol, on her client workstation, starts elocutor.exe from the server, opens a blank document screen, and types a short message.
2. Carol brings up the Voices dialog box from the Options menu, notices the default speaker is English-American Reed (Adult Male), changes to English-American Shelly (Adult Female) as the new speaker. Exhibit 36.2 shows voice characteristics[18] for each speaker.
3. Carol clicks on Reading in the same menu, chooses the option of reading every line while typing, and checks off the option of showing progress while reading every line.
4. While composing a short message, Carol listens to each line being spoken.
5. To double check, Carol uses the mouse to highlight the entire text, clicks the read-selection icon and listens to the speaker. Alternatively, Carol uses the Shift+Ctrl+RightArrow keys[19] to select an entire word, the LeftArrow key to place the cursor at the end of the word, clicks the read-selection icon and listens to the speaker.
6. Carol hears several mispronounced words. She fixes them one of three ways: spells the word literally, replaces the word with pho-

nemes, or adds the word to one of the dictionaries (special words, abbreviation or root dictionary). She repeats this process of converting text to speech and listens to the voice that now correctly pronounces all words.

7. Satisfied, Carol clicks the record-to-file icon to record the highlighted lines in the document, and saves it as message.txt.
8. The speech output is appended to the elocutor.wav[20] file.
9. Carol uses the RealProducer[21] to compress the .wav files in RealAudio formats,[22] which she, then, publishes under the Netscape and Microsoft Internet Explorer browser.
10. Carol brings up a Web page and sees a picture of Real Audio Player at the upper left corner. When the downloading completes, the player automatically begins to play. Carol clicks on the fast forward, rewind, stop, pause, or replay icon.
11. Carol begins a client e-mail program, sends a short message via a separate e-mail server to Bob, and asks him to check on her speech draft.
12. Bob gets his mail, starts Netscape or Microsoft Internet Explorer, brings up a Web page, clicks his mailbox, looks at the text, and listens to the speaker.
13. Bob returns a reply that he has added a sentence to the document.
14. Carol receives the message, starts eloqtalk.exe, reads the text, and listens to the text as it is being read. Satisfied, she switches to the browser, reloads the Web page showing the text, and listens to the speaker. She e-mails a note to Bob, and thanks him for the changes.

Now let's suppose Carol decides to translate the entire text document to French (that she does not know very well) and to have the browser speak it.

1. Carol starts a client translation program to translate message.txt to French and saves it as msgFR.txt.
2. Carol brings up elocutor.exe, and opens msgFR.txt.
3. Since the English-American Shelly (Adult Female) is the default voice, she wants to change it to a French female voice.
4. Carol brings up the Voices dialog box from the Options menu, and changes to Nathalie's voice (French Adult Female). Exhibit 36.3 gives voice characteristics for each speaker.[23]
5. Satisfied, she clicks the play icon to hear the entire document spoken.
6. Carol clicks the record-to-file icon, appending the speech output to a .wav file.
7. Carol passes the .wav file through the RealProducer to produce RealAudio files.

Exhibit 36.3 Female voice characteristics (American and French).

	American Female	French Female
Pitch	203	274
Speed	176	176
Volume	37640	26987
Head size	50	56
Pitch fluctuation	30	35
Roughness	0	0
Breathiness	50	50

CREATING THE ENVIRONMENT REQUIRED TO RUN A TEXT-TO-SPEECH SERVER

While Windows NT Server is seen as having a better support for threads and symmetric multiprocessor (SMP),[24] it is important to develop network performance assessment checklists to ensure the network for the TTS server is optimized. According to Winterspoon,[25] the following components need to be considered when optimizing the network:

- Workstation memory
- Transmission buffers
- Direct memory access (DMA) wait states and transfer speeds
- Disk drive seek times
- Disk controller transfer rates
- Number of active bus or SCSI devices
- Disk or disk array file fragmentation
- BIOS and VLSI chip sets and their efficiencies
- Workstation or server load capacities
- Software performance
- Client requests and server paging
- Database gateway efficiency
- Graphic screen refresh rate

While a fault-tolerant network system is ideal, the reality is that you need disk management tools as a preventive measure against server crashes. A data file, for instance, can be fragmented into small pieces scattered all over the disk. The smaller the piece and the larger the size of the disk, the more likely there will be wasted space. If the disk is not periodically optimized, excessive movements of a read/write head to fragmented pieces can wear out the head and cause a system crash.

To better control wasted space, use a disk partition tool to divide a disk into slices, and choose an optimal size for a partition you wish to create. The larger the partition, the more wasted space there will be for the short text documents. Given the correct partition size, partitioning improves performance by allowing you separate different groups of users and data. If one partition, for instance, ran out of space, the other partition would not be affected.

With multiple disks, you can have a smaller number of larger partitions on each disk. In a multiple disk arrangement, the first disk containing the operating system software and swap space is called the system disk, while the other disks may be called nonsystem or secondary disks. When you have more than one disk, you can modify the sizes of partitions on secondary disks without having to shutdown the system or reload the operating system software. Using multiple disks (and partitions) increases the I/O volume and throughput. With the I/O across multiple disks, you may avoid bottlenecks.

If you save a sensitive text document, edit it making it smaller than the original, then resave it, the chances are high that the residual data will be attached to the revised document. While this residual data are transparent to a user, it is not to a system administrator. With a disk management tool, the administrator can look at the residual data (and completely remove them).

Basically, a server[26] must have sufficient power, memory resources, and disk space to support the TTS system as well as related applications. The server must be running at all times, be on a very good UPS, and be able to reboot automatically upon restoration of power. There should be no other heavy duty applications and services on the same computer, such as an Oracle 8 Server. If the file and print sharing services are not heavily used, they can be installed on the TTS server.

The DNS server (included in the Windows NT 4.0 and higher) and even a firewall may run on the same server. Since FTP, Web, and mail servers will be heavily used in transmitting the documents and speech files, each should be set up separately from the TTS server. You should use Performance Monitor to ensure there is ample memory for avoiding excessive paging and that CPU utilization is no more than 90 percent. In addition, there should be (1) a mirror site that duplicates your master site, and (2) a backup at another site that can receive and temporarily hold TTS and related files until your site is back on-line should something happen to your site. Geographically isolated sites often work out reciprocal arrangements[27] to serve as backups for each other.

You must have a high speed modem to transmit quickly the text and speech files. Levels of service from a 56K modem to T1 are available from various ISPs. With a modem and a DNS server in place, use ping and tracert to check if the DNS operates correctly — particularly your local nodes and the nodes outside your IP domain.

INSTALLING AND CONFIGURING A TEXT-TO-SPEECH SERVER

Once all this underlying structure has been set up, the installation of a TTS server is not much more difficult than any other server types running heavy-duty applications.

There are three basic steps involved:

1. Create a Windows NT service account. This is just an NT account with minimum rights to run the server. The minimum right required is "Logon as a service." This account can be a domain account or a local account on the computer that will run the TTS system and related applications but rights must be granted on the local computer (log in as Administrator to your local computer, except on a Domain Controller, where this is not possible.).
2. Run the InstallShield setup programs on Windows NT to install the TTS software, Real Player, translation, OCR, browser acceleration[28] and other executables, install them in appropriate partitions, and start them running.
3. Use a browser to make sure the Web pages are talking. You may specify passwords and rights to groups of users accessing files and directories. Instances include granting passwords and access rights to the executives and denying them to certain users.

TESTING AND DEBUGGING YOUR NEW TEXT-TO-SPEECH SERVER

The first thing you should do in testing the server is to use the ping and tracert commands to check if the connection between your local system and the server is working properly. The ping command is used to test the TCP/IP connectivity, while the tracert command is used to determine how long it takes the IP packets to travel from your local system to a computer or a router and where on the network a problem is occurring.

If these two commands do not show any problems, start a client browser to test your TTS server, and download a Web page. When the downloading is complete, you should see a small Real Player box automatically turning on, and hear the speaker. If you use annotations to emphasize words, show emotions, or vary tones in the text document, you should be able to recognize, by ear, when the speaker is whispering, speaking a little louder, or getting emotional.

If you cannot hear the speaker at all, possibly a modem speaker has been set to turn off the volume. To raise the volume, click the speaker icon in the Windows status bar and use your mouse to move the slider to the maximum. You can then adjust the sound with the volume control on the Real Player box. Given the "no volume" setting for the modem, the speaker is automatically turned off when you reboot. If you still do not hear sounds after raising the volume, then you may have other problems that need to be debugged.

If the server has memory problems due, for instance, to excessive paging, use an option in the browser to clear the memory cache. If the problem persists, use the other option to clear the disk space cache assigned to hold temporary files from the Internet. As a last resort, reboot the client workstation and (or) server.

FUTURE DEVELOPMENTS

Current trends show that TTS technology will have the capability of translating from one language to another in both text and speech files. For instance, a TTS system of the future could translate a document in English to French and convert it to speak in Spanish. As client/server technology evolves, the number of speech channels will increase to accommodate a wide range of voices and tonal variations in, say, 50 languages. In addition, the annotations used to emphasize words or show emotions will be more automated. We hope to see a greater improvement in audio compression and streaming technology so sound files will consume less disk space, thus freeing up the space for other applications and server tasks.

SUMMARY

In 20 years, the TTS technology has evolved from a "choppy" voice to more natural, human-sounding voices. While the technology has greatly improved the voice quality, the issues of partitioning, fragmentation, wasted disk space, and residual data are still here. These issues must be kept in mind when you install, configure, test, and debug the server. Current trends, however, indicate that these issues will be resolved in future developments of more powerful TTS technology.

Notes

1. ETI-Eloquence: U.S. English, U.K. English, Castillian and Mexican Spanish, French, German, Italian, Canadian French, and Chinese.
2. QCS TTS server: English, Spanish, French, German, Italian, and Japanese.
3. IBM Talking Web Browser, developed originally in Japan, has been adapted to read English for the benefit of the visually impaired (and speech disabled).
4. Eloquence Technology, Inc., *An Introduction to Text-to-Speech Systems,* http://www.eloq.com/Edpap.htm, as of 12/9/98.
5. The concatenation approach puts together clauses from concatenated, recorded words.
6. According to Eloquence Technology, Inc., "the most publicized application for TTS is as a companion to the new SR [Speech Recognition] Dictation systems that have been introduced by IBM, Dragon Systems, Kurzweil (now Lernout & Hauspie). In addition, DECtalk software uses digital format synthesizer."
7. "Background on Eloquent Technology, Inc." http://www.eloq.com/etiOhst.html as of 12/9/98.

8. Hallahan William, I., "DECtalk Software: Text-to-Speech Technology and Implementation," *Digital Technical Journal* (electronic), updated 11 April 1996 (www.digital.com, as of 12/11/98).
9. ETI-Eloquence is applied as part of the IBM's voiceType speech toolkit.
10. It is not known, at this time of writing, if there are others using Radio Audio files to speak over the Web pages for the audience representing industrial, e-commerce, banking, and government sectors.
11. Visit http://www.real.com.
12. Eloquence Technology, Inc., "Interested in Text-to-Speech," www.eloq.com.
13. An uncompressed speech file, such as one containing "Hello, How are you, Bob?" can consume as much as 100K of disk space.
14. There are others, such as Quality Consulting Services, that provide TTS servers to make a wide variety of information available to every caller, such as the contents of an electronic mail message, the names or addresses of local dealers, or news updates. It makes this possible with software tools that it builds to expand the functionality of IBM's DirectTalk/6000 product offering.
15. ETI-Eloquence also runs on Windows 95/98, UNIX, and possibly other platforms.
16. "ETI-Eloquence SAPI Server," Release 4.0, 2 November 1998.
17. You may use, for instance, Internix to run UNIX applications on a Windows NT platform while you move to Windows NT from UNIX.
18. The *pitch, volume* and *breathiness* options are higher for the female voice than for the male counterpart.
19. See Microsoft Word Help for details on the use of keys to move to a character, word, paragraph, column, or object using shortcut keys.
20. In the demo Elocutor, all instances of Record are concatenated to elocutor.wav. To capture individual .wav files, delete elocutor.wav, record message.txt, rename elocutor.wav to message.wav. The full SDK allows you to control and name the individual recorded output files.
21. RealAudio Encoder is replaced by RealProducer, RealProducerPlus, or RealProducerPro.
22. The .wav files must be compressed, as they take up an enormous amount of disk space. For instance, an uncompressed short message, such as "Hello, What time is it, Carol?" consumes slightly over 100K of disk space. An uncompressed speech output from a long text document may take up more than 1.5 MB for each language.
23. The pitch, volume, and brightness are higher for the French female voice than for the American counterpart.
24. As of 12/28/98, ETI has completed a multithreaded server version with at least 14 speech channels.
25. Witherspoon, Craig & Coletta, *Optimizing Client/Server Networks,* Chicago, Illinois: IDG Books (Compaq Press), 1995.
26. For a small server or a workstation, ETI specifies 66 MHz as the minimum to run the TTS system. For ETI's multithreaded server version with a least 14 speech channels, a Pentium 200 is the minimum. Minimum requirements for Real Player Plus 5.0 are 75 MHz 16 MB for video and 66 MHz 8 MB for audio, and are higher for Real System G2. RealProducer runs on any Pentium class machine with at least 32 MB RAM.
27. Reciprocal arrangements should be part of contingency planning and disaster recovery plans.
28. One example is Real's PeakJet2000.

Chapter 37
CD-ROM Servers

Robert Brainard

So you want to build a CD-ROM server for your network. OK. So where do you begin and how do you proceed? This chapter discusses the various types of CD-ROM servers that are available and the items that you will need to consider when planning your deployment. The intent is to guide you through the myriad considerations involved in obtaining a CD server that is best suited for your environment and needs. The following topics are covered:

- CD-ROM networking benefits
- CD server types
- CD-ROM drive overview
- CD-ROM application considerations
- CD server product and vendor considerations
- CD technology advances

With an understanding of your options and considerations, you can arrive at the best solution for your network. You not only need to understand how to create a CD server, but also what your organization's needs, capacity requirements, user technical expertise level, and growth expectations for the future are. With a little planning, your CD server can grow with your organization without requiring a major overhaul next year.

First, let's review the benefits of networking CD-ROMs so that we can be sure of our objectives.

CD-ROM INTRODUCTION

CD-ROMs (Compact Disk Read Only Memory) have changed the way electronic information and software are distributed and accessed. CD-ROM is already the medium of choice for distributing software application programs, entire operating systems, vast knowledge bases, and information sets.

There is a wide array of information available on CD-ROM, including periodical collections updated monthly, technical documentation, state

0-8493-9823-1/00/$0.00+$.50

and federal statutes, government specifications, medical databases, insurance regulations, phone databases, catalogs, etc.

The adoption of the ISO-9660 CD-ROM format standard, which specifies how data must be arranged on the disk, was the catalyst that fueled the explosive growth of CD-ROM. Almost all CD-ROMs are written in this format, and almost every operating system and computer has software that can read this format, making CD-ROM one of the few universal formats that allow data to be read on all major computer platforms including DOS, Windows, Macintosh, OS/2, and UNIX. While most operating systems provide support for internal CD-ROM drives, few offer robust services for making those CD-ROM devices widely available over the LAN.

A CD-ROM can hold up to 650 MB (Megabytes) of information, the equivalent of 130,000 pages of text, or 450 floppy disks. Furthermore, just around the corner are DVD-ROMs (Digital Versatile Disks), which can deliver an astounding 4.7 GB all the way up to 17 GB on a single disk. DVD also promises to deliver additional rich media content such as audio and video.

CD-ROMs offer many advantages over traditional electronic storage and retrieval devices. They can store huge amounts of information, they are portable, lightweight, durable, and based on an international ISO standard file system. CD-ROMs are also a productivity enhancer, providing random and quick access to information, compared to tape storage solutions, which access information sequentially. CD-ROMs are also very inexpensive to produce and to purchase, as they leverage the economies of scale from the CD audio industry.

For software manufacturers, CD-ROMs are relatively inexpensive to produce, often costing less than one dollar per disk to manufacture in quantities. They are also easy to distribute because of their size. Because CD-ROM is such a cost-effective medium for distributing large amounts of information, software publishers can offer their applications (or suites) on one CD-ROM instead of ten or more floppy disks. This has two key benefits: installation of applications is much faster and it is also much cheaper for the publisher to distribute.

Consequently, CD-ROMs are everywhere, and they will surely make their way into your business operations. They have become the standard media on which a wide range of information is distributed. This information includes the following types of information:

- Knowledge bases particular to your industry
- Tax codes and forms
- Legal statutes and case histories
- Government specifications
- Company policies and procedures

- Encyclopedias
- Phone listings
- Periodicals and magazines
- Software applications such as Microsoft Office, Corel Draw, etc.
- Medical journals
- Mapping systems
- Multimedia
- Research data

WHY NETWORK CDS?

Assuming that your business has a network of some sort, you face a choice of whether or not to copy the CD-ROM information to your server hard disks, arm each user in your organization with his or her own personal copy of each CD-ROM, or finding a CD Server solution which enables you to share the CD-ROMs via your network infrastructure.

CD servers allow you to mount the CD-ROM media in CD-ROM drives connected to your network, and make those CDs directly available to users as network shares or volumes, complete with security access controls.

Most business users no longer have just one CD-ROM disk which they need to access; they need access to 5, 10, or even 20. Obviously, if internal CD-ROM drives are used to solve the problem of access, then users must constantly shuffle disks. They can only run one application at a time and they have to deal with managing and storing their disks. By networking CD-ROMs, these problems are solved, just like they were solved with conventional floppy-based applications.

Networks have become the vehicle to access information, and CD-ROM is becoming the vehicle to distribute information. Because of the benefits inherent in CD-ROM, almost all network users will need access to CD-ROMs in the coming years.

The benefits of networking CD-ROMs include easier management, installation, configuration and updates, and better security. They also offer cost savings in hardware and network software licenses, and, ultimately, higher user productivity and higher performance.

Networking allows multiple users to access a CD-ROM simultaneously. In addition, a single PC user can access multiple CD-ROM drives. The economic and productivity benefits can be significant. Eliminating the need for a CD-ROM drive at each PC can be a substantial savings in itself. Productivity gains come from not having to wait in line for information.

By having CD-ROMs connected to the network, hardware and software resources are usually located together. For instance, CD-ROM drives are typically grouped near the file server, or together somewhere in the work area, to allow for better control and administration of these resources.

CD-ROM software and applications that exist under the same controls already set up on the network allow for easy management of software, support, and issuing of upgrades. When CD-ROMs are not networked, network administrators are faced with updating each workstation with the same information or software in various places throughout a company. The job of tracking down those disks and issuing updates can be very time consuming. By integrating CD-ROMs into the network, administrators can easily accommodate changes and avoid many of these problems.

One of the often overlooked benefits of networking is in having CD-ROMs integrated tightly with the existing network operating system, i.e., Novell NetWare. Network administrators may then use familiar network management tools to access the CD-ROMs. The level of native integration that CD-ROM networking products offer may vary between network operating systems.

Because CD-ROMs are so portable and easy to use, and because they contain so much information, security can be a major concern. Often the CD-ROM contains mission critical information which must always be available for business operations. Access to the CD-ROM must be controlled to protect the disk from being accidentally ejected, causing users to become idle, losing productivity and, in some cases, valuable data.

Other reasons for restricting physical access to the CD-ROM are if the disk contains sensitive or confidential information, or if there is a risk of theft after hours. The highest level of security can be realized by networking CD-ROMs on the file server where disks can be locked away, access controlled, and activity monitored.

Hardware cost may also be a factor when considering networking CD-ROMs. Stand-alone hardware costs can add up quickly. Imagine a company with 50 users who all use the same CD-ROM software program. The company could purchase 50 internal CD-ROM drives, adapter interface (I/F) cards, and software drivers. Then install and configure them, fight IRQ/Port conflicts, etc. Or the company could attach one CD-ROM drive to the network, using a CD-ROM networking solution such as one of Microtest's DiscPort™ products. The CD-ROM software then appears to network users as a normal hard-disk-based application.

Another opportunity to reduce costs is software licensing. Generally, application network licenses are cheaper than the same number of single user fees. For instance, buying 50 copies of Corel DRAW! is more expensive than obtaining a network license for 50 users. Beware, however. Not all software is offered with network licenses. Make sure to read the license agreement before installing it onto the network. Networked CD-ROM applications are governed by the same license agreements and restrictions as normal hard-disk applications. Do not violate the CD-ROM license agreement. Most software publishers are issuing network licenses and enforcing them. Another hidden software cost savings includes not having to purchase CD-ROM driver software for every workstation, as most CD-ROM networking products include these drivers.

In most instances, networked CD-ROMs achieve higher performance than local internal CD-ROM drives. The file server processor and memory performance are usually much higher than most common workstations. Data and directory caching in file server memory provides network access to CD-ROM data faster than accessing a local CD-ROM drive. The file server memory, which is usually quite large, caches the data, so that data requests for information from the CD-ROM drive are often satisfied by a cache hit on the file server. These data are then returned directly via the network cabling to the workstation. The increased performance depends on many factors, including the processor speed of the file server, the amount of memory in the file server, the local workstation CPU speed and memory, the number of users on the network, the speed of the network cable, the amount of data being read from the CD-ROM drive, the amount of traffic on the network, and the type of application running.

The only perceived argument against utilizing a CD server is perhaps performance. We have found that, because most business CD-ROM titles are not multimedia based, but rather information resources that your users draw from to perform their jobs, performance is not an issue. Furthermore, with the advent of 10/100-mbps networks, bandwidth and throughput to user PCs have been dramatically increased, removing the most restrictive bottleneck.

CD SERVER TYPES

CD servers come in many shapes and styles. These styles are categorized as:

- Peer-to-peer CD sharing
- Fileserver CD extension software
- Work group CD server appliances
- Enterprise CD server systems

Understanding the types of CD servers that are available can help you make the best choice for your organization's needs.

Choosing the right architecture will help produce maximum benefits and minimize problems. The users' level of sophistication, knowledge, and usage patterns must all be considered when evaluating these architectures. In

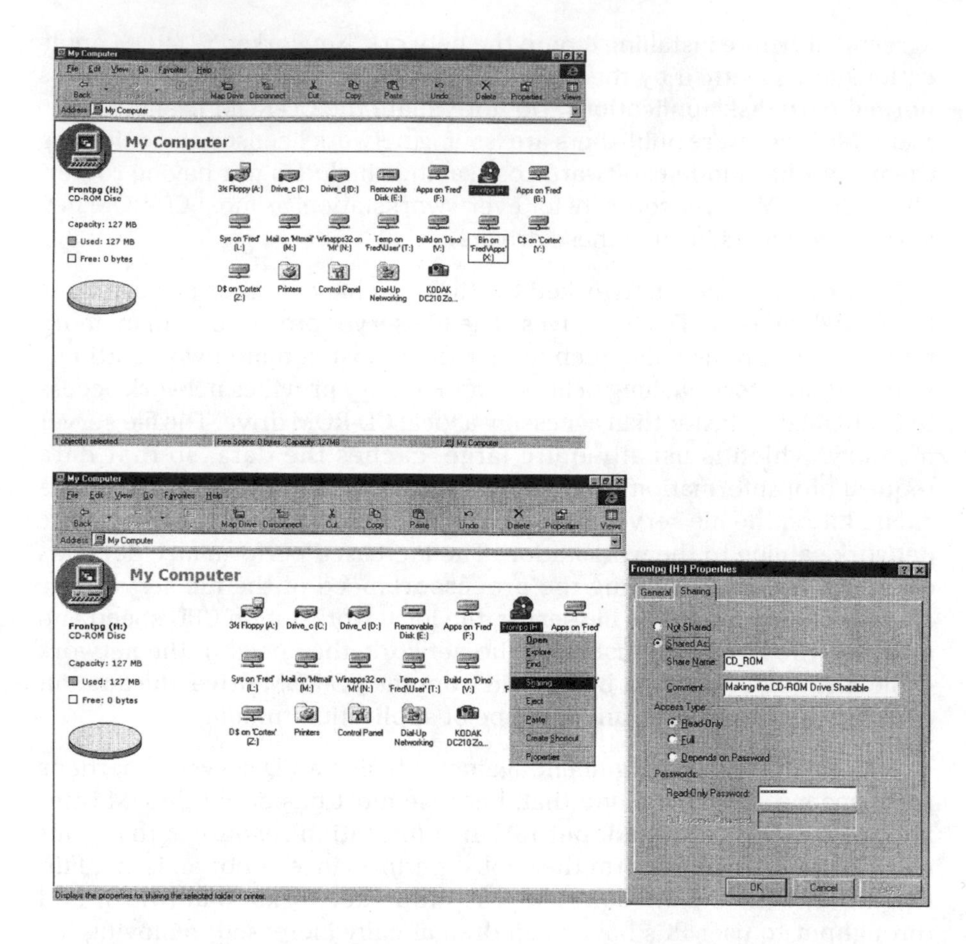

Exhibit 37.1 Windows 95 Explorer screen — setting CD drive to share.

many environments, more than one architecture may be effectively used at the same time in different parts of an organization.

Peer-to-Peer CD Sharing

For smaller organizations, the CD server can be as simple as a Windows 95 computer that has a CD-ROM drive, and where that CD-ROM drive (device) is made sharable by other users on the network. The computer that has the CD-ROM drive inside it is called the "host."

Windows 95 makes this very easy to do. You simply select the CD-ROM drive icon (see Exhibit 37.1) in the Windows 95 Explorer on the host computer and set its permissions to sharable and read-only. You may also assign

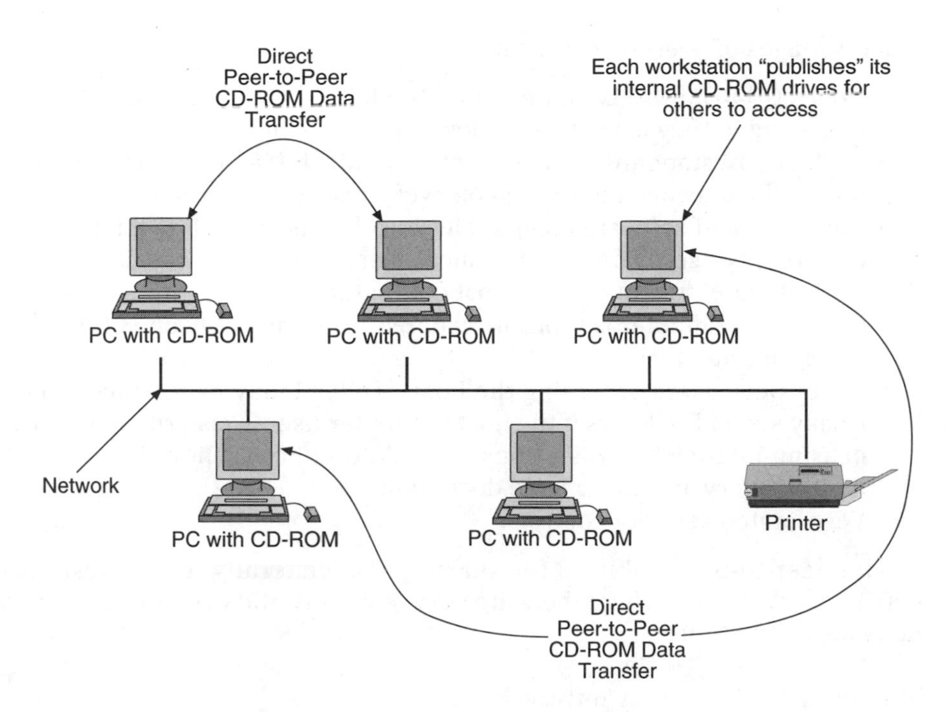

Exhibit 37.2 Peer-to-peer architecture.

---

it a password if desired for security reasons. Now other users on the network can access the CD disks that are inserted into this CD-ROM drive and execute the applications from the CD. Windows 95 shows this sharing capability by putting a "hand" icon under the CD-ROM drive icon on the host computer.

Windows 95 Explorer Screen Two Showing "Sharing" Icon. This model is often referred to as peer-to-peer (see Exhibit 37.2). This model works well when there is a very small number of users and also a small number (1 or 2) of CD-ROMs which need to be shared. Other operating systems such as Windows for Workgroups can also be set up to work in this manner. The peer-to-peer model has a few advantages (low cost being primary), but a number of distinct disadvantages. Advantages of peer-to-peer include:

- Low cost and ease of installation. Some operating systems, like Windows 95 and Windows for Workgroups, already provide this support.
- Each user does have control over his CD-ROM usage, but realize that this may also negatively affect other network users who are accessing that CD device simultaneously and may have their computers crash if the host computer user ejects the CD-ROM suddenly and without warning.

Disadvantages of peer-to-peer include:

- Host's CD-ROM use is limited. If the disk is ejected while others are accessing it, they may crash, or lose valuable work.
- Lack of any standard management or control. Management is often more difficult when distributed on every peer of the network.
- Hardware and software compatibility problems with CD-ROM drives, drivers, peer software, etc. You must install the necessary hardware and software drivers on each host computer.
- Each host computer is typically capable of sharing only one CD-ROM drive/disk at a time.
- Other peer users accessing the host CD-ROM may experience long (many seconds) delays if the host computer user is also running processing-intensive applications (e.g., Adobe Photoshop, Microsoft Access query, printing a large document, etc.).
- Very limited security controls.

The peer-to-peer solution for sharing CDs generally works best in smaller work groups, where there are very few users and CDs, and very little change over time.

Fileserver CD Extension Software

The second way in which you make CD-ROMs sharable on your network is by taking one of your existing file servers (NetWare, Microsoft Windows NT, or UNIX) and loading additional software on it to publish CD-ROM drives as server resources. You typically connect the CD-ROM drives and other CD devices directly to the file server (see Exhibit 37.3). This concept is pretty straightforward and easy to do. It does, however, require that your file server have extra capacity in terms of CPU performance, memory, and hard disk space so that the CD-ROM devices can be mounted and shared.

Most network operating systems (NetWare, Windows NT, and UNIX) contain a built-in support for CD-ROM drives and devices, but this is often very rudimentary. It seldom covers additional CD-ROM need such as:

- Support for CD multichangers and jukeboxes
- Support for newer DVD-ROM drives
- License management and monitoring
- Integrated security to the CD Level
- Performance optimization and advanced caching techniques

Also, CD server software adds additional features that make using and tracking CD resources faster and easier.

Advantages of file server extensions include:

- Leverages existing file server hardware and software costs

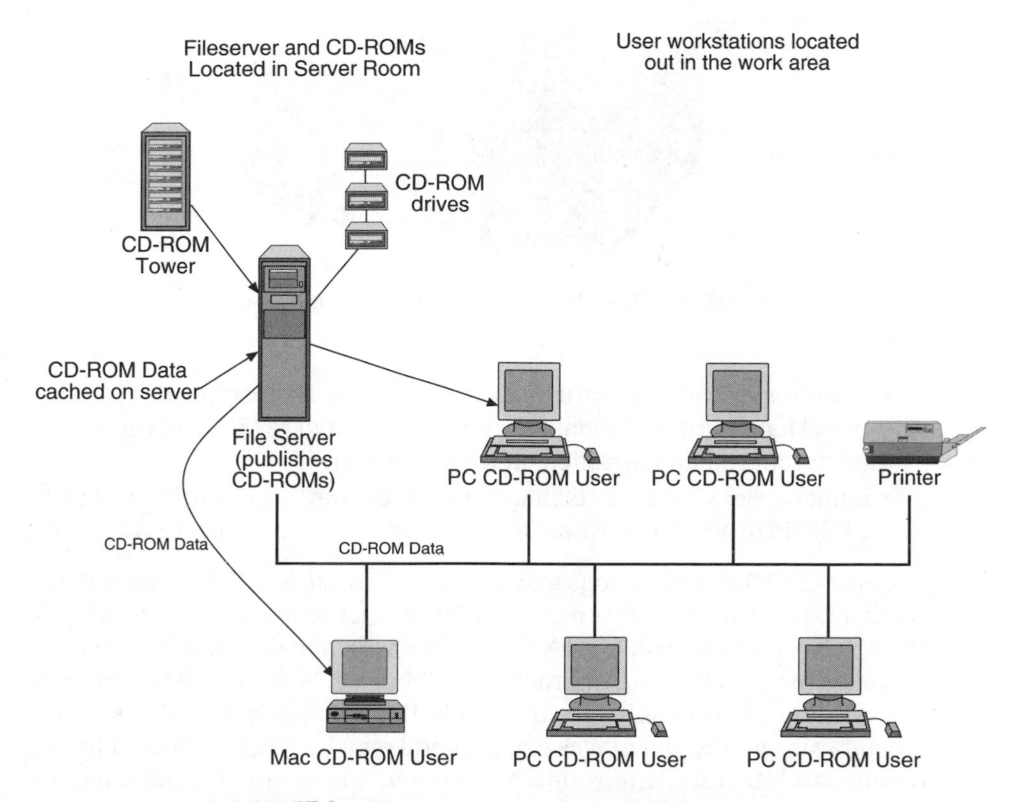

Exhibit 37.3 Fileserver extensions architecture.

- Offers high performance for CD-ROM access. Most file servers perform advanced caching to speed up CD-ROM (as well as hard disk) access
- Supports modest numbers of CD-ROM drives (many file servers can be extended to support 20 or more CD-ROM drives)
- Provides common, centralized management of CD-ROM drives, disks, and applications via the network, file server tools, and utilities
- Integrates data access from the CD-ROM drives, making CD-ROM data appear to users as more information on the file server
- Maximum physical security; CD-ROM disks and drives can be locked away in the server room with the file server

Disadvantages of file server extensions include:

- Burdens the file server with the additional load of servicing CD-ROM requests. Additional hardware, processing capability, memory, disk space, or other resources may need to be added to extend the file server.

Exhibit 37.4 Workgroup CD server appliance.

- Requires downtime of the file server to install these products; additional hardware (SCSI adapter cards) and software (SCSI drivers) must be purchased and installed on the file server.
- Limits access for user to change disks, as they do not have access to CD-ROM drives locked away in the server room (can also be a benefit).

Some CD-ROM applications may not run off the server if they are not network aware, or if they rely on calls which expect to access an internal CD-ROM drive, usually via MSCDEX.EXE, a TSR which loads on DOS to handle requests for data from an internal CD-ROM drive. Some CD-ROM networking vendors provide software that adds this compatibility. The specific advantages and disadvantages experienced may be greatly affected by the quality, architecture, and robustness of a vendor's specific implementation. Different vendors provide varying levels of network integration, functionality, ease of installation, support, etc.

Workgroup CD Server Appliances

The work group CD-ROM server is a network server, whose sole function is to publish and handle CD-ROM data requests. The dedicated CD-ROM server generally provides the highest performance and highest capacity for networking CD-ROMs. Most businesses today already use this model, having specialized database, e-mail, remote access, and other kinds of servers devoted to providing only one service to network users.

Work group CD server appliances come in many packaging form factors such as turn-key caching CD servers (see Exhibit 37.4) or network-ready plug-n-play universal CD-ROM towers (see Exhibit 37.5).

As shown in Exhibit 37.6, these products have an embedded server operating system that is specifically tailored and designed to serve up CD-ROM-based information. These products often come with an accompanying Windows management program or are manageable from a Web browser.

Exhibit 37.5 Workgroup CD network ready tower.

Windows CD Server Management Screen. Advantages of using work group CD-ROM server appliances include:

- Provides high peformance for work group-based access
- Does not burden other computers with CD-ROM serving demands
- Often offers universal client access to many different computers (Mac, Windows, OS/2, UNIX)
- Supports a modest number of CD-ROM devices, typically ranging from 7 to 28 CDs
- Offers common, centralized management of CD-ROM drives, disks, and applications via the network, file server tools, and utilities
- CD-ROM data appears to users as network shares or volumes. Users do not need to know they are accessing a CD-ROM drive
- Comes as a fully configured, network-ready system that can be up and running within minutes
- May offer advanced performance optimization and CD data caching capabilities along with rudimentary usage monitoring and load balancing

Disadvantages of dedicated CD-ROM servers include:

- Generally more expensive, although existing, unused 386 or higher PCs may be used

Work group CD-ROM server appliances can be very adaptable and useful for small and mid-size organizations as well as deployed in work groups within larger enterprises.

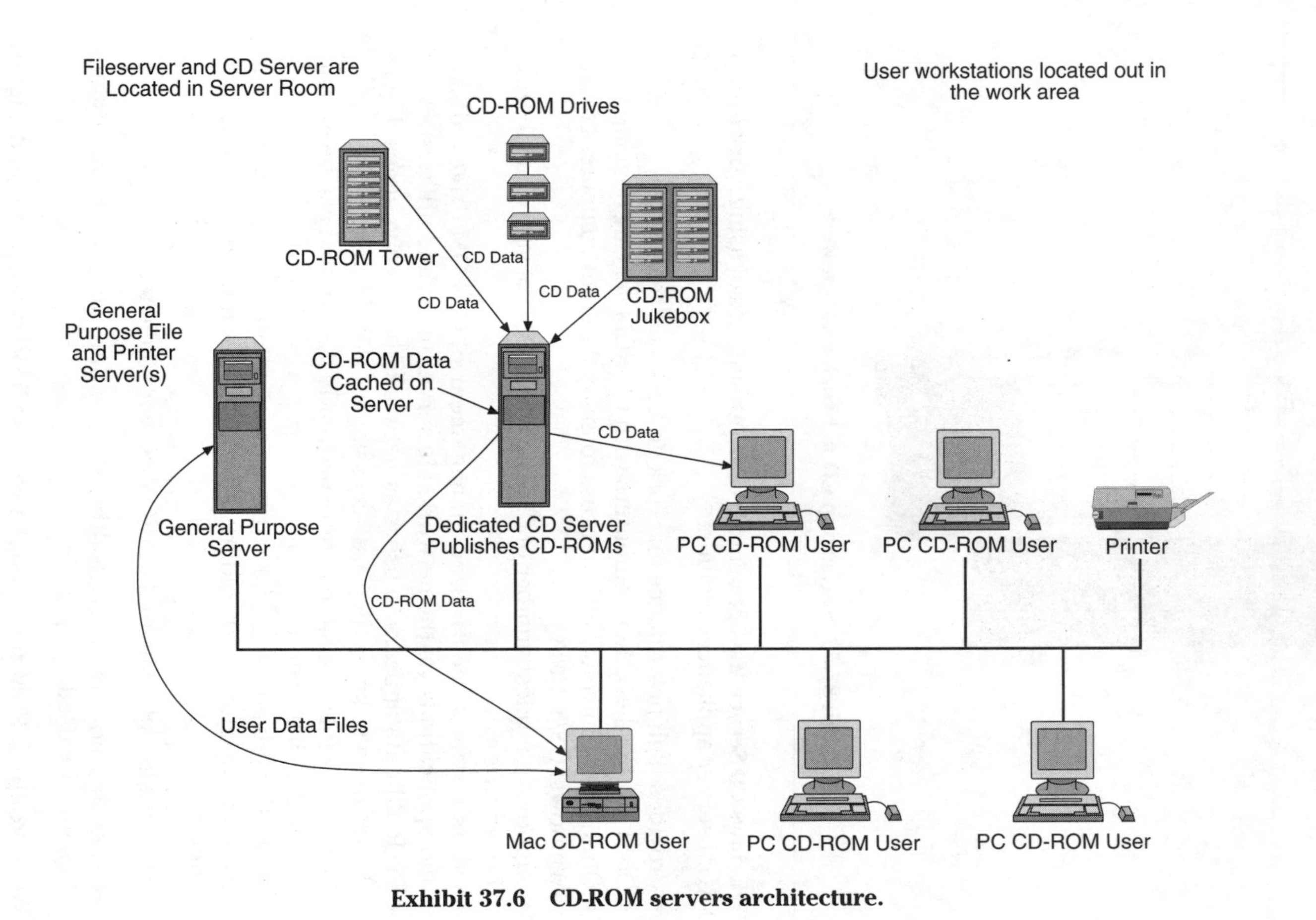

Exhibit 37.6 CD-ROM servers architecture.

Exhibit 37.7 Compaq-based CD server enterprise system.

Exhibit 37.8 Enterprise CD server rack system.

Enterprise CD Server Systems

The enterprise CD-ROM server is a high-performance, high-capacity network server, whose sole function is to publish and handle a large volume of CD-ROM data requests. The dedicated CD-ROM server generally provides the highest performance and highest capacity for networking CD-ROMs. Most businesses today already use this model having specialized database, e-mail, remote access, and other kinds of servers devoted to providing only one service-to-network user.

Enterprise CD server appliances come in many packaging form factors such as turn-key systems based on popular server hardware (such as the Compaq Server shown in Exhibit 37.7) or in rack-mount systems (shown in Exhibit 37.8).

As shown in Exhibit 37.9, these products have an embedded server operating system that is specifically tailored and designed to serve up large amounts of CD-ROM-based information. As with the work group CD Server appliances, these products often come with an accompanying Windows management program or are manageable from a Web browser.

Advantages of using Enterprise CD server systems include:

- Offers dedicated hardware and resources, providing the highest performance CD-ROM access
- Allows a large number, usually 100 or more, CD-ROM drives to be connected
- Supports higher-capacity CD-ROM jukeboxes, often handling up to 500 disks
- Existing production servers not burdened with servicing CD-ROM requests
- Often offers universal client access to many different computers (Mac, Windows, OS/2, UNIX)
- Offers common, centralized management of CD-ROM drives, disks, and applications via the network, file server tools, and utilities
- CD-ROM data appear to users as more information on the file server; users do not need to know they are accessing a CD-ROM drive
- Offers advanced performance optimization, caching, and other high capacity features such as fault tolerance
- Offers additional CD usage monitoring capabilities, load balancing, and license metering
- Completely integrated offering from one vendor can eliminate hassles of getting all software, drives, and hardware devices connected and functioning

Disadvantages of dedicated CD-ROM servers include:

- Generally the most expensive solutions for deploying a CD Server.

Choosing the Right Architecture

This section focuses on the issues that lead one to choosing the best architecture for their network. We will focus on file server extension software and dedicated CD-ROM servers.

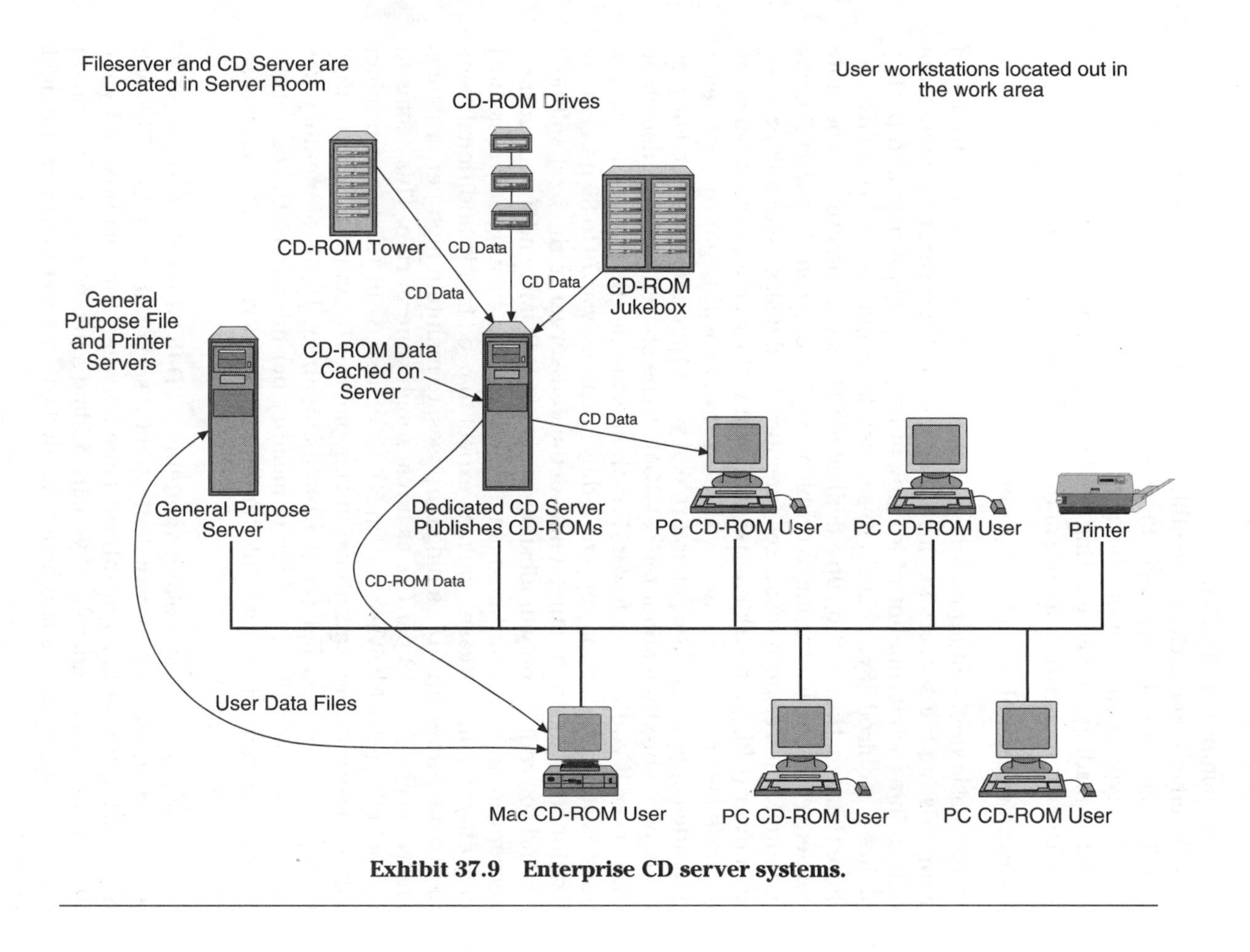

Exhibit 37.9 Enterprise CD server systems.

The choice of the appropriate architecture is determined from examining the following parameters for your expected installation:

- The number of CD-ROMs
- Whether jukeboxes will be utilized
- The number of users accessing the CDs
- The type of applications being run
- The loading on your existing production file servers
- The mission-critical nature of the CD-ROMs

We explore each of these in more detail.

The basic decision about whether to add CD-ROM services to one of your existing file servers vs. installing a dedicated server results from determining what amount of loading in terms of resources and performance is required. If you expect to access a large number of CD-ROMs, e.g., more than ten, then setting up a dedicated CD server can prove to be more cost-effective and less prone to problems in the long run. As each CD being mounted requires both resources (memory, hard disk space) and processing power (CPU, network bandwidth), you will reach a point at which an existing file server extended with CD services will begin to experience degraded operation. As a general rule, if you either currently need or plan to require more than about ten CD-ROMs to be shared, set up a dedicated CD server. If you require fewer CDs, then examine the current loading of your existing production servers. If they are already at 70 to 80 percent utilization, then again creating a new dedicated server is the best solution. Should you require only modest numbers of CD-ROMs, and if your current servers are not overburdened, adding them to the existing server is most cost effective. In this case, you may expect to need to add additional memory to the server for CD caching purposes, to maintain high performance. Figure on about 1 MB per CD, as a real rough general rule. This seems to apply both under NetWare and on NT. Of course, your CD server product may impose requirements above and beyond this to simply mount the CDs. Products on the market range from using about 16 KB of memory per mounted CD, to using 2 to 4 MB of memory per mounted CD. Look at the vendor's specifications carefully and understand your additional investment in memory.

Similarly, if you are looking at adding a CD-ROM jukebox to your network, plan on creating its own dedicated CD server at the outset. This will minimize any problems or interference between the jukebox and your other file servers. Generally, if you are adding a jukebox, you also plan on a larger number of CDs and hence the dedicated server again is the right choice.

Looking at the access side of the server, again a simple rule of thumb is that the more users you have, the more probability that you will need to create a dedicated server to sustain respectable throughput. As large numbers of users access CDs, having the dedicated server with sufficient caching resources (memory and hard disk) can make a big impact on overall response time for the users. Although it's hard to issue a general rule of thumb, if you plan on more than 50 users accessing the CDs, it's probably time to consider a dedicated CD server.

Similarly, if the CD-ROMs are data stream intensive, meaning that a lot of data is being read from the disks (e.g., multimedia, large graphics files, etc.), then you may need to set up a dedicated server. This again can be equipped to handle the throughput requirements to maintain a respectable response profile for your users.

Likewise, if your existing product file servers are near capacity, planning on a dedicated server right from the outset will save you frustration and costs in the long run. Like most network products, the total cost of installing and maintaining a CD server over several years is 10 percent product cost and 90 percent maintenance and support costs. Spend the 10 percent to create the dedicated server.

Finally, if your existing product servers must remain operational for mission-critical operation or, conversely, if your CD server is mission critical for serving up the CD content, separating the functions between file servers and dedicated CD servers will provide the most robust environment. Each server is then (reasonably) isolated from errors that may occur on the other, resulting in at least a partial operational capability for your organization.

Examination of your own network environment and user access patterns will help you make the best choice. Generally, it is less risky and costs less in the long run to set up a dedicated CD server from the outset, if you expect to require more than just a few CD-ROMs to be served. In the long run, you should experience fewer hiccups in operation and have greater productivity in the organization. The new CD server can also be incrementally added to your existing network, without affecting the operation of the current file servers, or require that they be brought down and serviced to add the CD service software. This can be a real benefit for mission-critical networks. The dedicated CD server can be set up, configured, tested, and later incrementally added to your backbone when your are satisfied with its reliability and operation.

CD-ROM DRIVE OVERVIEW

There are many different types, sizes, and speeds of CD-ROM drives. Drives come in both internal and external versions and utilize different interfaces such as ATAPI and SCSI. For CD server installations, some general guidelines apply. Since almost all drives that are networked use the SCSI or SCSI

II interface, we will limit our discussion to those drives. Drives which utilize the ATAPI interface are almost always found inside user desktop PCs.

First, let's define the various types of CD-ROM drives. Understanding the options can enable you to choose the most appropriate device. CD-ROM drives come in the following flavors:

- Single CD-ROM drives
- CD-ROM towers
- Multidisk changers
- Jukeboxes

Single CD-ROM drives are pretty straightforward. They can contain 1 CD-ROM, and that CD is exclusively available from that drive. These are much like the CD-ROM drives that come inside desktop computers, although usually purchased in an external case for network connection via a SCSI cable to a SCSI adapter. Using the SCSI interface, up to seven CD-ROM drives can be cabled (connected) together to form a chain. Each drive is then separately addressable by the CD server. CD-ROM drives provide a 1:1 ratio between disks and drives (laser mechanisms), resulting in 100 percent availability and maximum response performance to network user requests. Each CD is continually kept "spinning" inside the drive, ready to satisfy immediate response to data requests.

CD-ROM towers are banks of single CD-ROM drives which are internally cabled together, complete with a power supply, case, etc. CD-ROM towers are usually available in multiples of seven, ranging from 7 drives, 14 drives, 28 drives, 56 drives, and so on. The towers are usually purchased from a systems integrator or manufacturer, so that you simply install one connector to the tower, and the entire unit is connected to the CD server. Most towers will therefore have SCSI connectors enabling you to connect them to the CD server. Some towers can even come with an internal CD server preconfigured, so that there is a network tap out the back of the tower allowing direct connection to the network cabling. Delivered this way, the CD tower is actually a turn-key CD server. Other than packaging and simplicity of installation, subsequent operation is virtually identical to connecting a SCSI tower to an external CD server.

Multidisk changers are CD-ROM drives that typically hold between 2 and 18 disks. The CDs are usually loaded into cartridges where the disk changer has a small mechanical device internally which allows the different CDs to be swapped under the laser-read head. This allows one (SCSI) device to host multiple CDs. As a SCSI chain is limited to seven devices, using disk changers allows you to exceed seven CDs on the SCSI chain. Devices like this include the Pioneer DRM-624x, DRM-1804x, and Mountain. They are very similar to the home audio multiple cartridge CD players where you load six or seven disks, and then simply select which disk you want to play. Applied to CD-ROM, the disk changer automatically positions the appropriate CD under the laser-read head depending on the data request that was received from the CD Server. The CD Server software does need to know how to communicate with the disk changer, so that the particular disk addressing information is provided to the drive. The disk changers provide roughly a 6:1 or 7:1 ratio of disks to drives. Their main benefit is lower cost, where you wish to build a library of CD-ROM titles that are on-line for infrequent access. Purchasing and installing one disk changer is much less expensive than installing six CD-ROM drives, or a seven-bay Tower. The trade-off is lower performance (latency), as the disk changers usually have a 1- to 3-second "disk shuffle" time, to reposition to a disk that is not currently under the laser. Depending on the access profile of the CD within your organization, infrequently accessed disks can be successfully installed in disk changers, while providing a good balance of availability, price, and performance. Once a disk is positioned under the read laser in the mechanism, the performance of the disk-changer unit is virtually as fast as a single CD-ROM drive. The only performance hit that you incur is from the disk-swapping latency time. Because of this, you would not want to set up your network such that multiple users were continually accessing different disks in the changer, as the unit would start thrashing, constantly shuffling disks, and severely degrading throughput performance. Many CD server products recognize this potential problem and provide optimization techniques such as caching, prioritization, or read bunching to maximize the throughput from the changer device.

CD-ROM jukeboxes are high-capacity devices, typically storing from 100 to 500 CDs. These devices are much more expensive and typically range in price from $10,000 to $25,000. Since the CD-ROM jukebox can contain a large number of disks, it typically includes multiple CD-ROM drive readers. The jukebox then utilizes some sort of robotics mechanism to position various CDs into the readers. Since the readers are typically SCSI devices, jukeboxes usually contain anywhere from two to seven readers. This results in disk/drive ratios ranging from 25:1 to 100:1. As you can see, these devices are meant for building large CD-ROM libraries and are very valuable when you want to maintain a large number of CDs near on-line, in case you need access to them. Some of the CDs in the jukebox can be high-availability CDs, while others are there in case you need them. Examples of this include keeping application CDs in a jukebox, where they are accessible should you need to do an application install, or also a CD reference library that spans multiple CDs. CD server products vary greatly in their support for CD-ROM jukeboxes. As these devices are more complex, the CD server product must be designed with special features and capabilities, and must provide a special drive that knows how to track the disks in the jukebox and issue the appropriate robotics commands to the jukebox mechanism. CD server products also implement a wide range of techniques for mounting the CDs so that they are available on-line, and also have widely different means of optimizing the performance of the jukebox. For jukebox solutions, it is best to contact the CD server manufacturers directly and arrange for an evaluation copy. That way, you can see if it works in YOUR environment. Since CD-ROM jukeboxes are high-capacity devices, they are typically connected to dedicated CD-ROM servers, which can provide the performance and resources necessary to manage this large amount of information. Like CD-ROM towers, some jukebox models come complete with embedded CPU card and CD-ROM server software, so that they are directly network connectable. This way, the CD-ROM jukebox becomes a turn-key CD-ROM server itself.

In most real-world installations, a mixture of these devices will be installed, depending on the particular requirements. For instance, a jukebox may be installed, along with a CD-ROM tower. The high-availability and high-access CDs are placed in the tower, and the reference CDs are placed in the jukebox. Both devices are connected to the CD-ROM server, and network users can access disks from both devices simultaneously. Done well, the CD server software should even hide the location of the disks from the users. They don't care whether the CD is in a tower, jukebox, changer, or anything else. They simply want to run the application, get the data, and complete the job.

When planning your installation, map out the access patterns for the CDs that you expect to need, and plan for additional growth capacity in the future. Then check with the CD server vendor for specifics on support, features, performance, and pricing.

CD SERVER PRODUCT AND VENDOR CONSIDERATIONS

This section lists various questions that you should ask about the CD Server product and the vendor who offers that product. This information is meant to be a guide to assist you in obtaining the necessary information as you plan your installation.

What to Ask about the Vendor?

Choosing the right vendor is perhaps the most important element in your eventual success and peace of mind. Since your CD server will grow with your organization over time, and as CD technology is advancing rapidly, you need to choose a vendor that is progressive and will move technology forward with you. A few things to consider include:

- Is the vendor growing and following new technology developments?
- Is the vendor innovative?
- Is the vendor responsive to customer requests?
- Does the vendor provide a range of products to give flexibility?
- Is the vendor a market leader?

- Does the vendor have the necessary networking expertise, background, and history?
- Will the vendor nickel and dime me for features or accessories?
- How good is their technical support (do they have qualified personnel)?
- Is the vendor committed to CD-ROM networking technology?
- Is the company international?

What to Ask of the Product

As with the vendor, asking the right questions about the CD server product will minimize frustration and difficulty and result in the best solution for your organization. As the list of features provided by CD servers continually changes, contact the manufacturer for the latest information, and then compare the features against your needs now and in the future.

- How easy is the product to install, manage, and access?
- Is network down time required for installation?
- Does the product support the devices that I need?
- Can the CD-ROMs be managed from any workstation on the network?
- Is the product expandable and scalable?
- Are standards supported (ISO-9660, Macintosh HFS, etc.)?
- Can your workstations' internal CD-ROM drives be shared (if you have them)?
- How much server resources does the product take to support each CD-ROM?
- How quickly can the system mount and dismount CD-ROMs?
- Can disks be reinserted and automatically recognized and brought on-line by the system?
- Does the product migrate with my NetWare 3, NetWare 4, or Microsoft NT server software?
- Does it support the network protocols (IP, IPX, other) that you run?
- Does the product integrate seamlessly with my network operating system for security, management, access?
- How much resources does the product consume per CD-ROM?
- Is the user shielded from knowing which drive the CD-ROM disk is in today, or tomorrow after the administrator moves it to another (faster) drive?
- Does the users' workstation need a TSR to access CD-ROM?
- Must the user understand the complex CD-ROM dynamics on networks, or does the product hide these?
- Mapping and unmapping DOS drives.
- Mounting and unmounting CD-ROMs.
- Logging into and out of servers.
- Getting a disk on-line from near on-line in a jukebox.

ADVANCES IN CD TECHNOLOGY

The CD-ROM market is continually changing as new technologies are being rapidly developed. This section focuses on a number of areas that you should be aware of when planning your installation. If you expect that your organization will need these technologies in the future, make sure that you ask the appropriate questions from the CD server manufacturer so the support and features will be there when you need them.

Bigger, Better, and Faster CD-ROM Devices

CD-ROM drives are continually improving in terms of performance. Drives are measured in terms of throughput performance, relative to the original CD-ROM drives of several years ago. The original drives could deliver about 150 KB of data per second, and are referred to as "1X" drives. Today, a typical CD-ROM drive delivers "16X" or even "32X" performance. You will need to make sure that your CD-ROM server software and hardware can scale to these performance throughput levels if you plan to incorporate these advances in drive performance. The internal architecture and design of some CD server products can be a limiting factor in overall performance capabilities, when 10 or 20 of these high-performance drives are connected.

Likewise, CD-ROM jukeboxes are continually improving in reliability and performance and increasing in capacity. At the same time, price is declining rapidly, and as the number of disks used in your organization grows, you may find a CD-ROM jukebox in your future. Again, make sure that the CD-ROM server product can match your expected needs.

Other advances in interface design are also occurring. Today, almost all network connected CD-ROM drives, disk-changers, and jukeboxes utilize the SCSI II interface. While this interface meets the necessary capacity and performance requirements from today's drives, other interfaces are looming around the corner. These include:

- *SCSI 3:* A refinement, simplification, and standardization of the current SCSI II specification. This specification may also begin to standardize the commands used for CD-ROM jukeboxes etc.
- *Ultra SCSI:* An improvement on the standard SCSI interface capable of delivering twice the performance of SCSI II, ranging from 40 to 80 Mbps
- *FC-AL (Fiber-Channel Arbitrated Loop):* SCSI-type interface capable of delivering performance in the 100- to 200-Mbps range, while offering simplified ports, connectors, and cabling
- *Serial Storage Architecture (SSA):* A new high-speed serial interface capable of delivering 40 Mbps performance, again with simplified connectors and cabling

It is too early to determine if these interfaces will supplant or replace the SCSI II interface for CD-ROM devices. No CD devices can be identified as of today. They are identified here for awareness and planning purposes.

CD-Recordable

CD-Recordable (CD-R) drives allow you to write your own CDs for later use. The resulting CDs are fully compatible with CD-ROM drives and can be used just like a purchased CD-ROM disk. The CD-R drives use special "gold CD-R blank" disks, which can be purchased from major electronics stores and catalogs.

CD-R capabilities can be useful when you wish to archive or store information for later retrieval or for distribution to other sites, and when you wish to use the industry standard and widely accessible CD-ROM format to do this. Archiving the information on a CD-R disk, as opposed to tape, allows you to later read that information in any CD-ROM drive that is available, without requiring any special software or special hardware.

DVD-ROM — The Replacement for CD-ROM

DVD is a multipurpose technology suited to both entertainment and computer uses. As an entertainment product DVD will be used for full-length movies with up to 133 minutes of high-quality video (MPEG-2 format) and audio. The first in the DVD family of drives for computer use will be a read-only drive referred to as DVD-ROM. Subsequently, the family will be expanded with a one-time recordable version, referred to as DVD-R, and finally a rewritable version, referred to as DVD-RAM.

DVD-ROM drives will be used with computers as an alternative to CD-ROM, providing over seven times the storage capacity (4.7 GB). It is expected that the publishers of games, education, and entertainment software (which use multiple CD-ROMs) will offer their titles on a single DVD-ROM disk. With advanced video boards and high performance PCs, game and entertainment software will achieve new levels of capability using DVD technology. The DVD-ROM drive will read existing CD-ROMs and music CDs and will be compatible with installed sound and video boards. Additionally, the DVD-ROM drive will read DVD movie titles using an advanced (MPEG-2) video board — required to decode the high-resolution video format. DVD-ROM drives will ultimately be available from many manufacturers. The first drives, using a single-layer disk of 4.7 GB, became available during the second half of 1996 from several manufacturers, including Toshiba, Philips, Sony, and Hitachi. In 1997, dual-layer disks increased the disk capacity to 8.5 GB. In the more-distant future, it is planned to use double-sided, dual-layer disks, which increase the capacity to 17 GB.

Just as CD-R drives can record a disk that appears to a CD-ROM drive as a "pressed" disk, the write-once DVD-R drives will record a 3.9-GB DVD-R disk that can be read on a DVD-ROM drive.

Then the rewritable DVD-RAM drive will become available (this is similar to the unreleased CD-E drive). DVD-RAM drives will read and write to a 2.6-GB DVD-RAM disk, read and write-once to a 3.9-GB DVD-R disk, and read a 4.7/8.5-GB DVD-ROM disk. Also, it is expected that a DVD-RAM disk will be readable on both the DVD-R and DVD-ROM drives — a level of compatibility previously unseen.

The DVD-ROMs will eventually replace CD-ROMs. DVD-ROM provides for greatly enhanced storage capacities ranging from 4.7 GB to somewhere around 17 GB of data on a disk the same size as a CD-ROM. Issues remain as to backward compatibility, capacities, names, performance, etc. Eventually, it is expected that DVD-Recordable will also be available, allowing organizations to archive tremendous amounts of information on blank DVD disks.

DVD-Audio and DVD-Video are targeted more at the consumer market for delivery of music and video, respectively. The DVD-Video disks may intersect with computer networks at some point in the future for delivering MPEG-based video data, but that most likely will not occur for quite some time.

It is vitally important that you check whether your desired CD-ROM server software product can support DVD-ROMS, with their greatly increased capacities. For example, complete support for DVD-ROMs will require that your CD server run a 64-bit file system to support addressing up to 17 GB of data.

DVD-ROMs will be showing up more frequently, and you can expect to have to network them. As the amount of data is vastly higher than with CD-ROMs, CD-ROM server products need to utilize special technology to meld that data with the network operating systems. For instance, Novell NetWare was simply not designed to deal with removable media containing 17 GB of data. The CD-ROM server products must solve that incompatibility and provide a smooth and seamless migration for your organization.

100 MBPS Networks

In the quest for ever higher levels of performance, as more and more electronic data is shuffled around the network, organizations are moving to adopt 100-Mbps cabling technologies such as 100Base-T and ATM. Should your organization be moving in that direction, again make sure that the CD-ROM server software and hardware can provide the appropriate drivers, is compatible with 100-Mbps interface cards, and can deliver an increased level of performance over the 100-Mbps pipe. As each CD-ROM server product's performance level is different, they will respond differently when placed onto a 100-Mbps network. Instead of the network (10 Mbps) being the bottleneck, the CD server could become the bottleneck. A few questions today to the CD server manufacturer can help avoid a potential problem tomorrow.

The Internet, Intranets, and the World Wide Web

With the rise of the Internet and World Wide Web, and their rapidly increasing popularity, comes a new set of considerations for CD-ROM servers. The application of Internet technology in an organization's internal LAN is called an "intranet." As CD-ROMs are the most popular media on which to deliver rich media content and with the popularity of HTML as the standard for viewing rich media content, the marriage of CD-ROMs and HTML for intranet deployment is inevitable.

The considerations here are that if you plan on creating or deploying "Web-based CDs" which contain HTML-formatted data, make sure your CD server can seamlessly integrate these data into your organization's intranet LAN environment. This usually entails being able to manage your CD server resources from within a Web browser, and also that when an HTML CD is inserted in the CD server, that it is also immediately viewable by network users from within their favorite browser.

SUMMARY

We've looked at the benefits of networking the burgeoning amounts of information available on CD-ROM as well as the variety of architectures for sharing CD-ROM data. The format standards for CD-ROMs are stable and widely deployed. Speed access performance is continually improving and new technologies are being rapidly developed for CD-ROM drives. CD-ROM jukeboxes are also seeing improvements in reliability, performance, and increased capacity. At the same time, price is declining rapidly. When planning your installation, map out the access patterns for the CDs that you expect to need. Don't forget to plan for additional growth capacity in the future.

Chapter 38

Choosing and Equipping an Internet Server

Nathan J. Muller

The key to truly reaping the benefits of the Internet lies within the server. A server should provide ease-of-use and security and support such basic applications as e-mail and newsgroups. This article describes how to choose an Internet server and application software that best fit the needs of different users and organizations.

INTRODUCTION

The Internet is a global collection of servers interconnected by routers over various types of carrier-provided lines and services. It comprises more than 4 million hosts on about 100,000 networks in 160 countries. Approximately 30 million people have access to the Internet, a number that is expected to grow to 200 million by the turn of the century. The Internet comprises databases that have a combined capacity that can only be measured in terabytes — more information than has ever been printed on paper. It is accessed and navigated by PCs and workstations equipped with client browser software.

INTERNET AND INTRANET SERVICES

One of the most popular and fastest-growing services on the Internet is the World Wide Web, also known as WWW or simply "the Web." The Web is an interactive, graphically oriented, distributed, platform-independent, hypertext information system. Browser software such as Netscape Navigator and Internet Explorer make it easy for users to find information published on Web servers, which can be configured for public or private access. When configured for private access, companies can create Virtual Private Network or "intranets" to facilitate information exchange among employees, customers, suppliers, and strategic partners.

0-8493-9823-1/00/$0.00+$.50

Of all the services that can be accessed over the Internet, the Web holds the greatest promise for electronic commerce. Using catalogs displayed on the Web, customers can order products by filling out forms transmitted through e-mail. Often the transactions include buyers' credit card numbers, which require a secure means of transmission. Other electronic commerce applications include on-line banking and stock trading.

With a Web server, an organization can leverage Web technology for internal communication on an intranet by producing on-line documentation of corporate materials, automating salesforce activities, providing training-on-demand, or using data warehousing capabilities to analyze large amounts of data or complex data.

The factor driving these activities is the same in every case: providing users access to information. As companies use the Web to deliver new services, they need solutions that are capable of storing, managing, and organizing all of their existing data. Furthermore, these mechanisms need to tie into existing applications and be reliable, scalable, and open.

Early Web implementations focused on providing access to static data, mostly in the form of simple text and graphics. As Web-based interactions become more complex, the next step is the creation of real-world applications that can manipulate, input, modify, analyze, and apply this content to everyday tasks. The need for live, on-line applications that can manipulate dynamic, constantly changing data is driving the Web into the next phase of its evolution.

PLATFORM CONSIDERATIONS

The key to delivering services over the Internet is the server. The Internet is a true client/server network. Integrating into this client/server environment requires servers with strong connectivity capabilities suitable for high-traffic and mission-critical applications. The server must have ease-of-use functionality that allows corporate users to access information quickly and easily. The server must have security features that enable users to share confidential information or conduct encrypted electronic transactions across the Internet. Finally, the server must be able to support the many applications that have become the staple of the Internet, including electronic mail and newsgroups.

Processor Architecture

A high-performance server is a virtual requirement for any company that is serious about establishing a presence on the Internet. There are basically two choices of processor architectures: Reduced Instruction Set Computing-based or complex instruction set computing (CISC)-based. Reduced instruction set computing (RISC) processors are usually used on high-end

UNIX servers; CISC processors, such as Intel's Pentium Pro, are used on Windows NT machines. The performance of the Pentium Pro rivals that of Reduced Instruction Set Computing processors and costs less.

Because of the volume of service requests — sometimes tens of thousands a day — the server should be equipped with the most powerful processor available. The more powerful the processor, the greater the number of service requests (i.e., page lookups, database searches, and forms processing) the server will be able to handle.

SMP Servers. Servers with symmetric multiprocessing (SMP) enable the operating system to distribute different processing jobs among two or more processors. All the central processing units have equal capabilities and can handle the same tasks. Each CPU can run the operating system as well as user applications. Not only can any CPU execute any job, but jobs can be shifted from one CPU to another as the load changes. This capability can be very important at high-traffic sites, especially those that do a lot of local processing to fulfill service requests.

Some servers come equipped with multiple RISC or CISC processors. Users should be aware, however, that the added cost of a symmetrical multiprocessing server is not merely a few hundred dollars per extra processor. There are costs for additional hardware resources as well — such as extra RAM and storage space — that can add several thousand dollars to the purchase price. However, as needs change, users can upgrade SMP servers incrementally without having to buy a new system. In this way, performance can be increased and the original hardware investment can be protected. This requirement is especially critical in the rapidly evolving Internet market in which organizations want to implement new applications on their servers that require increasing database and search performance.

Operating System: UNIX vs. NT

When choosing a server, the operating system deserves particular attention. The choices are usually between UNIX and Windows NT. Although some vendors offer server software for Windows 3.1 and Windows 95, these are usually intended for casual rather than business use.

Most Internet servers are based on UNIX, but Windows NT is growing in popularity and may overtake UNIX in the near future. A Windows NT server offers performance and functionality comparable to a UNIX server and is easier to set up and administer, making it the platform of choice among developers of new sites.

Like UNIX, Windows NT is a multitasking, multithreaded operating system. As such, NT executes software as threads, which are streams of commands that make up applications. At any point during execution, NT's process manager interrupts (or preempts) a thread to allow another thread some CPU time. Also like UNIX, Windows NT supports multiple processors. If the server has more than one CPU, NT distributes the threads over the processors, allowing two or more threads to run simultaneously.

Fault Tolerance

If the server is supporting mission-critical applications over the Internet, several levels of fault tolerance merit consideration. Fault tolerance must be viewed from both the system's and subsystem's perspectives.

Site Mirroring. From the system's perspective, fault tolerance can be implemented by linking multiple servers together. When one system fails or must be taken off-line for upgrades or reconfigurations, the standby system is activated to handle the load. This is often called site mirroring. An additional level of protection can be obtained through features of the operating system that protect read and write processes in progress during the switch to the standby system.

Hot Standby. At the subsystem level, there are several server options that can improve fault tolerance, including ports, network interfaces, memory expansion cards, disks, tapes, and I/O channels. All must be duplicated so that an alternate hardware component can assume responsibility in the event of a subsystem failure. This procedure is sometimes referred to as a hot-standby solution, whereby a secondary subsystem monitors the tasks of the primary subsystem in preparation for assuming such tasks when needed.

If a component in the primary subsystem fails, the secondary subsystem takes over without users being aware that a changeover has taken place. An obvious disadvantage of this solution is that companies must purchase twice the amount of hardware needed, and half of this hardware remains idle unless a failure occurs in the primary system.

Because large amounts of data may be located at the server, the server must be able to implement recovery procedures in the event of a program, operating system, or hardware failure. For example, when a transaction terminates abnormally, the server must be able to detect an incomplete transaction so that the database is not left in an inconsistent state. The server's rollback facility is invoked automatically, which backs out of the partially updated database. The transaction can then be resubmitted by the program or user. A roll-forward facility recovers completed transactions and updates in the event of a disk failure by reading a transaction journal that contains a record of all updates.

Load Balancing. Another means of achieving fault tolerance is to have all hardware components function simultaneously, but with a load-balancing mechanism that reallocates the processing tasks to surviving components when a failure occurs. This technique requires a UNIX operating system equipped with vendor options that continually monitor the system for errors and dynamically reconfigure the system to adapt to performance problems.

Hot Swapping. Hot swapping is an important capability that allows the network administrator to remove and replace faulty server modules without interrupting or degrading network performance. In some cases, standby modules can be brought on-line through commands issued at the network management workstation or automatically upon fault detection.

Uninterruptible Power Supply. To guard against an on-site power outage, an uninterruptible power supply (UPS) can provide an extra measure of protection. The UPS provides enough standby power to permit continuous operation or an orderly shutdown during power failures, or to change over to other power sources such as diesel-powered generators. Some UPSs have simple network management protocol (SNMP) capabilities, so network managers can monitor battery backup from the central management console. For example, using simple network management protocol, every UPS can be instructed to test itself once a week and report back if the test fails.

INTERNET APPLICATION SOFTWARE

An Internet server must be equipped with software that allows it to run various Internet applications. Some server software supports general communications for document publishing over the World Wide Web. Often called a communications server or Web server, this type of server can be enhanced with software specifically designed for secure electronic commerce. Server software is available for performing many different functions, including implementing newsgroups, facilitating message exchange (i.e., e-mail), improving the performance and security of communications, and controlling traffic between the Internet and the corporate network.

Sometimes a server is dedicated to a single application such as e-mail, newsgroups, or electronic commerce. Other times, the server supports multiple Internet applications. The specific configuration depends on such factors as available system resources (i.e., memory, disk space, processing power, and port capacity), network topology, available bandwidth, traffic patterns, and the security requirements of the organization.

Communications Software

A communications server enables users to access various documents and services that reside on it and retrieve them using the hypertext transfer protocol. These servers support the standard multimedia document format — the hypertext markup language (HTML) — for the presentation of rich text, graphics, audio, and video. Hyperlinks connect related information across the network, creating a seamless Web. Client browser software is used for navigation. Some vendors offer servers preconfigured with these Internet protocols, allowing them to be quickly installed and put into operation.

A key service performed by any Internet server is the translation of complex Internet protocol (IP) addresses to simpler server domain names. When a user requests the uniform resource locator of a certain Web page, for example, the domain name server (DNS) replies with the numeric IP address of the server the user is contacting. It does this by checking a lookup table that cross-references server domain names and IP addresses.

For example, the domain name ddx might stand for "dynamic data exchange." This domain name might translate into the IP address 204.177.193.22. The translation capability of the Internet domain naming service makes it easy for users to access Internet resources by not requiring them to learn and enter long strings of numbers. To access the Web page of dynamic data exchange, the user would enter the uniform resource locator (URL) as http://www.ddx.com, which contains the domain name ddx.

Commerce Software

A commerce server is used for conducting secure electronic commerce and communications on the Internet. It permits companies to publish hypermedia documents formatted in hypertext markup language and to deliver them using hypertext transfer protocol (HTTP). To ensure data security, the commerce server provides advanced security features through the use of the secure sockets layer (SSL) protocol, which provides:

- *Server authentication.* Any secure socket layer-compatible client can verify the identity of the server using a certificate and a digital signature.
- *Data encryption.* The privacy of client/server communications is ensured by encrypting the data stream between the two entities.
- *Data integrity.* SSL verifies that the contents of a message arrive at their destination in the same form as they were sent.

As with other types of Internet servers, vendors offer commerce servers preconfigured with the protocols necessary to support electronic commerce.

News Software

A news server lets users create secure public and private discussion groups for access over the Internet and other TCP/IP-based networks using the standard network news transport protocol (NNTP). The news server's support of NNTP enables it to accept feeds from popular Usenet newsgroups and allows the creation and maintenance of private discussion groups. Most newsreaders are based on NNTP; some support SSL for secure communication between clients and news servers.

A news server should support the multipurpose Internet mail extension (MIME), which allows users to send virtually any type of data across the Internet, including text, graphics, sound, video clips, and many other types of files. Attaching documents in a variety of formats greatly expands the capability of a discussion group to serve as a repository of information and knowledge to support workgroup collaboration. Colleagues can download documents sent to the group, mark them up, and send them back.

Mail Software

Client/server messaging systems are implemented by special mail software installed on a server. Mail software lets users easily exchange information within a company as well as across the Internet. Mail software has many features that can be controlled by either the system administrator or each user with an e-mail account.

The mail software should conform to open standards, including hypertext transfer protocol (HTTP), MIME, Simple Mail Transfer Protocol, and POP3. MIME lets organizations send and receive messages with rich content types, thereby allowing businesses to transmit mission-critical information of any type without loss of fidelity. The SMTP ensures interoperability with other client/server messaging systems that support Internet mail or proprietary messaging systems with Internet mail gateways. The post office protocol version 3 (POP3) ensures interoperability with such popular client software as Zmail, Eudora, Pegasus Mail, Microsoft Outlook client, and most other Internet-capable mail products.

Proxy Software

To improve the performance and security of communications across the TCP/IP-based Internet, many organizations use a proxy server. This kind of software offers performance improvements by using an intelligent cache for storing retrieved documents.

The proxy's disk-based caching feature minimizes use of the external network by eliminating recurrent retrievals of commonly accessed documents. This feature provides additional "virtual bandwidth" to existing network resources and significantly improves interactive response time for locally attached clients. The resulting performance improvements provide a cost-effective alternative to purchasing additional network bandwidth. Because the cache is disk-based, it can be tuned to provide optimal performance based on network usage patterns.

The proxy server should allow dynamic process management, which allows the creation of a configurable number of processes that reside in memory waiting to fulfill hypertext transfer protocol (HTTP) requests. This feature improves system performance by eliminating the unnecessary overhead of creating and deleting processes to fulfill every hypertext transfer protocol (HTTP) request. The dynamic process management algorithm increases the number of server processes, within configurable limits, to efficiently handle periods of peak demand, resulting in faster document serving, greater throughput delivery, and better system reliability.

Firewall Software

An application-level firewall acts as a security wall and gateway between a trusted internal network and such untrustworthy networks as the Internet. Access can be controlled by individuals or groups of users or by system names, domains, subnets, date, time, protocol, and service.

Security is bidirectional, simultaneously prohibiting unauthorized users from accessing the corporate network while also managing internal users' Internet access privileges. The firewall even periodically checks its own code to prevent modification by sophisticated intruders.

The firewall gathers and logs information about where attempted breakins originate, how they got there, and what the people responsible for them appear to be doing. Log entries include information on connection attempts, service types, users, file transfer names and sizes, connection duration, and trace routes. Together, this information leaves an electronic footprint that can help identify intruders.

WEB DATABASE CONSIDERATIONS

Internet servers are the repositories of various databases. These databases may be set up for public access or for restricted intracompany access. In either case, the challenge of maintaining the information is apparent to IS professionals charged with keeping it accurate and up to date.

Vendors are developing ways to ease the maintenance burden. For example, database management vendors such as Oracle Corp. offer ways of integrating an existing data warehouse with the Internet without having to reformat the data into HyperText Markup Language. The data are not sent until a request is received and validated.

In addition, the server supports hypertext transfer protocol-type negotiation, so it can deliver different versions of the same object (e.g., an image stored in multiple formats) according to each client's preferences. The server also supports national language negotiation, allowing the same document in different translations to be delivered to different clients.

The database server should support the two common authentication mechanisms: basic and digest authentication. Both mechanisms allow certain directories to be protected by user name/password combinations. However, digest authentication transmits encrypted passwords and basic authentication does not. Other security extensions that may be bundled with database servers include secure hypertext transfer protocol (S-HTTP) and secure socket layer standards, which are especially important in supporting electronic commerce applications.

Maintenance and Testing Tools

The maintenance of most Web databases still relies on the diligence of each document owner or site administrator to periodically check for integrity by testing for broken links, malformed documents, and outdated information. Database integrity is usually tested by visually scanning each document and manually activating every hypertext link. Particular attention should be given to links that reference other Web sites because they are usually controlled by a third party who can change the location of files to a different server or directory or delete them entirely.

Link Analyzers. Link analyzers can examine a collection of documents and validate the links for accessibility, completeness, and consistency. However, this type of integrity check is usually applied more as a means of one-time verification than as a regular maintenance process. This check also fails to provide adequate support across distributed databases and for situations in which the document contents are outside the immediate span of control.

Log Files. Some types of errors can be identified by the server's log files. The server records each document request and, if an error occurred, the nature of that error. Such information can be used to identify requests for documents that have moved and those that have misspelled a uniform resource locator, which is used to identify the location of documents on the Internet. Only the server manager usually has access to that information, however. The error is almost never relayed to the person charged with document maintenance, either because it is not recognized as a document error or because the origin of the error is not apparent from the error message.

Even with better procedures, log files do not reveal failed requests that never made it to the server, nor can they support preventive maintenance and problems associated with changed document content. With a large and growing database, manual maintenance methods become difficult and may eventually become impossible.

Design Tools. New design tools are available that address the maintenance and testing issue by providing the means to visualize the creation, maintenance, and navigation of whole collections of on-line documents. Where traditional Web tools such as browsers and HTML editors focus on the Web page, these tools address the Web site, which may be either physical or logical in structure. These tools include a system to identify which pages are included in the site and another to describe how the pages are interconnected. The construction of a site is facilitated by providing templates for creating pages and scripts and linkage to tools for editing and verifying HTML documents.

In addition to offering high-level views of a site — either graphical or hierarchical — the design tools check for stale links (either local or remote), validate the conformance level of HTML pages, and make broad structural changes to the site architecture by using a mouse to drag and drop sections of the Web hierarchy into a different location.

Agents or Robots. Although design tools address document creation and maintenance at the site level, they do not comprehensively address the maintenance needs of distributed hypertext infrastructures that span multiple Web sites. This task can be handled by special software known as agents or robots. These programs can be given a list of instructions about what databases to traverse, whom to notify for problems, and where to put the resulting maintenance information.

For example, the agent or robot may be tasked to provide information about the following conditions that typically indicate document changes:

- *A referenced object has a redirected* uniform resource locator (i.e., a document has been moved to another location).
- *A referenced object cannot be accessed* (i.e., there is a broken or improperly configured link).
- *A referenced object has a recently modified date* (i.e., the contents of a document have changed).
- *An owned object has an upcoming expiration date* (i.e., a document may be removed or changed soon).

To get its instructions, the agent or robot reads a text file containing a list of options and tasks to be performed. Each task describes a specific hypertext infrastructure to be encompassed by the traversal process. A task instruction includes the traversal type, an infrastructure name (for later reference), the "top URL" at which to start traversing, the location for placing the indexed output, an e-mail address that corresponds to the owner of that infrastructure, and a set of options that determines what identified maintenance issues justify sending an e-mail message.

COMMON GATEWAY INTERFACE

An Internet server should support the common gateway interface (CGI), which is a standard for interfacing external applications with information servers, such as hypertext transfer protocol or Web servers. Gateway programs handle information requests and return the appropriate document or generate one spontaneously. With CGI, a Web server can serve information that is not in a form readable by the client (i.e., an structured query language [SQL] database) and act as a gateway between the two to produce something that clients can interpret and display.

Gateways can be used for a variety of purposes, the most common being the processing of form requests, such as database queries or on-line purchase orders.

Gateways conforming to the CGI specification can be written in any language that produces an executable file, such as C and C+. Among the more popular languages for developing CGI scripts are practical extraction and report language (PERL) and tool command language (TCL), both derivatives of the C language.

An advantage of using PERL and TCL is that either language can be used to speed the construction of applications to which new scripts and script components can be added without the need to recompile and restart, as is required when the C language is used. Of course, the server on which the CGI scripts reside must have a copy of the program itself — PERL, TCL, or an alternative program.

CONCLUSION

The client/server architecture of the Internet and its use of open protocols for information formatting and delivery make it possible for any connected computer to provide services to any other computer. With this capability, businesses can extend communications beyond organizational boundaries and serve the informational needs of all users.

The type of services that are available depends on the application software that runs on one or more servers. A server may be dedicated to a specific Internet application or multiple applications, depending on such factors as system resources and the specific needs of the organization. A careful evaluation of the hardware platform, operating system, and application software in terms of features and conformance to Internet standards ensures that the current and emerging needs of the organization and its users are met in an efficient and economical manner.

Section VI
Server Performance

The acquisition and operation of a server can represent a considerable expenditure of funds, especially when you note that an adverse level of performance can affect many employees. Thus, a common goal of organizations is to operate their servers as efficiently and effectively as possible. Chapters included in this section focus upon this area of server technology, providing information concerning the selection of an appropriate Web server network connection rate, the effect of images upon client/server operations, and the use of the Windows NT Performance Monitor utility program.

In Chapter 39 Gil Held addresses a key question often overlooked by organizations connecting a Web server to the Internet: what operating rate should service the connection? If the operating rate is too fast for the level of traffic, the WAN connection will be underutilized and the organization will be paying for unused or underutilized capacity. If the operating rate is too low, traffic bottlenecks will occur when remote users attempt to access the Web server, possibly resulting in lost sales or the inability of potential customers to access information necessary to make a purchasing decision. In his chapter, entitled "Selecting a Web Server Connection Rate," Held provides a methodology you can follow to select an optimum Web server WAN operating rate.

In the second chapter in this section, Held turns his attention to the effect of images upon client server operations. In Chapter 40, entitled "Working with Images in a Client/Server Environment," Held provides an overview of images, including data storage and transmission requirements. He also examines the potential use of different image file formats, cropping, and image conversion to enhance the efficiency of images stored and transmitted over a network.

The third performance-related chapter in this section focuses upon the Windows NT built-in Performance Monitor program. In Chapter 41, "Working with NT Performance Monitor," Gil Held illustrates how you can use this built-in utility program to determine if the server's processor is the culprit when users complain of sluggish server response and blame the network for the situation.

Chapter 39
Selecting a Web Server Connection Rate

Gilbert Held

If the operating rate of the Internet connection is too slow, anyone trying to access an organization's server from the Internet may get frustrated and terminate their access of information from the corporate Web server site. At the opposite extreme, if an organization's Internet connection operating rate exceeds the bandwidth required to support an acceptable level of access, you may be wasting corporate funds for an unnecessary level of transmission capacity.

As this chapter shows, with knowledge of the ways in which a Web server can be connected to the Internet, as well as knowledge about some of the transmission constraints associated with a Web server connection, it is possible to determine an appropriate Web server connection rate.

BASICS OF CONNECTING TO THE INTERNET

Exhibit 39.1 illustrates the typical method by which Web servers are normally connected to the Internet. A Web server resides on a local area network, with the LAN connected via a router to an Internet access provider. The Internet access provider has a direct connection to a backbone network node on the Internet, commonly using a full T3 or switched multimegabit data service (SMDS) connection to provide Internet access for a large group of organizations.

Although an Ethernet bus-based LAN is shown in Exhibit 39.1, in actuality any type of local area network that can be connected to a router (and for which TCP/IP drivers are available) can be used by the Web server. Thus, other common LANs used by Web servers include Token Ring and FDDI, as well as the numerous flavors of Ethernet, such as 10Base-T, 100Base-T, and 100VG-AnyLAN.

0-8493-9823-1/00/$0.00+$.50

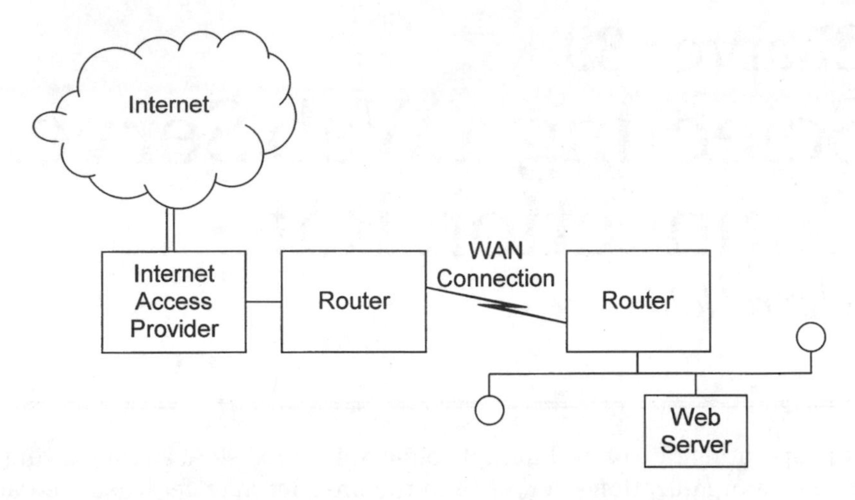

Exhibit 39.1 Web server connection to the Internet.

Analog vs. Digital Leased Lines

The actual WAN connection between the Internet access provider and the customer can range in scope from low-speed analog leased lines to a variety of digital leased lines. Only a few access providers offer analog leased line connection options, and the actual operating rate of the WAN connection is commonly limited to 19.2 or 24.4 Kbps, based on bandwidth constraints of a voice-grade analog leased line that limits modem operating rates. For digital leased lines, most Internet access providers recommend and offer 56 Kbps, fractional T1 in increments of 56 or 64 Kbps, full T1, fractional T3, and full T3 connectivity.

Connection Constraints

Although the WAN operating rate can constrain users from accessing information from an organization's Web server, another less recognized but equally important constraint exists — the traffic on the local area network on which the Web server resides. Although the focus of this chapter is on determining an appropriate WAN operating rate to connect a Web server to the Internet, it also examines the constraints associated with LAN traffic that affect the ability of the server to respond to information requests received from the Internet.

WAN CONNECTIVITY FACTORS

Three key factors govern the selection of an appropriate operating rate to connect a Web server to the Internet through a wide area network transmission facility. Those factors include:

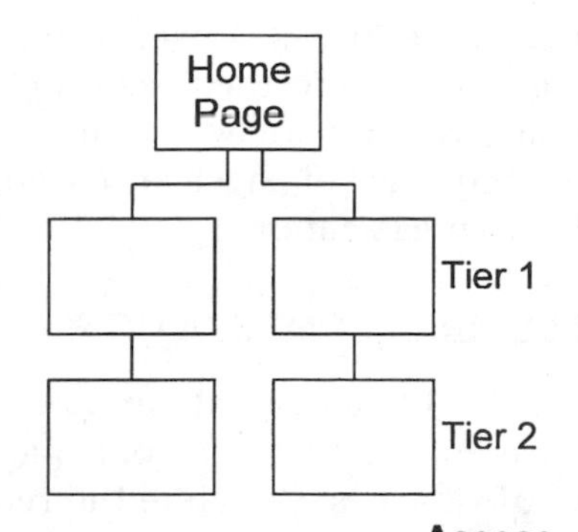

Data Storage	Bytes	Access Percentage
Home Page	12,500	40
Tier 1 Page	80,500	40
Tier 2 Page	60,500	30

Exhibit 39.2 Web page relationship.

- The composition of the Web pages residing on a server
- The types of pages retrieved by a person accessing the Web server
- The number of hits expected to occur during the busy hour

A typical Web page consists of a mixture of graphics and text. For example, a university might include a picture of “Old Main” on the home page in the form of a GIF file consisting of 75,000 bytes supplemented by 500 characters of text that welcomes Internet surfers to the university home page. Thus, this university home page would contain 75,500 bytes that must be transmitted each time a member of the Internet community accesses the home page of the university.

By computing the data storage requirements of each page stored on the Web server and estimating the access distribution of each page, it is possible to compute the average number of bytes transmitted in response to each Internet access of the organization’s Web server.

For example, assume an organization plans to develop a Web server that stores four distinct Web pages as well as a home page, providing Internet users with the ability to access two types of data from the home page. The construction of a two-tier page relationship under the home page is illustrated in Exhibit 39.2.

This example of computing an appropriate WAN operating rate is for illustrative purposes only. Although the Web home page is almost always the first page accessed, from the home page users typically access other server pages using hypertext links coded on the home page. Similarly, upon accessing different server pages, a user who wants to jump to other pages on the server is constrained by the links programmed on each page. Thus, the data transmitted in response to each page an Internet user accesses, as well as the sequence of pages accessed, will more than likely differ from organization to organization.

PERFORMING THE REQUIRED COMPUTATIONS

Assume that an organization has already determined that when Web pages are arranged in a tier structure, access to a home page at the top of the tier represents 40 percent of all accesses, while the remaining 60 percent is subdivided by remaining tiers. Furthermore, the organization's Web page structure is to be constructed in two tiers below the home page, with the data storage associated with each page to include text and graphics as well as the access percentage of each page, as listed at the bottom of Exhibit 39.2.

After determining the data storage required for each Web page and the distribution of Internet access by page, it is possible to compute the average number of bytes that will be transmitted from the Web server in response to each hit on the organization's server. (Here the term *hit* refers to access to a Web page on the server via the hypertext transmission protocol (HTTP) and a URL that represents a file stored on the server, which equates to the contents of a Web page.)

Using the information from Exhibit 39.2, the average data transmission rate resulting from a hit on the organization's server is computed as follows:

$$120{,}500 * .40 + 80{,}500 * .30 + 60{,}500 * .30 = 90{,}500$$

Thus, each hit on the organization's Web server results in a requirement to transmit 90,500 bytes of data from the server to the Internet via the WAN connection to the Internet access provider.

Hit Estimation

Perhaps the most difficult estimate to make is the number of hits that are expected to occur during the busiest hour of the business day. Access to an organization's Web server depends on a large number of variables, many of which are beyond the control of the organization.

For example, although a company can control advertising of its URL in trade publications, it may be difficult (if not impossible) to inhibit robot search engines from visiting the site, retrieving each page available for public access on the company's server, and indexing the contents of the server's Web pages. Once this index is placed onto the database of a search engine, access to the company's Web server can result from persons invoking a Web search using Lycos, Alta Vista, or a similar search engine.

Unfortunately, because of limitations associated with many search engines, forward references to an organization's Web server may not be relevant and can alter the distribution of page hits as many persons, upon viewing your home page, may click on the back button to return to the list of search matches provided by a search engine query and select a different match. If the organization was a tire distributor named Roosevelt Tires, for example, many Web search engines would return the home page URL in response to a search for the word "Roosevelt," even though the person was searching for references to one of the presidents and not for an automobile tire distributor.

Many Internet access providers can furnish statistics that may be applicable for use by an organization. A major exception to using such average statistics is if an organization is placing highly desirable information on the Web server, such as the results of major sports events like the Super Bowl or World Series as they occur. Otherwise, the information concerning busy hour hits can be considered to represent a reasonable level of activity that will materialize.

Returning to the estimation process, assume the organization can expect 660 hits during the busy hour. Although this hit activity may appear to be low in comparison to the tens of hundreds of thousands of hits reported by well-known URLs representing popular Web sites, during a 24-hour period you are configuring the operating rate of the WAN connection to support 24 × 660 or 15,840 hits, based upon a busy hour hit rate of 660.

According to statistics published by several Internet access providers, during 1995 the average number of hits per Web site, excluding the top 100 sites, is under 5,000 per day. Thus, if an organization is the typical business, college, or government agency, it may be able to use a lower WAN operating rate than determined by this example.

After determining the number of hits the Web site will support during the busy hour and the average number of bytes that will be transmitted in response to a hit, it is possible to compute other WAN operating rates. For this example, each hit results in the transmission of 90,500 bytes, and the WAN operating rate is sized to support 660 hits during the busy hour. Thus, the results obtained are

$$\frac{\text{660 hits per hour} * \text{90,500 bytes per hit} * \text{8 bits}}{\text{60 minutes per hour} * \text{60 seconds per minute}} = 132{,}733 \text{ bps}$$

LAN BANDWIDTH CONSTRAINTS

Based on the preceding computations it would be tempting to order a 192 Kbps fractional T1 as the WAN connection to the Internet access provider because the next lower fraction of service, 128 Kbps, would not provide a sufficient operating rate to accommodate the computed busy hour transmission requirement of 132,733 bps. However, before ordering the fractional T1 line, the business needs to consider the average bandwidth the Web server can obtain on the LAN it is connected to.

If the average bandwidth exceeds the computed WAN operating rate, the LAN will not be a bottleneck so it would not need to be modified. If the average LAN bandwidth obtainable by the Web server is less than the computed WAN operating rate, the local area network will function as a bottleneck, impeding access via the WAN to the Web server. This means that regardless of any increase in the operating rate of the WAN connection, users' ability to access the organization's Web server will be restricted by local traffic on the LAN.

If this situation should occur, possible solutions would be segmenting the LAN, creating a separate LAN for the Web server, migrating to a higher-speed technology, or performing a network adjustment to remove the effect of a portion of the local LAN traffic functioning as a bottleneck to the Web server.

Determining the Effect on Local Traffic

To illustrate the computations involved in analyzing the effect of local traffic, assume the LAN shown in Exhibit 39.1 is a 10 Mbps 10Base-T network that supports 23 workstations and one file server in addition to the Web server, resulting in a total of 25 stations on the network. This means that on the average, each network device will obtain access to 1/25 of the bandwidth of the LAN, or 400,000 bps (10 Mbps/25).

However, the bandwidth of the LAN does not represent the actual data transfer a network station can obtain. This is because the access protocol of the network will limit the achievable bandwidth to a percentage of the statistical average.

For example, on an Ethernet LAN that uses the carrier sense multiple access collision detection (CSMA/CD) protocol, collisions will occur when two stations listen to the network and, noting an absence of transmission, attempt to transmit a frame at or near the same time. When a collision occurs, a jam signal is transmitted by the first station that detects the high voltage resulting from the collision, causing each station with data to transmit to invoke a random exponential backoff algorithm.

This algorithm generates a period of time the network station delays attempting a retransmission; however, the frequency of collisions, jams, and the invocation of backoff algorithms increases as network utilization increases. For an Ethernet LAN, network utilization beyond a 60 percent level can result in significant degradation of performance, which can serve as a cap on achievable transmission throughput. Thus, the average bandwidth of 400,000 bps previously computed should be multiplied by 60 percent to obtain a more realistic level of average available bandwidth obtainable by each station on the LAN to include the Web server.

In this example, the Web server will obtain on the average 240,000 bps of LAN bandwidth (i.e., 400,000 bps × .6). Since the average LAN bandwidth obtainable by the Web server exceeds the computed WAN operating rate, no adjustment is required to the LAN. If the Web server were connected to a Token Ring LAN, the average bandwidth of 400,000 bps should be multiplied by 75 percent, since a Token Ring LAN does not have its performance seriously degraded until network utilization exceeds 75 percent.

MAKING WEB PAGE ADJUSTMENTS

As described thus far, network managers need to consider the LAN bandwidth obtained by the Web server as well as the WAN operating rate to effectively select a WAN connection method to an Internet access provider. When computing the WAN operating rate, it is important to note that the rate depends on the following factors:

- The number of hits expected to occur during the busy hour
- The storage in bytes required to hold each page (which represents data that has to be transmitted in response to a page hit)
- The distribution of hits on the pages that are placed on the company's Web server

The first and third factors are obtained by an estimation process. However, a company has a high degree of control over the composition of its server's Web pages, and this fact can be used as an effective tool in adjusting the WAN connection's ability to support the estimated number of hits expected during the busy hour.

Because initial access to a company's Web server is through its home page, that page will have the highest distribution of hits on the company server. Thus, if the estimate of busy hour traffic is low, it is possible to increase the selected WAN operating rate to support additional hits by reducing the transmission associated with each home page hit. Methods include replacing GIF images with their equivalent JPEG images that require less storage, cropping images to reduce their data storage requirements, or eliminating all or some images on the home page.

RECOMMENDED COURSE OF ACTION

The selection of an appropriate WAN operating rate to connect a corporate Web server to the Internet depends on three key factors, of which two — the expected number of hits during the busiest hour and the distribution of hits per server page — can only be estimated. This means that the WAN operating rate's ability to service expected traffic will be only as good as two traffic-related estimates.

However, by planning ahead the organization can adjust the third factor — the data storage on the server's home page — and obtain the flexibility to alter the selected WAN operating rate to support additional hits during the busy hour.

By following the methodology presented in this chapter, network managers and others involved in corporate Web page creation will be able to remove a large degree of the guesswork associated with connecting a Web server to the Internet. In addition, they should be able to adjust rapidly to capacity of a WAN connection to support additional Web server hits if such an adjustment shouid become necessary.

Chapter 40
Working with Images in Client/Server Environments

Gilbert Held

Advances in monitor display technology coupled with the development of multimedia software and increased capacity of disk drives are just some of the several factors contributing to a rapid increase in the use of images in client/server environments. Today it is common to find pictures of employees in a personnel database, images of houses and interiors in a real estate database, and the results of CAT and MRI scans in a hospital patient database.

Although the adage "one picture is worth a thousand words" contributes to the increased use of images, the storage and transmission of images can rapidly diminish network resources. Using images in client/server environments has to be carefully planned so that the resultant application is effective both in terms of cost and of operations. If it is not, the potential adverse effect to network performance will decrease the productivity of all network users — those using image-based applications as well as those using other client/server applications. Inappropriate use of images can also result in inefficient use of disk storage, necessitating costly disk drive upgrades or the acquisition of a hierarchical storage management system, whose use introduces delays as migrated files are moved back to disk storage on the receipt of a user access request.

Appropriate use of images is therefore vital to enhancing user productivity and the effective use of network bandwidth and storage.

AN OVERVIEW OF IMAGING

Two basic types of images are used in computer applications: raster and vector. A raster image consists of a grid of equally sized sources, referred to as pixels, with a color element for each pixel. Raster images include

0-8493-9823-1/00/$0.00+$.50

those taken with a camera and scanned photographs. In comparison, vector images are a collection of geometric shapes that are combined to form an image and are recorded as mathematical formulas. Computer-aided design (CAD) drawings represent an example of a vector image.

Vector data cannot reproduce photo-realistic images. As a result, most computer-based applications require pictures of persons, places, or things using raster-based images. This discussion of techniques for using images effectively in a client/server environment is therefore limited to raster-based images.

Storage and Color Depth

In a raster-based image a color is associated with each pixel. That color can vary from a simple black or white denotation to the assignment of one color to each pixel from a palette of more than 16 million colors.

The color assignment to a pixel is based on the use of one or more bits per pixel to denote the color of each pixel, a technique referred to as the color depth of the pixel. For example, a black and white raster image would use one bit per pixel to denote the color of each pixel, with each pixel set to 1 to denote black and 0 to denote white. For a 16-level gray scale raster image, each pixel would require 4 bits to indicate the pixel's gray level (24 = 16). The following table indicates the correspondence between the number of bits per pixel (i.e., color depth) and the maximum number of colors that can be assigned to a pixel. Note that the use of 24 bits per pixel is referred to as true color and represents the maximum number of colors the human eye can distinguish when viewing a raster image.

Bits per pixel (color depth)	Maximum number of colors
1	2
2	4
8	16
16	32,768 or 65,536 (depends on format)
24	16,777,216

Data Storage

The amount of storage required for an image depends on its size, resolution, and color depth. The size of an image is its vertical and horizontal size, typically expressed in inches or millimeters. The resolution of an image is the number of pixels per inch or per millimeter, and the color depth represents the number of bits per pixel required to define the color of each pixel in the image. Thus, the total amount of data storage can be expressed in bytes as follows:

$$\text{Data storage} = \frac{\text{length} * \text{width} * \text{resolution} * \text{color depth}}{8}$$

The computation of data storage can be illustrated through the example of using 3.5 × 5 inch photographs in a visual database. A scanner is used to digitize each photograph using a resolution of 300 dots per inch. Then, without considering the effect of the color depth of the scanned images, each photograph would require 3.5 in. × 5.0 in. × 300 × 300/(8 bits/byte) or 196,875 bytes of storage.

The following table compares the data storage requirements for a 3.5 × 5 inch photograph scanned at 300 dpi using different color depths:

Data storage (bytes)	Color depth
196,875	1 bit (black and white)
393,750	2 bits (4 colors/4-level gray scale)
787,500	4 bits (16 colors/15-level gray scale)
1,575,000	8 bits (256 colors/256-level gray scale)
3,150,000	16 bits (32,768 or 65,536 colors)
4,725,000	24 bits (16,777,216 colors)

In examining the data storage required for the 3.5 × 5 inch photograph, note that the maximum number of colors supported based on the indicated color depth is indicated in parenthesis to the right of the color depth. As indicated by the entries in the table, the use of color significantly affects the data storage required for an image. This effect governs not only the number of images that can be stored on a visual database, but also the ability of users to work with stored images. Concerning the latter, the memory requirement of a workstation to display an image is proportional to the amount of storage the image requires. Thus, physically large images scanned at a high resolution with a high color depth may not be viewable, or only partially viewable at a time, on many workstations.

Color-Depth Trade-offs

Careful selection of a color depth appropriate to the particular image-based application supported can result in significant savings in data storage and the time required to transmit images. For example, in a personnel database, a color depth of one or a few bits would probably be sufficient for pictures of employees. Note from the preceding table that the use of black and white images requires 196,875 bytes to store a 3.5 × 5 image, whereas the use of a 24-bit color depth results in a data storage requirement approximately 24 times greater. This means that the use of black and white pictures of employees could reduce the data storage requirements of a personnel database containing employee images by a factor of 24. For an organization with hundreds or thousands of employees, these savings could translate into a significant amount of disk storage becoming available for other applications, or the reduction in equipment required to support an image-based application.

Although few image applications use black and white, significant savings in data storage and transmission time can be achieved by selecting an appropriate color and color depth. In the real estate field, for example, the use of digital cameras is expanding. Many real estate professionals now take pictures of their listings and enter them into a central database for viewing by clients or other members of the organization. Although most digital cameras can support a 24-bit color depth, use of that color depth does not provide any appreciable viewing difference of a home, room, or swimming pool over the use of an 8-bit color depth. Thus, selecting an 8-bit color depth can reduce data storage of color images by a factor of three from the default 24-bit color depth used by many digital cameras.

Data Transmission

Images also affect client/server operations because of transmission time. To illustrate the effect of image storage on transmission time, the example of a 3.5 × 5 inch photograph stored on a server connected to a 10Base-T Ethernet local area network (LAN) is used. There are 39 workstations and one server connected to the LAN for a total of 40 network devices.

Ethernet is a shared access network, meaning that at any one time only one user can transmit or receive information. Thus, although each device can transmit or receive data at 10 Mbps, on the average, the devices obtain the use of 10 Mbps/40 or a 250-Kbps data transmission capability. Assuming the photographs were stored using a 24-bit color depth, each image would require 4.75 MB of data storage. Storage would actually be slightly more than that amount because images are stored using special file formats that add between a few hundred to approximately a thousand bytes of overhead to each file. Using a storage of 4.75 MB and an average transmission capability of 250 Kbps, the time required to download the image from a server to a client workstation would be 4.75 MB × 8 bits per byte/250 Kbps or 152 seconds. Thus, on the average it would take almost 2.5 minutes to download each photograph.

Displaying the image on a monitor would result in a slight increase in time because the server would have to access and retrieve the file containing the image from its disk storage system, and an image display program operating on the workstation would require some time to display the image. In any event, 2.5 minutes is not an appropriate waiting period for a guard attempting to verify personnel entering a building, a real estate broker attempting to show a client a series of pictures of homes, or a doctor attempting to view a previously performed MRI scan of a patient.

MANAGING STORAGE AND TRANSMISSION REQUIREMENTS

Several techniques have been developed to manage the storage and transmission challenges associated with the use of images in a client/server environment. They are presented in the sections that follow.

File Format

One of the most effective methods for reducing image storage requirements and transmission times is to use an appropriate file format when storing the images. Today, most imaging programs support a wide range of raster file formats, such as the CompuServe GIF, Joint Photographic Experts Group (JPEG or JPG), Truevision's TGA, Aldus' TIF, and Microsoft Windows's BMP. Some of those file formats store scanned images on a pixel by pixel basis, using one to three bytes per pixel for storing the color depth. Other file formats include a built-in data compression feature that compresses the scanned image before storing it.

Although it is tempting to select a file format with a built-in compression capability to reduce data storage and the transmission time required to move images from a server to client workstations, the various compression methods have important differences. For example, CompuServe's GIF file format uses the Lempel Ziv Welch (LZW) lossless compression method, whereas Aldus' TIF file format supports six different types of compression, including the JPEG lossy compression method.

A lossless compression method is fully reversible and does not result in the loss of any details on an image. In comparison, a lossy compression method groups blocks of pixels and considers two blocks to be equivalent for compression purposes if they differ by only one or a few pixels. Thus, decompression can result in the loss of a few pixels per block of pixels. Lossy compression can significantly reduce the data storage requirements of images. Although this compression method would not be suitable for storing some images, such as a chest X-ray where every pixel is important, it is highly effective for storing pictures of personnel, homes, and other objects for which the loss of a few pixels does not significantly alter the image's usability.

To illustrate the potential differences in data storage obtained through the use of different file formats, the same photograph was scanned several times and stored each time with a different format. The first scan was accomplished specifying the TIF file format without compression. The resulting file required 2.4 MB of data storage. Next, the same image was stored using the CompuServe GIF file format, which includes built-in LZW compression. The resulting data storage required for the photograph was reduced to 1.8 MB. Although a reduction of 600 KB is significant, LZW compression is fully reversible and does not represent the best method of compression for storing many images.

To illustrate the potential of data storage reduction that can be obtained through the use of lossy compression, the photograph was rescanned two additional times. The first time the image was scanned using the JPG file format specifying least compression, which results in two blocks of pixels being considered as equal if they only differ by one pixel. The resulting scan required 1.365 MB and yielded a reduction of 435 KB in storage from the GIF file format. Next, the image was scanned and saved using the JPG file format specifying moderate compression, where two blocks of pixels are considered to be the same even if they differ by up to four pixels per block. As a result of using a moderate level of lossy compression, the amount of data storage was reduced to 163 KB — a reduction of more than 2 MB in the storage of the image using a noncompressed file format. The table that follows provides a comparison of the data storage required for storing the same photograph using four different file formats:

File Format	Data Storage
TIF no compression	2.4 MB
GIF LZW compression	1.8 MB
JPG least compression	1.365 MB
JPG moderate compression	163 KB

If each file was stored on a server on the previously described Ethernet LAN, it would take 1.365 MB × 8/250 Kbps or approximately 44 seconds to retrieve the image stored using a JPG file format and least compression. In contrast, the retrieval of the image stored using a JPG file format and moderate compression would require 163 KB × 8/250 Kbps or approximately 5.2 seconds. Thus, the use of a moderate level of lossy compression can significantly reduce client retrieval time as well as server data storage requirements.

Cropping

Another effective mechanism for reducing image storage and transmission requirements is cropping. Cropping eliminates portions of an image that are not applicable to the application. For example, if images are required for an employee database, a large portion of the background of the scanned photograph can be cut out.

File and Image-Type Conversions

Two additional techniques for enhancing the use of images in a client/server environment are file and image-type conversions.

File Conversion. Many times software provided with a scanner or digital camera limits the user to storing images with one file format. Thus, the use of an image management program that provides a file conversion capability may provide the ability to significantly reduce the data storage for an image. By converting from one file format to another as well as specifying the use of an appropriate image compression method for the application, users can significantly reduce the storage requirements for an image.

Image-Type Conversion. The second conversion method involves a change in the image type. Here image type is used to denote the use of color depth with an image. For example, assume a photograph is scanned using 24-bit true color. Three bytes would then be used to store color information for each pixel. If the photograph is being used in a personnel file, the image could probably be converted to 16 color requiring 4 bits per pixel or 256 color requiring 8 bits per pixel to store color detail information. Thus, converting a one-inch 300 dpi image requiring 90,000 pixels from true color to a 256-color image type would reduce its data storage requirement to 180,000 bytes.

CONCLUSION

Although the use of images in applications has the potential to enhance user productivity, it also can adversely affect network bandwidth and server storage availability. Effectively using images in network-based applications requires the careful consideration of the use of different image file formats, compression methods, cropping, and image type. By carefully considering each of these image-related features, the characteristics of the image can be tailored to the application. Doing so can significantly reduce the amount of storage required for the image as well as the transmission time from servers to client workstations.

Chapter 41
Working with NT's Performance Monitor

Gilbert Held

One of the more valuable administrative tools built into both Windows NT Workstation and Windows NT Server is the Performance Monitor program. Although the design goal of Performance Monitor is to assist network managers and LAN administrators in performing capacity planning operations, this program can also be used to facilitate a variety of troubleshooting activities, including network troubleshooting. Since the best way to obtain an appreciation for Performance Monitor is by its use, we will examine its basic operation through the use of a few screen displays. However, prior to doing so, let's first obtain an overview of the general capabilities of this graphical tool and potential differences between its implementation in Windows NT Version 4.0 and Version 5.0, along with information on how it is implemented in the beta release of Windows NT 5.0.

OVERVIEW

Performance Monitor is a graphical tool which provides both Workstation and Server users the ability to measure the performance of their computer along with other computers on a network, with the latter capability based upon a user on one computer having a valid account on another computer accessed via a network.

Performance Monitor is accessed through the Start-Administrative Tools sequence. As an alternative for those who still like to use command-line entries, you can enter the command perform at the command prompt. Both Windows NT Version 4.0 and Windows NT Version 5.0 support both methods of invoking Performance Monitor.

Once Performance Monitor is active, you can use it to view performance metrics for objects on your computer or another computer on the network, create alerts with predefined thresholds that are invoked when the thresholds are exceeded, and export data from charts, logs, and Alert Logs maintained by Performance Monitor to spreadsheet or database programs for

0-8493-9823-1/00/$0.00+$.50

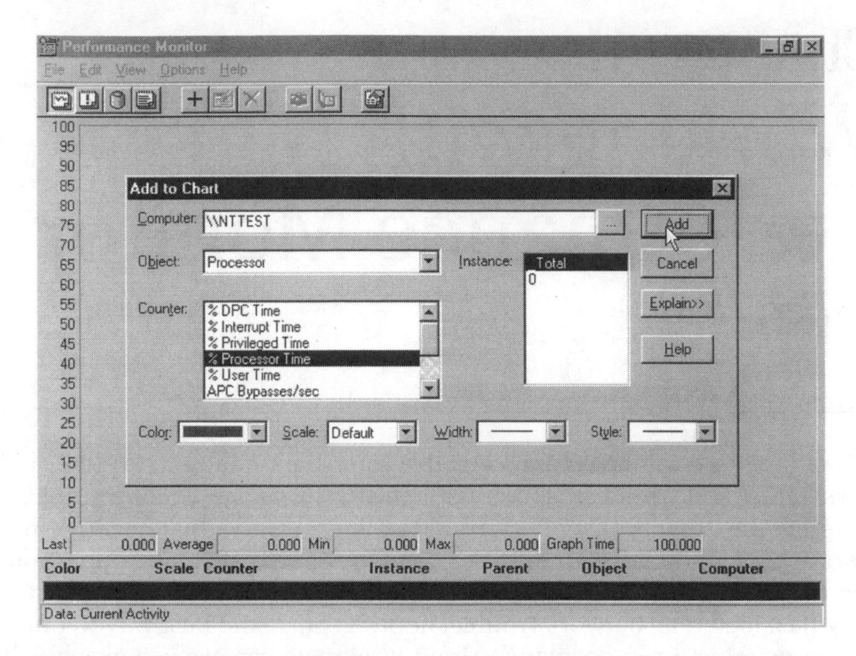

Exhibit 41.1 Using performance monitor's "add to chart" dialog box.

manipulation. Our discussion of Performance Monitor is applicable to both Windows NT Version 4.0 and the beta release of Windows NT Version 5.0. Where appropriate, I will indicate differences between the two.

USING THE GRAPHIC INTERFACE

Upon selecting Performance Monitor from the Administrative Tools bar, or entering the appropriate command-line entry, Performance Monitor displays a screen with a graphical area that is blank. That blank area is used for charting the performance of selected objects and object counters.

Exhibit 41.1 illustrates the initial blank Performance Monitor screen in the background of the figure. The foreground shows the dialog box labeled *Add to Chart* which is displayed as a result of clicking on the plus (+) symbol icon on the toolbar of the background screen. Since the Add to Chart dialog box is the key to the use of Performance Monitor to display metrics concerning the operation of your computer or another computer on the network, let's examine the entries in this dialog box.

The computer entry provides you with the ability to chart the performance of one or more objects on your computer or a different computer that is connected to your network. The object, counter, and instance entries provide you with a mechanism to specify exactly what is to be monitored or charted.

In Windows NT, an object represents a mechanism used to identify a system resource. Objects can represent individual processes, sections of shared memory, and physical devices. Clicking on the downward pointing arrow associated with the object window results in a display of a series of objects to include memory, processor, paging file, and physical disk. Under the beta version of Windows NT Version 5.0 several new objects were added to Performance Monitor, including a network object and a logical disk object. For each object, Performance Monitor maintains a unique set of counters that, when selected, produce statistical information. This is illustrated by the selection of the Processor object in Exhibit 41.1. Note that the counter window illustrates six counters associated with the Processor object. In actuality there are more than six Processor counters, since you can scroll through additional counter entries that are not directly visible in the Counter window. For illustrative purposes, we will select the % Processor Time counter since that counter indicates the percentage of elapsed time that a processor is busy executing a non-idle thread. When attempting to isolate poor server performance, it is important to determine if the bottleneck is the server or the network. By examining the value of the % Processor Time counter over a period of time, you can obtain valuable insight as to whether or not the server's load is causing the bottleneck.

Note also the window labeled Instance in Exhibit 41.1. Each object type can have one or multiple instances. For example, a multiprocessor capable server could have multiple processors installed on its system board. This means that a mechanism is required to provide you with the ability to examine each processor. That mechanism is obtained by the window labeled Instance. Some object types, such as Memory and Paging File, do not have instances. If an object type has multiple instances, each instance will produce the same set of statistical information since the counters are the same for each instance. That is, counters vary only by object type.

The lower portion of the Add to Chart dialog box provides you with the ability to specify or customize the manner by which counters are charted. You can select a particular color, scale, line display width, and line style for your chart. If you do not select any particular value or set of values, Performance Monitor will cycle through a default set of values if you plot multiple counters. In doing so, Performance Monitor will assign different default values to each counter you wish plotted, enabling you to easily distinguish one plot from another.

Exhibit 41.2 illustrates the initial plotting of the % Processor Time counter on the author's NT computer. Note that at the time the screen was captured, the % Processor Time counter value had a maximum of

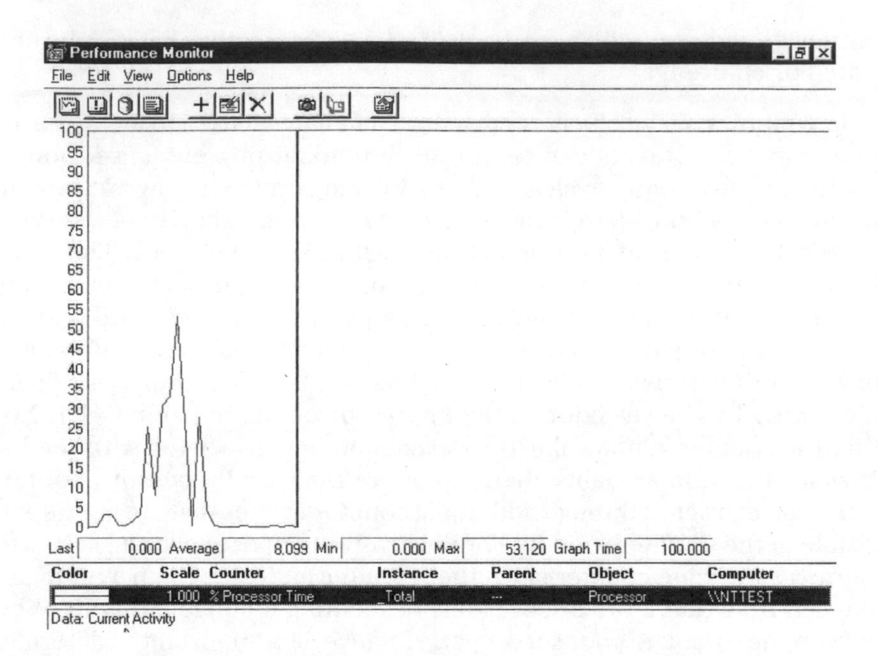

Exhibit 41.2 Observing the % Processor Time utilization.

slightly over 53 percent, a minimum value of 0, and an average value of 8.099 percent. During the monitoring interval, this author executed several programs to ascertain their effect upon processor performance which resulted in the spike of activity shown in Exhibit 41.2. Through the use of Performance Monitor in a similar manner, you can determine the effect of different workloads being placed upon the processing capability of an NT-based system. This in turn provides you with a mechanism to determine if you should consider either upgrading a single processor system to a more powerful microprocessor, assuming that it is upgrad able, or adding one or more processors to a multiprocessor-based computer. Thus, the use of Windows NT Performance Monitor can be a valuable tool for capacity planning purposes.

WORKING WITH MULTIPLE OBJECTS

Very often it is difficult to determine the culprit for poor performance. This is due to the complexity of modern computers and networks for which many areas of activity interact upon each other, resulting in the possibility that a hardware, software, or network bottleneck could be adversely affecting the level of performance, resulting in complaints from remote users. Recognizing the necessity to provide a mechanism to view the values of multiple counters, the designers of Performance Monitor permit you to simply specify additional counters for display.

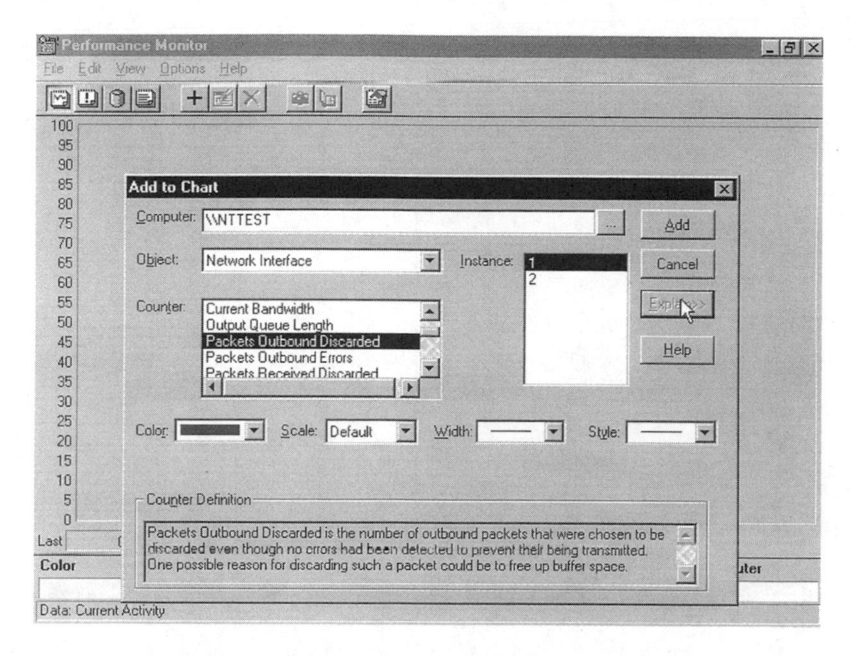

Exhibit 41.3 The beta version of Windows NT 5.0.

Exhibit 41.3 illustrates one of several new objects and related object counters added to the beta version of Windows NT 5.0. In Exhibit 41.3 the Network Interface object is shown selected. Several additional objects added to the beta version of Windows NT Version 5.0 that are not included in Version 4.0 include TCP, IP, and IPX, providing an enhanced ability for network capacity planning operations.

In Exhibit 41.3, note that the Instance window shows the values 1 and 2. This is because the computer has two network adapters installed. Through the use of the Add to Chart dialog box, statistics could be obtained for either or both network adapters. Also note the counter definition at the bottom of the dialog box. One of the more helpful features of Performance Monitor is the ability to click on the Explain button which results in the display of a short definition of a highlighted counter. This can be an extremely useful feature due to the large number of counters associated with many objects, as well as because of the use of counter names that, upon occasion, are difficult to relate to a particular activity. Both Windows NT Version 4.0 and the beta version of Windows NT 5.0 include the Explain button capability.

Exhibit 41.4 illustrates the charting of two counters. Although the illustration is shown in black and white, making discrimination between the

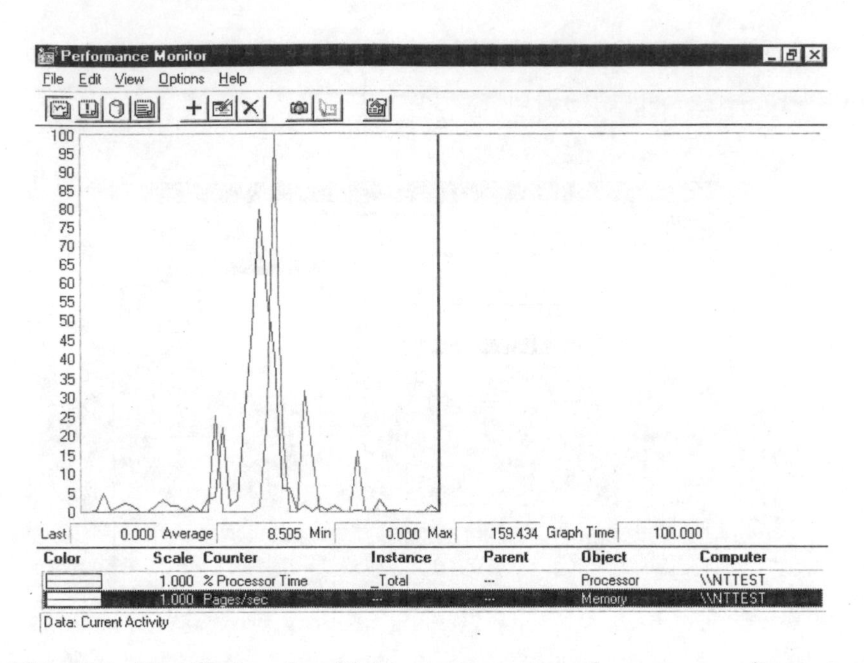

Exhibit 41.4 The display of multiple counters on a performance monitor screen.

two counter charts difficult, on a color monitor the plotting of each counter is in a different color which facilitates viewing the charting of multiple counters. However, if you closely examine Exhibit 41.4, you can still differentiate between the two plots. To do so you would first examine the color indicated for each counter which shows the plotting style used for plotting. Doing so you will note that the pages/sec counter display uses a dotted line. Although a bit difficult to discern in black and white, the use of different styles enables you to discriminate among plotted objects. To make it easier to differentiate among multiple plots, you can use the width window previously shown in Exhibit 41.1 to adjust the width of the lines used to plot different counters. In addition, you can select more pronounced styles, such as dots, dashes, and other symbols, to better differentiate one plot from another.

Because there is only one row in Performance Monitor to indicate the last, average, minimum, and maximum values for a plotted counter, a mechanism is required to select the display of such data when multiple counters are displayed. That mechanism is obtained through scrolling a highlighted bar among the counter information displayed at the bottom of the Performance Monitor display. For example, in Exhibit 41.4 the highlighted bar is shown placed over the Pages/sec counter previously selected from the Memory object. Thus, the values in the last, average, minimum, and maximum

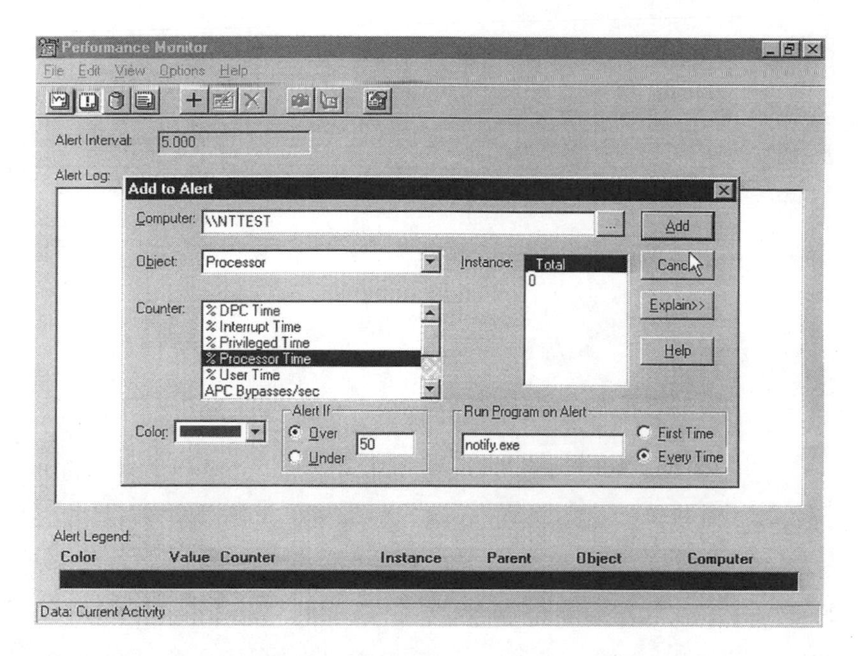

Exhibit 41.5 Creating an alert that will execute the program Notify.exe when the % Processor Time exceeds 50 percent.

windows are associated with the Pages/sec counter. If you move the highlighted bar upward over the % Processor Time counter, the previously described windows would reflect values for the % Processor Time counter.

USING ALERTS

While Performance Monitor provides a valuable visual display of data concerning previously defined counters, most people have more important things to do than sit before a console for a long time waiting for an event that may never happen. Recognizing this fact, Performance Monitor includes a built-in alert facility which enables users to set predefined thresholds. Once those thresholds are reached, Performance Monitor will automatically place an appropriate entry into the Windows NT Alert Log. In addition, if you desire, you can enter the name of a program that will be automatically executed when an alert occurs.

Exhibit 41.5 illustrates the use of Performance Monitor's Add to Alert dialog box. In this example the % Processor Time counter for the Processor object is shown selected. In the box labeled "Alert If," the Over button is shown selected and the value "50" is shown entered in the Alert If window. This means that an alert will be generated if the % Processor Time counter value should exceed 50 percent.

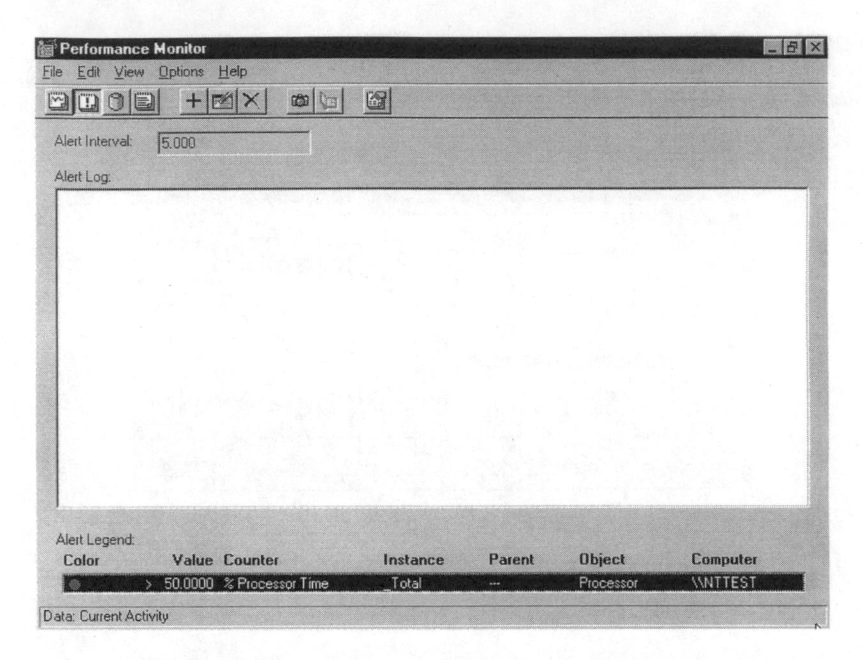

Exhibit 41.6 The performance monitor alert log.

In the lower right portion of the Add to Alert dialog box is a window with the label "Run Program on Alert." Note the entry "notify.exe." This entry results in the automatic execution of the previously mentioned program when the % Processor Time value exceeds 50 percent. In addition, since the button labeled "Every Time" is shown selected, this means that the program will be executed each time the previously defined alert occurs.

The program notify.exe represents an e-mail shell that was developed by this author to transmit a predefined message to technicians. You can also obtain commercial programs that can generate a paging message, send an e-mail, and even dial a telephone number and generate a predefined voice message. Through the use of such programs combined with the Run Program on the Alert window, you can alert your staff to different conditions where human intervention may be necessary. For example, by setting an alert to be generated upon the occurrence of a predefined number of network transmission errors, you would alert networking technicians to check the network. Doing so may enable potential problems to be noted and fixed prior to such problems' adversely affecting network users or causing a network failure if left unchecked.

In concluding this chapter on the use of Windows NT, Exhibit 41.6 shows the Performance Monitor's Alert window. The Alert Log when the screen display was captured is empty since it was extremely difficult to exceed a 50 percent processor time level of utilization to coincide with a screen capture operation. If one or more alerts occurred, they would be time-stamped and listed in sequence of occurrence in the Alert Log window.

SUMMARY

Performance Monitor provides an easy-to-use statistical reporting capability which also enables you to define alert conditions. Through the use of this built-in utility program, you can note the growth or decrease in the use of system resources and plan hardware and network upgrades accordingly. Since Performance Monitor is included with each copy of Windows NT, it provides an ideal mechanism to identify and correct bottlenecks, plan for increased processor and network workloads, and troubleshoot a variety of computer and network-related problems. Thus, network managers and LAN administrators that support Windows NT clients should become conversant with the capabilities of this utility.

Section VII
Security

Today we live in a world filled with inquisitive and unscrupulous people. While many urban and rural areas are not safe for walking late in the evening, a similar situation occurs electronically when you consider connecting an organizational network to the public Internet. That is, without the protection of router access lists and firewalls, the network connection may not be safe. Thus, we would be remiss if we did not consider a variety of security issues when connecting organizational networks to the Internet.

In this handbook we are primarily concerned with technology associated with servers. Thus, while important, the role of router access lists and firewalls is beyond the scope of server-related topics to be covered. However, regardless of the efficiency of router access lists and firewall features, such devices are not infallible. In addition, they do not provide protection against the possibility of disgruntled or curious personnel on the network attempting to view information that they should not be allowed to access, enabling or disabling employee or customer accounts, or performing functions that would not be conducive to the operation of the organization. This means that it is extremely important to review a number of server security policies and to ensure that those applicable to your organization are efficiently and effectively implemented.

In this section you will find four chapters that discuss server security issues. The first chapter, which is Chapter 42 in this handbook, provides an overview of server security functions through an examination of server functions and the relationship of those functions to security. Authored by Jon David, this chapter, entitled "Server Security Policies," looks at access control, logging, backup, communications, and other server security-related topics to provide a comprehensive list of policies you may wish to consider.

In Chapter 43, "Security and Resource Sharing in an NT Environment," Gil Held examines the method by which NT shares are created and uses this information as a foundation to explain how to examine this file sharing technique to facilitate security. Held explains how to display information concerning NT shares and how you can control access to shared files as well as how to immediately disconnect suspicious people that appear to be attempting to share the use of files you do not want them to access.

A second chapter oriented toward Windows NT is focused upon access security. In Chapter 44, "Windows NT Access Security," Gil Held describes how you can use the features built into the operating system to control access to both workstations and servers. In this chapter Held covers the use of NT's User Manager for Domains facility, covering how to examine user properties, enable and disable group memberships, restrict utilization to defined days and hours of the day, and develop and enforce different account policies. The information provided in this chapter is similar to its title in that it informs you of the details you need to know to control access to your organization's NT-based computers.

One of the given facts associated with Internet surfing is the high level of probability that an employee will acquire a virus that may wind up on the corporate server. This means that anti-virus software is high on the priority list of products most organizations seek to acquire. In concluding this section Chapter 45, "Evaluating Anti-Virus Solutions Within Distributed Environments," which was written by the staff of Network Associates, provides a detailed methodology for evaluating the acquisition of anti-virus software. In the chapter we are first acquainted with some of the more common types of viral threats and counteractions to those threats. This forms the basis for describing and discussing a set of anti-virus features that can be used for the acquisition of this important software product.

Chapter 42
Server Security Policies

Jon David

INTRODUCTION

Local area networks (LANs) have become the repository of mission-critical information at many major organizations, the information processing backbone at most large organizations, and the sole implementation avenue for IP efforts in smaller concerns. The growing importance of LANs — the integrity and confidentiality of data and programs on them, their availability for use — demands proper security, but LANs have historically been designed to facilitate sharing and access, and not for security. There is a growing pattern of interconnecting these networks, further increasing their vulnerabilities.

The Internet has similarly become an integral part of day-to-day operations for many users, to the point that business cards and letterheads often contain email addresses, and a large number of organizations have their own Internet domain, organization-name.com. The worldwide web (WWW) is an extension of the Internet, actually an additional set of functions the Internet makes readily available. It is gaining in popularity at a very fast rate, to the point that it is now common to see even TV advertisements cite Web addresses for additional information or points of contact (e.g., www.news-show.com, www.product-name.com, etc.). Today, even with the Web still in its infancy, there is much Web commerce, e.g., the display and purchase of merchandise. While LANs come from a background where relatively little attention was devoted to security, the RFCs (Requests for Comment, i.e., the specifications to which the Internet conforms) specifically state that security is not provided and is therefore the sole responsibility of users. The Internet is rife with vulnerabilities, and the Web adds a further level of risks to those of the Internet.

Although servers are integral parts of various types of networks, this chapter will deal with LANs, not the Internet or the Web, or any other type of network. The Internet and the Web are individually and together important, and

0-8493-9823-1/00/$0.00+$.50

servers are particularly critical components (with PCs through mainframes being used as Web servers), but it is felt that most of you reading this book will be LAN oriented. The exposures of both the Internet and the Web differ significantly from LAN vulnerabilities in many areas, and deserve separate (and extensive) treatment on their own.

THE NEED FOR SERVER SECURITY

For a long time information — and its processing — has been a major asset, if not *the* major asset, of large organizations. The importance of information is even reflected in the language we use to refer to what originally was a simple and straightforward function: What was once known as computing became electronic data processing (EDP), and is now information processing; when we look for expert guidance in this field, we seek people skilled in information technology (IT); in our contemporary electronic world, both the military and commercial segments fear information warfare (IW).

The information that we enter into, store on, and transmit via our computers is critical to our organizations, and in many cases it's critical not just for efficiency, profit, and the like, but for their very existence. We *must* keep prying eyes from seeing information they shouldn't see; we *must* make sure that information is correct; we *must* have that information available to us when needed. Privacy, integrity, and availability are the functions of security.

LANs are a key part of critical information processing, and servers are the heart of LANs. The need for proper server security is (or at least certainly should be) obvious.

Server/NOS vendors do not help the situation. As delivered, servers are at the mercy of the "deadly defaults." Since security tends to be intrusive and (or) constraining in various ways, servers "from the box" tend to have security settings at the most permissive levels to make their networks perform most impressively.

THE NEED FOR SERVER SECURITY POLICIES

The media have been very helpful in recent years in highlighting the importance of proper information security. While they certainly haven't been on a crusade to make large operations more secure, the many security breaches experienced have made good copy and have been well publicized. Since pain is an excellent teacher, and successful organizations endeavor to learn from the pain of others, the publicizing of various breaches of information security has made everyone aware of its importance.

Successful organizations endeavor to remain successful. If they recognize a need (vs. merely a nicety), they endeavor to treat it. "Go out and buy us some," "What will it cost," and the like are frequently heard once the need for security is recognized. Unfortunately, security is not something you go out and buy. It's something you plan, and it's something you work on ... when planning it, when creating it, and when living with it.

Security policies are a prerequisite to proper security. They provide direction; they treat all areas necessary for proper security; and, possibly most important because it's so rarely recognized, they provide a means for consistency. Without direction, completeness, and consistency, security can always be trivially breached. If your security efforts concentrate on securing your servers, yet you don't tell users not to have stick-on notes with their passwords on their monitors, your security policies are deficient; if you require server changes be made only at the server console, yet allow anyone other than duly authorized administrators to make such changes, you've again missed the boat in terms of security policy. And, when networks that are 100 percent secure in and of themselves can each compromise the others via inconsistencies in their respective security types if they are interconnected (and interconnection has been a hot item for some time), having components with proper security is no longer enough; you must have consistent security set forth in your policies.

(*Warning*: Your policies should fit *your* operational environment and requirements. It is unlikely that the policies of even a similar organization will be best for you in every area. This does not mean that looking at the policies of other organizations cannot be of help to you — if they are good policies, of course — in terms of suggesting things like the types of areas to be treated, but you need to do what is right for you, not what may or may not have been right for somebody else.)

POLICIES

Servers are parts of networks, networks are parts of information processing structures, and information processing structures are parts of organizational operations. While this chapter deals only with server security policies, all other security areas *must* be dealt with in an organization's full security policies statement. A single security breach of any type can, and often does, compromise all operations. (If, for example, visitors were allowed to enter a facility unchallenged, and if nodes were left operational but unattended — during lunch periods or whatever — the best server security policies in the world would readily be defeated.)

The statements of policy set forth below are generic in nature. Not all will apply, even in modified form, to all servers, and many, if not most, will have to be adapted to specific operations. They are, though, most likely better than those you're likely to get from friends, and should serve as a good start for, and basis of, proper server security policies for your particular situation. For convenience, they are grouped in functional areas.

One area, and possibly the most critical one, will not be covered: the LAN security administrator. Your security cannot be any better than your administrators make and maintain it. You require the best possible personnel, and they must be given the authority, and not just the responsibility, to do whatever is necessary to provide the proper server — and network — security. Too often we see "the Charlie Syndrome." LANs come in, administrators are needed, Charlie is free, so Charlie is made the system administrator. Why is Charlie free? Well, in all honesty, it's because Charlie isn't good enough to be given anything worthwhile to do. What this means is that, rather than having the best people as system administrators, the worst are too frequently in that position. System administration should not be a part-time assignment for a secretary!

SERVER FUNCTIONS

Access Control

- The server shall be able to require user identification and authentication at times other than logon.
- Reauthentication shall be required prior to access to critical resources.
- File and directory access rights and privileges should be set in keeping with the sensitivity and uses of the files and directories.
- Users should be granted rights and privileges only on a need-to-know/use basis (and not be given everything except the ones they're known not to need, as is very commonly done).

Encryption

- Sensitive files should be maintained in encrypted form. This includes password files, key files, audit files, confidential data files, etc. Suitable encryption algorithms should be available, and encryption should be able to be designated as automatic if appropriate.
- For any and every encryption process, clear-text versions of files encrypted must be overwritten immediately after the encryption is complete. This should be made automatic, effectively making it the final step of encryption.

Logging

- Audit logs should be kept of unsuccessful logon attempts, unauthorized access/operation attempts, suspends and accidental or deliberate disconnects, software and security assignment changes, logons/logoffs, other designated activities (e.g., accesses to sensitive files), and, optionally, all activity.

- Audit log entries should consist of at least resource, action, user, date, and time and, optionally, workstation ID and connecting point.
- There should be an automatic audit log review function to examine all postings by posting type (illegal access attempt, access of sensitive data, etc.), and for each posting type. If a transaction threshold (set by the LAN administrator) for any designated operation exception is exceeded, an alarm should be issued and an entry made in an action-item report file.
- The audit file should be maintained in encrypted format.
- There should be reporting functions to readily and clearly provide user profiles and access rules, as well as reports on audit log data.

Disk Utilization

- As appropriate to their sensitivity, ownership, licensing agreements, and other considerations, all programs should be read-only or execute-only, and (or) should be kept in read-only or execute-only directories. This should also apply to macro libraries.
- Users should be provided with private directories for storage of their non-system files. (These include files that are shared with other users.)
- There should be no uploads of programs to public areas; the same is true for macros and macro libraries.

Backup

- The availability of the LAN should be maintained by the server scheduling and performing regular backups. These backups should provide automatic verification (read-after-write), and should be of both the full and partial (changed items only) varieties. All security features, including encryption, should be in full effect during backups.
- Both backups and the restore/recovery functions should be regularly tested.
- Backups should be kept off-premises.
- Automatic recovery of the full LAN and of all and individual servers (and workstations) must be available.

Communications

- Communications (i.e., off-LAN) access should be restricted to specific users, programs, data, transaction types, days/dates, and times.
- An extra layer of identification/authentication protocol should be in effect (by challenge-response, additional passwords, etc.) for communications access.
- All communications access should be logged.
- All communications access messages should be authenticated (using MACs, digital signatures, etc.).

- The password change interval should be shorter for communications access users.
- Stronger encryption algorithms should be used for communications access users.
- Any and all confidential information — passwords, data, whatever — should be encrypted during transmission in either or both directions for all communications access activities.
- Encryption capabilities for communications should include both end-to-end and link encryption.

Server Access and Control

- There shall be no remote, i.e., from other than the console, control of any kind of the server, and there shall similarly be no remote execution of server functions.
- All server functions must be done from the console. This specifically excludes access via dial-in, gateways, bridges, routers, protocol converters, PADs, micro-to-mainframe connections, local workstations other than the server console, and the like.
- All administrator operations (e.g., security changes) shall be done from the console.
- Supervisor-level logon shall not be done at any device other than the console.
- If supervisor-level use of a device other than the console becomes necessary, it shall be done only after a boot/restart using a write-protected boot diskette certified as "clean" (this implies that such diskettes are readily available, as they should be for even stand-alone PCs), or from tape.
- There shall be no user programs executed at the server by user (i.e., remote or local workstation) initiation.
- There shall be no immediate workstation access to the server or to any server resource(s) following a diskette boot at the server.
- All communication among and between nodes must be done through the server. There shall be no peer-to-peer direct communication.
- There shall be no multiple user IDs (UIDs)/passwords logged on (i.e., the same user on the system more than once at a given time). There should further be the ability to suspend the active user session and (or) issue alarms should this situation occur.

General (Node) Access Control

- Both a user ID and a password shall be required by servers for a user as part of logging on.
- The server should be able to identify both the workstation and workstation connection point at logon.

- All files (programs and data) and other resources (peripheral equipment, system capabilities) should be able to be protected.
- All resource access should be only on a need-to-know/need-to-use basis.
- File access control should be at file, directory, and subdirectory levels.
- File access privileges should include read, read-only, write (with separate add and update levels), execute, execute-only, create, rename, delete, change access, none.
- Resource access should be assignable on an individual, group, or public basis.

Passwords

- There should be appropriate minimum (six is the least, eight is recommended, more is better) and maximum (at least 64, more is better) lengths. (Longer "passwords" are usually "pass-phrases," e.g., "Four score and 7 years ago.")
- Passwords should be case sensitive.
- There should be a requirement for at least one upper case character, one lower case character, one numeric, and one alphabetic character to be used in user-selected passwords. For high-security access, this should be extended to include one non-print (and non-space) character.
- There should be computer-controlled lists of proscribed passwords to include common words and standard names, and employee/company information as available (name, address, social security number, license plate number, date of birth, family member names, company departments, divisions, projects, locations, etc.). There should further be algorithms (letter and number sequences, character repetition, initials, etc.) to determine password weakness.
- Passwords should be changed frequently. Quarterly is a minimum, monthly is better. High security access should have weekly change.
- There should be reuse restrictions so that no user can reuse any of the more recent passwords previously used. The minimum should be five, but more is better, and eight is a suggested minimum.
- There should be no visual indication of password entry, or even of password entry requirements. This obviously prohibits the password characters from echoing on the screen, but also includes echoing of some dummy character (normally an asterisk) on a per-character basis, or use to designate maximum field length.
- New passwords should always be entered twice for verification.
- LAN administrators, in addition to their password(s) with associated supervisory privileges, should have an additional password for "normal" system use without supervisory privileges.

(*Note:* There are password test programs to allow automatic review and acceptance/rejection of passwords. These are usually written in an easily ported language, typically C, and can be readily structured to implement whatever rules the security administrator feels are appropriate. They're used between the password entry function and the password acceptance function already in place, so only proper passwords get used by the system.)

Physical Security

- All servers should be as secure as possible in keeping with their sensitivity.
- Servers should be physically secured in locked rooms.
- Access to servers should be restricted to authorized personnel.
- Access to the server area should be automatically logged via use of an electronic lock or other such mechanism as appropriate.
- The room in which the server is kept should be waterproof and fireproof.
- Walls should extend above the ceiling to the floor above.
- Water sprinklers and other potentially destructive (to computers) devices should not be allowed in the server room.
- The server console should be kept with the server.
- Servers should have keylocks.
- Connection points to servers should be secured (and software-disabled when not in use) and regularly inspected.
- All cabling to servers should be concealed whenever possible. Access to cabling should be only by non-public avenues.
- All "good" media practices — encryption of sensitive information, storage in secure locations, wiping/overwriting when finished, etc. — should be in full effect.

Legal Considerations

- Programs that by license cannot be copied should be stored in execute-only or, if this is not possible, read-only directories, and should be specifically designated as execute-only or read-only.
- Concurrent use count should be maintained and reviewed for programs licensed for a specific number of concurrent users. There should be a usage threshold above which additional concurrent access is prohibited.
- Access rules should be reviewed for all programs licensed for specific numbers of concurrent users.
- Appropriate banner warnings should be displayed as part of the logon process prior to making a LAN available for use.
- Appropriate warning screens should be displayed on access attempts to sensitive areas and (or) items.

Other

- There shall be no unauthorized or unsupervised use of traffic monitors/recorders, routers, etc.
- There should be a complete formal and tested disaster recovery plan in place for all servers. This should include communications equipment and capabilities in addition to computer hardware and software. (This is of course true for full LANs and for the entire IP operations.)
- There shall be no sensitive information ever sent over lines of any sort in cleartext format.
- Servers should require workstations that can also function as stand-alone PCs to have higher levels of PC security (than those PCs that are not connected to a LAN). Workstations that operate in unattended modes, have auto-answer abilities, are external to the LAN location (even if only on another floor), and (or) are multi-user should have the highest level of PC security.
- Workstation sessions should be suspended after a period of inactivity (determined by the LAN administrator), and terminated after a further determined period of time has elapsed.
- Explicit session (memory) clean-up activities should be performed after session disconnect, whether the session disconnect was by workstation request (logoff), by server initiative (such as due to inactivity), or by accident (even if only temporary, as might be the case with a line drop).
- In cases where session slippage tends to occur (such as line drops), or in instances where service requests require significant changes of access level privileges, reauthentication should be required.
- Unused user IDs and passwords should be suspended after a period of time specified by the LAN administrator.
- Successful logons should display date and time of last logon and logoff.
- There should be the ability to disable keyboard activity during specified operations.
- The integrity of data should be maintained by utilization of transaction locks on all shared data — both data files and databases.
- The integrity of data and the availability of data and the entire LAN should be maintained by specific protections against viruses and other malicious code.
- All security functions and software changes/additions should be made only from the server and only by the LAN administrator.

Higher-Level Security

While the above capabilities will be significantly more than most LAN servers would find appropriate, there are still more sophisticated security features that are appropriate to LANs with high risk/high loss profiles. For the sake of completeness, rather than suggest the typical reader will have even an academic interest in them, major ones are set forth below.

- Access to critical resources should require reauthentication.
- Access to critical resources should not only authenticate the user, but further verify the correctness of the workstation in use, the connection point of that workstation, and the correctness of the day/date/time of the access.
- Message sequence keys should be used to detect missing or misordered messages.
- After a failed logon attempt, the server should generate an alarm, and be able to simulate a proper logon for the failed user (to keep this user connected while personnel go to the offending workstation).
- After excessive access violations, the server should generate an alarm and be able to simulate a continuing session (with dummy data, etc.) for the failed user (to keep this user connected while personnel go to the offending workstation).
- Traffic padding — the filling in of unused transmission bandwidth with dummy pseudo-traffic — should be used to prevent transmission patterns from being readily detected (and thereby making it easier to "trap" valid information).
- Multiple (at least two) LAN administrators should be required for all potentially critical server changes. (These might be adding a new user, altering an existing users' rights and privileges, changing or adding software, and the like.) For example, one administrator could add a user, but only from a list internal to the computer that a second administrator created. This means that any deliberate breach of security by an administrator would require collusion to be effective.
- LAN administrators should have separate passwords for each individual server function they perform, the rights and privileges associated with that password being the minimum necessary to do the specific job for which it is being used.
- The server should be fully compatible with tokens, biometric devices, and other such higher-security access control products and technologies.
- The server should be able to do automatic callback for any and all communications access.
- To improve the availability of the LAN, it should be fault tolerant. Multiple (shadow) servers, disk mirroring, and the like should be in place.
- There should be a file/system integrity product in regular and automatic use to alert the administrator to any and all server changes.
- Sophisticated authentication methodologies should be in place to ensure not only the contents of a message/request, but also the source. Message authentication codes (MACs) and digital signatures are viable means to certify contents, and public key/private key

(commonly known as RSA-type) encryption provides acceptable source verification.

- Backups should be made to an off-LAN facility. This could be an organizational mainframe, a service bureau or whatever. With this "store and forward backup," recovery media is immediately away from the server.
- Servers should be compatible with biometric devices (fingerprint, retinal scan, palm print, voice, etc.) for user verification.

Caveats

Just as seat belts, air bags, and other automotive safety devices merely make it less likely you will be seriously injured in an accident, and certainly do not guarantee your not being involved in one, so too does computer security merely lessen the chances your systems will be misused, lessen the likelihood of damages associated with certain common incidents, make it more likely to promptly discover and limit any misuse and (or) damages, and make it easier to recover from various types of accidents and misuse.

No realistic computer security "can't be beaten," and this certainly includes server security. Proper server security will make networks much more difficult to compromise, and can make it not worth an intruder's while (in terms of anticipated cost to break in vs. expected return as a result of a breakin) to even attempt to break into a properly secured network.

With servers viewed as being in the hands of "experts" (which they often are, of course), many, if not most users rely exclusively on server security for total protection and do not practice proper security as part of their operations. Server security, and even full network security, is not a substitute for other types of security; your security policies must reflect this.

Teeth

The best policies in the world will fail if they're not enforced. ("Thou shalt not print your password on a stick-on note and post it to your monitor" sounds good, but people still tend to do it. If you don't make sure that they don't, or take proper corrective actions if they do, your policies are little more than a waste of paper.) Your policies should have teeth in them. As appropriate, server, as well as all other, security policies should contain monitoring and enforcement sections.

Because operational environments are often in a virtually continuous state of change — new equipment and users, changing capabilities, rights and privileges, etc. — you should regularly review your server (and full) security to make sure it continues to be in agreement with your server security policies.

Similarly, untested server security may only be security on paper. Since even the most qualified of personnel can make mistakes in creating server security policies and (or) in implementing them, your security should be tested to see that it really works as intended, and conforms to your server security policies. This should obviously be done when you design/develop/install your security, but should also be done on a reasonable periodic basis (quarterly, yearly, whatever). Such tests are usually best done by outsiders, since truly capable personnel often aren't available on staff, and employees often have friends to protect and personal interests in particular operations.

CONCLUSION

LANs have become critical processing elements of many, if not most, organizations of all sizes, and servers are the hearts of LANs. As the frequent repository of highly sensitive, often mission-critical, information, proper security is of prime importance. Without proper security policies, security is unlikely to succeed, and policies have to be in place to allow the appropriate security to be designed and installed. Adequate security can be obtained by companies willing to work a bit at it, and the work must start with proper security policies, and must continue by seeing that security continues to conform to existing security policies. The key element by far in LAN security is the LAN administrator; for all purposes, and in spite of whatever products you may purchase, the quality of security will be in one-to-one correspondence with the abilities and efforts of this person.

Chapter 43 Security and Resource Sharing in an NT Environment

Gilbert Held

The ability to enable remote users to share resources on NT workstations and servers represents one of the major benefits of the use of this operating system. By enabling employees in an organization to share access to drives, folders, and files, you can significantly enhance productivity. However, you also provide the potential for others with access to your organization's network to either intentionally or unintentionally read or modify existing data as well as store new files whose contents could have an adverse effect upon the operation of your organization. Thus, it is important to understand the options associated with Windows NT resource sharing as they control access to shares as well as govern the level of security associated with network-based resources.

Because an understanding of the method by which shares are created facilitates an understanding of different levels of data security associated with resource sharing, we will examine in detail the creation of a shared folder on one computer and its access via another computer. In doing so we will refer to a series of screen images that illustrate the actual use of the Windows NT operating system to create shares. We will also discuss and describe methods that control access to shared resources, providing different levels of security concerning the ability of network users to access shared resources and, once accessed, their ability to manipulate data.

OVERVIEW

Windows NT provides users with the ability to share access to the contents of an entire drive or folder over a network. Although shares operate only at the drive and folder level, the operating system provides several methods to control access down to individual files.

0-8493-9823-1/00/$0.00+$.50

Because sharing is integrated with the operating system's database of user accounts and passwords, it is possible to restrict sharing to a specific user, a group of users, or several predefined user groups. In addition, because sharing is also integrated with the Windows NT file system, it is also possible to assign a variety of permissions to shared resources. However, your ability to do so depends upon the file system used to create a volume. If a share resides on a volume created using the original File Allocation Table (FAT) file system that dates to the original IBM PC, shared resource permissions will be restricted to a series of basic permissions. In comparison, shares residing on a volume created through the use of the more modern NT File System (NTFS) can be assigned additional file-by-file permissions due to the enhanced level of capability of that file system.

SHARE CREATION

An appreciation for the security of data associated with resource sharing is best facilitated by creating a shared resource. Thus, in this section we will create a shared folder and examine how access to data in the folder can be controlled.

The creation of a shared resource requires you to first select a drive or folder icon. Once that is accomplished you would right-click on the previously selected icon to display a list of options associated with the drive or folder.

Exhibit 43.1 illustrates the selection of the folder icon labeled PKZIPW located in the folder PKWARE and the resulting pop-up menu displayed as a result of right clicking on the selected icon. You can consider the selected folder to represent a subfolder or, for die-hard DOS fans, a subdirectory.

From the pop-up menu displayed as a result of right clicking on an icon, you can invoke sharing in one of two ways, selecting either the Properties or the Sharing menu entries. Selecting the Properties menu entry results in the display of a dialog box labeled "Properties," prefixed with the name of the previously selected icon. That dialog box contains two tabs as illustrated in Exhibit 43.2. The first tab is labeled "General." The contents of that tab are displayed in the foreground and provide information about the selected icon, such as its location, storage size, contents, creation date, and any assigned attributes. The second tab, labeled "Sharing," is located in the background of the display and provides you with the ability to share the resource as well as control access to it.

If you select Properties from the pop-up menu, shown in Exhibit 43.1, you would then have to select the Sharing tab shown in Exhibit 43.2. In comparison, if you select the Sharing entry in the previously shown pop-up menu, the Sharing tab would be directly displayed. Thus, the selection of the Sharing entry from the pop-up menu can be considered a small shortcut for

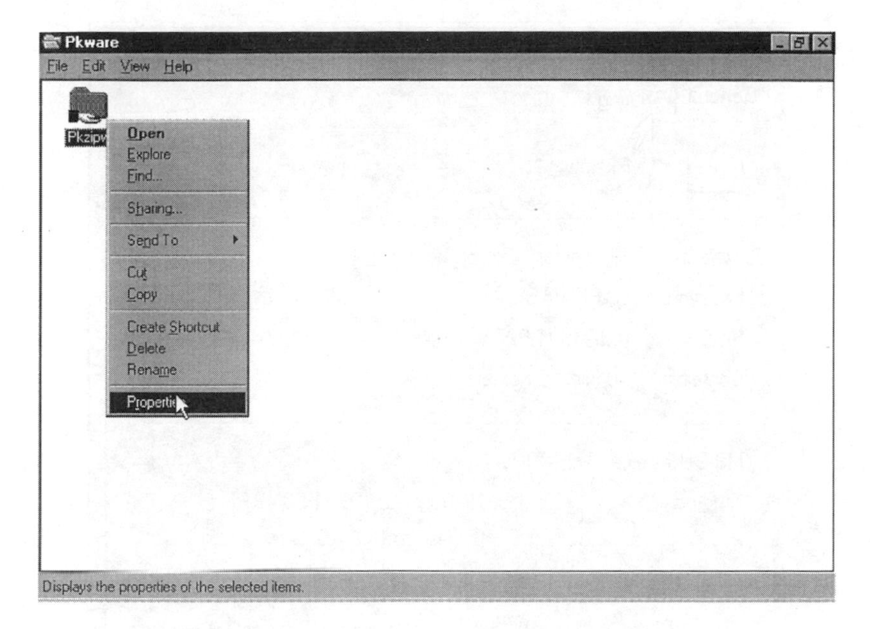

Exhibit 43.1 Creating a shared directory.

accessing the sharing facility for a Properties dialog box for a selected drive or folder icon.

Because we wish to create a share we would either click on the tab labeled Sharing in Exhibit 43.2 or select "Sharing" from the pop-up menu entry shown in Exhibit 43.1. Either action will result in the display of the Sharing tab for the Properties menu of the selected icon.

Exhibit 43.3 illustrates the display of the Sharing tab for the previously selected PKZIPW folder. Several entries in the tab display were either by default placed by the operating system into an appropriate box or entered by the author of this article and will now be described as we take a tour of the entries.

When you initially select a non-shared drive or folder the button labeled "Not Shared" will be shown selected. When you click on the Shared As button the name of the folder will by default be placed in the box associated with the label "Shared Name." You can enter a share name up to 80 characters in length; however, you should probably consider a much shorter name length, especially if your organization has Windows 95-based computers that will access the share. This is because a Windows 95 system cannot see a share name more than 12 character in length. In addition, DOS and Windows for Workgroup users are restricted to the 8.3 DOS file naming

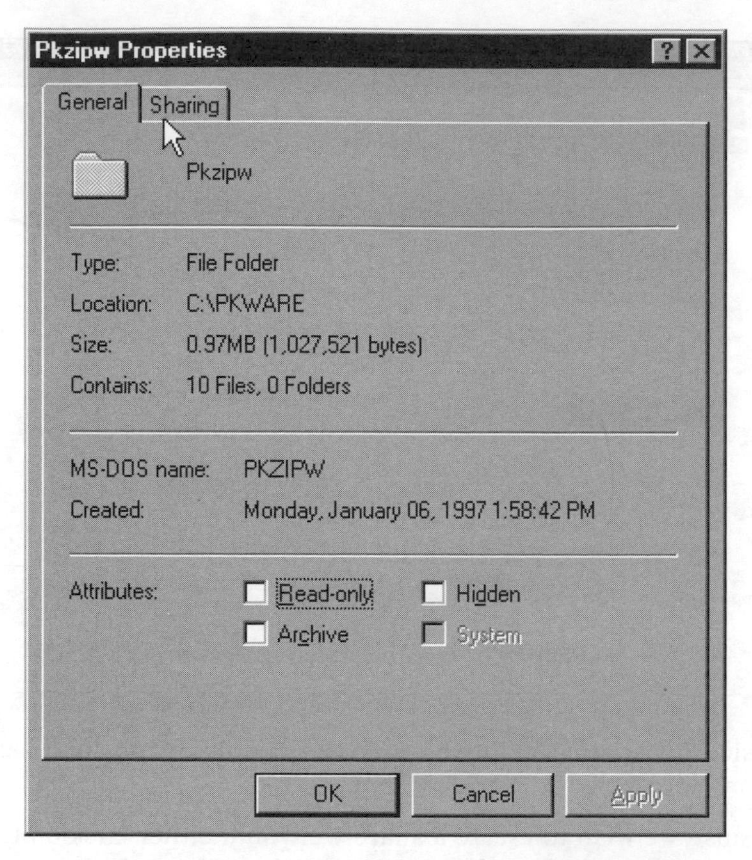

Exhibit 43.2 A folder or drive properties dialog box.

convention, so when in doubt as to who will access a share it's best to use the lowest common denominator of naming restrictions.

The actual composition of a share name can include all letters and numbers and the following special characters: $ % ' - _ @ ~ ! ( ) ^ # and &. If you use long filenames you can also include six additional special characters: +,: = [and]. One additional item that is worthy of mention concerns the use of the dollar sign ($) as a suffix to a share name. Doing so hides the share from direct viewing by other network users when browsing Network Neighborhood and provides a bit of privacy. However, if improper permissions are assigned to the share, other users that know its universal naming convention (UNC) address can access the data in the share.

Note the down arrow to the right of the box for the Share Name. As we will shortly note, the operating system provides the ability to share a

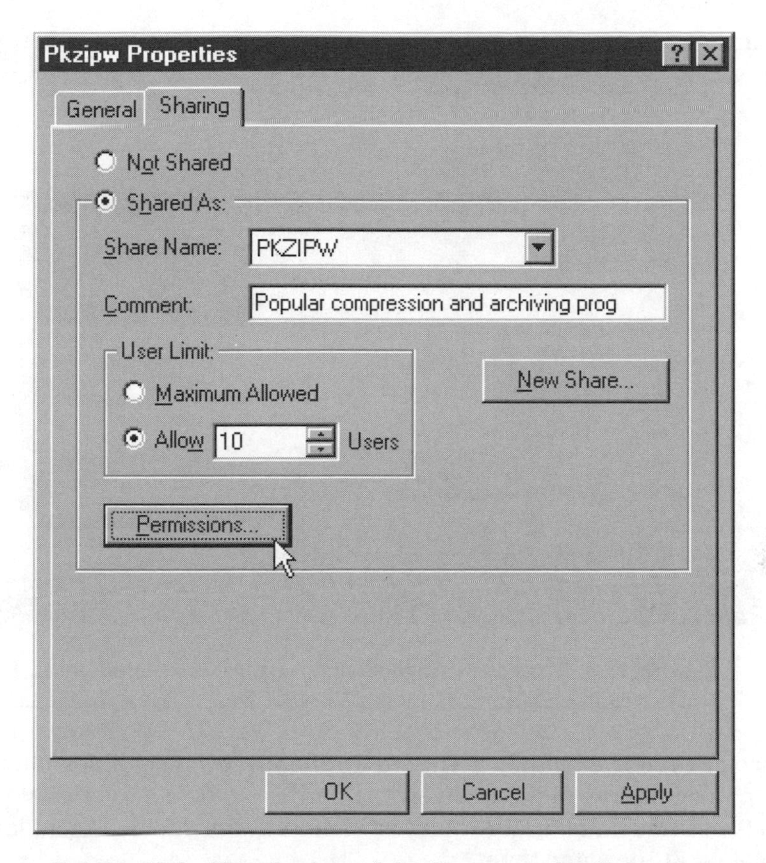

Exhibit 43.3 The sharing tab in the properties dialog box.

resource under multiple names, allowing you to tailor access and permissions to each share name. Another word of caution is in order concerning share names. Once you name a share, the only way to rename it is to disable it and then re-enable it using a new name.

Continuing our tour of the Sharing tab, the box labeled "Comment" provides you with the ability to add a comment concerning the share. The User Limit area provides you with the ability to control the number of users that can access the share at one time. While this option does not restrict access based upon any permissions it's valuable to consider from both a license and performance perspective. That is, its use enables you to remain in compliance with certain software product licenses that restrict the number of users that can use the product at any one time. In addition, this feature permits you to consider the movement of data across a network and indirectly control both data flow on the network and its effect upon network

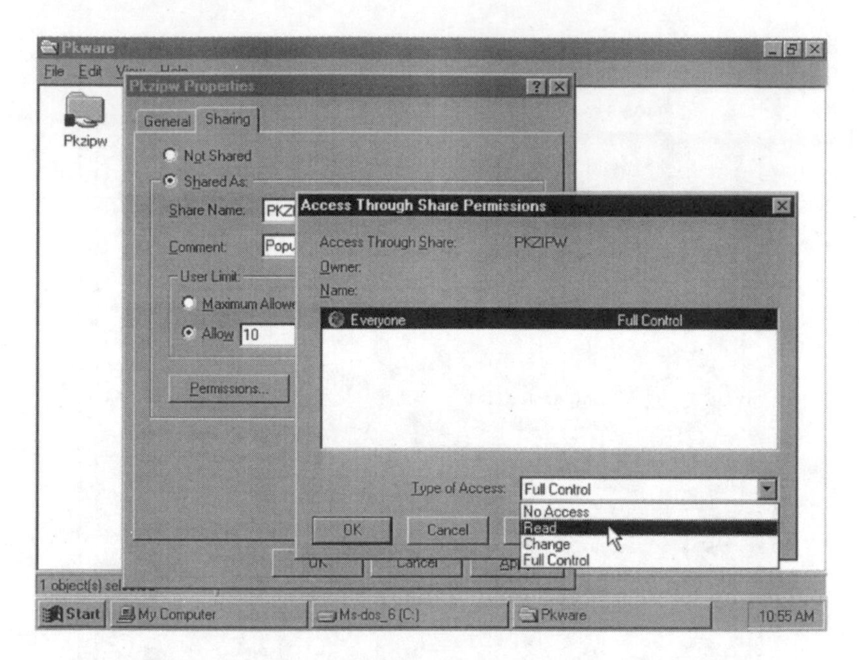

Exhibit 43.4 Access through share permissions dialog box.

performance. For example, a shared application that enables clients to backup the contents of their drives to a disk array on a server could bring the network to a halt if a large number of employees decided to initiate the application at the same time.

SETTING PERMISSIONS

Returning our attention to the Sharing tab displayed in Exhibit 43.3, note the button labeled "Permissions." This button provides you with the ability to control access to the share both via users that have accounts on the computer as well as by setting permissions that control access to the data in the share.

Exhibit 43.4 illustrates the result of clicking on the Permissions button, the display of a dialog box labeled "Access Through Share Permissions." At the lower right portion of Exhibit 43.4 you will note a drop down menu which lists the four types of access control that can be placed on a shared drive or folder — No Access, Read, Change, and Full Control.

No Access, as its name implies, prevents access to the share, even if a user belongs to a group that has access to the share. Read enables the user to view data and run files that are programs. Change adds the ability to modify data in a file or delete files. Finally, Full Control adds the ability of a user to change permissions and take over ownership; it should be assigned with caution. Another item that deserves attention is the relationship of share access permissions to any previously established file permissions. Because the Windows NT file system is part of the operating system's security module, file system permissions override share permissions. That is, if you previously associated a read-only permission to a file but the share for a folder in which the file resides was set to Change, a network user would be limited to reading the contents of the file or running it if it was a program. Thus, it's also important to note the file permissions within a share to ensure that their settings, along with the type of share access you set, provide the intended result.

Hidden from view by the drop-down menu in Exhibit 43.4 are buttons labeled "Add" and "Delete." Those buttons provide you with the ability to control individual users and groups of users that will have access to the share you are creating. By clicking on the Add button the operating system links the share creation to its User Manager for Domains, allowing you to add users or user groups. By default when you set up a share, access is assigned to the group labeled "Everyone," with full control assigned to this group. In this author's opinion this is a dangerous default and should be carefully examined prior to finalizing the share.

Exhibit 43.5 illustrates the display resulting from clicking the Add button to add individual users and user groups with access to a share. Note that the resulting display resembles the display of the Windows NT Administrative Tool, User Manager for Domains, for, as previously noted, it's linked to that facility. Thus, you can assign access to a share to any previously established Windows NT account.

MULTIPLE IDENTITIES

Two powerful capabilities included in Windows NT's sharing facility are the ability to create multiple-named shares as well as to create more than one share at a time. If you click on the button labeled "New Share" shown in Exhibit 43.3, a dialog box appropriately labeled "New Share" will be displayed. An example of this dialog box is illustrated in Exhibit 43.6. The use of this dialog box provides you with the ability to assign a new name and description as well as user access limits and permissions to the same share.

Although you may initially have reservations concerning the creation of a new name and attributes for an existing share, from a security perspective this capability provides an additional level of control. For example, you might wish to create a share and provide different levels of access to different groups. However, if some groups require only temporary access to the

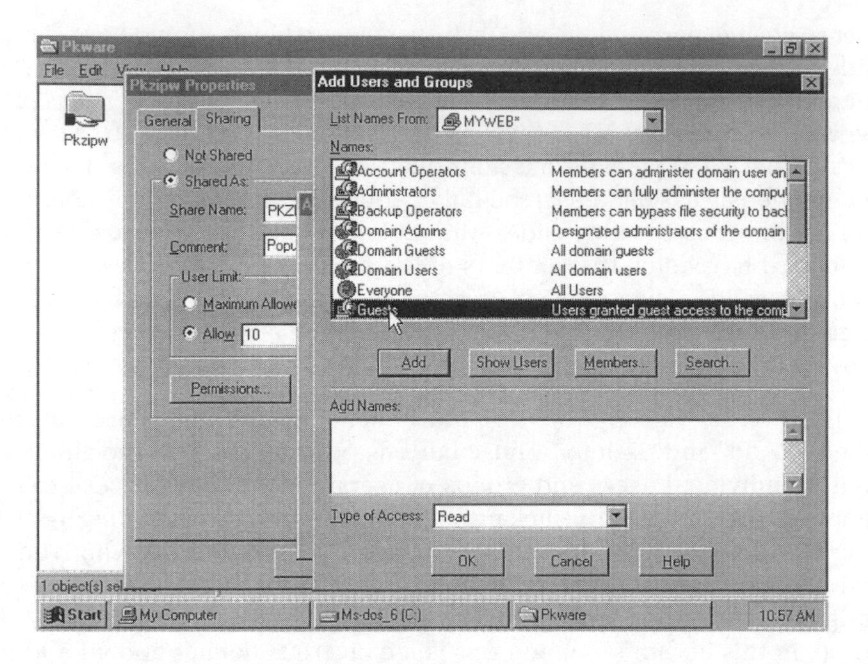

Exhibit 43.5 Sharing's link to the administrative tool 'User Manager for Domains.'

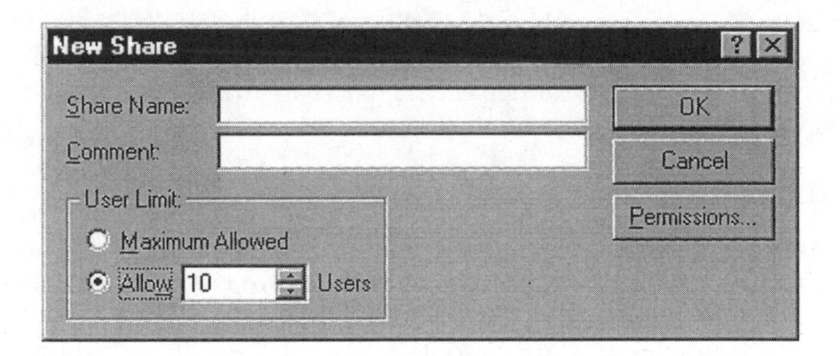

Exhibit 43.6 The new share dialog box.

---

share it might be easier to assign them to a new name which can be easily deleted when their access requirement expires.

CONSIDERING SUBFOLDERS

In the example we are using, we created a share for the folder PKZIPW located in the PKWARE folder. If you create another share and associate it

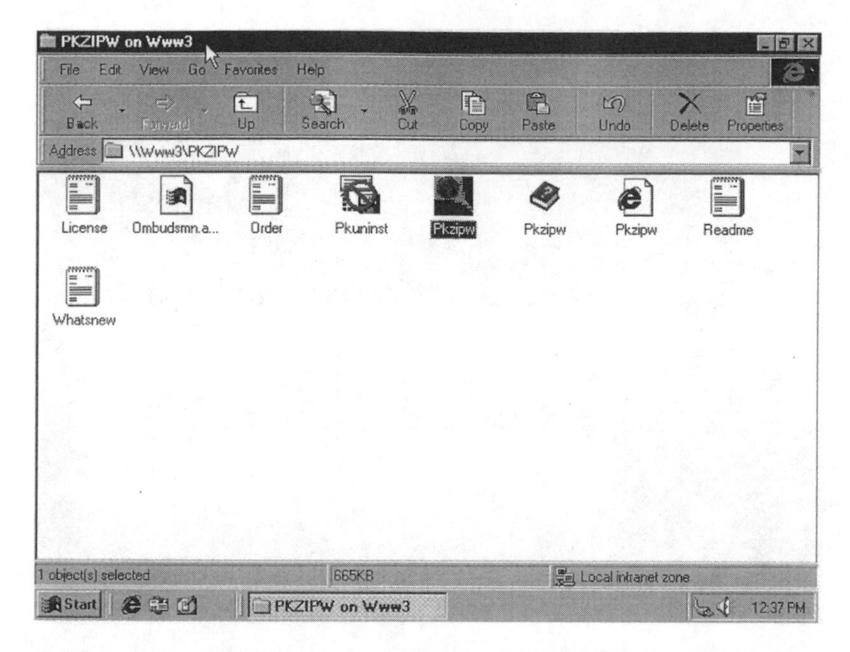

Exhibit 43.7 Viewing the shared folder from a Windows 2000 client.

with the higher level directory (PKWARE) you can segment access to a directory and one or more of its subdirectories, an important concept to note. For example, if you create a share named PKWARE, then anyone who connects to that share can also see the contents of the subdirectory PKZIPW, which may or may not be your intention. Thus, you should carefully consider the directory structure of the computer prior to setting up shares. In some instances it may be advisable to move a folder to a new location prior to sharing it, especially if those sharing the folder have no need for accessing files located in one or more subdirectories currently located under it.

OPERATIONAL CONTROL

Exhibit 43.7 illustrates the access of the shared folder PKZIPW from a Windows 2000 computer, explaining the display of Microsoft's Internet Explorer as the browsing facility. This screen display resulted from first selecting Network Neighborhood, then accessing the computer WWW3 to obtain access to the shared folder.

Many network managers and administrators can be lulled into complacency after they set up one or more shares. Thus, it's normally a good idea to periodically examine the User Sessions display to determine who is using the shares established on a computer.

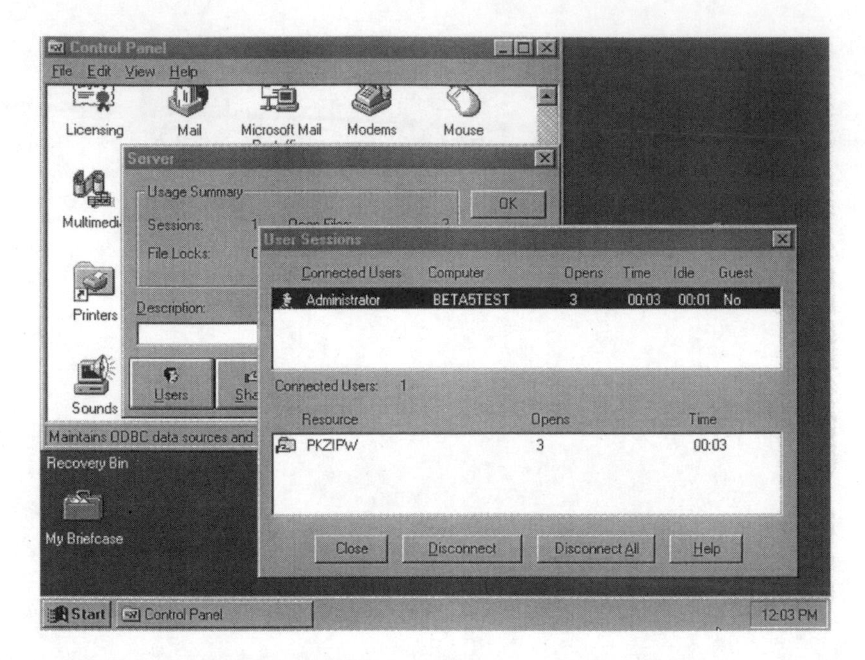

Exhibit 43.8 The server icon in the control panel.

You can display information concerning the use of shares through the Windows NT Control Panel, first selecting the Server icon in the Control Panel. That action provides you with the ability to display a variety of computer usage data. In Exhibit 43.8 the User Sessions display is shown in the foreground, the result of pressing the Users button in the "Server" dialog box shown in the background. Note that in this example the display indicates the user accessing the share as well as the share resource being accessed. Also note that if you detect a problem you can immediately disconnect the user prior to changing his ability to access the share via its configuration screen.

SUMMARY

Sharing provides an important productivity tool which enables employees to gain access to common programs and data files. However, sharing cannot be done in a vacuum and requires careful consideration of the users and user groups that require access to shared data. In addition, you must consider existing file permissions for all files in the share in conjunction with the type of access you assign to users. Finally, you need to consider the location of shared data and the content of any subdirectories under the folders you wish to share. By properly planning you can share data in a safe and reliable manner without having a security loophole that can harm your organization.

Chapter 44
Windows NT Access Security

Gilbert Held

Windows NT, scheduled to be marketed as Windows 2000 in its pending release, represents a popular workstation and server operating system designed from the ground up to support a variety of networking protocols. Similar to any modern operating system, Windows NT has a range of security features either imbedded directly in the operating system or supported by the ability of software developers to link their products to the operating system. Because the proverbial front door to an operating system is through its access manager, and the one built into Windows NT is used by many third party products, a detailed understanding of the operation of its User Manager for Domains and associated features of the operating system can considerably facilitate an understanding of the techniques that can be used to control access to NT-based computers. Thus, the focus of this article is upon obtaining an appreciation for the use of the Administrative Tool User Manager for Domains and the various operating system facilities that can be used to further control access to an NT-based computer as well as provide information concerning its use.

Because Windows NT is a graphical user interface (GUI) based operating system, any examination of its operation can be facilitated by the use of screen images that illustrate certain operations. Thus, in this chapter a series of screen images captured from an NT Server will be referenced to illustrate the operation of certain operating system features and facilities.

OVERVIEW

Windows NT security is based upon the creation of user accounts. User accounts can be placed into groups to facilitate assigning various levels of permission or denying them access to any computer resource.

Exhibit 44.1 illustrates the display of the initial User Manager for Domains screen for the server named MYWEB. The top portion of the screen display shows previously defined user accounts, while the lower

0-8493-9823-1/00/$0.00+$.50

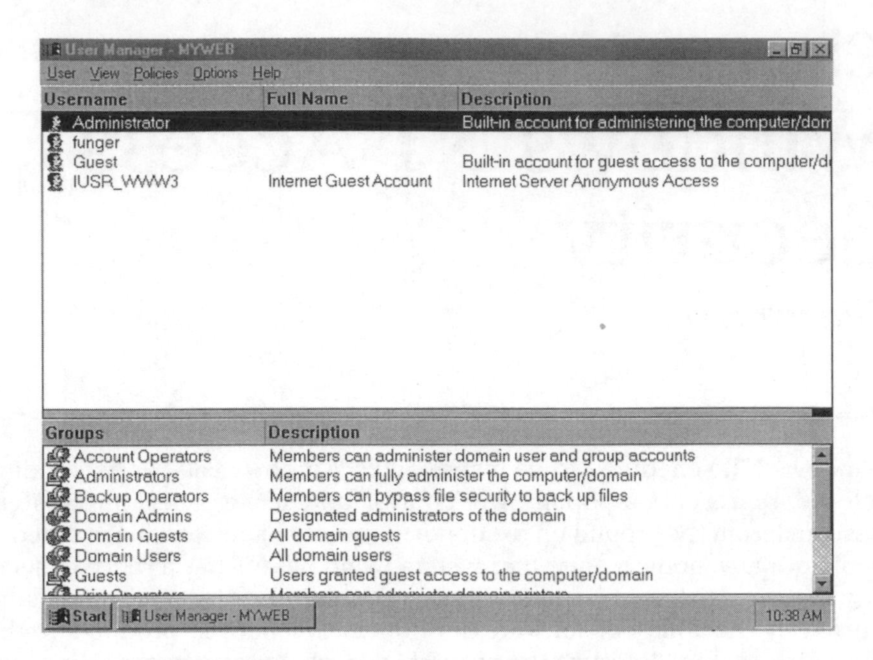

Exhibit 44.1 The user manager for domains screen.

portion of the display shows a portion of the available groups supported by the operating system. By default, both an Administrator and Guest account are established on both a Windows NT workstation and a Windows NT Server when a user installs the operating system. The Administrator account is by default named Administrator, which as we will soon note provides hackers with a valuable target to attack. Similarly, the Guest account represents another predefined account many hackers view as an opportunity to attempt to gain access to your NT-based computer.

The screen illustrated in Exhibit 44.1 is accessed from the Start Menu by first selecting the Program Menu. You would then select Administrative Tools, and then User Manager for Domains from the Administrative Tools menu. Similar to most Windows NT screens, the menu bar shown at the top of Exhibit 44.1 provides users with the ability to select many options associated with the display. One option that should be on your priority of items to consider is the Rename entry in the User entry in the menu bar. By selecting the Administrator account, shown highlighted in Exhibit 44.1, you can use the Rename option to rename the account. Because that account has full access to all computer facilities, it is a frequent target of hackers who commonly attempt to gain access to the account by a dictionary attack. That is, they typically obtain an electronic dictionary and run the contents of the dictionary against the Administrator account in an attempt to break into the account. If successful, they obtain access to everything on your computer, a potential occurrence that can cause an administrator to have nightly insomnia.

If you rename the Administrator account you should do so following common guidelines that are normally associated with the protection of username passwords. That is, you should consider including one or more numerics or valid special characters, such as the asterisk (*) as the prefix, suffix, or mid-separator for alphabetic letters to preclude a successful dictionary attack. For example, the use of "cat" and "dog" as passwords are relatively simple to overcome since they are in every English dictionary. Similarly, "catdog" and "dogcat" could be overcome with a little effort by a hacker that creates a program to use word combinations. However, the password "dog4cat*" would be much more difficult to break. Using the preceding as a guide you might wish to consider renaming the Administrator account so that it includes alphanumerics, making it much more difficult for a hacker to attack.

USER PROPERTIES

In addition to renaming the Administrator account it's important to consider the setting of user properties for each account established on an NT-based system. To do so you can either highlight a username and select the Properties entry from the User menu in the menu bar, or simply double-click on a username. Either action will result in the display of a "User Properties" dialog box, similar to the one illustrated in Exhibit 44.2 for the Guest account on the computer MYWEB.

In examining the User Properties screen displayed in Exhibit 44.2, you will note that in addition to being able to assign and verify the entry of a password the screen includes five rectangles, referred to as check boxes, one of which is currently shown shaded. The shaded checkbox is associated with the legend "Account Locked Out" and will be activated only if the account lockout policy for the account is activated and the number of bad logon attempts exceeded a predefined threshold. If this occurs, the checkbox can be selected to remove the lockout and enable the account user to gain access to the computer. Otherwise you cannot access the check box. Because it is shaded, many NT administrators that are new to the job incorrectly assume that the shading means the account cannot be accessed. To the contrary, you cannot lock an account with this check box and must instead select the disable option.

To illustrate how the selection of check box entries can cause confusion, this author purposely selected a combination that does not make sense. In the example illustrated in Exhibit 44.2, you will note that the password never expires yet the account is disabled. This illustrates that the operating system

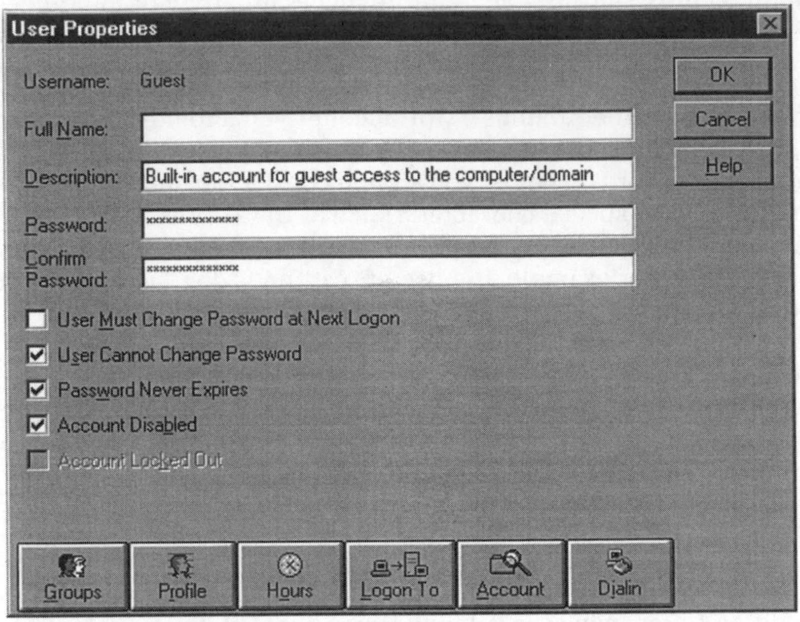

Exhibit 44.2 The user properties dialog box.

does not prompt you to carefully consider different check box settings and you should do so with a degree of caution. For example, the idea behind having a Guest account is to provide people who do not have an established account on a computer with the ability to access the computer, usually providing them with a minimum set of rights concerning the ability to perform certain operations. This relieves the Administrator of the requirement to set up individual user accounts. However, because the name Guest is well known to hackers, it's probably advisable to use a different name for establishing this type of account. Now that we have an appreciation for the potential use of the User Properties display, let's turn our attention to the buttons located at the bottom of the display which further control the ability of users to perform computer-related operations.

GROUP MEMBERSHIP

Pressing the button labeled "Groups" at the bottom of the User Properties display previously illustrated in Exhibit 44.2 results in the display of a dialog box labeled "Group Membership." An example of this dialog box for the Guest user account is illustrated in Exhibit 44.3.

In examining the Group Membership dialog box shown in Exhibit 44.3 note it is subdivided into two windows. The window on the left lists the

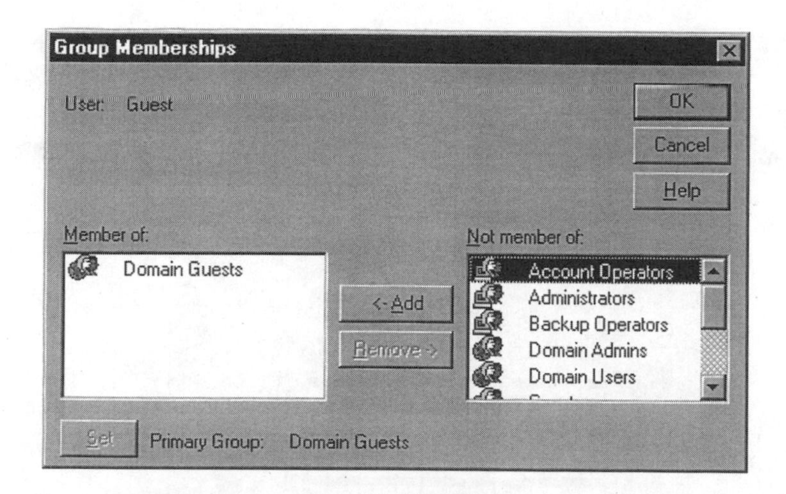

Exhibit 44.3 The group membership dialog box.

groups the user is assigned to, while the window on the right portion of the dialog box lists the remaining groups that are defined but of which the user is not a member. Due to the enhanced ability of certain group memberships to perform different computer related operations, it's important to carefully consider the assignment of group memberships. For example, it would more than likely be totally unwise to assign membership of the Guest account to the Administrators group.

USER ENVIRONMENTAL PROFILE

The second button located at the bottom of the User Properties dialog box, shown in Exhibit 44.2 is labeled "Profiles." Selecting it results in the display of a dialog box labeled "User Environment Profile," which provides the ability to specify a user profile path, login script location, and home directory. This option is primarily an administrative tool as it governs the execution of any predefined programs (login script name) and the location where a user will be placed on the computer upon logon (home directory). Because file permissions and user rights actually control the ability of users to access and modify data once they gain access to a computer, the User Environmental Profile can be viewed as more of a user convenience mechanism than having an association with computer security.

RESTRICTING HOURS OF UTILIZATION

Continuing our examination of the use of the buttons located at the bottom of the User Properties dialog box, let's select the button labeled "Hours."

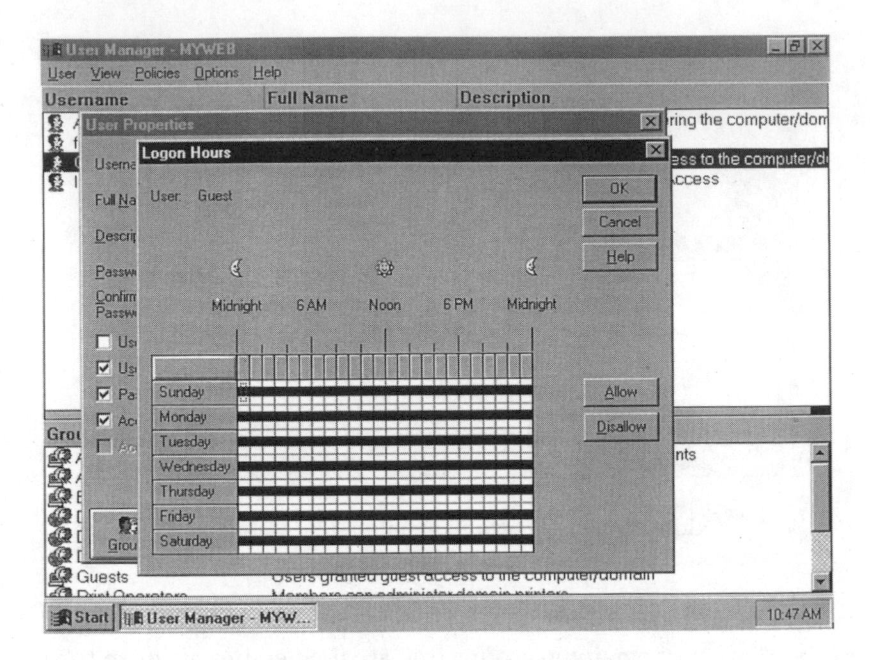

Exhibit 44.4 The logon hours dialog box.

Doing so results in the dialog box labeled "Logon Hours," which is illustrated in Exhibit 44.4.

The purpose of the Logon Hours dialog box is to provide administrators with the ability to control the periods when different user accounts have access to the computer. Although the primary use of the Logon Hours dialog box is to permit unfettered predefined periods for backups and system maintenance, this capability can also be used to enhance the level of security of the computer. For example, if you permit certain accounts you may also wish to consider restricting their hours of operation to periods of time when an administrator is onsite. Therefore, any problem or alerts generated by certain thresholds being exceeded can be investigated by appropriate personnel without undue delay.

CONTROLLING ACCESS WITHIN A DOMAIN

Within a Windows NT domain you can have a large number of workstations. Because many organizations locate most workstations in offices and other restricted areas, while other workstations are installed in lobbies and other public areas, the ability to specify the workstation or workstations from which users can log onto their account represents another security feature you may wish to consider. You can access this Windows NT feature through

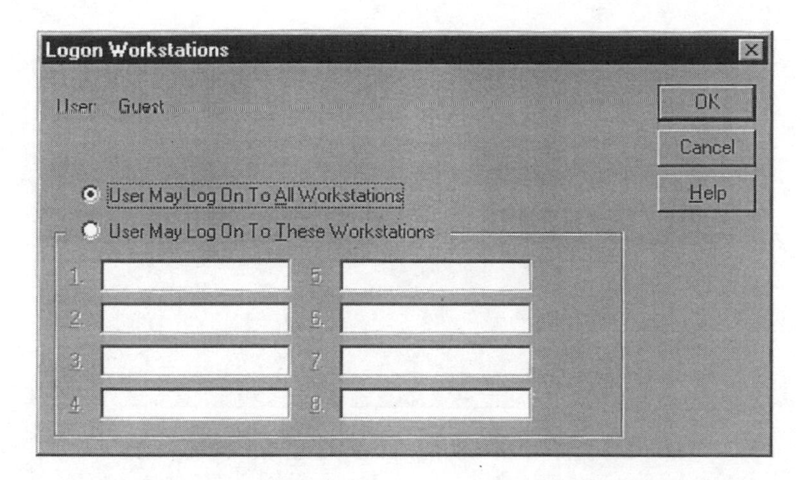

Exhibit 44.5 The logon workstations dialog box.

the Logon To button shown at the bottom of the User Properties dialog box illustrated in Exhibit 44.2.

Exhibit 44.5 illustrates the Logon Workstations dialog box for the user account Guest. By default the setting for this option is set, as shown, to "User May Log On To All Workstations." In actuality, the title of the two options shown in Exhibit 44.5 are in error and many times can cause a bit of confusion. The first option heading, "User May Log On To All Workstations," means that the user or user group selected can log on *from all* workstations in the domain. The second option heading, "User May Log On To These Workstations," means the user or user in a selected user group can log on only from up to eight workstation names you specify.

ACCOUNT CONTROL

There is an old adage that nothing lives forever, and user accounts are no exception. Because employees are periodically assigned to different projects that may require access to different data, it's common for administrators to set up a new account for the user with a different series of permissions. When you know the duration of an assignment or hire a contractor for a set period of time, you should consider assigning an expiration date to their accounts.

Exhibit 44.6 illustrates the Account Information dialog box for our well-used Guest account. The default setting for this option is a never ending account, which in many instances may result in the ability of a former employee to access the computer long after he or she departed the organization. Although a good administrator is responsible for maintaining user

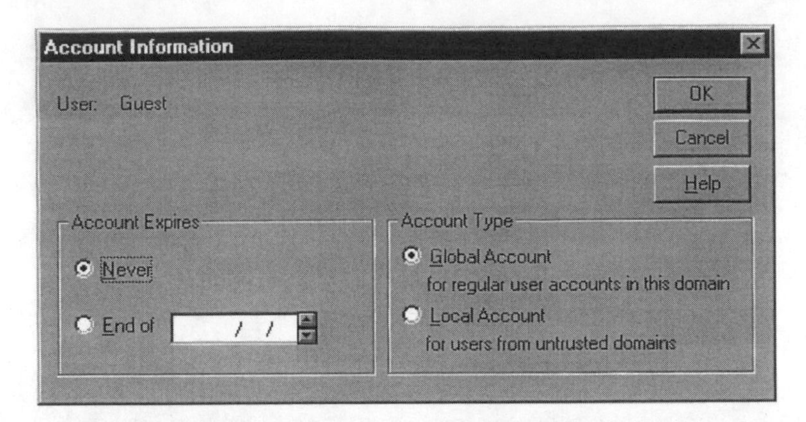

Exhibit 44.6 The account information dialog box.

accounts, if he or she is not informed by personnel of the departure of an employee it may be difficult to know when to remove an account. Thus, if you place an expiration date on an account you can override the failure of appropriate paperwork to reach the computer administrator.

In examining Exhibit 44.6 note that by selecting the button labeled "End of" you can define an expiration date which will render the account useless. This capability can be especially appealing to administrators that have to provide access for a large number of contractors, such as those being hired by many organizations to fix the Y2K problem.

DIAL-IN ACCESS CONTROL

If you intend to provide remote access via dial-in you would select the button labeled "Dialin" shown at the bottom of Exhibit 44.2. The selection of the Dialin button results of the dialog box labeled "Dialin Information," shown in Exhibit 44.7. Note that by default accounts do not have dial-in permission and the check box labeled "Grant dialin permission to user" must be selected to enable the user to access the computer via a modem connection. Also note that there are three call-back options you can consider. The first is no call-back. The second is set by the caller when he or she first dials into the computer, while the third option permits you to define the telephone number the computer will call after receiving a call from the particular user. By setting a call-back you can ensure dial-in connections are from a verified telephone number, enhancing the security associated with dial-in access.

Earlier in this chapter we briefly discussed the lockout option in the dialog box labeled "User Properties." In addition to the lockout option, there are

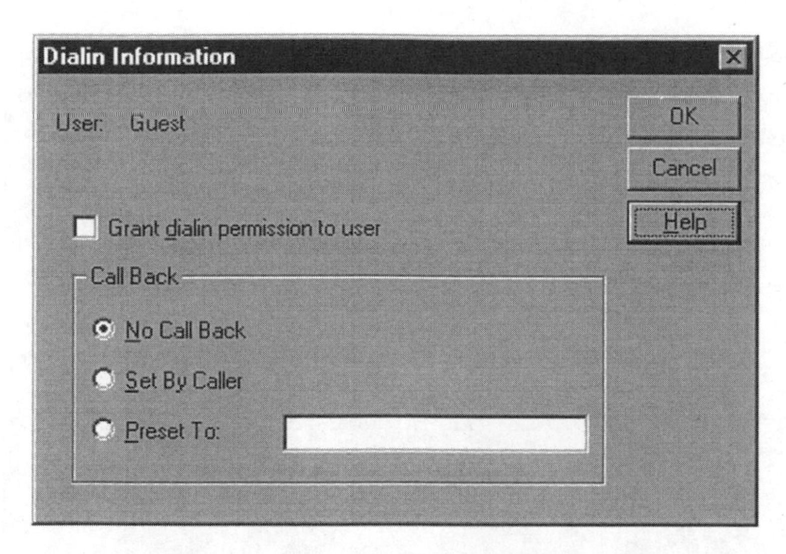

Exhibit 44.7 The dialin information dialog box.

definable user rights that can govern the user's ability to perform certain computer-related functions. In concluding this chapter, we turn our attention to these topics as well as examine the auditing capability of the operating system.

CONTROLLING THE ACCOUNT POLICY

Although the dialog box labeled "User Policies" provides a basic capability to assign and change passwords, to set password restrictions and account lockout requires the use of the dialog box labeled "Account Policy," which is illustrated in Exhibit 44.8. To access this dialog box you would select the Account option in the Policies menu, located in the main User Manager for Domains screen display. That display for the computer MYWEB is illustrated in Exhibit 44.1.

In examining the Account Policy dialog box shown in Exhibit 44.8 note that the top half of the display provides four password related options you can consider. You can set a predefined expiration date in terms of the number of days from the present date, define when the user can change his or her password, define the minimum password length and its uniqueness. Concerning the latter, many people, including this author, like to rotate through a series of passwords. By setting the fourth option to a high number you can force a person to stop using an easily recognizable series, such as the names of family members. However, in examining the four password-related

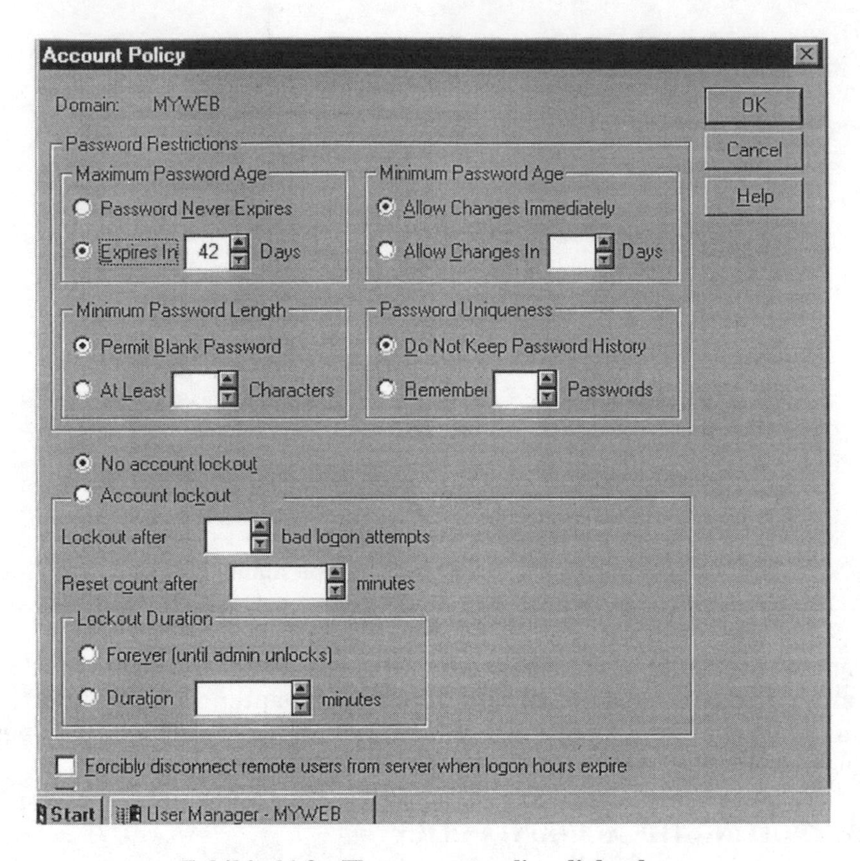

Exhibit 44.8 The account policy dialog box.

entries, note there is nothing to preclude a person from entering common language names or objects as passwords. To obtain a higher level of password security to assist your organization in warding off a dictionary attack, you should consider the acquisition of a third-party password security add-on. As an alternative, you can enable the account lockout facility shown in the lower portion of Exhibit 44.8.

Through the use of the account lockout feature you can disable access to an account after a predefined number of invalid logons occurred. Setting this option is advised as it disables the account in the event a hacker attempts to run the contents of an electronic dictionary against an account.

Two additional security related options, located in the Policies menu of the User Manager for Domains screen, control user rights and auditing. Thus, in concluding this article on Windows NT access security we would be remiss if we did not focus our attention on each.

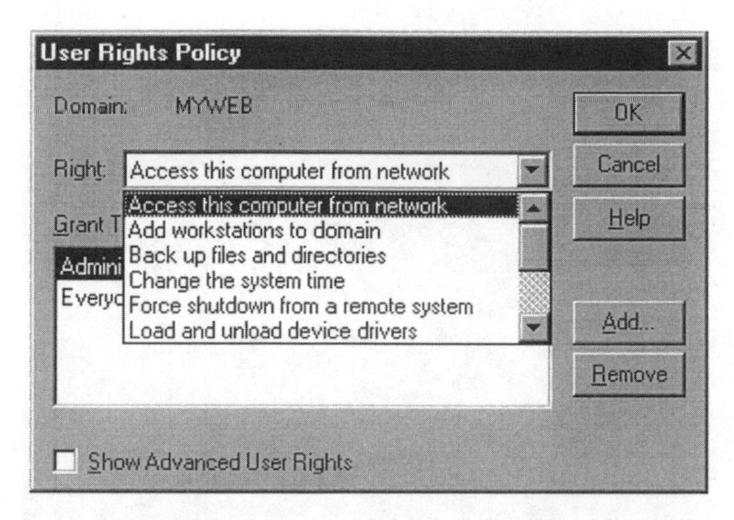

Exhibit 44.9 The user rights policy dialog box.

USER RIGHTS

Exhibit 44.9 illustrates the dialog box labeled "User Rights Policy." This box is displayed by selecting the User Rights option from the Policies menu on the User Manager for Domains screen. In Exhibit 44.8 the drop-down menu illustrates a portion of assignable rights. By default certain rights are assigned to each user group. Selecting a predefined right displays the user groups that have that right. Then you can use the buttons labeled "Add" and "Remove" to alter the user groups associated with a particular right. For example, currently only users that are members of user groups Administrators and Everyone have the right to access the computer from the network. By using the buttons labeled "Add" and "Remove" you could alter this default setting to satisfy a particular requirement of your organization.

AUDITING THE COMPUTER

Perhaps one of the most overlooked capabilities of the Windows NT operating system is its audit capability. By selecting one or more events to be audited you can ensure their occurrence will be logged in the built-in Event Viewer security log. In addition, third-party products generate an alarm upon the occurrence of a predefined Event Viewer log entry, allowing you to be notified of potential or actual security problems through the operating system audit policy capability.

Exhibit 44.10 illustrates the dialog box labeled "Audit Policy." This box is the result of selecting the Audit option in the Policies menu from the User Manager for Domains main screen, illustrated in Exhibit 44.1.

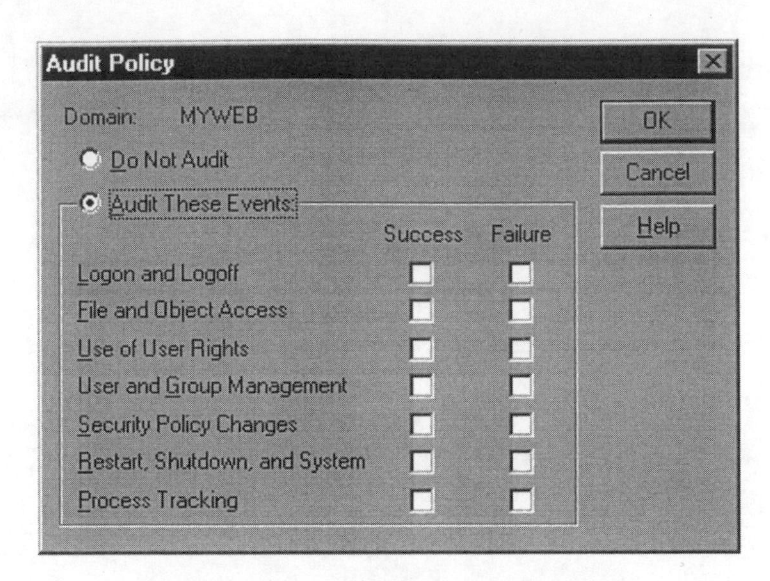

Exhibit 44.10 The audit policy dialog box.

By default auditing is disabled and represents a potential security problem: if someone attacks your computer the absence of auditing may result in the attack's going unnoticed until a serious problem occurs. As a minimum most organizations should audit logon and logoff failures, as doing so can inform you of an attack on an account. Similarly, the auditing of the failure of the correct use of user rights can indicate that an employee is either attempting to do something they should not be doing or is confused about what they are trying to accomplish. In either event the employee is probably a candidate for a visit from the network administrator.

SUMMARY

Windows NT represents an important operating system that many organizations are putting to use. Like most operating systems the ability of users to gain access to resources controlled by the operating system is governed by a variety of access controls. By understanding those controls, default settings you should consider changing, and effective use of different policy options you can secure access to your NT systems.

Chapter 45

Evaluating Anti-Virus Solutions Within Distributed Environments

Network Associates, Inc.

Since effective anti-virus products are available, it appears that organizations have found the establishment of protective measures to be confusing, disruptive, and possibly too expensive. Indeed, the decision process is complex, especially in distributed multi-user, multiplatform environments (which typically undergo almost constant evolution). More important, it would seem that no two product comparison reviews offer similar assumptions, results, or conclusions.

As a leading provider of anti-virus software for networked enterprises, Network Associates proposes that the following evaluation process will enable decision makers to achieve cost-effective enterprisewide protection and policies.

DISTRIBUTED SECURITY NEEDS

As in the evaluation and purchase of any other category of software, the difference in needs between individual PC users and networked users is great. In the case of virus infection, the very proliferation of networks and increasing mobile users is at once both a source of the problem and an obstacle to solving it.

Industry estimates for the introduction of new viruses/strains range from 100 to 300 per month, so anti-virus measures are important even for stand-alone PCs, which inevitably "contact" virus potential via modem use or sharing and borrowing data and programs from other users via disk or network.

0-8493-9823-1/00/$0.00+$.50

The creation of a new virus is a deliberate act (one often aided by multiple resources readily available to perpetrators). However, the introduction of viruses into corporate networks is most often innocent — employees take work to the home PC or exchange files via the Internet etc., bring it back to the network, and have no idea their files, programs, or computers were infected.

In a networked enterprise, anti-virus strategies must take into account further complications as follows:

- Multiple system entry points for viruses
- Multiple platform constraints on compatibility
- Resource requirements for adequate protection
- Need for integrated support efforts
- Need to minimize "epidemic" potential
- Multilevel reporting of network/system activity
- Logistics of installation across enterprise
- Ongoing implementation and maintenance costs
- Need for companywide policies, overlaying political considerations

The above factors may seem obvious, but it is important to review them, as their perceived difficulty has contributed to today's state of affairs — in which anti-virus products and strategies are seriously underutilized and network vulnerability increases daily. Currently, viruses already infect 40 percent of all networks.[1]

EVALUATION REASONING

The key evaluation criteria for choosing anti-virus solutions are detection, performance, features, and price. However, in choosing products to meet specific organizational needs, some typical assumptions about these factors need to be challenged in order to make the best decision.

Detection

Isn't the product that detects the greatest number of viruses always the best?

Not necessarily. The rate of virus proliferation is such that every vendor's current detection numbers are constantly changing. And although researchers have identified as many as 12,000 to 15,000 viruses, the same few hundred viruses continue to account for a vast majority of infections. Therefore, a sacrifice of other key criteria for the sake of a higher detection "count" may be a costly mistake.

Performance

Isn't the best product the one that works fastest and demands the least of system resources?

Maybe and maybe not. Performance statistics cited by vendors and reviewers, no matter how responsibly they were formulated, can only be taken at face value unless every aspect of the tested environment happened to be identical to that of the evaluating organization.

Therefore, experiments performed in one's own network environment are necessary, albeit time-consuming. Finally, virus detection rates, *measured at various performance configurations,* may add further meaning. What real difference will a few seconds (or milliseconds) make if damaging viruses were missed?

Feature Set

The product with the most features wins, right?

By now, most evaluators of technology products are aware of the features trap, but it is still important to analyze claimed features very closely. Some "features" are simply characteristics and may be beneficial to some environments while downright harmful to others. And it is not enough simply to decide whether or not a feature is needed by the enterprise. It should be attempted to assign *weighted importance* to each feature, and then weight test results accordingly (guidelines follow later in this chapter), before a given feature should be considered key to the decision.

Price

Wouldn't it be great if the lowest-cost product is also the best?

Definitely. But of course, there is far more to cost analysis than initial purchase price, especially with anti-virus software. Considerations that affect basic pricing are license types, platform support, terms of purchase, and maintenance structure. Further, cost of ownership considerations include: internal and external cost of user support and training, cost of future changes, and ultimately, the perhaps incalculable cost of *inadequate* protection. (American businesses lost $2.7 billion due to viruses in 1994 alone.)[2]

When the evaluation team understands the basic pitfalls of relying too heavily on any one of the above factors, it is ready to move on. The next step is to establish the framework of tests and analyses in which all leading products will compete.

TODAY'S VIRUS PROFILE

Prior to recommendations for testing methodology, this overview of today's active virus types may be helpful.

Boot vs. File Viruses

IBM-compatible and Macintosh viruses (which constitute the majority of network infections) fall into two basic categories: "Boot" and "File." *Boot* viruses activate upon system start-up and are the most common type. They infect a system's floppy or hard disk and then spread (by replicating and attaching) to any logical disks available. *File* viruses are actually programs that must be executed in order to become active, and include executable files such as .com, .exe, and .dll. Once executed, file viruses replicate and attach to other executable files.

Four Special Challenges

Particularly troublesome virus classes today are known as *Stealth* (active and passive), *Encrypted, Polymorphic,* and *Macro.*

Stealth viruses are difficult to detect because, as their name implies, they actually disguise their actions. *Passive stealth* viruses can increase a file's size, yet present the *appearance* of the original file size, thus evading Integrity Checking — one of the most fundamental detection tactics. *Active stealth* viruses may be written so that they actually attack installed anti-virus software, rendering the product's detection tools useless.

The challenge of *encrypted* viruses is not primarily one of detection, *per se*. The encryption engine of this type of virus masks its viral code — making *identification,* as opposed to detection, more difficult. Therefore, detection and prevention of recurring infections is harder even with frequent anti-virus software updates.

The *polymorphic* category has grown considerably, presenting a particular detection challenge. Each polymorphic virus has a built-in mutation engine. This engine creates random changes to the virus's signature on given replications.

Recently, *macro* viruses have gained much notoriety and growth. A macro virus is a set of macro commands, specific to an application's macro language, which automatically executes in an unsolicited manner and spreads to that application's documents. Since macro virus creation and modification is easier than other viruses and since documents are more widely shared than applications, they pose the most significant new threat.

COUNTERACTION TECHNOLOGY OPTIONS

As the threat and number of viruses grew along with the conditions so conducive to their spread, so did the sophistication of the efforts to combat them. There are now five major virus detection methods:

- *Integrity Checking* — Based on determining whether virus-attached code modified a program's file characteristics.

- *Downside Examples* — Checksum database management, the possibility of registering infected files, inability to detect passive and active stealth viruses, no virus identification.
- *Interrupt Monitoring* — Looks for virus activity that may be flagged by certain sequences of program system calls, after virus execution.
- *Downside Examples* — May flag valid system calls, limited success facing the gamut of virus types and legal function calls, can be obtrusive.
- *Memory Detection* — Depends on recognition of a known virus' location and code while in memory.
- *Downside Examples* — Can impose impractical resource requirements and can interfere with legal operations.
- *Signature Scanning* — Recognizes a virus' unique "signature," a pre-identified set of hexadecimal code.
- *Downside Examples* — Totally dependent on currency of signature files (from vendor) and scanning engine refinements, also may make false-positive detection in valid file.
- *Heuristics/Rules-based Scanning* — Recognizes virus-like characteristics within programs for identification.
- *Downside Examples* — Can be obtrusive and can cause false alarms, dependent on currency of rules set.

Usually, all five techniques can be used in real-time or on-demand, and on both network servers and workstations. The number of virus types and their many invasion and disguise methods, combined with the five counter-action methods' various strengths and weaknesses, has resulted in the fact that today all effective products leverage a combination of methods to combat viruses. This field is by no means limited to known methods, however. Constant involvement in virus counteraction steadily increases the knowledge base of anti-virus software vendors, enabling refinement of these methods and the development of entirely new ones for the future.

TESTING OBSTACLES

Detection Testing

The Test Parameter seems simple: send a library of files and viruses through the client/server host and calculate the detection percentage of each competing product.

Obstacles to this process, and to a valid analysis of test results, are numerous. Evaluators should attempt to answer all of these questions before proceeding with detection testing:

- Where will we get the viruses? Should test files consist of only one library of viruses or multiple sources?
- Do the sample viruses replicate?
- What total number of viruses will we need for a valid test?

- What mix of virus types will reflect real-world best in testing (i.e., macro, boot vs. file viruses, polymorphics, encrypted, passive or active stealth, etc.)?

Performance Testing

For each product being evaluated, a library of files and viruses (the same used in Detection Testing) is sent through the client/server host, while both speed of operation and host degradation are calculated.

Obstacles to this test include:

- Determining the number, sizes, and types of control files to be scanned for viruses
- Determining all control conditions of the testing platform, i.e., what is "normal"? (Evaluators should average multiple test runs for both workstations and servers to produce a standard set of values.)
- How will we assure accurate timing metrics?
- The overriding concern in performance testing — that test conditions be as real-world as possible

Features Testing

A list of all product feature claims must be compiled. Decide which features the organization needs, tabulate the scores, and compare competing products.

Obstacles to features comparison are

- Certain features will appear to be common to all products, but how comparable are their actual implementations?
- Not all implementations will be suited to all needs — what is really required to take advantage of a given feature, and is it really worth it?
- If a "standard" feature set (such as may be derived from Novell and NCSA certification) is the basis for comparison, in what ways does each product go beyond that standard set?
- Verifying the depth of analyses behind each claim (does the vendor provide empirical evidence?)

All features, common and unique, must be analyzed as to their true importance within a defined description of the evaluating organization's needs and desires. Inevitably, when evaluators adhere to a methodology of assigning a numerical weight of importance to each feature, certain features are revealed as far more critical than others, and those critical features may be very different from one company to another.

Additional Obstacles

The three key factors that can be tested (detection, performance, and features) share several *common* obstacles as well. In preparing to test each factor, evaluators should also assess:

- To what extent statistical inferences meet actual objectives
- Whether the latest releases of all products have been obtained for testing (remember that monthly signature file and regular engine updates are a *de facto* standard in the anti-virus software industry)
- To what extent the products can be *normalized* (for "apples to apples" comparison)
- In which tests should products also be *maximized* (letting the "oranges" stand out where they may, and finding a reliable way to quantify their value)

TESTING PREPARATION

Virus Sample Configuration

Evaluators should begin researching sources of testable viruses early in the process. Sources are critical to achieving reliable tests, and it may prove time consuming to obtain adequate numbers of the virus types determined necessary for testing.

Here are some considerations regarding virus sample size and type:

- How many viruses constitute a large enough sample? Is 1,000 out of a potential 7,000 (14 percent) enough? (Obtain samples in your environment)
- One standard deviation (66 percent of total universe) is potentially 5,600 viruses — possibly overkill and almost certainly too costly to fully test across competing products for both detection and performance analyses. (Again, most virus infections involve the same 200 viruses over and over.)
- Boot, file, and macro viruses should probably all be tested, and boot virus tests can demand significant time and resources. Because of that demand, the evaluation weight appropriate to such viruses should also be determined during sample design.
- Use pure viruses, which are those obtained from initial samples that have successfully replicated twice (avoid droppers). Example of pure virus model:

  virus 1, infected file 1, infected file 2 → validation

- Polymorphics, unto themselves, require a fair-size sample by their very nature. Example of preparation for polymorphic testing:

  Take polymorphic virus, replicate 100 times;
  take an infected file of that sample and replicate 100;
  take an infected file of that sample and replicate 100...
  Repeat until sample of at least 2,000 is achieved.

- From polymorphics, sample library should capture each of a given major type such as SMEG, MTE, etc.
- Tests must be set up so that file viruses can infect different file extension types, as appropriate (.com, .exe, .DLL...).
- Due to the fact that macros can be easily modified, it is suggested that one include only macro viruses deemed most likely to be found in your environment.

It should be clear from the above that collection of the virus test library alone is a fairly extensive project. Various on-line services and bulletin boards may be good resources, as are some independent testing bodies. Make sure to research how the sample library of viruses was obtained; is it possibly too vendor oriented?

Anti-Virus Product Configuration

Establish a cut-off date by which vendors must provide you with the latest releases appropriate to your platform(s), along with all appropriate documentation. Also make sure vendors provide virus signature files that are equally up to date. Take time for a general review of all competing products prior to testing (remember to build this time into the initial project schedule).

Testing Platform Configuration

In configuring a network or stand-alone platform for testing, and during actual detection and performance tests, it is critical to maintain a controlled environment. This environment should resemble actual workplace conditions as much as possible.

Today's anti-virus products must safeguard network servers, workstations, stored data, and remote users, so it is important that all tests must be run consistently across the same devices. Those devices must also be installed with the same "typical" software in all categories.

TESTING PRACTICES — DETECTION

Normalized vs. Maximized Protection

Normalization of the products being tested will tend to provide more dependable results, in terms of extending their meaning for real-world conditions. All products should be capable of performing as closely as possible on identical operations. Prior to testing, this required level of protection must be determined.

Maximization testing allows each product to perform to its full capability, with no set baseline or limit in the number of viruses that can be detected or the time allowed for detection. Evaluators may then calculate weighted averages for each product's maximized protection tests in order to compare them to each other and to normalized results.

TESTING PRACTICES — PERFORMANCE

Timed Tests

In timed tests, the biggest variable will be the server (and workstation) disk cache. Therefore, the first test should record performance without any anti-virus protection. Again, adequate timing mechanisms should already have been established. A timer program or even a stopwatch may be used. Below is a timer program example:

timer, map, timer off
timer, copy c:\test f:\sys\test, timer off

For statistical reliability, each test needs to be run at least five times; then the lowest and highest values are removed and the remainder averaged. Also, products being tested must share a common configuration for accurate results.

Throughout the testing process, one might double check that file and virus sample libraries are identical in each test. Evaluators must also down the server or workstation after tests of each product to clean the cache in preparation for the next product's test.

Resource Utilization Tests

Resource utilization testing actually began with the first Timed Test — in which the system's performance was measured without any virus protection installed. Now, with anti-virus products installed in turn, the effect of each on overall performance may be measured.

A fundamental test is how long it takes to launch programs from the file server. Windows (consisting of over 150 files) and Syscon are good samples to test.

Testing larger files will allow the server copy overhead delays, giving the virus scanner more time to meet outstanding file scans. Be sure to test files both read from and written to the file server.

Again, every aspect of the test environment must be controlled. To prevent interference, other users cannot be allowed to log on to the test environment. Product configurations must be similar, tested on identical computers, without cache or other bias.

Performance, in terms of speed, was measured with a timer program or stopwatch. Resource utilization of each anti-virus product is best judged using Novell's monitor. One can conduct resource utilization tests on workstations by implementing similar tests.

TESTING PRACTICES — FEATURE SET

Tabulation

For each product under consideration, list all the features claimed within the product package and any accompanying documentation. To make features comparison easier, use a common format and standard terminology on all product feature lists.

Feature Validation/Implementation

Validate all the apparently desirable feature claims by following documented procedures. If a feature does not seem to work as described, check the documentation again. Throughout the validation process, take detailed notes on each feature's implementation, particularly at any points which seem poorly documented or cumbersome.

Critical and highly desirable features, of course, should receive the closest analysis of their implementation and performance as promised.

Feature Rank and Weight

Each feature should be weighed as to its ability to meet requirements, and then ranked in relation to comparable features from competing products. After eliminating any products which have important implementation barriers (discovered during feature validation), determine the weighted average among remaining products.

Using Exhibit 45.1 as a guide, summarize each product's performance/delivery of desired features for both client and server platforms.

COST OF OWNERSHIP — PRICING

Throughout the software industry, product pricing is affected by four main issues that may differ greatly from one enterprise to another: type of license, platform support, maintenance agreement, and terms of purchase.

With anti-virus software for networks, one aspect is perhaps simpler than other applications — there should be little fact finding required to answer the "who needs and really uses this software?" question. The entire network requires protection, and every workstation, whether on-site or remote, must be viewed as a potential entry point and considered vulnerable to viruses that might originate from other entry points.

License Types

Several different licensing schemes are commonly available, and most vendors offer some degree of choice between concurrent use, per-node, per-user, home users, etc. Purchasers must assess the enterprise's overall network

Exhibit 45.1 Feature set weighted average.

Server Platform Support Price (5 High...1 Low)	**Client Platform Support Price (5 High...1 Low)**
Detection	**Detection**
General	General
Polymorphic	Polymorphic
Macro	Macro
Scanning	**Scanning**
Performance	System Resources
Options	Requation
Capabilities	Performance
	Capabilities
Administration	**Ease of Use**
Manage/Security	Installation
Console/Interface	User Interface
Flexibility/Notification	Flexibility/Update Distr.
Client Support	Security Control
	Notification/Reporting
	Virus Encyclopedia
Client/Server Integration	**Support**
Auto Install/Update	Updates/Accessibility
Notify Log	Documentation/Package
Isolate/Enforce	Technical Support
Support	
Documentation/Package	
Technical Support	

configuration, including authorized home users, to decide on the type of license that will provide adequate coverage at the lowest cost. Organizations should also consider their own growth potential and investigate vendor procedures for adding users and nodes as needed, with a minimum of disruption to network operations.

Platform Support

Another pricing factor that will vary from product to product is support of multiple platforms, which is often necessary in today's network environments. As prices might vary depending on the platforms required, buyers must confirm pricing specifics for all current platforms (preferably in native mode) as well as the cost associated with users migrating from one system to another. Availability and pricing variations, if any, for platforms the company may consider adding should also be checked.

Maintenance

Maintenance agreements usually fall into one of two basic plans:

- Base price plus stated cost per user per year (how much covers current users and how much for additional users in the future)
- Percentage of overall contract per year (platform by platform)

If a vendor can offer either type of maintenance plan, evaluators must compare any stated differences in degree of support, response time, etc., and then calculate actual cost estimates based on known platforms and user base over time. In evaluating maintenance plans, buyers should also consider each arrangement's potential impact on internal administrative resources.

Terms of Purchase

Because of its complexity of implementation, enterprisewide network software, including anti-virus software, is most often purchased on a contract basis, intertwined with support and maintenance agreements.

A typical arrangement is an annual contract including maintenance, renewable each year. Longer-term contract periods, such as two to four years, should provide added value to the buying organization in terms of lower buy-in price, added support, simplified administration, etc.

TOTAL COST OF OWNERSHIP — SUPPORT

For a large distributed network, the initial price of new software is only the beginning of a careful cost-of-ownership exploration, especially with anti-virus solutions. Investment in an anti-virus system protects all other software investments and invaluable corporate information, yet it also necessarily adds complexity to the overall network operation. Without adequate support, this additional "layer" can disrupt network operations, resulting in hidden added costs.

Anti-virus software, least of any category, cannot just be installed and left alone, so evaluators must thoroughly analyze how competing products can satisfy the following support considerations.

Signature, Engine, and Product Updates

To remain effective, even the best package must include frequent signature file and scanning engine updates in addition to product operational updates, because of the constant flow of new viruses/strains.

While most well-known anti-virus software vendors will certainly produce updated virus signature files, their methods, quality assurance, and timeliness of distributing the files to customers vary. Physical distribution, in the form of disks accompanied by documentation, is one method. Even if physical distribution is done via express courier, however, electronic distribution via file download is the faster (and if anything, lower-cost) method.

Training, Education, and Consulting

In terms of logistics and cost (included or not in contract, on-site vs. vendor site, depth of staff included, etc.) training and education services will vary from vendor to vendor. Key network management and support staff should probably obtain direct training from the vendor.

Further training of the next personnel levels might be accomplished in-house, but organizations should weigh the cost and time required of their own internal resources against the cost of outside training (which may prove a relative bargain). The availability of the vendor to perform flexible consulting services may also improve implementation time and reduce costs.

General and Emergency Services

For ongoing maintenance services, proposals should detail what the vendor will provide as part of the contract, how often, how they may be accessed/ordered, and the availability of any special features.

Emergency services are a key aspect of anti-virus software maintenance. Customers need to know in advance the vendor's emergency-basis policies (and additional costs if any) for analysis of suspect files, cure returns, and cure guarantees. Often, if it sounds too good to be true, it is. What are the guarantees backed with?

VENDOR EVALUATION

Surrounding all the tests, analyses, and qualitative considerations about product features and performance are key questions about the vendor companies being evaluated. Decision-makers should thoroughly investigate the following aspects of each competing company.

Infrastructure

Assuming that vendors under strong consideration offer desirable technology, buyers should not neglect to research the *business* side of each vendor. A company's resources and the way it uses them are important issues for customers pondering a fairly long-term contract for critical anti-virus software and services. Simple and practical questions to be answered are how many, and of what experience level are the support, development, quality assurance, and virus research people employed by the vendor? And what is the size and scope of its user base?

The vendor's economic health should also be assessed, as a struggling company will have difficulty servicing many large clients and supporting monthly ongoing product development.

Technologies

The company's product development track record, in the context of continued organizational strength, is an indicator of how well it will support future development. Have its new product releases been significant improvements or departures from standard technology of the time? Have updated releases been truly enhanced or mere "fixes"? Have those releases arrived on schedule, or is the company usually running behind?

Beyond its handling of individual products, the vendor's grasp of technological trends is important, too. Does it recognize issues about the future that concern its customers? In proposing anti-virus solutions, do its representatives ask knowledgeable questions that reflect this ability?

Relationships

Technological strength alone or excellent service without robust technology cannot provide the best anti-virus solution for a complex distributed environment. Gathering this information about a potential vendor's infrastructure *and* technological strength will give a clearer picture of its ability to fully support the relationship.

SUMMARIZED CRITERIA

In today's large network environments, the task of choosing an enterprise-wide anti-virus solution involves a multidisciplinary evaluation of many factors. Products and vendors that are strong in certain areas may have weak links in others that are of particular importance to a given organization. Ultimately, the best solution is the one that provides the best overall balance of these factors in relation to your needs:

- *Detection* — consistently high detection rates, tested repeatedly over time on appropriate virus samples
- *Performance* — ability to provide security with minimal impact on network operations
- *Administration* — practicality of central and distributed management, with convenient support
- *Reporting* — mechanisms for communication and measurement of virus incidents
- *Reliability* — stability of the product, vendor, and ongoing support
- *Total Cost of Ownership* — as measured in real dollars and impact on other resources, initial and long-term

- *Vendor Considerations* — ability to sustain key relationship based on technological expertise and healthy infrastructure

REAL-WORLD EXPERIENCES

Research of products and vendors combined with analysis of data from customized tests provides a strong foundation for anti-virus software evaluation. However, decision makers should not neglect the highly valuable resource represented by peers from other, similar organizations.

Ask about their experiences in obtaining viruses to test, equipment utilized in tests, whether they normalized tests, and their method of weighting results. The fight against viruses is a relatively popular area in which all involved can benefit from shared experiences.

The anti-virus software industry can provide more, too, by formulating guidelines and supporting more in-depth education. Vendors should offer their own tests to supplement user-customized tests, and assist with virus libraries for sample configuration.

Anti-virus software certification by the National Computer Security Association (NCSA) serves both users and vendors by providing measurable industry standards. It is an ongoing process of independent "blind" testing of anti-virus product detection.

Over 30 participants from several different countries, including major software vendors and independent researchers, contribute to NCSA's effort to catalog all viruses, as they become known. Currently, NCSA is in the process of recertifying products against new, tougher standards for virus detection.[2]

CONCLUSION

To choose the anti-virus solution that best suits the needs of an organization, it is vital that the evaluation team have a thorough understanding of the following issues:

- The threat of virus exposure today
- Detection technologies currently available
- Trends in viruses and anti-virus countermeasures
- Functional requirements, present and future
- Environment constraints, present and future
- Pros and cons of various testing methodologies
- Evaluation methodologies and resources

It is hoped that this paper has enhanced potential evaluators' knowledge of these issues, in addition to providing a sound methodology for discovering solutions that satisfy their organizations' needs well into the future. We are prepared to provide further assistance to your evaluation effort.

References

1. Computer Security Association, *Virus Impact Study*, 1994.
2. National Computer Security Association, *NCSA Anti-Virus Certification Scheme*, November 1997 Edition, NCSA Worldwide Web Site.

Network Associates, Inc. currently has 60 percent worldwide unit market share for anti-virus software, according to IDC Research. Millions of individual and corporate VirusScan customers and Network Associates agents in over 65 countries enable us to analyze more viruses first and meet demands to resolve incidents quickly. This tremendous installed user base and exhaustive testing, including monthly NCSA and VSUM virus library certifications, provide Network Associates with hundreds of suspect virus files per month.

Consistent virus detection and removal require a sizeable staff of skilled and experienced virus researchers, engineers, and support teams. We are constantly expanding our capabilities — recruiting expert virus researchers, streamlining support mechanisms, and designing groundbreaking virus extrapolation tools. This customer-driven effort delivers significant monthly updates and the best reviews in the industry.

Network Associates is dedicated to strengthening the industry's number-one anti-virus solution and our customers' security. We are confident that you will find Network Associates to be the premier vendor to meet your virus and security needs.

Note: All company and product names, registered and unregistered, are the property of their respective owners.

Section VIII
Server Add-Ons and Troubleshooting Methodologies

This concluding section discusses two key server-related topics. The first topic recognizes the fact that many people add a variety of hardware and software products to their servers to obtain an additional level of functionality and processing capability. Because such products are added to servers, they are commonly referred to as server add-ons.

While we normally prefer to have an optimistic view of client/server computing, reality tells us that upon occasion problems will occur. When server-related problems occur, knowledge about appropriate troubleshooting techniques and methodologies can considerably facilitate returning equipment to an operational state. Thus, we will conclude this final section with two articles focused upon troubleshooting.

In Chapter 46, the first chapter in this section, Paul Davis turns our attention to the cost of software which, if unchecked, can literally place an organization in the proverbial poor house. In his article, titled "Software Management: The Practical Solution to the Cost-of-Ownership Crisis," Davis describes the use of software management tools which enable IT managers to address the costs of managing a networked enterprise. Davis describes and discusses software licenses, training and technical support, and network administration. He also provides an example of how one large company used software management to reduce their software license expenditures by nearly 80 percent.

Chapter 47 recognizes the fact that inventory control represents another method to control cost. In his chapter "Inventory — Cornerstone Technology for Managing the Network," Roy Rezac provides us with an appreciation of the rationale to manage network devices. He first illustrates the considerable cost difference between managed and unmanaged computers. Once this is accomplished, Rezac introduces us to inventory control operations and includes a comprehensive table that can be used for evaluating different inventory programs. This table lists over 120 items, covering a wide range of hardware and software add-on products.

Because software distribution can represent a considerable effort for large organizations, a number of software distribution tools were developed to facilitate this activity. In Chapter 48, "Software Distribution," Steven Marks covers this timely and relevant topic. Marks examines the components of software distribution systems and illustrates how they can be used to facilitate the installation of software across a network.

Chapter 49 represents the first of several chapters that examine hardware-specific add-ons. In that chapter Jeff Leventhal turns our attention to one of the most critical components of a server, its storage subsystem. In his chapter entitled "Fault Tolerance Protection and RAID Technology for Networks: A Primer," Leventhal examines several vendor products as well as describes in detail the operation, advantages, and disadvantages associated with different popular RAID levels. This chapter is followed by a chapter covering another critical server-related activity — the backup and restoration of the contents of a server's hard drives. In Chapter 50, "Network Data and Storage Management Techniques," Laurence D. Roger describes different media rotation schemes as well as methods you can consider for monitoring and managing storage. The third hardware-related chapter in this section covers a vital product whose operation is critical in many locations that experience summer brownouts and blackouts. That product is the uninterruptable power supply (UPS), which is discussed in Gil Held's chapter, "Acquiring and Using a UPS System," Chapter 51. Held reviews common electrical power problems and discusses how a UPS operates. Once this is accomplished, Held turns his attention to describing different UPS parameters to facilitate its acquisition and includes a worksheet you can use if you need to acquire this server add-on hardware device.

As previously mentioned at the beginning of this section introduction, the second topic to be covered concerns troubleshooting methodologies and techniques. In concluding this section you will find two chapters, authored by Kristin Marks, which provide us with information on this topic. In Chapter 52, "Troubleshooting: Resolution Methodologies," Marks discusses several methods, such as documentation and a spare parts inventory that can facilitate returning equipment to an operational state. In Chapter 53, "Troubleshooting: The Server Environment," Marks reviews environmental issues that can adversely affect server performance and provides a insight into physical security issues we should consider to obtain a reliable level of operations.

Chapter 46 Software Management: The Practical Solution to the Cost-of-Ownership Crisis

Paul Davis

THE BALLOONING COSTS OF MANAGING A NETWORKED ENTERPRISE

Since their introduction in the late 1970s, personal computers have reshaped the way employees spend their time and redefined the role of the corporate information technology (IT) department.

Each year, as the power of PCs has increased, their purchase price has gone down. Corporations have been willing to spend on desktop hardware and software with the increasingly common conviction that the returns in user efficiency more than justify the investment.

But recent studies of the cost of ownership of desktop personal computers have brought to light a sobering truth: that the *real* costs of owning PCs are much greater than previously suspected. The Gartner Group estimates that each year "a networked PC costs about $13,200 per node for hardware, software support and administrative services, and end-user operations."[1]

These estimates have fueled concern among the people responsible for corporate networks, from system administrators to senior executives. They want to know whether these previously "hidden" costs are justified and how to reduce them. Many organizations are taking a closer look at their software expenditures. This is because software drives PC ownership costs — not only as an expense itself, but also as the main reason for hardware purchases and upgrades, support calls, etc. Until recently, however,

0-8493-9823-1/00/$0.00+$.50

the lack of a clear understanding of cost of ownership was exacerbated by the lack of effective tools to measure and evaluate software usage, so that informed decisions can be made.

This chapter will discuss how the emergence of software management tools is enabling IT managers to gain awareness of real software usage and to dramatically reverse the trend of rising costs of ownership. It will also suggest some ways to get started on implementing an effective software management program.

WHAT IS SOFTWARE MANAGEMENT?

Software management is a set of practices that optimize an organization's software purchasing, upgrading, standardization, distribution, inventory, and usage. Good software management means a company is using its software assets in a manner that optimally supports its business activities, without spending any more than necessary. It helps companies save money, comply with license agreements, and gain new levels of network control and efficiency.

While system managers have been performing more labor-intensive methods of software management for years, today the most effective way to achieve optimal performance is by using a new class of software applications that fall into the categories shown in Exhibit 46.1.

REDUCING OWNERSHIP COSTS, ONE-BY-ONE

With software management tools, IT managers can systematically address the key costs of managing a networked enterprise.

Cost #1: Software and Hardware Purchases and Upgrades

Manage Software Licenses for Optimal Efficiency. The advent of networks made it possible for hundreds of users to run applications stored on a server simultaneously. But single-user license agreements, which require that each piece of software run on only one PC, made the practice of sharing applications on a network illegal. Concurrent licensing was introduced to address this dilemma and soon became the industry licensing norm.

Concurrent licensing allows users to share applications over a network, and requires the purchase of licenses equal to the greatest number of people using a particular application at any given time. But since there has been no accurate way to determine usage patterns for any given piece of software, the simplest way to ensure license compliance has been to purchase a license for every potential user (see Exhibit 46.2).

Software management helps businesses realize the potential cost savings of concurrent licensing. With metering software, system managers can

Exhibit 46.1 The key activities of effective software management.

Activities	Advantages	The Old Way	The New Way
Metering — the measurement of software usage and the cornerstone of software management. Metering reveals which applications are in use by which users and how often.	• Reduce software costs by more effectively allocating resources • Stay in compliance with license agreements • Plan for upgrades according to true user needs • Determine training requirements	Rely on user-supplied information about usage. In addition to the work involved, the problem with this method is that users may inaccurately report their usage.	Metering software provides a fast and ongoing profile of software usage across the enterprise. Better metering packages also reduce software expenses dramatically by optimizing the use of licenses.
Inventory — the assessment of how much and which software is installed at individual workstations.	• Reveal what software is installed on which machines • Ensure license compliance • Detect new installations of unapproved applications	Visit each workstation frequently to see which software components are in use. This method is labor intensive and impractical, as users frequently download new software from the Internet, from home, and from other remote sites.	Inventory software automatically gathers information over the network in the background, without interrupting users' work flow. (Inventory packages vary greatly in their ability to "see" all desktop applications.)
Distribution — the installation and upgrading of software across networks.	• Efficiently ensure that the right versions of software applications and information resources are available to designated users	Visit each desktop to physically install the new software.	Distribution software automatically copies and installs software over the network in the background from a server — and bases installation on what users really need.
Reporting — the collection, integration, and presentation of data on software usage, inventory, and distribution.	• Build a foundation for decision making and planning • Share software management information among collaborative groups • Compile data across multi-site enterprises for administrative efficiency and group purchasing discounts	Manually input data from various sources, integrate it, and organize it.	Good software management packages can generate a variety of reports on the use of information technology assets. By integrating reporting and control, they can automatically optimize many aspects of IT asset management, and can implement policies without manual intervention.

Exhibit 46.2 A show of hands — the cost of informal polling.

Consider the case of a large U.S. public utility that wanted to evaluate which users it should upgrade to a new version of a popular database program.

The system manager sent an e-mail to all employees to find out which ones considered themselves "frequent" users of the current version of the software.

Eighty-three percent said they frequently use the application. To be safe, the decision was made to purchase the software upgrade for 100 percent of the utility's employees. They also planned to upgrade the RAM on every employee's PC, as the new version of the application required 8 megabytes more than the old.

Later, the organization was given the opportunity to perform an audit of actual software usage. They discovered that only 13 percent of employees were actually frequent users of the software. If they had known this before they made the purchase, they could have saved 87 percent of the upgrade costs.

monitor usage over time and purchase the least number of licenses necessary to satisfy user demand. In many cases, companies can reduce the number of licenses they purchase by 40 to 80 percent or more through the use of metering software.

Another benefit of software management tools is their ability to monitor different types of licenses. A system manager may decide, for instance, that he's going to purchase dedicated licenses of Microsoft® Word for the most active users of the application and concurrent licenses for the remainder of the user pool. The most effective software management tools allow him to clearly see which users should be assigned which license type. Establishing and maintaining an efficient license mix ensures that every user has access to the software she needs, when she needs it, and that software costs are kept to a minimum.

Plan Ahead for Software Purchases and Upgrades. The standard operating mode for today's system manager is reactive. The complexity of making a great variety of software and hardware elements work together — combined with ever-increasing user demands for assistance — leaves little time for active planning. At the same time, those who plan for software and hardware needs do end up saving time and money. By knowing usage patterns, a system manager can forecast needs, budget more accurately, and take better advantage of quantity discounts.

Today's software management tools provide the information needed to plan effectively with very little effort on the system manager's part. The result is that users are better served and resources are deployed more efficiently. Exhibit 46.3 illustrates how one large U.S. corporation used software management to cut software license expenditures by nearly 80 percent.

Stay Legal. While not typically associated with cost of ownership, audits of companies suspected of illegal software use can cost a great deal of time,

Exhibit 46.3 How one large U.S. corporation used software management to cut software license expenditures by nearly 80 percent.

Application	Licenses required without software management	License cost without software management	Licenses required with software management	License cost with software management	Savings by using software management
Lotus Freelance	543	$171,045	95	$29,925	$141,120
Micrografx ABC Flowchart	36	$10,800	15	$4,500	$6,300
Microsoft Access	502	$147,588	95	$27,930	$119,658
Microsoft Excel	718	$211,092	140	$41,160	$169,932
Microsoft Project	297	$128,304	50	$21,600	$106,704
Microsoft Word	1,088	$319,872	280	$82,320	$237,552
Total	**3,184**	**$988,701**	**675**	**$207,345**	**$781,266**

money, and aggravation, not to mention bad publicity. Many company executives don't realize that they are responsible for all applications run on company equipment, regardless of whether the applications are launched from the server or from clients, or whether they are company-approved or not.

With the proliferation of networks, it became much more difficult to monitor and comply with software license agreements. At the same time, the penalties for not doing so have increased. For example, in the U.S. in 1992, Congress instituted criminal penalties for software copyright infringement. Violators of license agreements can now be fined up to $250,000 and sent to jail for up to five years. And according to the Software Publishers Association (SPA), as of 1996 more than $16 million in fines had been levied against companies in violation of software licensing agreements.

Software management not only helps to ensure license compliance; the reports available with software management tools can act as proof of compliance if a company is ever questioned.

Cost #2: Training and Technical Support

Reduce Training Expenses. Appropriately directed training leads to more productivity and fewer help desk calls. This is the rationale that leads companies to make large investments in training resources. Software management can help make those investments pay off by revealing which users should be trained on any given application.

For example, if a large corporation has standardized on one spreadsheet application, software management can reveal small populations of employees using other products. IT can then train those employees on the company-approved software and reduce help desk support for the unapproved applications. This also reduces the expense of keeping support personnel and trainers up-to-date on seldom-used applications.

And software distribution allows the controlled rollout of software to only those users who have been trained. This also reduces help desk costs because the majority of help requests for a given piece of software occur when the user is first learning how to use it. Similarly, with software management, corporations can determine the most frequent users of any given application and train them first.

Cost #3: Network Administration

Control the Software Invasion. On any given day in a large corporation, employees may load hundreds of unapproved software applications onto their desktops. While many of these applications are from home, an increasing number comes from the Internet, where it's easy to download new software with the click of a mouse.

In addition to the obvious effects on employee productivity, for the system manager this influx of software is a nightmare. It often leads to frozen machines and embarrassed employee calls to the help desk. And at its worst, it causes serious problems on the network.

Software management tools constantly monitor which applications are loaded and allow system managers to respond quickly to installations of unapproved applications.

Monitor the Effectiveness of IT Applications. IT departments spend a lot of time and money developing custom internal applications. But they may not know how effective their software is in meeting their objectives. Software management reports can give system managers the information they need to either improve the applications or discontinue their development efforts. Software management not only saves development time, but also helps companies more effectively plan for hardware needs (which are often quite expensive for internally developed applications).

Putting License Agreements to the Test. Enterprise license agreements are a common way for organizations to purchase software. Manufacturers of popular business applications (word processors, spreadsheets, databases, etc.) and office suites offer discounts to companies that agree to purchase an individual license for each PC on the network. These agreements are touted for their ability to save the customer money and ensure legality. But without software management tools, they may be only partially successful in accomplishing either goal.

REAL COST SAVINGS REQUIRES REAL USAGE INFORMATION

When creating an enterprise license agreement, the software manufacturer uses its own formulas to determine how many licenses are needed for a given customer. These formulas are standards they use with all of their customers and are not based on actual software usage.

Through software management, system administrators can be more informed customers. They can find out exactly how many licenses they need, rather than depending on external formulas. With this information, they are in a much better position to negotiate the most advantageous license agreement and the best price possible.

Partial Legality Means Only Partial Peace of Mind

It's easy to believe that by signing an enterprise license agreement, an organization's license-compliance worries are over. This may be true for the one application or set of applications covered by the agreement. But today's enterprise employs a wide variety of applications from many different manufacturers. And employees have the ability to install any software on their desktops.

When it comes to staying legal, true peace of mind for the system manager is only possible with the certain knowledge that *all* of the applications being used across the organization are being employed legally. Software management can supply that knowledge.

The First Step in Establishing a Software Management Program — Measure Usage

The best way to start managing software assets more efficiently and to reduce cost of ownership is to measure current usage accurately. This can be done with a software metering application in what's known as a software audit. A software audit provides a snapshot of exactly how a company uses its software and where that company stands with regard to compliance.

An effective software audit goes far beyond a simple inventory (such as that provided by many audit tools). Rather, it continually tracks which workstations on a network access which applications over the course of the audit, to reveal usage patterns.

By tracking current usage, a system manager can assemble a wealth of previously unknown information, including:

- Unauthorized applications in use
- Versions of applications in use
- The locations of all software run on users' PCs
- The actual users of specific applications
- Peak usage levels of individual applications

Exhibit 46.4 Network computers: the ultimate answer to cost of ownership?

The network computer (NC), an intranet-based "thin client," is being widely touted as an answer to rising costs of ownership. It offers a lower initial purchase price, reduced support costs, and excellent extensibility. However, the cost-conscious system manager may choose to not standardize on the new machines exclusively. This is because doing so requires a large investment in changing the network infrastructure. Expanded network bandwidth, server upgrades, and new applications are just a few of the expenses involved in establishing a viable NC environment. And while many are attracted to the NC paradigm because of the promise of ease of management, effective tools have yet to be developed.

Software management applications can help the system administrator determine which users are the best candidates for NCs and which ones need the desktop processing power and storage available with a PC.

And no matter what mixture of PCs and NCs an organization chooses to deploy, software management tools will offer cost-saving benefits in the form of usage information, inventory, license management, and more.

- The optimal location(s) to install software
- Exactly how much software is needed

With the results of a software usage audit in hand, the next step in establishing effective software management is to set priorities and make decisions about how to structure access to the organization's software assets. One of the most important of these decisions is which software management tools offer the features to help implement an organization's strategies.

Here are some questions to ask when evaluating a software management product:

- Does it measure usage of applications launched from both the desktop and the server?
- Does it provide universal metering — the ability to see software on any network operating system?
- Does it ensure legality by detecting launches of all applications, whether or not they have been designated for monitoring by the system administrator?
- How comprehensive is its database of recognized applications?
- Is it easy to install, administer, and support?
- Are its various components well integrated?
- Does it operate in the background, allowing users to remain productive?
- Are its available reports truly useful? Do they help you make good decisions?
- Does the manufacturer of the software offer not only products, but also the ongoing support and expertise you need to develop and ensure the success of your software management strategies?

SUMMARY

In today's technology-centered workplace, software is an essential employee tool. The way an organization acquires, uses, and manages software directly affects its short- and long-term success. Until now, however, there has been no way to effectively determine whether software resources are being used efficiently.

Thanks to the new class of software management applications, system administrators now have access to the information and control they need to significantly increase productivity and improve their bottom line.

References

1. *TCO: The Emerging Manageable Desktop,* The Gartner Group, October, 1996.

Chapter 47
Inventory — Cornerstone Technology for Managing the Network

Roy Rezac

Inventory is the cornerstone of any effective network management solution. Without an accurate inventory of network systems and components there is no way to manage a network efficiently, regardless of the size. Recent downsizing trends and headcount capping procedures have forced many administrators to support more end users than ever before. Proactive management with a practical tool set can compensate for these personnel constraints. Inventory is the beginning of proactive network management in every environment.

THE COST OF THE NETWORK

Due to the rapid deployment of the PC network at all levels of the corporation, the migration of mission-critical applications and data to the network has become a necessity. The Gartner Group research firm estimates the management of a PC environment can reduce costs by up to 50 percent over five years, while costs in an unmanaged PC environment will rise 20 to 50 percent over the same period (see Exhibit 47.1).

"To appreciate the enormity of the potential benefits, consider that organizations with 5,000 desktops can achieve savings ranging from $1.5 million to $13 million in the first year after implementing an effective enterprise-wide IT asset management system, according to Gartner Group analysis."

0-8493-9823-1/00/$0.00+$.50

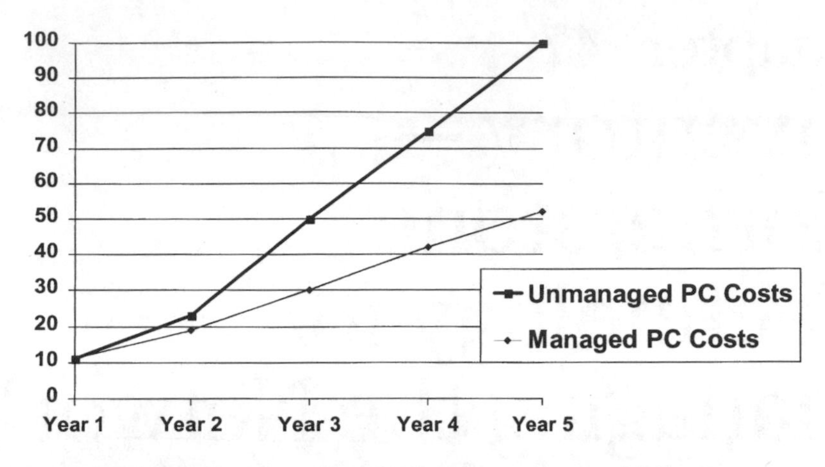

Exhibit 47.1 Unmanaged PC costs vs. managed PC costs.

"IT" MANAGEMENT

The dispersion and rapid change of network assets make it difficult to identify what is actually on the network. You can't manage it if you don't know what it is. Auditing has unveiled some interesting results. As reported in *InfoWorld* (Poor Asset Management is Breaking the Bank, by Doug van Kirk) several companies simply guess. One company thought it had 15 users per printer, while an audit indicated a ratio of 1 user to every 1 printer.

An accurate inventory can eliminate redundant purchasing, minimize the cost of upgrades, migrations, moves and changes, dramatically decrease troubleshooting time, and most important, reduce user downtime. In short, an accurate inventory impacts a company's bottom line.

The core of many network management problems is the lack of detailed knowledge about the network and how it changes over time. Understanding hardware, software, and configuration details of every PC in your network lets you plan for inevitable product upgrades, changes, and obsolescence, ensures corporate standards are maintained, and simplifies system maintenance.

The Gartner Group estimates only 12 percent of the total costs of PC software are direct capital costs. The remaining 88 percent are labor-related costs: end-user operations, administration, and support (see Exhibit 47.2).

AN IDEAL INVENTORY

There are two ways to inventory your network. You can walk around with a clipboard and pencil and record your findings or implement an automatic inventory process which allows the network to inventory itself.

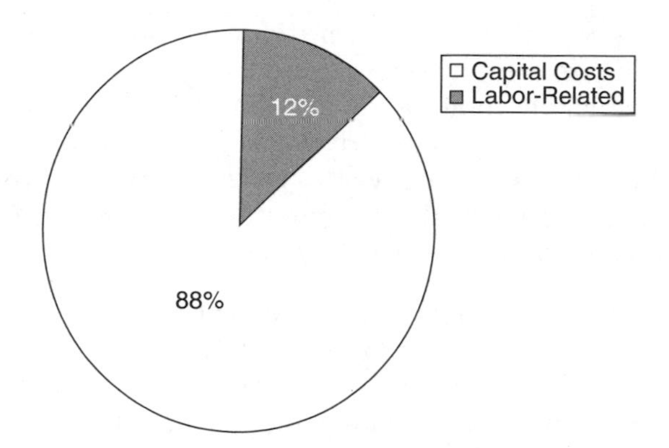

Exhibit 47.2 Cost of PC software ownership.

According to the Gartner Group, an Inventory/Asset Management program can reduce your costs of ownership anywhere between $200 to $700 per year per machine. Since inventory programs cost between $5 and $25 per desktop, the payback period is 1 to 2 weeks, and the return on investment (ROI) is an estimated 3000 to 4000 percent. According to a 1996 study conducted by Techknowledge Research Group, the network management solution being given the highest consideration for implementation is network inventory (78 percent).

Definition of Key Terms

Inventory. Collecting data about your network.
Asset Management. The ability to use that data to reduce costs, increase productivity, and effectively plan your systems for the future.

An inventory program should collect information about every networked system regardless of platform or specific server configuration to include both hardware and software data collection and store this data in an easy-to-use, centrally located database. Flexible query and reporting processes and integration with other network management tools, like software distribution and license metering, provide the foundation for proactive network management.

KEY ASPECTS OF AN INVENTORY PRODUCT

There are several key aspects of an inventory product. If the product can't satisfy these key aspects, it should not be considered a viable inventory solution.

Breadth of Inventory

It is rare to find a homogeneous network running a single desktop operating system or network operating system in the corporate world. Even environments that adhere to strict purchasing standards and commit resources to only three or four system models still contain internal component variations. A whole room of identical model number server PCs can have different size or types of hard drives, memory, and configuration data. A reliable product allows you to inventory every system on your heterogeneous network. More specifically, an inventory program should recognize a variety of hardware and software components.

Inventory programs should be able to detect as many different components as possible, as listed in Exhibit 47.3. Only with breadth of identification do you have assurance that you can accurately support, troubleshoot, and manage any component that may appear on your network. The network is not a static object. As soon as you have everything strategically set up, it can change. Purchasing an inventory program from a supplier who will keep the product updated to support new hardware and software platforms is essential.

Exhibit 47.3 Table for evaluating an inventory package.

Does it	Yes	No
Collect hardware data for:		
Windows 3.x workstations		
Windows 95 workstations		
Windows NT workstations		
OS/2		
DOS workstations		
Macintosh workstations		
NetWare Servers		
Windows NT Servers		
How many hardware components inventoried:		
CPU Types		
Intel		
NEC		
Cyrix		
Common Device Types:		
EISA		
MCA		
SCSI		
IDE		
ATAPI		
PCMCIA		
PCI		
Does it store CMOS Disk information for recovery		
Does it collect user id		
Does it collect NIC id		

Exhibit 47.3 (continued) Table for evaluating an inventory package.

Does it track each system independently of user/nic id?		
Does it track PC resources such as:		
DMA		
I/O ports		
Interrupts		
Shared memory		
Does it provide fields for storing and displaying:		
Asset tags		
Serial numbers		
Warranty information		
Can you query all data collected ex. All 486 processors?		
Does it adhere to DMTF standards:		
DMI		
MIF		
Service layer		
Does it support the Compaq Intelligent Manageability initiative?		
Software Inventory Parameters		
Does it accurately inventory software on:		
Windows 3.x workstations		
Windows 95 workstations		
Macintosh workstations		
Windows NT workstations		
Novell File Servers		
Windows NT Servers		
Is the master software file comprehensive?		
Is software vendor information provided?		
What criteria is used for identifying software?		
File size		
Date		
Supporting files		
Can you add titles to the software master for internally developed applications?		
Can you print the software master for review?		
Can it identify application versions?		
Can it scan for and report unknown files		
Network Details		
Does it collect key network details for each NOS on your network?		
Does it provide configuration information like:		
Shell versions		
LSL version		
IPX version		
SPX version		
VLM type		
VLM version		

Exhibit 47.3 (continued) Table for evaluating an inventory package.

Does it track network parameters like:		
Number of packet burst buffers		
Large Internet packets		
First network drive		
NET.CFG files		
TCP/IP information		
Vendor		
Version		
Address		
IP address		
Subnet mask		
File Collection		
Does it collect:		
AUTOEXEC.BAT		
CONFIG.SYS		
SHELL.CFG		
NET.CFG		
LANMAN.INI		
DOSLAN.INI		
PROTOCOL.INI		
WIN.INI		
SYSTEM.INI		
Any text file?		
Can it collect any file with a		
Minimum size (collect any text file smaller than 30 bytes)		
Maximum size (bigger than 1k)		
With a specific date		
On any drive		
On a specific search path		
Recurse subdirectories		
Is there a limit on how big collected text files can be?		
Can you view any collected file and copy or paste to other applications		
Administrative		
Is there support for displaying administrative data?		
Can you edit administrative data?		
Can you create new user-defined fields		
Can you create user-defined screens		
Can you report on administrative data?		
Scheduling		
Can you schedule hardware and software collection independently?		
Do schedule options include:		
Every login		
Every day		

Exhibit 47.3 (continued) Table for evaluating an inventory package.

Every week		
Monthly		
Quarterly		
On next login		
On demand		
Change Alert Notification		
Can you be notified via:		
25th line message		
Windows pop-up box		
Microsoft Mail & Exchange		
cc:Mail		
Reporting		
Can you easily filter data and print quick reports		
Are standard reports included like:		
Buggy Pentiums		
Windows compatibility		
Can you customize reports including font, color, page layout?		
Can it generate group subtotals (how many 486s, Pentiums, etc.)		
Database Support		
ODBC level 2 compliance		
Ship with SQL database and ODBC driver		
Sybase SQL Anywhere		
NOS Support		
Novell NetWare		
Banyan Vines		
Microsoft LAN Manager		
Windows NT Server		
Windows for Workgroups		
IBM LAN Server		
DEC Pathworks		
Agents		
Do collection agents self install?		
Are TSRs required for any data collection?		
Are 16 and 32-bit native collection agents included?		
Integration		
Software		
Remote control		
Virus protection		
Backup		
Other		

Change Alerting

When personnel changes or offices are redesigned, equipment is moved and applications revised. Migration to new versions of core applications for a new feature may be required. For example, operating systems change. In order to maintain your support relationship with the vendor, you may have to upgrade. This means that new hardware must be purchased to support new operating systems.

Inventories must also continue to identify all components being installed onto your network. In some environments this constitutes thousands of unique components. When changes occur, inventory programs should inform the administrator of unauthorized changes. Inventory programs that alert you to a change enhance inventory functionality. Alerting can occur in a variety of ways including broadcast messages, e-mails, and even notification via your beeper.

Performance Issues

While this robust inventory takes place, your users should not notice any degradation in network access or performance. In an attempt to bypass annoying network events, a user may interfere with the inventory collection process. Most users are literate enough to execute the CTRL-Break command to halt a process and inadvertently meddle with your inventory. Effective collection of workstation data needs to be tamper resistant.

Ease of Installation

Ask any network administrator how much time he or she has for installing and configuring new software and the response will invariably be, "10 minutes."

Networks vary in the number of attached nodes. Corporate networks range from a couple hundred to several thousand network-attached PCs. Implementing an automatic inventory across many desktops should not be a time-consuming task. Spending days and weeks installing an inventory program is frequently so daunting that many companies just can't free the resources to complete the installation.

Consequently, easy installation of inventory collection agents on every platform on the network stands as another key feature, with a self-installing agent as the ideal. As users log in, the inventory program should check if the agent is installed on the workstation. If not, the install should automatically occur followed by a collection scan. This prevents nodes from "disappearing."

Since your network will change and grow, openness and scalability are important factors in an inventory product. Once the data are collected, they should be stored in an "open" database. An open database has a published set of APIs (application programming interfaces) that allow any programmer to write add-ons like report writers to the stored data. The most popular set of APIs is ODBC for SQL database servers. ODBC (open database connectivity) is almost exclusively used on Windows workstations.

Using an open database lets the administrator extend the database schema, add fields, or store data about systems in a variety of ways. Openness provides easy access to other applications like an interface to your fixed assets' application. Openness is achieved through ODBC or by using a common file format like .DBF. A client/server database system, such as those using ODBC, is scalable. A .DBF-based system, which processes the data at the workstation exclusively, generates traffic and slows as the files grow.

Benefits of scalable databases increase directly with the size of the network. Data may need to be collected on a variety of subnetworks, or geographically dispersed LANs and then analyzed at headquarters for enterprisewide management decisions.

Reporting

The point of an inventory program is to provide administrators with useful information. Data are translated into information through useful and practical reports. Inventory packages need several predefined, or vendor-supplied reports. Standard support questions should be easily answered through existing reports. Custom reports should be almost as easy to create. These reports should cover such standard data as disk space usage, too little memory, operating system versions, and configuration changes. Each environment has unique ongoing report requirements. Administrators should be able to create and save reports to suit their needs and save them for frequent use.

Integration with Other Tools

Inventory is the foundation of other network management tasks and tools. For example, effective software distribution is closely related to inventory. In addition, licensing can be an integrated component of effective network management. Staying current and maintaining license compliance are tasks supported by inventory programs.

Most network management tasks are dependent upon accurate inventory data. Troubleshooting requires identifying hardware and software configurations. Integration with other network management tools, such as remote control and help desk, are also essential for the proactive network administrator.

Standards

Future technology changes require that software adapt and adhere to standards. The desktop management task force (DMTF) has more supporters than any other standards organization in this arena. The most widely implemented standard from the DMTF is DMI (desktop management interface). Network-attached desktop components can provide information to management applications, like inventory programs, through the DMI standard format and interface. Support for DMI is widespread, and the DMTF continues to garner support from the vendor community. Choosing an inventory package that adheres to these standards will provide technical longevity.

If your corporation owns Compaq systems, then you should choose an inventory program that supports the Compaq Intelligent Manageability Initiative. Compaq is committed to providing network management vendors with easily accessible data about systems and components in a standard format. This program has a smaller impact than DMI because it revolves around a single vendor — Compaq. Nevertheless, any help for efficient management should be embraced.

Evaluation

When you evaluate inventory products look for these key features:

- Support for all operating protocols, servers, and systems on your network
- Broad, accurate component recognition capabilities
- Ease of installation
- Flexible, scalable, and open database
- Easy and flexible reporting capabilities
- Integration with other management tools like software distribution and licensing
- Adherence to standards

Using these guidelines, you can select an inventory package that is adaptable to the changes in your network environment.

The Ideal Inventory Process

Electronic inventory packages follow different architectures. In order to adhere to the key criteria, select a package with an architecture that implements:

- Ease of collection
- Highly detailed data for hardware and software
- Robust database management
- Low performance impact on the network

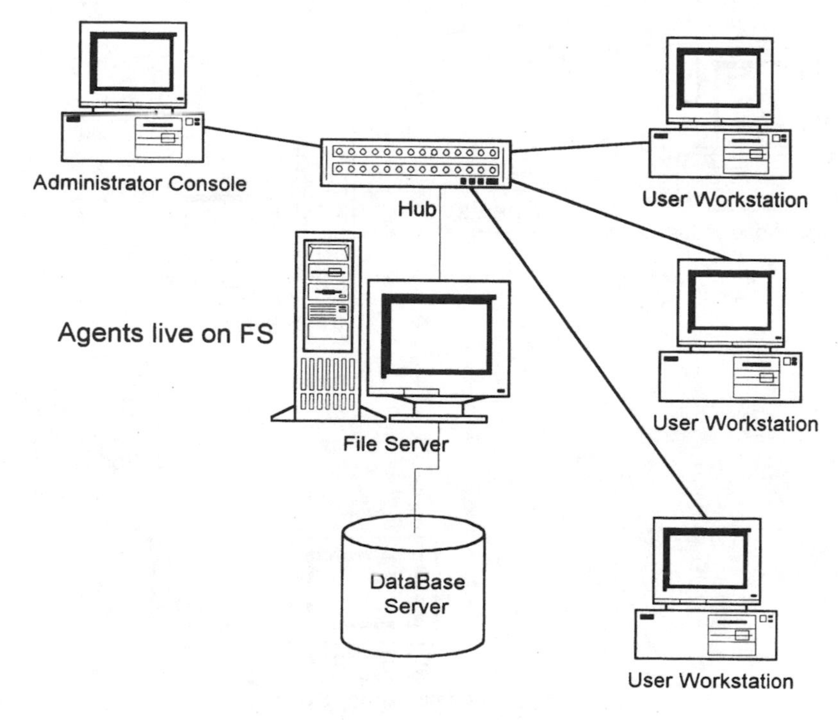

Exhibit 47.4 Architecture.

- Easy-to-use reporting
- Strategic alliances with operating system vendors, hardware vendors, and participants in the DMTF

INVENTORY PROGRAM INSTALLATION

The network exists to service the users, not the network management programs. In order to keep performance delays to a minimum, it is more efficient to run the inventory database on a system other than the file server. This allows complex querying and reporting, which trigger a lot of processing, to be run off-line, minimizing traffic on the network (see Exhibit 47.4).

As the enterprise grows, so should the database. Multiple smaller databases may maximize performance; however, doing so shouldn't compromise the viewing of data throughout the entire network. Central management staff should be able to analyze data by merging or rolling up the data to a single access point. The database structure becomes even more important as other network management tools like software distribution and metering programs are implemented. Only an open standard industry database will provide forward compatibility and technological longevity.

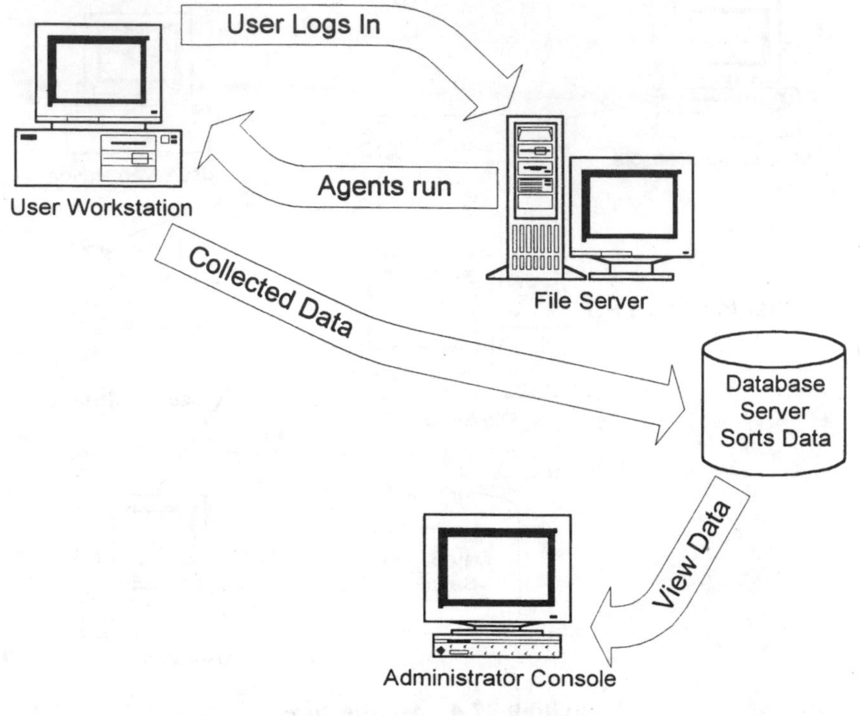

Exhibit 47.5 The collection process.

Agents generally run as executable programs called from login scripts. If, however, the collection agent can scan from code resident on the workstation such as a .DLL on a Windows workstation and use the login process for transporting the scanned data back to the database, the collection process runs faster and with less delay (see Exhibit 47.5).

The console is the access point to the inventory data; therefore the administrator's console application should exploit the powerful workstation most administrators have at their desks. Thirty-two-bit consoles running on fast processors with lots of system RAM provide the horsepower necessary to readily access the variety of data administrators need to analyze.

In order to stay competitive, any inventory vendor should develop close relationships with the major hardware and software vendors. Within the network operating system market, Novell and Microsoft are leaders. Strong relationships between operating system vendors and management vendors should allow nearly concurrent release of new operating systems and management tools. Without good communication between the operating system vendors and the management utility vendors, network upgrades are slow to occur. This hurts both the OS and management vendors.

PRACTICAL INVENTORY USES

Walking Hardware

There are several scenarios that exemplify the need for accurate and current hardware information. One of the most frequent is the "walking hardware syndrome." Yesterday there was plenty of virtual memory space on the 500-MB hard drive, but last night the 500-MB hard drive went home with the user. Administrators most frequently find out about walking hardware through support calls from users.

Network cards, RAM chips, hard drives, network cards, and ROM drives have been known to take extended vacations from corporate systems. Troubleshooting these problems is effortless with a strong daily hardware inventory. Each and every "it worked yesterday" call should be addressed first with a check of the inventory change log. Here, hard drive type and memory can be retrieved and future "hardware walks" can be prevented.

Migration Planning

Migration planning always begins with an accurate inventory of the existing network. The recent push to 32-bit platforms on the desktop has escalated the need for accurate inventories. Every migration to a 32-bit platform should begin with building a query that includes the minimum standard components required by a new operating system. This will allow you to formulate purchasing needs and budget requests. Saving the query report will let you monitor your upgrade progress.

Application Roll-Out

Any application roll-out requires an understanding of existing software configurations. Software inventories that include configuration files like AUTOEXEC.BAT, CONFIG.SYS, WIN.INI, PROTOCOL.INI, etc. are necessary for painless software upgrades. Too often, administrators distribute software without checking the configuration of the target system and cause the new software to crash. This leads to excess downtime for the user and stressful reinstalls for the administrator.

System crashes caused by hard drive errors frequently require an administrator to reconfigure the hard drive type in CMOS. A good inventory program will have this information readily accessible in the database. Any system should be able to be recreated from a complete hardware and software inventory. Hardware configuration, network protocol configuration, startup files, and resident applications are all part of a complete inventory. With these data in hand, rebuilding and reconfiguring any system becomes an easy step-by-step process.

Tracking Administrative Data

Administrators need to track data that are external to network systems like warranties, asset tags, and serial numbers. Because these are inventory data, a means to include it in the inventory database is critical. The ability to manipulate tracked data like any other piece of inventory is just as important.

Software Inventory. Software inventory is a particularly thorny process. Installed software needs to be identified by more than just the filename of the executable. Many applications today use an .EXE file as a loader to call the rest of the application code with nontraditional executable file extensions like .OVLs and .DLLs. Unless a software inventory searches for all the files that compose an application, you will be unable to identify what is installed on the network.

Unknown files should not be ignored. Just because they aren't recognized by a software inventory does not mean that they don't require support, upgrades, and configuration management. The most common example of this is internally developed applications. There is no way inventory vendors can identify these applications, yet the network administrator still needs to manage them. Unknown files should be flagged along with the information display. Industrious administrators will add these custom applications to the software master list so that they can be automatically identified in the future.

Software Version Tracking. Intelligent software inventories also track application versions. Filtering the database to analyze the variety of software and the even greater variety of application versions helps administrators comply with licensing agreements and plan software budgets more accurately.

Tracking Mobile Users. One common problem with inventory programs is their reliance on user names or network card IDs for tracking each system's records. Users tend to login from a variety of systems. If an inventory program relies solely on a user name, it will appear as though your more mobile users change their systems quite frequently.

Network cards occasionally walk from system to system. This is especially true of laptop and notebook computers that log in from a variety of PCMCIA cards alternating with docking systems. Therefore, network card ID numbers can lead to the same confusion as relying on user login names. One way to overcome this is for the inventory program to create a unique ID number for each system stored on the hard drive combined with user name and network card ID. This is a feature especially important to companies with many mobile users.

Easing Help Desk Support. Troubleshooting desktop problems begins at the help desk. Yet, before the troubleshooting process can begin, help desk personnel need accurate information about the configuration of the users' system. The help desk provides users with a means to communicate what's not working on their system. Referencing the desktop inventory record should provide the help desk with the configuration.

Attaching the inventory record to a trouble ticket supplies support staff with the additional information required to accurately resolve complex problems.

Change tracking is another inventory feature that help desk personnel can use to solve problems. "What changed last?" is the most frequently asked question by experienced troubleshooters. A quick check of the change log, or alert log will lead to problem resolution. The help desk then becomes a streamlined user support system, instead of a bottleneck of unanswered calls.

SUMMARY

There are many more practical troubleshooting and management techniques that a robust, full-featured inventory program facilitates. When evaluating inventory packages, build a list of key features that you require.

If you can only implement one network management tool, make it an inventory program that supports all your network operating systems, provides extensibility and scalability so that it can grow with your network to make sure it includes proactive features like change alert notification, and adheres to the most popular standards. This is your insurance that the inventory program will keep pace with operating system technology.

Wasting precious time and corporate resources in an attempt support the unknown platforms and applications is not necessary. Higher end-user productivity is an achievable goal when downtime is cut short by proactive management. A network administrator's sanity can be sustained with robust, comprehensive tools.

Chapter 48
Software Distribution

Steven Marks

The rapid development of automated software distribution systems is being driven by the explosive growth of PC networks. The systems under management have quickly grown both in size and in complexity, beyond the ability to manage by traditional means. Support organizations have grown tremendously, over the past few years, in an attempt to gain control over growing networks but have failed to return the flexibility needed to provide competitive advantage.

The move from DOS to Windows NT and NetWare networks, over the past few years, has led to a number of changes for managing PC workstations. No longer could we simply install an application on a file server, point a batch file toward it, and consider the installation done. Now, Windows applications require each application to be individually integrated into the Windows operating system. Windows itself has required a rearchitecture where installations require software be installed on the users' workstation directly. The move from centrally installed and managed applications to workstation-based installations has caused a world of problems for systems administrators — one that cannot be solved by adding more staff or developing better procedures alone.

The ever-increasing size of PC networks has also created its own problems and complexity. The sheer size of many networks today makes manual change management too labor intensive to be viable. The lack of alternatives has created a management void that has left many network managers uncomfortable, but with few alternatives.

The result has been that many Windows installations become stagnant, with versions of applications growing increasingly out of date. Administrators' ability to manage PCs has eroded as systems have grown and expanded. Systems are increasingly vulnerable to problems caused by user-installed software. Support costs increase as users tinker with their workstation configurations. There is no way to manage the size and complexity of modern PC installations without a means of automating the processes of installing and maintaining workstation configurations.

0-8493-9823-1/00/$0.00+$.50

COMPONENTS OF SOFTWARE DISTRIBUTION

Software distribution systems have developed out of the need to manage what has become unmanageable. A software distribution system consists of a number of modules, rather than a single system, which interact to provide a means of managing user workstation configurations. The system automates the tasks of installing, configuring, and updating user workstation configurations. The most basic elements are a hardware/software inventory system and a distribution mechanism. Many manufacturers include a number of other modules which assist in the distribution/management function, such as remote control of user PCs or software metering. But two basic modules, inventory and distribution, form the basis of the management system.

The process begins with inventory. Inventory has become a truly ubiquitous function. Many manufacturers offer inventory programs either as stand-alone or as modules as part of a larger suite. However, merged with distribution capabilities, inventory functions take on a new importance in the management of PC networks.

Hardware inventory has become very sophisticated, particularly as a result of efforts to standardize information gathering by the Desktop Management Task Force. The DMTI spec has allowed inventory packages to collect very detailed information about the contents of a PC, including hard disk utilization, memory and interface cards installed, and even the condition of the battery. Some PCs, notably those from Compaq and Hewlett-Packard, even include the ability to read the machine's serial number. Intel's recent efforts to include data gathering firmware within Pentium Pro systems carries the collection even farther. These newest systems include monitoring of real-time environmental factors, such as temperature and voltage.

The hardware inventory function is important to the distribution process to determine if a machine has the necessary resources to receive a specific distribution. Before installing new software it is important to make sure there is adequate disk space for the installation. Also, distributions are sometimes based on specific hardware types. For instance, we might want to install only a processor intensive application on machines with adequate processing power.

Software inventory collects information about files stored on the local hard disk of a PC. The program scans the disk and collects the names of files. The software inventory system compares the file names to a table stored within the program that relates specific file names with known applications to compile a list of applications installed on the PC. Users are able to add to the list so the inventory system can track applications not included in the built-in table.

The software inventory provides vital information for managing licenses and controlling access to applications. The distribution process allows us to control the software installed on PCs. Software also serves as a condition for distributions. For instance, we want to install an upgrade only on machines that currently have the previous version of an application.

Inventory on its own is a valuable management tool. Hardware inventory provides valuable information. The basic information provides asset information about the equipment owned by the organization, information it would be difficult or expensive to collect manually. This information is also valuable in planning upgrades. It becomes easy to see which systems require an upgrade with a simple query of the database to find all systems which do not have sufficient disk capacity, processor speed, or memory. The hardware inventory also serves a security function to find hardware configurations that have changed, either by the removal of a board or the installation of unapproved products.

Software inventory serves a similar function. The database created by the inventory function allows us to keep a record of applications installed on users' drives. We can keep track of the number of installed copies of an application. Regularly generating reports on installed software aids compliance with license agreements. The software inventory also allows searches for unapproved software users may have installed on their PCs.

A valuable function of software inventory is the ability to track changes of specific files. User changes to system configuration files can cause significant problems. For instance, a user installs an application that modifies their configuration files and reduces available memory on their PC. They later try to load a large spreadsheet and receive an insufficient memory error. Support personnel could spend a long time searching out the source of the problem, without knowing the configuration files have changed. The ability to track specific files can save valuable time in resolving problems and result in a more stable system as a result.

THE DISTRIBUTION MODULE

The distribution module of the system automates the task of applying changes to workstation configurations. The kind of changes necessary to manage sophisticated operating systems such as Windows 95 or NT must be much more sophisticated than in earlier operating systems. Distributions can be divided into two categories — those that simply write a new system and those that modify an existing system.

The simplest type of distribution is a full workstation installation. All the pieces of the operating system and applications are rolled up into a single distribution which was on the workstation before the installation was run. This provides a simple means of preparing new workstations for users. The alternative, before software distribution, would have been to run the setup program for the OS and each application and then run each application to configure specific settings in each.

CONFIGURATION MANAGER

An interesting variation of a full software distribution manager is a product from Intel called Configuration Manager. The Configuration Manager, a hardware-based server, allows distribution of full workstation configurations for a variety of operating systems — Windows 95 and NT, DOS, and OS/2.

The Configuration Manager is primarily designed for new, system board-based firmware Intel has developed as part of its work with the Desktop Management Task Force. An option allows booting to the Configuration Manager, rather than loading the operating system from the local drive, when a PC is booted. A menu provides the reconfigured choices of configurations. Administrators can set up multiple configurations for each operating system. For instance, a system may have choices set up for separate Windows NT configurations for Engineering and Marketing and a Windows 95 system for Sales. Each can have different layouts and application groups in addition to different operating systems. After selecting a choice, a full configuration is performed on the PC. The system allows booting from a floppy if the PC does not have the Intel firmware to autoboot to the Configuration.

What is unique about the Configuration Manager approach is that it allows installation of a variety of operating systems from a single system. Other software distribution systems are limited in the operating systems they can install. Windows-based software distribution systems, for instance, may not be able to install OS/2 operating systems.

There are several limitations to the Configuration Manager approach, however. First, it is hardware based. This makes it prohibitively expensive for small installations. The hardware, at this point at least, is proprietary, which some people object to. Second, the Configuration Manager only allows for a full workstation installation. This is important to an organization with a large number of workstations to install. However, the greatest value of software distribution is in allowing ongoing change management to prevent installations from becoming stagnant and out of date. The Configuration Manager does not allow these types of changes.

Software-based software distribution systems can be used, for the most part, to perform the same distributions. Most can run the OS setup program to install the operating system or can install a suite of applications if the OS is already installed. Few are able, however, to change one operating system for another easily. For instance, users can sometimes install Windows

NT over a previous OS/2 installation with some preinstallation work, sometimes a significant amount of work. The Configuration Manager approach allows full installation without even logging in to a network.

Initial installations are a significant problem for managers of large installations, but the ongoing management of additional applications and upgrades has been, until now, an intractable problem. Additional capabilities are necessary to automate the process of installing new applications, adding icons to the desktop or changing user permissions. Without these capabilities, a distribution system cannot be of maximum benefit. Versions of existing systems with the ability to perform extensive editing of systems files are just starting to appear, because of the newness of operating systems such as Windows 95 and NT.

Software distribution systems are immensely valuable in returning the ability to make changes to user workstation configurations as needed. Whether the changes are new applications, upgrades, new drivers, or bug fixes, the result is increased stability and a more supportable network. Software distribution packages provide a scripting language to move and manipulate files to make these changes possible. To be useful, the language needs to be powerful and flexible enough to recreate the steps one would perform while sitting at the user's workstation, either manually or when running an application setup program. A distribution package needs to be able to copy files to the local drive, delete specified files and directories, and modify selected files. The program does this by way of a specialized scripting language, which provides a straightforward, if not simple, way to perform these functions.

To modify files the program needs to be able to search for specified text strings. Script commands need to allow the user to search specific files for text and delete the line, or append text before or after. For Windows environments, the distribution system needs to be able to modify not just the text INI files, but also edit registry files to add install applications and icons (see Exhibit 48.1).

One problem in scripting a software distribution script is deciphering exactly what happens when a Windows setup program is run. Since the setup program is designed to automate the installation of the application, the steps it takes are not visible. Software distribution systems take several approaches to this problem.

One way to handle it is to take a snapshot of a standard installation and monitor the changes made by the setup program. A distribution can then be designed to create the same effects. This can be very time consuming but allows the greatest control over the distribution.

Exhibit 48.1 Norton administrator script for installing Windows 95.

Job Actions

Display Message...	Starting Job #1 "Windows 95 Installation"
Execute File...	md c:\dossave
Copy/Replace Files...	C:\dossave
Execute File...	md C:\master
Execute File...	md C:\master\windows
Execute File...	md C:\master\windows\msnds
Execute File...	md C:\master\winsvc
Copy/Replace Files...	C:\MASTER\WINDOWS
Execute File...	G:\PROGRAMS\BATCH\WINCHK.BAT
Copy/Replace Files...	C:\
Execute File...	C:\win311.exe -d -o C:\
Copy/Replace Files...	C:\
Execute File...	C:\WINDOWS\WIN G:\PROGRAMS\WIN95\CABS\W95INST3.EXE

Some programs automate this step, monitoring and recording the changes Setup makes and automating the process of recreating this on the user's workstation. This is a faster process but allows less control.

Other programs skip this step entirely and provide a means for running the application's setup program directly on the user's workstation as if you were sitting in front of that computer. A distribution script is written which runs the setup program from the server.

Using distribution to run setup programs is not without pitfalls. It is still necessary to write a distribution script. While it is not necessary to write a script to perform the actual installation, it is necessary to simulate the input from the user the program is expecting. This is not always easy. Sometimes it is necessary to hold a response until the setup program actually asks the question. Trigger the response too soon and the setup program hangs, waiting for a user response. If the timing is not set right, or something else untoward happens, the installation may hang.

The advantage of this approach is that the setup program is often very smart. It may analyze the existing environment and modify the installation procedure based on other installed software. Having the software distribution system run the application setup program instead of analyzing the setup program's effects on the system and duplicating it with a detailed distribution script has several benefits. It saves time. There is no need to analyze the steps the setup program takes, assemble the files, then write and debug the distribution script. You also gain the benefit of any custom analysis the setup program performs.

Default settings must be set up in some way. In some applications it might be possible to create a macro; for others, copying a configuration file may be enough. But, in most cases, the best way is to edit the SETUP.INF file for the application.

The Microsoft Setup program is a very sophisticated tool. It includes a parameter file which is a scripting program on its own. This file contains settings for the specific application and really contains all the information needed for the generic setup program to work.

Determining when to run distribution scripts is where the distribution feature and inventory come together. Most often, distributions are run based on a hardware or software condition. For instance, we may want to upgrade everyone with a certain version of an application. To do this, we first must find out what versions of the application people are running and who the specific people are who need the upgrade. We might also have a new version of a driver and want to find which users have a specific network card and issue them the upgrade. The steps in performing a distribution, then, are to use the inventory system to determine who should be given the installation, determine what steps must be scripted to perform the upgrade, and then combine the two to schedule who should receive a specific upgrade.

A good software distribution system should also provide some safeguards for the process. A check should first be made to make sure there is adequate disk space available to complete the installation before starting. There should also be the ability to reverse or undo any distribution just in case of some problem.

Policing systems for unlicensed or prohibited software is one area where the inventory and distribution functions really come together. When all applications were stored on the network, administrators had more control over users' actions. The problem has grown as software installations were forced to shift to user workstations. This is a growing area of concern for firms interested in managing software assets and protecting themselves from liability from software audits. The problem of maintaining only approved software on systems continues to grow as more people have computers at home.

The ramifications can be grave for managers charged with maintaining the integrity of networks. For instance, when the shareware game Doom first came out, a network mode allowed people to play against each other over the local area network. Unfortunately, the program was very inefficient with network bandwidth and caused massive, difficult-to-isolate slowdowns.

Software distribution systems allow administrators to regain control over the integrity of their networks. The solution is simple, using the combination of features in the software inventory and distribution modules.

First a query is made to the inventory module to identify all workstations with Doom installed and then a distribution can be designed to erase the files and directories on the users' workstations. It is even possible to send a message letting the user know this was done.

This same combination allows administrators to take fast, decisive action in case of a virus attack or similar threat, for instance, the Microsoft Word macro virus. The same procedure, described above, can immediately wipe this problem from all users' machines. The alternative would be a long, manual process which might not succeed in finding the problem in all places.

PUSH VS. PULL DISTRIBUTIONS

We need flexibility in sending distributions to user workstations. An application can take a long time to install if we are trying to distribute large applications. Forcing distributions on busy users will only create problems for the organization. As a result, two different distribution mechanisms have been designed, which can be used together to tailor distributions which allow some freedom for the user but still ensure delivery of important changes.

The more obvious distribution is called a push. A push distribution forces the changes onto the users' PC either with notification or without. If we were only taking control of the workstation for a few moments, it might not make much of a difference; however, a new application can take a significant amount of time. If we try to perform an installation when someone is sitting down at their computer to get something out quickly it would interfere with their work process. A user trying to make some last minute changes to a presentation, for instance, would not be pleased with a 15-minute delay while your software distribution systems commandeers their machine, no matter how helpful they may find the new software.

A push distribution is especially useful for critical changes. If an update to correct a critical system function is available or if a bug fix for a template is developed, we want to deliver these changes as quickly as possible. Whenever a change will effect system stability, we want to deliver these changes in the minimum amount of time.

The alternate distribution, for noncritical or large changes, is a pull. This is the most common type of distribution. The pull distribution gives the user a choice in when the distribution is to run. Since we cannot determine what a user will be doing at any time, it is usually better to offer them a choice. The pull distribution asks permission, and gets the users' approval before the process is run.

A pull distribution system allows us to query the user before sending a distribution. We can, for instance, open a box on the user's screen with the message "There is an upgrade from Word 7.0 to Word 7.5 which needs to be installed on your PC. The installation will take approximately 15 minutes. May I install it now?" The user can then respond yes or no. This way distributions are always installed at the users' convenience. Each time they log in they will be prompted for the distribution again, giving them another opportunity for the installation.

But there are no convenient times for some users, as anyone who has managed a network of any size knows. To some users everything they do is critical, and they are so busy there is never a good time for system maintenance. If we give these users a choice, they will reject our pull distributions reflexively. Some even develop a reflex action, replying no, without even seeing what is being sent, whenever they see the distribution request box. Worse, over time, these users will build up a backlog of distributions. When they finally do agree to receive distributions, they may find themselves swamped.

In most cases, distributions will be performed as a combination of both push and pull. With a pull distribution mechanism in place we can specify to offer a pull distribution *n* number of times. Say we set our distribution threshold for three times. The user can reject the distribution for a maximum of three attempts. We can design our distribution to automatically force a distribution, a push, after the threshold is reached. So, after our user rejects our three pull distributions, they will have the distribution delivered on a push the next time the distribution is sent. Combining push and pull methods in the same distributions allows us to offer users flexibility while maintaining control over the stability of the network.

As software distribution systems develop, distribution will not be limited to the local area network. Companies have already started to offer software distributions over the Internet. These are almost always pull distributions to prevent undue intrusions on users. Pointcast has been doing this for some time. Pointcast is an innovative news distribution system. They provide client software which runs on your PC and pulls down news stories from their servers over the Internet. They also built in a software distribution system into their software.

As an innovative, quickly developing product, they had the problem of wanting to enhance their software on an ongoing basis. Using the traditional method of having users select and download the latest software from their Web site would not have given them the level of control they needed to continually update their software. Since the software automatically connects to their site for new news stories, they built in a mechanism where the software also reports its version number.

If the software is not the latest revision, the user is offered the opportunity to upgrade: a pull distribution. If they answer yes, then the software is automatically uploaded and installed by the Pointcast application. If users reject opportunities to upgrade their software, then eventually they will find their software too out of date to communicate with the Pointcast server.

Pointcast used their software distribution ability to make information about the 1996 Summer Olympics available for a limited time. Users found a new section of information about the Olympic events and results available as the Olympics began. Shortly after the Olympics, this section was removed. As more vendors incorporate software distribution systems within their applications, we will continue to see innovative changes in the way software updates are performed.

Eventually, built-in software distribution agents may make periodic software updates obsolete. Instead, vendors will create products which evolve over time, with features and new abilities appearing on a regular basis or available as needed for additional cost. This, of course, will create new challenges for systems administrators, help desks, and trainers at the same time it solves the problem of periodic updates software distribution systems were designed to alleviate. In the trade, we refer to this as irony.

Networks with a large number of users and servers have an additional problem. Setting up the same distribution on server after server removes some of the automation we had hoped to achieve from the distribution process. More sophisticated distribution systems provide a mechanism to provide levels of distribution. Staging servers can be set up which then distribute the distribution to other servers, which then are either staging servers for other servers or distribute to workstations.

Staging servers are especially effective for disbursed organizations on wide area networks. A server in a central location can stage to other servers in different geographic locations. These servers can then distribute to all other servers in their location. This minimizes the traffic created on the network to perform the distribution, while automating the process of getting the distribution to all locations.

Many distribution packages include modules for other, related modules to create a distribution suite. These applications are not necessary for the distribution process directly, but expand the management capability of the package.

One common addition is a remote control utility. Remote control utilities are a very common system management tool. The program allows all screen and keyboard activity to be shared, either through a local area network or dial-up connection, with a second PC. The ability to manually control a user's PC allows the system to be used for problem resolution.

Typically, a call comes into the help desk from a user experiencing problems with his PC. The technician examines the tracked files log to see if there have been any changes made to critical files on the PC. The log shows changes have been made to one of the configuration files on the PC. The user admits installing an application on the PC, but needs the program to get some work out. Rather than activating a distribution to send the standard configuration file, the technician takes control of the user's PC and modifies the file to correct the problem.

Because of the interaction among modules in a package, many vendors have built a management console to integrate the modules together. This makes it easier to move among modules during the resolution process. This way, a technician can gain access to problems reported by the anti-virus module or examine a system performance module to solve network problems.

Another common addition is a software metering module. Like inventory, metering has become a fairly ubiquitous application. Metering allows control of a limited number of software licenses. The metering program keeps track of how many people are using a specified application at any one time. When the number of users reaches a preset limit, additional users are locked out until some quit the application. This allows organizations to minimize the software expense by purchasing fewer licenses, while maintaining compliance with manufacturer's license agreements.

Software metering products are not without problems, however. Windows-based operating systems make it easier to switch among applications. It is not necessary to exit an application in order to use another. Quite often users will fail to close an application when they start working with another. In this case, the first application would not be freed from the metering system for use by another person.

Some manufacturer's licensing agreements do not permit concurrent usage at all. It is important to read the specific licensing agreement. Some agreements specify use by a named person or installation on only one computer. More and more vendors are realizing the value of Enterprise licensing. Allowing a company to license a program for any number of users removes the need for software metering.

A specific problem for software metering products is software suites. A suite is a collection of applications, from a single vendor, which share a licensing agreement whether they do or do not share functionality. Most suite licenses state that using any application in a suite uses the suite license. So, for instance, using the spreadsheet program of a suite uses the suite license. Another user cannot use the word processor of a single copy of the suite without violating the licensing agreement.

HOW DO YOU KNOW IF YOU CAN USE SOFTWARE DISTRIBUTION?

Not every organization is suitable for software distribution. There are two factors you can use to judge your organization's suitability as a candidate for software distribution — size and level of standardization of workstations.

Creating distributions takes time. There is little benefit to automating the process if it does not save time over doing it manually. If there is not a sufficient number of users, then manual means of managing software might be adequate for managing user desktops.

There is little need for software distribution in networks of 100 to 200 nodes. There may still be issues controlling user installing applications on their own, but, in general, smaller networks are managed easily enough without resorting to the expense of an automated distribution system.

The second reason software distribution may be unsuitable is lack of standardization. Software distribution will not work if there are too many different workstation configurations. If you want to build a distribution to upgrade even something as simple as users' network drivers, you first need to know where they reside. If user workstations have these drivers installed in a variety of different directories, then you will need to create multiple distributions to do the same job. It is time consuming to document different workstation configurations, and create, test, and run each version of the same distribution.

You might embark on a standardization program as groundwork for implementing a software distribution system. However, this is often not as easy as it sounds. The problem is not lack of standards; all organizations have standards. The problem is most organizations have had so many standards over time, it looks as if there are no standards.

Legacy applications often foil attempts at standardizing workstation configurations. It can be difficult to standardize systems when there are many old applications where the file locations are hard coded.

The mere size of a network often defeats grand efforts to standardize. The time it takes to analyze all applications on a 10,000 node network, design a new configuration that suits all or most users' needs, and reconfigure all workstations is often more work than an organization can sustain.

Software distribution offers a toolset that makes it possible for large organizations to regain control of PC desktops lost when Windows replaced DOS on the desktops of users around the world. The present accelerated rate of development in both hardware and software systems requires a systematic approach to desktop management to be effective.

A well-implemented software distribution system can provide an organization a powerful tool to organize and maintain control of a large distributed network. It provides the tools necessary to reduce the cost of rolling out new workstations. Of even greater value, software distribution provides a foundation for creating a management system which controls the user environment while allowing upgrades and environmental changes creating a stable and flexible environment. The result is increased stability and a more supportable network, whether the changes are new applications, upgrades, new drivers, or bug fixes. Over time the ability to make these kinds of changes is of immeasurable value to any organization.

Chapter 49 Fault Tolerance Protection and RAID Technology for Networks: A Primer

Jeff Leventhal

According to a recent *Computer Reseller News*/Gallup poll, most networks are down for at least two hours per week. The situation has not improved for most companies in the past three years. If an organization has 1,000 users per network, this equals one man-year per week of lost productivity. Even if a network is a fraction of that size, this number is imposing. For nearly a decade, many companies responded by deploying expensive fault-tolerant servers and peripherals.

Until the early 1990s, the fault-tolerant label was generally affixed to expensive and proprietary hardware systems for mainframes and minicomputers where the losses associated with a system's downtime were costly. The advent of client/server computing created a market for similar products created for local area networks (LANs) because the cost of network downtime can similarly be financially devastating. Network downtime can be caused by anything from a bad network card or a failed communication gateway to a tape drive failure or loss of a tape used for backing up critical data. The chances that a LAN may fail increases as more software applications, hardware components, and users are added to the network.

This chapter describes products that offer fault tolerance at the system hardware level and those that use fault-tolerant methods to protect the integrity of data stored on network servers. The discussion concludes with a set of guidelines to help communications managers select the right type of fault-tolerant solution for their network. This chapter also discusses RAID (redundant array of independent [formerly "inexpensive"] disks)

0-8493-9823-1/00/$0.00+$.50

technology, which is used to coordinate multiple disk drives to protect against loss of data availability if one of the drives fails.

DEFINING FAULT TOLERANCE

PC Week columnist Peter Coffee noted the proliferation of fault tolerance in vendor advertising and compiled a list of seven factors that define fault tolerance. Coffee's list included safety, reliability, confidentiality, integrity, availability, trustworthiness, and correctness. Two of the factors — integrity and availability — can be defined as follows:

- Availability is expressed as the percentage of uptime and is related to reliability (which Coffee defined to be mean times between failures) because infinite time between failures would mean 100 percent availability. But when the inevitable occurs and a failure does happen, how long does it take to get service back to normal?
- Integrity refers to keeping data intact (as opposed to keeping data secret). Fault tolerance may mean rigorous logging of transactions, or the capacity to reverse any action so that data can always be returned to a known good state.

This chapter uses Coffee's descriptions of availability and integrity to distinguish between products that offer fault tolerance at the system hardware level and those that use fault-tolerant methods to protect the data stored on the network servers.

Availability

This is the proliferation of hardware products with fault-tolerant features attributable to the ease with which a vendor can package two or more copies of a hardware component in a system. Network servers are an example of this phenomenon. Supercharged personal computers equipped with multiple power supplies, processors, and input/output (I/O) buses provide greater dependability in the event that one power supply, processor, or I/O controller fails. In this case, it is relatively easy to synchronize multiple copies of each component so that one mechanism takes over if its twin fails.

Cubix's ERS/FT 11. For example, Cubix's ERS/FT II communications server has redundant, load-bearing, hot-swappable power supplies, multiple cooling fans, and failure alerts that notify the administrator audibly and through management software. The product's Intelligent Environmental Sensor tracks fluctuations in voltage and temperature and transmits an alert if conditions exceed a safe operating range. A hung or failed system will not adversely affect any of the other processors in the system.

Vinea Corp.'s StandbyServer. Vinea Corp. has taken this supercharged PC/network server one step further by offering machines that duplicate any server on the network; if one crashes, an organization moves all its users to its twin. Vinea's StandbyServer exemplifies this process, known as mirroring. However, mirroring has a significant drawback — if a software bug causes the primary server to crash, the same bug is likely to cause the secondary (mirrored) server to crash also. (Mirroring an iteration of RAID technology, which is explained in greater detail later in this chapter.)

Network Integrity, Inc.'s LANtegrity. An innovative twist on the mirrored server, without its bug-sensitivity drawback, is Network Integrity's LANtegrity product in which hard disks are not directly mirrored. Instead, there is a many-to-one relationship, similar to a RAID system, which has the advantage of lower hardware cost. LANtegrity handles backup by maintaining current and previous versions of all files in its Intelligent Data Vault. The vault manages the most active files in disk storage and offloads the rest to the tape autoloader. Copies of files that were changed are made when LANtegrity polls the server every few minutes and any file can be retrieved as needed. If the primary server fails, the system can be smoothly running again in about 15 seconds without rebooting. Because all the software is not replicated, any bugs that caused the first server to crash should not affect the second server.

NetFRAME Servers. The fault tolerance built into NetFRAME's servers is attributable to its distributed, parallel software architecture. This fault tolerance allows the adding and changing of peripherals to be done without shutting down the server, allows for dynamic isolation and connection of I/O problems (which are prime downtime culprits), distributes the processing load between the I/O server and the central processing unit (CPU), and prevents driver failures from bringing down the CPU.

Compaq's SMART. Many of Compaq's PCs feature its SMART (Self- Monitoring Analysis and Reporting Technology) client technology, although it is limited to client hard drives. If a SMART client believes that a crash may occur on a hard disk drive, it begins backing up the hard drive to the NetWare file server backup device. The downside is that the software cannot predict disk failures that give off no warning signals or failure caused by the computer itself.

DIAL RAID FOR INTEGRITY

In each of the previous examples, the fault tolerance built into the systems is generally designed to preserve the availability of the hardware system. RAID is a technology that is probably the most popular means of ensuring the integrity of corporate data. RAID (redundant arrays of independent disks) is a way of coordinating multiple disk drives to protect against loss of data availability if one of the drives fails. RAID software:

- Presents the array's storage capacity to the host computer as one or more virtual disks with the desired balance of cost, data availability, and I/O performance
- Masks the array's internal complexity from the host computer by transparently mapping its available storage capacity onto its member disks and converting 110 requests directed to virtual disks into operations on member disks
- Recovers data from disk and path failures and provides continuous I/O service to the host computer

RAID technology is based on work that originated at the University of California at Berkeley in the late 1980s. Researchers analyzed various performance, throughput, and data protection aspects of the different arrangements of disk drives and different redundancy algorithms. Exhibit 49.1 describes the various RAID levels recognized by the RAID Advisory Board (RAB), which sets standards for the industry.

The redundancy in RAID is achieved by dedicating parts of an array's storage capacity to check data. Check data can be used to regenerate individual blocks of data from a failed disk as they are requested by the applications, or to reconstruct the entire contents of a failed disk to restore data protection after a failure.

The most common forms of check data are a mirror (identical copy) of user data and shared parity, which involves appending mathematical code to data bits for later comparison, matching, and correction. Different combinations of mapping and check data comprise distinct RAID levels.

Striping

Of the six well-defined RAID levels, as listed in Exhibit 49.1, three are commonly used. Level 1 uses mirroring for data protection and may incorporate striping. Striping refers to the location of consecutive sequences of data blocks on successive array members. Striping balances I/O load, thereby increasing performance.

Levels 3 and 5 both use parity for data protection and almost always incorporate striping. RAID levels 3 and 5 use different algorithms for updating both user data and check data in response to application write requests.

In a RAID level 3 array, the disks are physically or logically synchronized, and each contributes to satisfying every I/O request made to the array (i.e., parallel access). In a RAID level 5 array, the disks are allowed to operate independently (i.e., independent access) so that in principle the array may satisfy multiple application I/O requests concurrently.

Exhibit 49.1 RAID levels.

	Description	Benefits	Disadvantages
RAID 0	Disk striping; data are written across all drives, multiple disk drives.	Storage is maximized; tolerance features good performance and low price.	Has virtually no fault.
RAID 1	Disk mirroring; data are copied from one drive to the next.	Data redundancy is increased 100 percent; has fast read performance.	Slower write performance, but twice the disk drive capacity; more expensive.
RAID 2	Spreads redundant data across multiple disks; includes bit and parity data checking.	Has no physical benefits.	Has high overhead with no significant reliability.
RAID 3	Data striping at a bit level; requires a dedicated parity drive.	Has increased fault tolerance and fast performance.	Is limited to one write at a time.
RAID 4	Disk striping of data blocks, requires a dedicated parity drive.	Has increased fault tolerance and fast read performance.	Slower write performance; not used very much.
RAID 5	Disk striping of both data and parity information.	Features increased fault tolerance, efficient performance, is very common.	Write performance is slow.

Some RAID levels are theoretically faster than others, but in many situations the existing hardware technology does not always enable these performance enhancements to be realized. Other factors that are significant in overall system performance include the combinations of the disk drive, the host adapter, the tuning of the operating system, and how these components function together.

Parity

In RAID level 3, parity information that is saved to one designated tape drive can be used to regenerate data from a failed drive or tape media (see Exhibit 49.2).

RAID 5 offers improved storage efficiency over RAID 1 because parity information is stored rather than a complete redundant copy of all data. The parity information is essentially a number determined by adding up the value of all the bits in the data word. Parity requires some amount of overhead, ranging from 50 percent on RAID 1 to somewhat less than 20 percent on RAID 5. The result is that three or more identical drives can be combined into a RAID 5 array, with the effective storage capacity of only one drive sacrificed to store the parity information. Therefore, RAID 5 arrays provide greater storage efficiency than RAID 1 arrays.

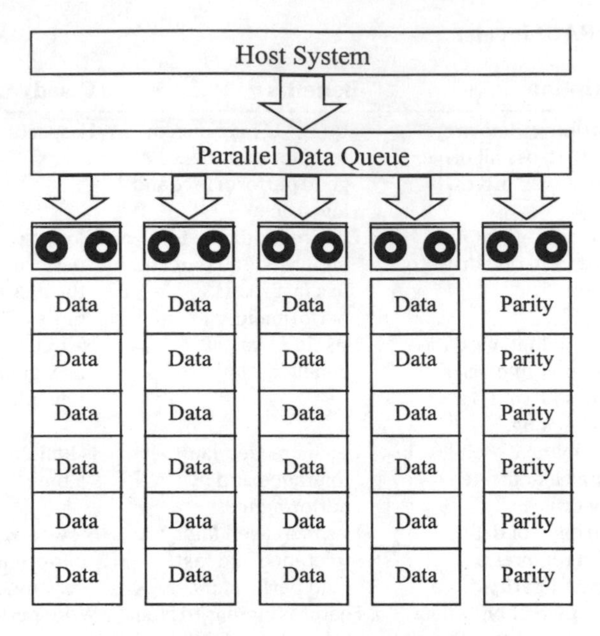

Exhibit 49.2 Example of RAID 3.

RAID 5 is an implementation in which parity information is striped across all the configured drives. This method can increase the array throughput by separating the parity information across all drives. This is the preferred method when using transaction or database processing (see Exhibit 49.3).

Data blocks and parity blocks are striped to the drives or tapes in a stair-step, or barber-pole fashion, allowing for full restoration even if a disk or tape is lost or damaged. In such an event, the data and parity blocks on the remaining drives or tapes contain enough information for the software to extrapolate the "lost" data.

Mirroring

Modern operating systems that are built for the enterprise, such as Windows NT 4.0, provide both RAID 1 and RAID 5 fault-tolerance protection. RAID Level 1 (disk mirroring) simultaneously streams data to two hard drives (or tape devices) throughout the entire job, not just in the event of the hardware failure. Both drives are considered to be one drive by the Windows NT software (NT's fault tolerance driver is called FTDISK.SYS).

Disk mirroring creates a duplication of partition data onto another physical disk. Any partition, including the boot or system partitions, can be mirrored. This strategy protects a single disk against failure.

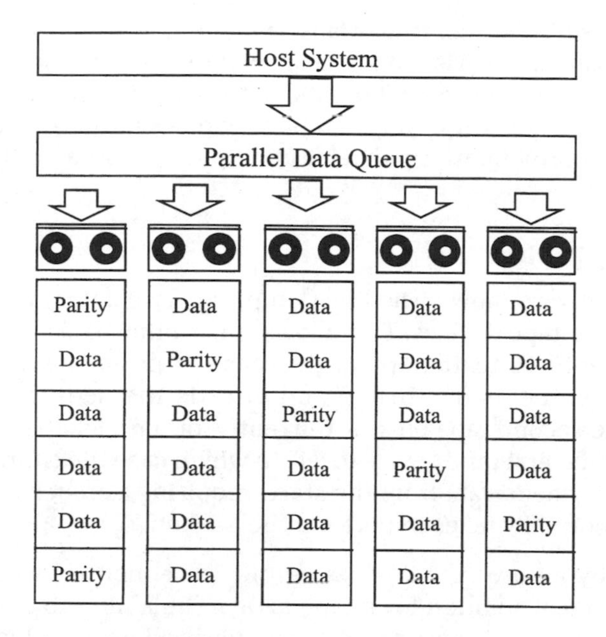

Exhibit 49.3 Example of RAID 5.

Disk mirroring is generally appropriate for small LANs, as its initial cost is limited to only two disk drives. As the need for more network storage capacity grows, mirroring may become more expensive per megabyte than other forms of fault tolerance because only 50 percent of the disk space is being used. However, with the advent of low-cost integrated drive electronics (IDE) (IGB for under $300, for example), disk-mirroring controllers and large mirrored RAID drives could become viable alternatives, in workstations and small work groups, to the faster, more powerful small computer system interface (SCSI)-based arrays — provided users do not need the full performance of the SCSI systems.

For extra security, an entire RAID system can also be mirrored, but of course that doubles the cost. How communications managers determine whether mirroring or striping is best for their enterprise depends on their confidence in the technology and how much they are willing to spend.

Mirror drive sets provide a modest performance increase for reading because the fault tolerance driver can read from both members of the mirror set at the same time. However, there is a slight performance decrease when writing to a mirror set because the fault tolerance driver must write to both members simultaneously. When one drive in a mirror set fails, performance returns to normal because the fault tolerance driver is working with only one partition.

If a mirror set's disk controller fails, then neither drive of the mirror set will be accessible. It is wise to install a second controller in the computer so that each disk in a mirror set has its own controller. This arrangement is called a disk duplex. Duplexing is a hardware solution to fault tolerance and should improve performance. It is helpful to add a duplicate power supply as well.

Tape Backup Devices

Cheyenne Software (now owned by Computer Associates) was the first to offer RAID 5 for tape devices. This allows the routine backup window to be met (albeit with 10 to 15 percent decrease in performance when both drives are treated as one drive) and affords the user the security of archiving the second tape off-site. Currently, other backup software products wait for the first tape device to fail, in which case the entire backup job must be performed again from the start, requiring as much as double the time that it normally takes to perform the backup.

Because by nature tape devices employ a sequential access method, RAID 5 is an ideal solution for a tape array. The ability to deliver several parallel data streams to an array of tape devices allows for a highly scalable transfer performance. Although tape devices possess very intense error detection and correction algorithms, the user is still left vulnerable to mechanical or tape cartridge failure that could render a critical restore of data unobtainable. A RAID 5-based array of tape devices provides an economical way to protect the backup/restore session data.

RAID level 5 (data striping) is only supported with three or more disk drives in NT, and three tape drives with Cheyenne's tape software. If one of the drives is destroyed, a complete restore of the data set can still be made from the remaining tapes. If less than three tape drives are used, the data striping will occur, but without fault tolerance. The location of the parity information is moved from drive to drive in a barber-pole fashion. This is extremely important for tape RAID systems, where tape drives compress information.

SUMMARY

When selecting a network's RAID level, the choice between the RAID 1 and RAID 5 depends on the level of protection desired and an organization's budget. The major differences between disk mirroring and striping with parity are performance and cost.

Generally, disk mirroring offers better I/O performance. It also has the advantage of being able to mirror the boot or system partition. Disk striping with parity offers better read performance than mirroring; however, the need to calculate parity information requires more system memory and can slow write performance. The cost per megabyte is lower with striping because disk utilization is much greater. For example, if there are four disks in a stripe set with parity, the disk space overhead is 25 percent, compared to 50 percent disk space overhead for disk mirroring.

An effective means of achieving fault tolerance is to combine RAID 1 disk mirroring with RAID 5 data striping on the same computer. Consider mirroring the system and boot partitions and protecting the rest of the drive with stripe sets with parity.

Chapter 50
Network Data and Storage Management Techniques

Lawrence D. Rogers

One of the fundamental services provided by a multi-user computer system has always been access to shared data. The shared data, once counted in megabytes, are now usually counted in gigabytes. Protecting and managing its storage is one of the most important tasks of the LAN administrator. To protect the data and manage storage requires a combination of hardware protection, operating systems services, and software tools. There are two main problems — maintaining data integrity and storage management.

Data integrity means that systems data are always correct or can be corrected. Data integrity systems provide the ability to recover data in the face of hardware, software, and human error or from direct physical or programmatic attack. Error correcting codes (ECC) and fault tolerant disk systems can ensure immunity to all but the most disastrous hardware failures, and corrections are performed, for the most part, invisibly. Recovery from software overwrite failures and human errors, however, relies on keeping multiple copies of files and file histories. Disaster recovery (including recovery from direct physical attack) is usually achieved by keeping copies and file histories at separate secure locations. In the latter cases some human intervention is required, if only to make choices during the recovery process. Protection from programmatic attack requires an adequate security system and the use of virus detection or prevention programs with a file copy and history system for when serious security breaches are committed.

Storage management means optimizing the use of systems storage media with respect to both cost and performance. Moving infrequently accessed data from expensive primary storage to less expensive storage media improves access performance and makes primary storage space available for the most commonly used data. Defragmenting disks improves

0-8493-9823-1/00/$0.00+$.50

performance by reducing access times. Compressing files frees disk space. Balancing the load across multiple disk drives and servers to prevent simultaneous overutilization and underutilization of disks in the same system smoothes performance. These and other techniques are commonly applied to storage systems to optimize performance and cost.

This chapter discusses tools and techniques used for maintaining file histories and for managing the location of files in the storage system and briefly touches on related disaster recovery techniques. The chapter does not discuss fault tolerant systems or hardware technology for ensuring first-level data protection, nor does it discuss antivirus software or security or many other related topics. Each is beyond the scope of this chapter.

MAINTAINING FILE HISTORIES

Backup and Restore

An important technique for ensuring that data are not lost is to maintain a version history for each file. This is done by periodically copying files to a different media, such as tape or optical disk. This is usually referred to as backing up files, because it involves making backup copies of files in case an original is lost or destroyed. The versions of a file available are determined by the frequency with which these copies are made and how long the copy is kept. The two main components of a backup strategy are the type of backup performed and the media rotation strategy. This chapter first defines some different types of backup and then discusses how these can be combined with various rotation schemes.

Full Backup. A full backup (i.e., a backup of all files) made on a regular basis allows a network administrator to recover versions of all files from the system. The available versions of any given file are determined by the interval between backups. Periodic full backups have the advantage of containing all the systems data on one backup medium, or at least a minimal number of media. This can help speed recovery if a full volume must be restored. Due to the size of a full backup, and hence the time required to perform one, it may be difficult or impractical to execute full backups as often as desired. Properly combining full backups with partial backups can still give reasonable protection.

Incremental Backup. An incremental backup captures all files that have been modified since the last full or incremental backup. It requires less time to perform because less data are being copied. To restore a full volume requires two steps, first restoring from the last full backup tape and then applying each incremental backup. Generally a full backup is done each weekend and incrementals each weekday.

Differential Backup. A differential backup captures all files that have been modified since the last full backup. This means that a Monday differential captures Monday's changes after a weekend full backup. A Tuesday differential captures Monday's and Tuesday's changes. Differentials done later in the week require more space and time than incrementals, but a restore requires only the full backup restore and the latest differential.

Modified Full Backup. A modified full backup captures all files, except those for which three unmodified copies already exist on three different media. In other words, if a file has been copied three times each to a different medium, and it has not changed, it will not be backed up again. This reduces the time to do a full backup because files that do not change are not backed up more than three times.

If tape rotation schemes are used, precautions must be taken to see that needed copies are not overwritten. Modified full backups done on the weekend, with incrementals or differentials during the week, will ensure that the file history has copies of all files in their most recent state. Weekend backup time will be shorter, but the files will be spread across many different media. The big disadvantage is restoring a full volume or even a directory because there is a high probability the required files will be stored on many different tapes.

Continuous Backup and Transaction Tracking. The techniques described previously allow a system to be restored to its state at the time of the last backup. How often a backup is done depends on how critical the data are. A company's financial files might get backed up twice a day, while routine correspondence might be copied only every two days. Some data are so critical that a business cannot afford to lose any of it. For these instances, a continuous file history can be recorded. Often these types of data are found in on-line database systems. Most serious database systems maintain a separate log of transactions. There are various reasons for doing this that include backing out transactions to correct errors.

This log can also be effectively used for providing continuous backup. If a consistent backup of the database is made, and then the transaction log is backed up frequently, a system can be restored to the time of the last transaction captured by restoring the backup and then reapplying each of the transactions.

It is good practice, then, to store the transaction log on a disk other than the one the data reside on, and to have it backed up frequently or even duplexed. When a full backup of the database is performed a new log file should be started. The log file itself should then be backed up repeatedly, as often as after every transaction is written to it. This same technique can be applied to important non-database files by acquiring a system that supports continuous backup. These systems basically capture a baseline backup of the files of interest and then create all the writes to those files. Using this technique, a file can be restored to the time of the last write.

Tracking the File History and Recovering Files

Knowing what media to use when recovering a whole volume is straightforward. As described above, the last full backup is restored and then the necessary partial backups are applied, depending on what kind of backup strategy has been employed. Usually the media are labeled, and this is all that is necessary.

To recover individual files or groups of files, however, the communications manager must know what media are being used. Even if a recent version is desired, knowing which media it is on is not always obvious, and finding it without the help of the file history system can be tedious. It is important for file history or backup systems to keep an on-line database that tracks all versions of files and where they are stored. Various approaches to finding these files can then be supported.

When the primary file system is used, a file is referred to by its unique path name (i.e., the volume, directory, and file name used for storing it). Well-designed history databases support this same view and their relationships. Users can then browse the backup file much as they would browse the directory of the primary file system.

When a particular file is selected, each of its versions and their location can be displayed. A common type of users' request is "Recover my spreadsheet; it was called BUD... something. I had it a few months ago." To find such a file, locating it by name or type is a good approach. The file history system should support searching by file name, by extent, and permit the use of wildcards (e.g., those supported in the DOS and Novell operating systems directories).

Another approach is to locate the file by the save set to which it was copied. A save set is a grouping of data that was all copied to a backup media together. Most backup systems that have on-line data support tracking the files as they were stored. Tracking a file by set allows looking for a file as it existed at a particular time.

MEDIA ROTATION SCHEMES

So far this chapter has not discussed how long each copy of a file is to be kept. In the previous discussions it has been implicit that when a copy has been made it is henceforth always available. This is not usually necessary, or cost effective; it would mean always buying new storage media to replace the ones that fill up. When a file has been superseded by later versions it is, after some period of time, usually not required.

Keeping too few copies of files, however, can seriously diminish the security of the data, and so trade-offs are almost inevitable. The speed with which files can be restored, the amount of time available for backup, and the amount of data also enter into the equation.

Various schemes for reusing media are employed, both as a cost-saving measure and to keep system clutter (e.g., database size and number of tapes) under control. When media are reused, previously backed-up data are overwritten. The file history database must be updated to reflect the media's new contents. Most backup databases remove all the files stored on a tape when the user indicates that a tape is to be reused. The administrator should know four things in order to make appropriate decisions about media rotation:

- The periods each day when the network is fairly dormant (the backup window)
- The amount of data on the network disks
- The expected time that would be available to the manager to perform a full system restoration
- How far back in time the file history must go

Father–Son Rotation Schemes

An example would be a small manufacturing concern whose network has 120 nodes, two servers, and a total of 2 GB on the server disks. The network is running three shifts, five days a week, and there is only one hour each night during the week to perform routine backup. The ability to restore data is crucial, but only four hours is available in which to do it. Three to four weeks of data history is sufficient. In this case, the eight-tape rotation schedule shown in Exhibit 50.1 might be chosen.

The eight-tape schedule provides for one full backup on Fridays, augmented with four incremental backups Monday through Thursday. The Friday tape is on a separate subschedule of four tapes. During week one, tapes one through five are used. During week two, tapes one through four are followed by tape six, and so on until tape eight is used. Then the cycle repeats. This is ideal because the daily incrementals can be done during the 60-minute window each night (which is enough time to back up at least 300 MB of changed data), and for the Friday backup there is plenty of time before the next shift starts.

The only penalty this arrangement has is restoration time. In a worst-case scenario (i.e., restoring an entire volume), it would be necessary to load the last full backup tape, followed by every subsequent incremental

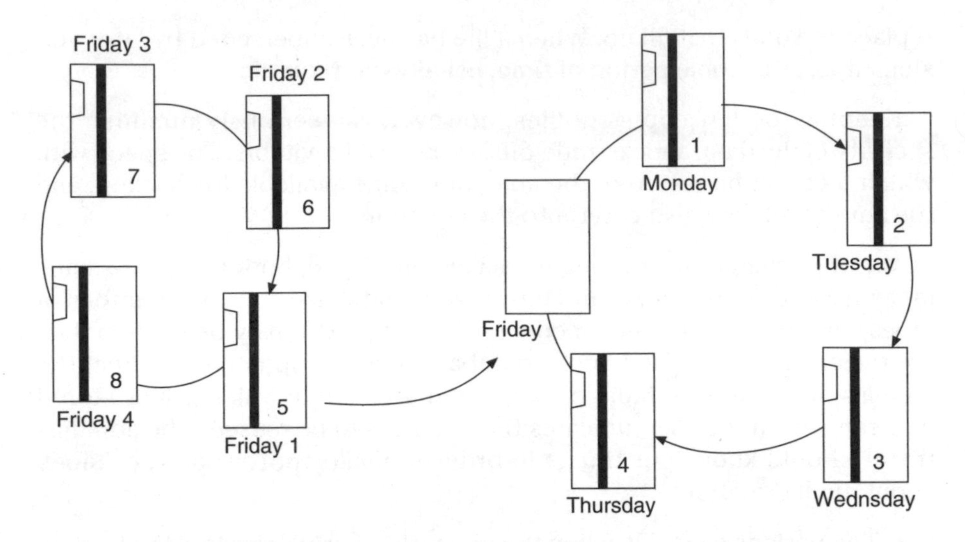

Exhibit 50.1 A father–son tape rotation schedule.

tape, to ensure that the latest version of every file is restored. If more time is available at night for doing backups and less for restore, differential backups would be done in place of the incrementals. If a longer file history is required, more tapes would be included in the weekly, father tape rotation.

Many other father–son schemes can be constructed. Performing two full backups a week on Tuesday and Friday, with increment on Monday, Wednesday, and Thursday, will increase confidence and restore times but will take more time at night twice a week. Keeping three weeks of history will now require 11 tapes. Performing a full backup every night is the safest method of securing data and requires the least restore time. However, it requires the most time each night and will cost as much in tapes as a father–son scheme that has one father tape being made each week, that covers the same history period.

Grandfather–Father–Son

Grandfather–father–son schemes increase the history period, but leave more holes in it. Suppose the company decides that it must keep three to four months of history. The grandfather–father–son scheme uses incremental or differential backups each weekday night. Four tapes are rotated for these. Each weekend for three weeks (Friday 1, Friday 2, Friday 3) a full backup is done, and once every four weeks a monthly backup is performed. A new monthly tape is used each month (i.e., each four-week period) for three months (Monthly 1, Monthly 2, Monthly 3) before the rotation starts

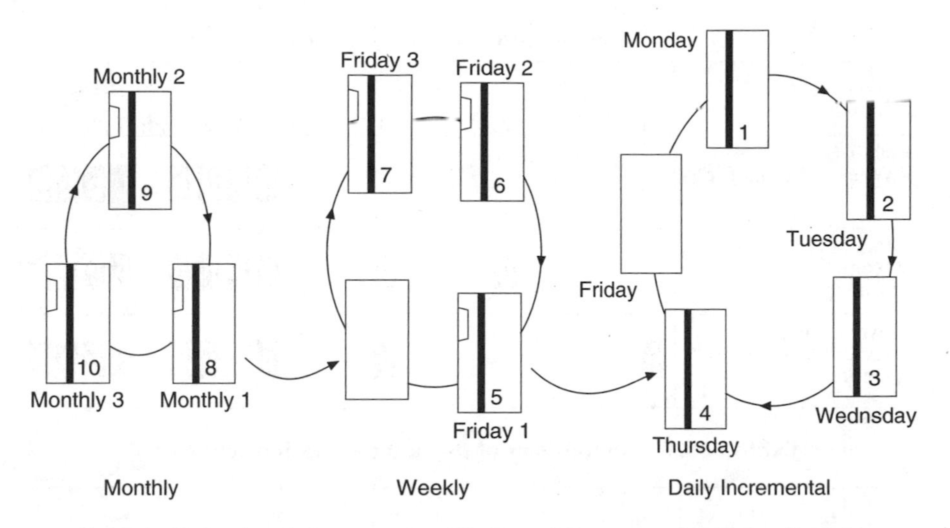

Exhibit 50.2 An eight-tape grandfather–father–son rotation schedule.

Exhibit 50.3 Tower of Hanoi tape rotation.

Tape 1 is used weeks	1	3	5	7	9	11	13	15	17	... 31
Tape 2 is used weeks	2	6	10	14	18	22	26	30		
Tape 3 is used weeks	4	12	20	28						
Tape 4 is used weeks	8	24								
Tape 5 is used weeks	16	32								

over (as shown in Exhibit 50.2). Data can be recovered from any day in the most recent week, any of the previous four weeks, plus data that are up to 12 weeks old in four-week increments. To increase the number of months in the history, more monthly tapes can be added.

Tower of Hanoi Tape Rotation

If economy and a long history period are required, the Tower of Hanoi algorithm can be used (at the expense of increasing the complexity of backup operations) for father weekly full backups. The usual four-tape son rotation of weekday incremental or differential backups is still used, as in the schemes discussed earlier. This scheme (named for a child's toy) uses weekend tapes at exponentially growing intervals and gets up to 2n — I weeks of coverage for n tapes. Five tapes, therefore, can give up to 16 weeks' coverage, as shown in Exhibit 50.3.

The rotation is then repeated. Just before Tape 5 is reused at week 32, its data are 16 weeks old, Tape 4's data are 8 weeks old, Tape 3's are 4 weeks

Age of Backups Currently on Tape

Tape Rotation Scheme	Weeks																Days				
	16	15	14	13	12	11	10	9	8	7	6	5	4	3	2	1	F	T	W	T	M
Father-Son 4 Weekly, 4 Daily Tapes													■	■	■	■		■	■	■	■
Grandfather-Father-Son 2 Monthly, 2 Weekly 4 Daily					■				■				■	■	■	■		■	■	■	■
Tower of Hanoi 5 Exponential 4 Daily	■								■				■		■	■		■	■	■	■

Exhibit 50.4 Comparison of three tape rotation schemes.

old, Tape 2's are 2 weeks, and Tape 1's are 1 week old. This is approximately the same coverage as grandfather–father–son with one less tape. Simplicity is probably worth the cost of an extra tape. A weakness of this scheme is the high use and hence potential failure of Tape 1. Exhibit 50.4 illustrates the length of file history, the gaps in the file history coverage, and the number of tapes required for these different tape rotation schemes.

MANAGING STORAGE

Monitoring Volume Use

Storage requirements are likely to change over time. An administrator should monitor the data storage devices to determine when additional storage is needed. Usually, filing a volume to 70 to 80 percent of its capacity is a comfortable range. This depends on the size of the volume and the type of data stored on it. More free volume space should be maintained if the data are subject to temporary files causing rapid growth spurts, and less is required if the data are largely stable. Deciding when a volume is full means understanding the mix of programs that will run and their storage profiles.

For example, the month-end financial program may use many intermediate temporary files and overflow the disk once a month unless this is anticipated. Understanding the storage profile of disks is key. Performance is also influenced by a lack of available disk space. Full disks are usually fragmented — that is, the space allocated is scattered across the disk rather than localized. The last files written are likely to be the ones most often accessed, and these are the ones that are most heavily fragmented.

Disk fragmentation causes head movement even when clever caching algorithms are used, resulting in a decrease in performance. Keeping sufficient disk space open allows the system to maintain performance. Adding disks is not necessarily the answer to dwindling free space. Archiving dormant files (i.e., files that have not been accessed in some predetermined period of time) will probably free up sufficient disk space.

Archiving

Many data are transitory in nature, being active for a short period of time and then dormant for months or even years. Still, data are a valuable commodity and it is impossible to determine when they will be required. Archiving is a technique wherein a particular data set is moved from on-line primary storage to near-on-line or off-line archival storage media and then deleted from primary storage. This differs from maintaining a file history (backup), which captures copies of files that are still active and on-line so that they can be retrieved if they are needed. Before data can be considered safely archived and removed from primary storage, copies have to be made on multiple media (e.g., three different tapes or two optical platters).

Whether archival storage is near-on-line, off-line, or off-site usually depends on the age of the data. Data might be automatically archived to a near-on-line device after they have been dormant for three months, and then taken off-line after six months. Off-site data storage is usually a disaster recovery measure, although sometimes it is convenient simply to save physical space.

Archival storage is intended to be long-term storage, so it is important to match the life expectancy of the media to the required data availability. The life span of tape depends on the environment in which it is stored (a computer room environment is recommended) and the care taken with it. Frequent retensioning, or spinning the tape from end to end, will keep it from sticking to itself.

With proper care, a 10-year life span is practical. Media vendors make available guidelines on storing tape. Optical disk is a newer technology than tape and is more expensive. Less is known about its durability, however, less environment control is necessary than for tape, and a practical life span of 20 to 30 years is probable.

An archiving system should provide an easy method for locating archived data. Each tape, optical disk, or other media used for archival storage must be titled and labeled for easy retrieval. This label should be both electronic, as data on the media, and physical, attached to the cartridge, tape, or whatever medium is used, in a visible location. The physical label should bear a name, archive data, and general description. The electronic label should duplicate the physical label plus provide information on the contents of the media (e.g., directories and save sets).

The same database that is used for tracking file history for backup should be used for tracking archived files, and the same search techniques should be available. Archived files are most frequently searched by the file name or by browsing the directory tree of archived data, rather than by the time that they were archived.

Whenever data are moved from one device to another, as with archiving, it is crucial to move all directory information (name space) and security (operating system attributes) with it. The removable media used for archiving should be protected by password, physical location, or both. The number of people with physical access to archived media should be kept at a minimum to further protect the privacy of the information contained in the archives.

Combining Archiving and Backup

Archiving and backup activities are similar, and it is convenient to share the file history database between them. It is important to remember that they serve different purposes and that archived files are intended to be kept for long periods of time, while backup files are maintained until they have been superseded and then discarded. Media that are used for archiving are never overwritten, while data integrity tapes are reused on a specific media rotation schedule.

The simplest and safest way to combine these two operations is to let them share a media database, but they should be kept on separate physical media and these media should be treated differently. Backup time and media cost can be reduced by combining the techniques. The modified full backup scheme described earlier reduces the time taken for a full backup by not capturing files that have been copied to three different archival media. As noted earlier, this saves backup time at the expense of restore time.

If performing modified full backups is important, it is possible to go one step further and combine backup and archiving on one media. On each weekend backup session the first thing written to a tape is the set of dormant or stable files that have not been captured three times. Then active files are written to tape. The next time this tape is used, any newly dormant or stable files would be appended to the first set of archived files, overwriting the active files. Then active files would be written. This process continues until the tape is full, at which time it is permanently stored offsite. Exhibit 50.5 illustrates this combined strategy.

Data Migration

Archiving data is an explicit action performed by a user or LAN administrator when files have been determined to be dormant. Sometimes the system

Exhibit 50.5 Combining archive and backup on one tape.

Tape After First Use

Dormant and Stable Files (First Use)	Active Files (First Use)	Unused Tape

Tape After Second Use

Dormant and Stable Files (First Use)	Dormant and Stable Files (Second Use, Overwrites Active)	New Active Files (Second Use)	Unused Tape

will take this action according to previously published system rules. Files are removed from the primary system and the names are removed from the primary file system directory tree. Retrieving them is also an explicit act. This requires locating the desired file(s) in the file history database and then restoring them. Archiving is performed as a routine practice for maintaining files over time and to keep them from cluttering expensive primary storage.

Data migration provides an automatic method of moving data from primary to near-on-line storage in order to maintain more free primary storage. This is done automatically according to aging rules such as those discussed for archiving (e.g., files that have been dormant for three months would be moved to near-on-line storage). Exhibit 50.6 illustrates this concept.

With data migration, however, file names are not removed from the original name directory and will still appear to a user browsing those directories as if they are on-line. When a user attempts to access a file, the file is restored from near-on-line to primary-on-line storage. The user or LAN administrator does not have to do any explicit moving of data or tracking of files. These are performed automatically by the data migration system.

Data migration must make use of near-on-line media or retrieve times become too long, and system time-outs become a potential problem. Typically, near-on-line storage employs high-capacity, rotating, removable media (e.g., optical or Bernoulli disks). Before installing such a system, it is important to make sure it is matched to the environment. Automatic data migration is similar to providing virtual storage for the primary disk system. The rules for when files are moved and when and how they are restored are sensitive to various virtual memory afflictions such as thrashing.

Using Hierarchical Storage

Data migration, archiving, and backup can be planned as one logical, harmonious set of actions. Archiving and data migration are used for managing storage. When using both these methods, the communications

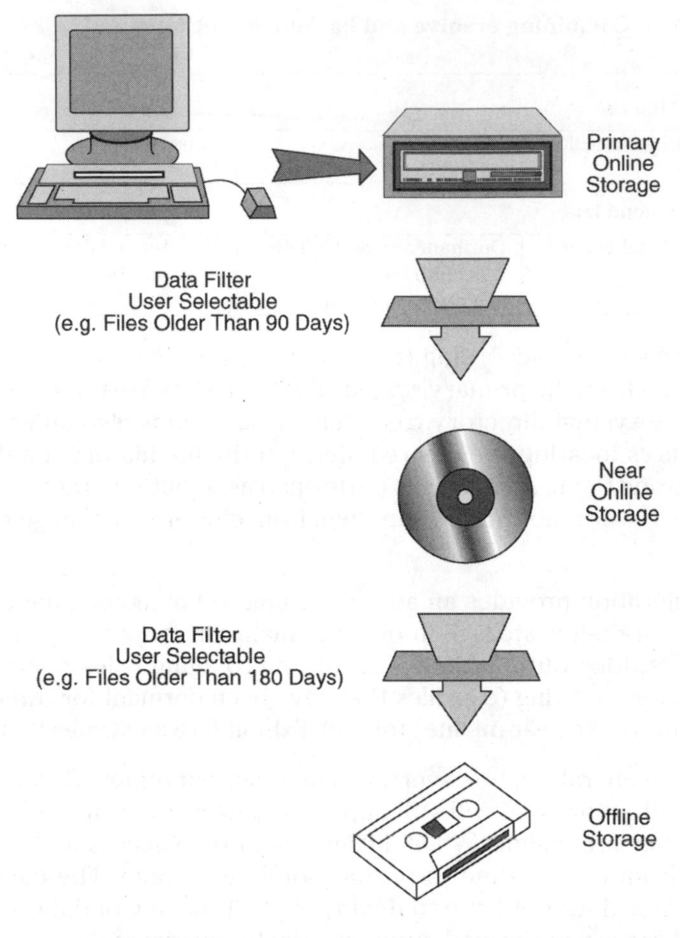

Exhibit 50.6 Schematic of data migration.

---

manager would set up the data migration using, for example, a near-on-line optical disk jukebox. The communications manager might archive from there to an autoloading tape device and eventually move the archive files off-site. If the archives must be kept for a long time, then specific optical media might be used for the permanent archives. Weekly and daily backups would be performed to the tape autoloader, with media being rotated from there off-site according to a schedule. The file history system would be used for tracking all copies of the files.

Such a system is called *hierarchical,* because data move down the hierarchy from primary on-line storage to fast-access near-on-line storage, to a slower-access near-on-line storage, to off-line and off-site storage. If all these levels of storage are to be managed, it has to be done automatically or the LAN administrator will be devoted solely to this task.

DISK FRAGMENTATION

Fragmentation has been a problem with storage systems since the introduction of direct access rotating media. When the file system needs to write data to a disk, it searches for the first available block and starts writing to this block. If the block is smaller than the amount of data that needs to be written, the storage algorithm will break up the data and store the remainder in another open block on the disk.

If, for example, a 1-MB file is being written to disk, the disk space found may be a 500-KB open block. The first half of the file is written to this area and then the system continues seeking. The next free space may be 300 KB, and more of the file is written. Finally, 200 KB of open space is located and the remainder of the file is laid down on disk. The single file has been fragmented into three pieces.

If these three fragments are physically far apart on the disk, the read head will have to move to retrieve them. Moving the disk head is a relatively slow mechanical operation; thus fragmentation adversely affects performance.

Normally, a file system will try to place all information on the disk contiguously. However, after files have been written and erased, holes are left. It is unfortunate that as disks fill up they are more prone to fragmentation, and the newest, most active files are usually the most severely fragmented. The effect of fragmentation can be minimized by systems employing efficient storage allocation techniques and disk caching. Administrators can help by ensuring that ample free space is available so the system has room to manage the storage. Even so, sometimes defragmentation is required. Defragmentation can be accomplished best using specialized utilities that have been designed for the task. These utilities rewrite data into continuous blocks to reduce disk arm movement.

An alternate method for defragmenting a volume involves the deletion and restoration of all data. When data are laid down on a fresh, unused volume, they are written contiguously; therefore, the whole volume can be backed up, the data deleted, and then a complete restore done. This should not be undertaken lightly. But with a proper full volume backup system it can be a fairly painless operation. It would be prudent to undertake multiple full backups and some restores to test the media and system before undertaking this form of defragmentation.

DATA COMPRESSION

Another technique for making better use of disk space is to compress the data stored on it. This can be done automatically in the system software or in hardware on the disk controller. The penalty paid is in performance, but this can be minimized.

One software technique is to compress only dormant files and to perform the compression automatically in off-peak hours. The data then can be expanded automatically when it is referenced. Since dormant data are referred rarely, there is little performance overhead. This technique is very similar to data migration except that data are not moved off-line, they are simply compressed. This type of data compression may be built into future operating systems.

There are many different compression algorithms, and usually they are suited to compressing different types of data. An algorithm that is designed for compressing image data may achieve ratios as high as 50:1; a good algorithm designed for compressing English text may achieve ratios of 8:1. General algorithms applied to data without regard to type usually average only about 2:1 (extravagant claims by vendors aside). How much space this technique can reclaim is dependent on the type of data and the compression algorithm. Ordinarily this is less than half the disk space.

DISASTER PROTECTION

In the broadest sense, the topic of disaster recovery is well beyond the scope of this chapter. However, the chapter does focus on reconstructing a file system, assuming it has been partially or completely destroyed.

A general disaster recovery principle states that if a disaster occurs, the more accurately an environment can be reproduced the faster the system will be back up and running. The fewer changes that need to be made in bringing a system back up, the easier it will be.

Disaster recovery companies dealing with mainframes specialize in keeping duplicate environments (called hot sites) running at a separate secure location so that recovery time is absolutely minimal in the event of a disaster. A more modest position is to make sure that the environment can be reconstructed quickly.

Preparing a disaster plan is an important part of deciding what is the correct level of protection for any particular company. The first step is to have documentation that accurately catalogs the hardware environment. This means documenting the type of computer, storage media, add-on hardware, devices, network connections, and so on, as well as model numbers, time of purchase, the level of upgrade, in short, a complete journal of all the hardware for each server that has to be restored. The same applies to software. Maintaining a database (which must be backed up regularly) is a good way of keeping this information, but hard copy reports should be kept off-site in the event of the database's destruction.

The second step is to have a carefully thought out backup plan that includes off-site tape rotation. This implies installing a good set of software utilities and making disciplined use of those tools. It also includes the manual task of transporting tape off-site. When global networks are faster and less costly this too will be automated, but at the moment a manual system is the most cost effective.

The first step in rebuilding the system after a disaster is to reconstruct the hardware as precisely as possible. This will mean replacing whole machines or faulty components and performing hardware system tests to ensure proper operation. After restoring the hardware environment, the network operating system has to be installed, at least the minimum system sufficient to support the backup software and hardware and the file system, including any volume partitions and logical drives. Workstation-based backups require a properly configured workstation and server support for both communications and the file system. The last step is to restore the file system to the most recent state backed up. This means restoring from the most recent full backup and subsequent incrementals or differentials, according to the backup scheme. Particularly important are the security attributes of the users, groups, work groups, and file servers. If the security and file attribute information cannot be restored, then all users, groups, and other objects specific to that server must be recreated from scratch. After the hardware, the system and data files, and their attributes are restored the server should be tested and reviewed before putting it on line. If it is put on-line with incorrect data or systems and customers begin using it in this configuration, there will be trouble reconciling the changes made against this restored system with the old system when its correct version is finally restored (and almost certainly it will be necessary to go back and do the correct restore). As a part of the normal backup scheme, weekend backups at least should be rotated off-site.

For more security, all backups should go off-site nightly. Fires, water damage, earthquakes, hurricanes, and other disasters happen more frequently than might be expected. Disasters also can happen to user workstations on the network. In most environments there is at least some local storage. This storage is subject to the same risks as centralized storage. Fortunately, it also can take advantage of many of the same protective strategies. Most of today's network data management tools provide some method of backing up and restoring local workstations. Critical local data should be treated just as server data are treated.

SUMMARY

This chapter has focused on some important techniques for preserving the integrity of data and managing system storage. There are many variations on the themes presented here and the terminology used by different tool and system vendors may vary slightly. This chapter should make it possible to sort through the different options available and choose and operate the ones that are most appropriate for a given network.

Chapter 51
Acquiring and Using a UPS System

Gilbert Held

The data center manager must be familiar with the operation and use of various types of equipment to operate the data center effectively and efficiently. Possibly the most important type of equipment, however, is an uninterruptible power supply (UPS) system. Originally designed to provide battery backup in the event primary utility-provided power was lost, today's modern UPS system performs additional functions beyond the scope of the traditional UPS. The acquisition of a UPS system, therefore, involves the consideration of a number of features beyond battery backup time.

First, the data center manager must understand the various power problems, their effects on data processing, and how UPS systems can be used to alleviate those problems. This information serves as a base for developing a checklist, which facilitates the equipment-acquisition process.

After the type of UPS system is chosen, the data center manager must perform a second, critical step: correctly size the system to support the organization's data center environment. Otherwise, the manager may not have an appropriate amount of time to correctly bring down a mainframe or minicomputer system during a power outage or be able to continue operations if the battery of the UPS system is connected to a local motor generator.

This chapter focuses on the power protection features incorporated into many modern UPS systems as well as computational methods to determine the capacity of such systems to satisfy the operational environment. Highlighted are the power-protection features that should be evaluated during the UPS selection process as well as the methods used to obtain a system with sufficient capacity to satisfy the organization's data center operational requirements.

0-8493-9823-1/00/$0.00+$.50

Exhibit 51.1 Common electrical power problems.

Power Problem	Typical Cause	Potential Effect
Brownout or Sag	Start-up power consumption of electrical motors or voltage reduction by electric utility	Reduction in the life of electrical equipment
Blackout	"Act of God" or accident	Loss of data being processed; possible disk crash
Spike	Lightning or resumption of power after a power failure	Damage to hardware, loss of data
Surge	Completion of an electrical motor cycle, such as an air conditioner	Stresses to equipment's electrical components, resulting in premature failure of a service
Noise	Electromagnetic interference caused by equipment, lightning, or radio interference	Data corruption

POWER PROBLEMS

Inhabitants of the Northeast U.S. are familiar with the effects of a summer heat wave — when the local electric utility company lowers the voltage level during peak electrical consumption periods. Dimmer lights, television screens, and computer monitors and slightly lower operating grates of motor-driven elevators are but a few of the effects of brownouts. Also known as voltage sags, this situation results from the start-up power demands of different types of electrical motors. Brownouts can adversely affect computer systems, causing electrical devices to fail as a result of repetitive occurrences. Although brownouts and sags are common, they are not the only electrical power problem the data center manager can expect to encounter.

Exhibit 51.1 lists five of the most common electrical power problems, including the previously discussed brownout or sag. Although most, if not all, managers should be familiar with the term *blackout,* which represents the total loss of electrical utility power, the remaining events in the exhibit may require a degree of elaboration.

A spike represents an instantaneous increase in voltage level resulting from lightning or when utility power is restored after an electrical outage. The effect of a voltage spike can literally fry computer equipment, causing the destruction of hardware as well as the loss of any data stored on the hardware, both in memory and on disk.

A power surge represents a short-term increase in voltage that over a period of time stresses electrical equipment until a point of failure is reached. The most common cause of a power surge is the cycling of equipment, such as air conditioners, refrigerators, and similar machinery for which motors are turned on and off on a periodic basis. When such equipment is turned off, extra voltage that was flowing to operate the equipment is dissipated through the power lines in the office or home.

The last power problem listed in Exhibit 51.1, noise, represents electromagnetic interference (EMI) and radio frequency interference (RFI). Both EMI and RFI result in noise that disturbs the smooth, alternating sine wave generated by the electrical power utility. Depending on the amount of noise, the effects range from no adverse effect on equipment and data to the corruption of data when noise reaches a level that precludes equipment from operating correctly.

An appreciation for the types of power problems is critical before understanding how UPS systems operate. The description of UPS operation in the next section provides information that can, in turn, be used to evaluate UPSs during the equipment-acquisition process.

UPS OPERATION

Early UPS systems were developed as a backup power source and simply consisted of a battery charger, battery, and invertor. The invertor converts direct current (DC) to alternating current (AC) and is commonly referred to as a DC-to-AC converter.

A UPS can be used in an online-without-bypass mode of operation, in which it always provides power to equipment, or in a standby mode, in which a transfer switch is used to provide UPS power in the event utility power fails. Exhibit 51.2 illustrates the configuration of UPS systems used in an online-without-bypass mode of operation and in a standby online mode of operation.

When used in an online-without-bypass mode of operation, the UPS accepts and uses raw electric utility power to charge its battery. In actuality, the battery illustrated in each example in Exhibit 51.2 represents a bank of batteries whose number and charge capacity vary. The number and capacity are based on the amount of electrical power required by the organization both in the event that primary power fails and for the duration for which battery backup power must be furnished during that failure. Because raw electrical power is first used to charge the battery, the effect of any surge, spike, or brownout is isolated from equipment that obtains power from the online UPS system shown in Exhibit 51.2.

In the standby mode of operation, raw electrical utility power is directly output through the transfer switch to equipment when utility power is present. Only when power fails does the transfer switch sense the loss of power and switch the connection to battery-supplied power. The key problem associated with this UPS configuration is it does not provide protection

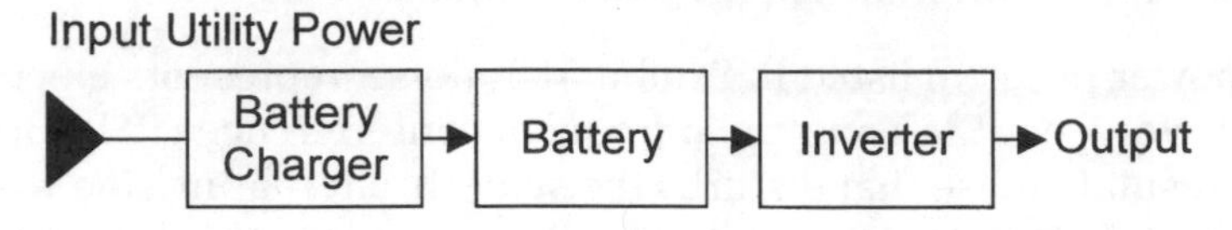

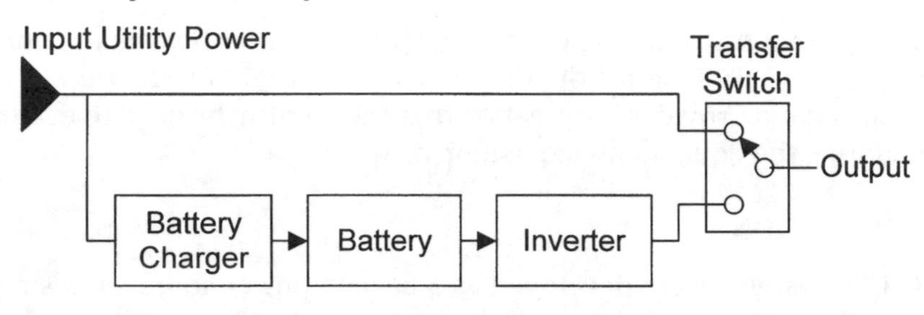

Exhibit 51.2 Basic UPS configurations.

to equipment from dirty electric power, corrupted through utility company-induced brownouts, natural causes, or man-made problems.

Because of this deficiency, the standby mode has been supplemented by the use of separate surge protectors and power line filters. The surge protector is used to block or chop off peak voltage spikes and smooth out reduced voltage surges, while the power line filter is used to reshape the power sine wave, eliminating the effect of sags and noise. Some UPS manufacturers are incorporating surge-suppression and power line-filtering capabilities into their products, resulting in the surge suppressor and power-line filter operating on raw electrical power that bypasses the UPS when electrical utility power is operational. Exhibit 51.3 illustrates a general-purpose UPS system configuration that protects equipment from utility power irregularities as well as provides a redundant backup power source.

COMMON VARIATIONS

The general-purpose UPS configuration illustrated in Exhibit 51.3 forms the foundation for all modern UPS equipment. Batteries used to store electrical power for use when primary power fails are both bulky and costly; therefore, by itself a typical UPS is sized to provide no more than 30 minutes to one hour of backup power, typically a sufficient duration to provide an

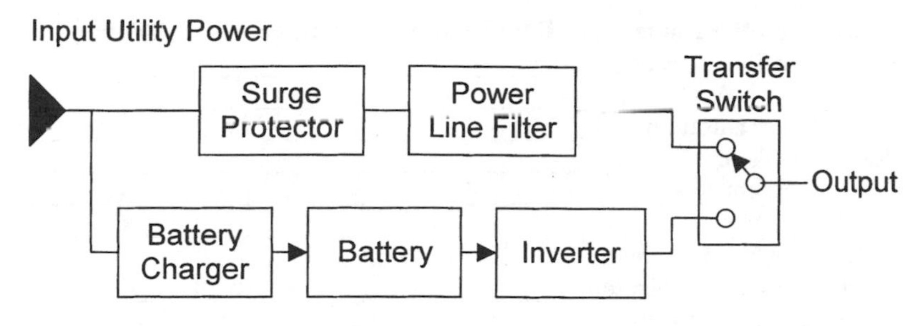

Exhibit 51.3 General-purpose UPS system configuration.

orderly shutdown of a mainframe or minicomputer in the event primary power fails. Recognizing that many organizations need to continue computer operations for prolonged periods of time once primary power fails, a common technique to meet this need is to use a diesel generator to supply power to the UPS when primary power fails. This can be accomplished by a transfer switch, which changes the UPS battery charger power source from the utility company to the diesel generator. Then, once diesel-generated power flows into the UPS system, the battery backup capability is limited only by the organization's ability to keep the generator operational.

SIZING A UPS SYSTEM

Sizing a UPS system is the process by which the data center manager determines the battery capacity required to operate equipment for a desired period of time after primary power is lost.

To size a UPS, the manager first makes a list of all computer equipment as well as lights, air conditioners, and other equipment required for continuity of operations. Next, the manager determines the voltage and amperage requirements for each device and other equipment, such as interior lighting, that may be on the circuit that the UPS is to protect. Multiplying voltage times amperage determines the volt amps (VA) requirement for each device. (Some types of equipment specify their power consumption in watts (W); that number can be converted to VA units by multiplying watts by 1.4.) The sum of the VA requirements for all resources requiring power is the minimum VA capacity required in a UPS.

The VA rating measures only the power load, however, and does not consider the duration for which battery power at that load level must be supplied. Therefore, the data center manager must determine the period of time during which the organization will require battery backup power. For many organizations this normally equals the time required to perform an orderly computer shutdown plus a small margin of safety time. For example,

Evaluation Parameter	**Requirement**	**Vendor A**	**Vendor B**
Maximum Line Current			
Capacity in VA			
Load Power Duration			
Recharge Time			
50 percent			
100 percent			
Operating Environment			
Operating Temperature			
Relative Humidity			
Dimension			
Weight			
Transfer Time (Milliseconds)			
Surge Response Time (Milliseconds)			
Noise Suppression (dB)			
Warranty			
Cost			

Exhibit 51.4 UPS operational parameters worksheet.

some IBM mainframes require 20 minutes for an orderly shutdown; in this case, a 30-minute battery backup at peak load is usually sufficient. Because total power provided by a UPS battery backup is measured in volt-ampere hours (VAH), the VA requirements are multiplied by .5 to determine the VAH capacity the battery backup must provide.

UPS EVALUATION

Although it is important to ensure a UPS is correctly sized to satisfy organizational operational requirements, it is equally important to examine other UPS operational parameters. For example, a product from one vendor may provide a higher level of protection from dirty electric utility power than a competitive product, offers a longer warranty, or has another feature which can satisfy organizational requirements in a more cost-effective manner than a competing product.

To facilitate the UPS evaluation process, Exhibit 51.4 contains a checklist of UPS operational parameters for consideration. Although several of the entries in the exhibit were previously discussed or are self-explanatory, some entries require a degree of elaboration.

The recharge time contains two entries, 50 percent and 100 percent. Although not a common occurrence, power does sometimes fail, is restored, and fails again; in this case, the ability to obtain a quick partial recharge of UPS batteries can be more important than determining the time required to obtain a complete battery recharge.

The operating environment entries refer to parameters that govern where the UPS system can be located. Some organizations construct a separate building or an annex to an existing building to house a large-capacity UPS system. Other organizations either use an existing location within a building or modify the location to support the installation and operation of the UPS. Due to the weight of lead acid batteries normally used by UPS manufacturers, the floor-load capacity of the intended location of the batteries should be checked. This is especially important when considering the use of battery shelf housing, since the weight of a large number of batteries mounted in a vertical shelf concentrated over a relatively small area of floor space can result in sagging floors or, worse, the collapse of the floor.

The transfer time entry refers to the time required to place an offline UPS system online when utility power fails. If a load sharing UPS system that is always active is installed, the transfer time is quoted as zero; however, there is always a small delay until battery backup power reaches the level required by equipment and lighting connected to the UPS. Thus, many times the term *transfer time* is replaced by *transfer to peak time*, which is a better indication of the time required to fully support the backup power requirements. As might be expected, the quicker this occurs the better, and better performing equipment requires the least amount of time. Similarly, the quicker the built-in surge suppressor operates, the better. Thus, a UPS requiring a small amount of time to respond to a power surge condition is favorable.

The last entry in Exhibit 51.4, noise suppression, indicates the UPS's ability to filter out EMF and EMI. Noise suppression capability is measured in decibels (dB), and a larger dB rating indicates an enhanced ability to remove unwanted noise.

RECOMMENDED COURSE OF ACTION

Dependence on raw electric power can result in a number of problems ranging in scope from brownouts and blackouts to spikes, surges, and noise. Since the investment of funds used to acquire data processing equipment can be considerable, the acquisition of a UPS system can save an organization many times its cost simply by protecting equipment from harm. In addition, the ability to continue operations by using battery backup power provided directly from the UPS system or indirectly from a diesel motor generator can be invaluable for many organizations. Because there can be considerable differences between vendor products, the data center manager should carefully evaluate the features of different UPS systems. By doing so and installing a UPS system correctly sized to satisfy the organizational requirements, the data center manager may avoid equipment failures that result from power problems as well as permit continuity of operations on an orderly computer shutdown in the event of a blackout.

Chapter 52
Troubleshooting: Resolution Methodologies

Kristin Marks

Since computers were invented, mankind has lived in fear and fascination of these devices. In the entertainment media, computers have been elevated to a god-head status, imbued with an advanced and slightly sinister intelligence. In reality, they are just machines which execute programmed instructions. These machines require maintenance and proper operating environments, just like a car or lawn mower. When networks lack these care-taking measures, they eventually cease to function. This chapter makes some suggestions on how best to resolve server and network problems.

The issues we'll look at include problem isolation techniques, issues related to the environment of our networks, and physical security. Problem isolation is a science. When a problem hits, there are two basic ways we can attack it. We can try our best guesses, based upon knowledge, experience, and luck ... or we can follow a logical pattern of testing to locate and resolve the problem.

UNDERSTANDING TROUBLESHOOTING TECHNIQUES

Troubleshooting skill is something all networkers need. Good troubleshooting is closer to an art than a science, though. Good practice of troubleshooting involves thorough network documentation, sound knowledge of networking concepts, product knowledge, logic, and intuition. When a problem hits, we need to be able to employ all of these skills to effectively resolve the problem and prevent its reoccurrence.

Documentation

Few areas of server management are as neglected as the *Docs*. Often, it is rationalized that since we put the server together and maintain it, we are

0-8493-9823-1/00/$0.00+$.50

already imbued with a thorough knowledge of its components, operation, and needs. Therefore, documentation is redundant and time wasting. Wrong. Good documentation will show you things you may have never otherwise known about your server, and it will save considerable time in the long run.

Even if we did know everything about the network, new team members and outside contractors would be unaware of important facts. New personnel would be clueless as to where things are located, and where the skeletons are buried. Speaking of skeletons, how about problem records? Sound Docs incorporate records of the things which have happened, and their resolution. There are several areas of documentation which need to be recorded for posterity. They are

- Network and server
- History and users
- Resources

The Network. The network is the core of the documentation. The information in this section details several different aspects of the physical and logical network. We need to know where everything is, how many there are, how the components are related, and who uses them.

Maps. Maps tell us several important things: where things are and their relationship. Two kinds of network maps are useful to maintain: logical and physical.

Logical maps, as depicted in Exhibit 52.1, are topology overviews, which focus on the internetworking devices connecting the networks. The purpose is to establish the relationship between devices and show the flow of data. These sorts of maps are not very good for giving someone directions on how to find a file server, but they can help us to locate network bottlenecks, plan expansion, and discern the general source of a particular problem.

Physical maps, like the example shown in Exhibit 52.2, show us exactly where things are located spatially. These are the maps we need to maintain so that things can be found quickly. Information which would go into a physical map includes a blueprint (even if you have to draw it by hand), showing room names and locations, wiring between rooms, and cable specifications.

Networks grow and change, but all too often the physical map is never updated. Think of the few minutes it takes to update it each time you make a change as you would a deposit in a bank. When the system starts to give you headaches, and it crashes or slows down, you can withdraw those minutes in the form of saved hours, if and only if you have kept the plan up to date.

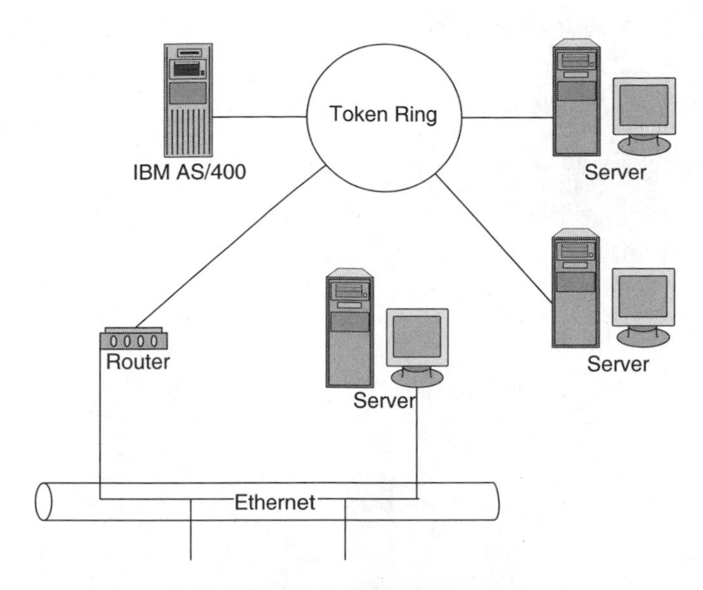

Exhibit 52.1 A logical map.

Inventory. An equipment inventory is another vital element for good documentation. We might be tempted to think that this only means the spare parts inventory we keep *just in case*, but that's only a small part. We also need to know how many clients are attached to the networks, how many servers exist along with their component configurations, internetworking devices, as well as the model and version of the attached hardware.

Clients. On the client side, the inventory is concerned with basics. How many are there, and what kinds of network cards do they have? This makes for useful planning information. At this stage we are not really concerned with the contents of CONFIG.SYS, AUTOEXEC.BAT — just how many there are and where they are physically and logically located.

Servers. Inventory of the production and test servers should include detailed information about each of the servers, its location, make, model, operating system, hard disks, LAN cards, and other peripherals. Along with file servers, we should also maintain appropriate information about fax servers, print servers, and the like. There are several third-party tools available to help you record settings and other key data about your networks.

Internetworking Devices. Repeaters, concentrators, bridges, routers, and gateways should be well documented. The information needed includes

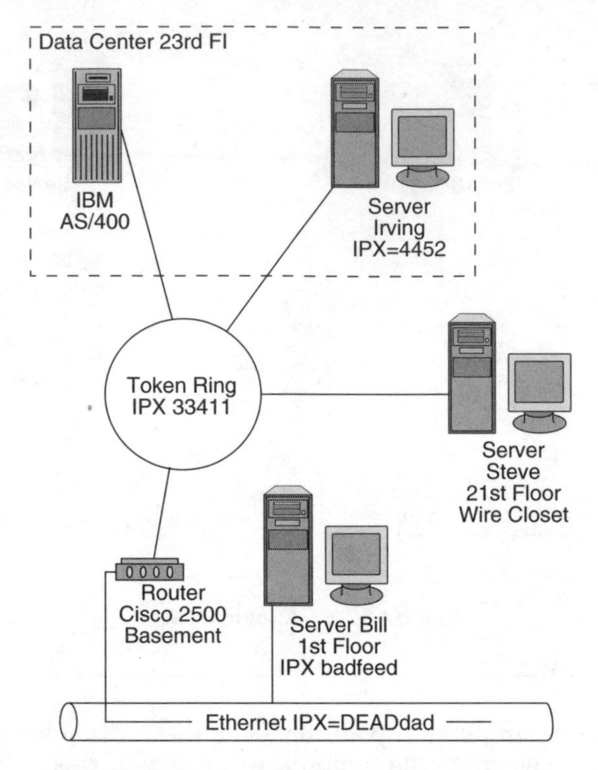

Exhibit 52.2 A physical view of a network.

---

vendor, model, location, and general connections. For example, a bridge might list

- Port A connects to east wing LAN
- Port B connects to west wing LAN
- Port C connects to 64 KB circuit #11122233333 and links network to headquarters office.

Spares. The spares inventory is one of the most critical lists an administrator will have. This list will tell us what options are available when something breaks. Some people wonder if they need a spare inventory if they have a maintenance contract. The answer depends on what kind of contract it is, and how patient your users are. Many maintenance contracts specify the items covered, response time, and feeds. A few go into more detail and require the maintenance company to keep a supply of certain components in their stock. The reverse is also true. Some contracts specify that the customer must keep the stock.

Exhibit 52.3 A sample of useful spare parts inventory.

Item	Comments
System unit, with motherboard	Should be at least equal in speed to current servers. The number and types of slots available need also match current servers in production use.
RAM chips	Make sure they all are the same speed and will fit all production servers.
Disks	Also must meet or exceed current production configurations. Factor into this the interface (SCSI, IDE, ESDI) must also match.
Disk controllers	Same as for Disks above.
Keyboards	Usually one spare will do. Server keyboards are seldom used and therefore rarely break.
LAN cards	At least one extra for each vendor and model combination in use.
I/O ports	Keep a few for replacements and growth.
Connectors, transceivers, and other attachment devices	Troubleshooting and growth require having several for each type of connector.
Cabling	Extra cables, again for each type of interface to be used. Keep enough on hand for troubleshooting and growth.
Concentrators, hubs	Keep at least one spare of each kind that you have in use.

If you have mission-critical service going through your network, then the users won't be patient while you start to fill out purchase orders and start calling for prices on a dead network component. The rule should be that spares inventories are needed.

If your server's network card dies and you have the same model in your workstation, but no spares, swap out yours. You won't get any less work done, and the others will be up and running much sooner.

So, what should be included in this inventory? Extras for all production pieces of software and equipment. Exhibit 52.3 contains a listing of normal equipment spares to have on hand.

As you can see from Exhibit 52.3, a heterogeneous network with equipment from numerous vendors is the most expensive type for which to maintain a spares inventory. Limiting the number of vendors you deal with may cost you a bit in terms of purchase pricing, but it can save a fortune when the spares inventory is factored in.

Do yourself a big favor and do not put questionable pieces of equipment into spares inventory. If an item is doubtful in terms of its ability to be placed into production use, scrap it, or tear it apart and use it for education. You don't want to be in a position of having to rely on it when things go bad. Troubleshooting is hard enough without adding new factors in the form of questionable equipment.

Clients

General Client Data. General data about client devices are useful for reviewing widespread problems. It is very difficult to maintain an accurate database of all client configurations. This is especially true of networks where the clients number in the hundreds or thousands. There are packages available which can help you to maintain these data. WinLAND from Seagate is an example. It allows an administrator to have access to key configuration data about any client connected to the LAN.

Special Situations. Every network, large and small, has some special users and special systems which are never touched because they are so sensitive. Perhaps there is some nonstandard software running on it. Whatever the case, you're afraid to touch it, and when that user calls, you prepare an updated resume before making that "house call."

Document these cases with great detail. Configuration files, directory structure, memory managers, applications ... all of it. Some day that station will need attention. Being able to research the current set up is critically important. One last benefit of good Docs is being able to put the system back together, should you or the user clobber it with changes or upgrades.

Change Order Log. A change order log is a lost art in most networks. In many mainframe environments, the change order log has been a standard procedure for years. You simply did not leave before recording what work you did, and why it was done. The basic goals of a Change Order Log are to keep trend information on servers, and to discover if we broke one thing while attempting repair on another. It adds accountability to changes made to the network. This log, like regular backups, is a procedure which never quite caught on with LANs. But, you can probably add several years to your expected life span if you implement a simple Change Order Log. The elements of a good change order log are shown in Exhibit 52.4.

A good rule of troubleshooting is there is no such thing as a coincidence. It often happens that when we make some change to a network, the following day some problem pops up in another area. Being true optimists, we seldom suspect our own labors which were geared toward eliminating problems as the causes of yet another.

If you view a situation logically, there is nearly always a strong case for suspecting the last change as the cause of current problems. After all, if the network was running for days and weeks, and suddenly a problem arises after a change, there is more than ample reason to look at the change first.

Exhibit 52.4 The elements of a change order log.

A basic description of the change.	This should be a quick description of the nature and scope of the work.
Who did the work.	In case of a problem, we need to talk to the person who made the change.
Why was it done.	What motivated the change. Perhaps the goal was to eliminate a problem; perhaps to improve performance.
What components were touched.	An important item. In case a problem is later detected on the network, we need to know what components were handled, or swapped.
When was the work started and completed.	This helps bracket problems which start as a result of change or occurred during the changes.

Always suspect the last changes made to a network. A change order log is an invaluable tool to help us zero in on high-probability suspects when we seek to resolve a problem.

Problems. Whenever a problem occurs, we always seek to repair it. Yet one more issue remains: do we learn from it? In order to learn from a problem, a record has to be maintained. If we do not log problems, then we'll likely go through all the discovery and resolution steps from scratch each time the same problem occurs. This leads to two problems. The first is that we duplicate effort and delay the resolution by going through the whole discovery process again. The second issue is that the problem may keep coming back. Our goal should be to eliminate problems, not to keep fixing the same ones over and over. That is where a problem log comes in.

Baseline Data. Keeping a baseline of network data helps us to resolve some of the trickier problems which may trouble us and our networks. Baseline data include information on server utilization, connections typically in use, file storage used, and LAN traffic. In order to gather these data, we need to employ tools designed specifically for the job. Several vendors offer solutions including Seagate, Intel, and HP.

Escalation Procedures. Not all problems can be solved by one person, no matter how skilled they may be. Escalation applies largely to who gets involved in solving a problem. As we look at troubleshooting from a high level, there are three basic levels of escalation. Often troubleshooters are assigned jobs and responsibilities according to the levels described in Exhibit 52.5.

User Data. This documentation need not be long, but should include some information which will help you characterize the LAN users. An example of this would be as follows:

Exhibit 52.5 The three escalating levels of tech support.

Level 1	A Level 1 person checks to see if correct procedures were followed. If they cannot resolve it, they categorize the basic problem area (network, server, user applications, printers) and call in a Level 2 who works in that area.
Level 2	Level 2s use standard troubleshooting tools in their field of endeavor and try to determine the cause of the problem. Their job is to find and use standard procedures and tools to attempt a fix. If they cannot resolve the problem, a Level 3 is called in.
Level 3	Level 3s are usually few in number and employ nonstandard tools and techniques to discover a problem and resolve it. Their areas of expertise are often fairly narrow, but within that area they are highly specialized.

> The engineering department uses file servers named ENGXXX. Their primary application is AUTOCAD and their bandwidth requirements are very high. We expect their storage requirements to double each year through 1998. As of September 1999, they are using approximately 1 GB of storage.
>
> The Accounting and Sales departments use the servers called ADMINXXX. They are users of vertical application located on servers which only they log into. They also use word processing and spreadsheet programs. Current storage in use is 1.2 GB, with 20 percent expected annual growth.

Resources. Some last items to track in the network documentation are the resources we can employ to resolve problems. The resources include vendor sales, internet URLs and tech support lines, reliable consultants, frequently used telephone numbers, and other technical support within your company. Make it a standard procedure to place new entries in the list as you come across them. Include business cards in the list. In no time at all you'll have a very respectable list.

Networking Concepts. Knowing how things work, theoretically and in actual use, is crucial. In order to know how things work, we need to know design capabilities and limitations. For example, 10BaseT imposes a 100-meter cable length limitation. When this limitation is exceeded, attenuation of signals and excess collisions may occur. Without knowing this, it would be hard to solve a problem caused by a new network attachment. It is important that you know the systems you now use, as well as have an understanding of those coming soon.

PRODUCT KNOWLEDGE

Knowing the products, software and hardware, which are employed on your networks is key to troubleshooting problems. The knowledge spans not only the *book* information, but also the *down-in-the-trenches* type stuff that only research and experience yields. Product knowledge breaks down into two basic areas:

- Standards/published information
- Secrets of the trade

Standards and published information consists of knowing how things are supposed to operate. Getting to know this information is a daunting task which takes years. Truthfully, the learning is never complete. But it is absolutely worth the investment in time and effort. Knowing how things are supposed to work can do two things for you: it can keep the network out of trouble by making sure it is assembled and installed correctly in the first place. Second, you have known standards by which to measure current performance and operation.

Secrets of the trade is a topic most techies love to build on. This category of knowledge consists of things learned through the *school of hard knocks*. Most complex systems have back doors, short cuts, and fixes which never seem to make it into the manuals.

Internet forums and newsgroups in which people from around the world gather electronically to discuss issues related to servers and networks are also helpful. There are volumes of files which can be downloaded and numerous discussion areas which are of tremendous benefit. Experts from all across the world exchange information there.

LOGIC

Being able to follow a logical, disciplined approach to problem solving is another necessary capability for good troubleshooting. Some companies test their job applicants for skills in these areas when considering a new hire. It is that important. If we fail to follow set procedures and act systematically, our efforts at troubleshooting may end up causing more problems than they solve. As with many things in life, there are exceptions to every rule. You'll also need intuition and good guess work.

REASONING

On its surface, philosophy may not seem to be deeply related to network troubleshooting. But if we look a bit closer, we can garner some tidbits which can help us to properly frame our minds when trying to solve a problem. Reasoning breaks down into two basic categories, inductive reasoning and deductive reasoning.

Deductive reasoning is attempting to answer a question based upon having a set of data and logically attempting to find an answer which fits what we have observed. This sort of reasoning would be used if we had an entire cabling system that failed for some unknown reason; all the users cannot get to the server and network monitors are reporting many errors. Here we have much data and we need to find the problem (or problems) which can account for these conditions. Either start at one end and work forward or start at the other and work backward. Do not alternate between the two. Be consistent. Don't start at one end, test, and then move back to the other.

Inductive reasoning comes into play when we have a small amount of data and need to discover more about the environment that surrounds and caused the problem. This is the sort of reasoning that comes into play when we have a part that has failed for no apparent reason. Swapping does not get the system back on line. All we know is that this item continues to fail. From there, inductive reasoning demands we discover more about the environment surrounding the system and find links.

On the television show *Star Trek,* there is a famous character named Mr. Spock. He comes from a planet called Vulcan, where emotion is discouraged and logic is the rule for life and daily living. He has a photographic memory and is always able to state the current time of day without consulting a timepiece. Yet, in spite of this intellectual prowess, Mr. Spock is unable to solve many problems. The reason is that he is shackled by his inability to make a guess ... something which leaps beyond logic.

Guessing correctly when trying to fix problems is a useful gift, often required, when trying to resolve a problem quickly. Following strict troubleshooting guides and problem isolation techniques should not be overdone. This is not to say we can throw caution to the wind and use a dart board as our method of problem discovery. Rather, we are recommending that you consider the realities of time and other constraints when approaching a problem.

TROUBLESHOOTING STEPS

Regardless of what your job is, or the industry in which you work, the steps toward problem resolution listed in Exhibit 52.6 apply very well.

Once a server is in production, we usually do just two things with it: administration and troubleshooting. This chapter has focused on resolution methodologies. Good troubleshooting is more than having lots of knowledge and a willing attitude. It also means being disciplined and placing yourself in a good position to not only remedy problems, but also to anticipate and prevent problems.

As you have seen in this chapter, 90 percent of troubleshooting is best done before the problem occurs. We can react to problems or anticipate them. Documenting all past network problems and changes in their appropriate logs, knowing the standards of our systems as well as the specifics of the

Exhibit 52.6 Summary

Investigate	First employ diagnostics. Discover as much information as possible within reasonable time constraints. Make good use of the Docs. It pays to approach any problem from a position of knowledge. Counterbalance the time spent diagnosing with the urgency of the matter and the ability to use brute force solutions. That is, if any easy swap out of parts will solve an extremely urgent problem, do it. Spend time later isolating the exact point of failure.
Plan	Write down a systematic approach to implementing a proposed fix. It is easy to get lost in a quagmire of trial and error.
Execute the plan	Stay close to the plan. If it fails, go back to diagnostics and make another plan. A plan can only be deemed successful if the problem is no longer present in your opinion and in the opinion of the people who use the network.
Record the results	Open your documentation log and record the problem and resolution in the Problem Log, and the changes you made in the Change Order Log.

products we have purchased is an ounce of prevention that's worth weeks of overtime.

We can attack problems fastest if we maintain good network documentation (especially the maps), keep spares, have the appropriate tools handy to respond when a problem arises, and apply sound logic to isolating the failure. Anyone who has been through a network failure knows what a life-changing experience it is. Learn from the mistakes and problems of the past and be ready for them. Have the necessary assets available to solve the issues which are likely to come up and be ready to approach them all with a calm mind, ready to make a reasonable plan of attack, flavored with a few good guesses.

Chapter 53
Troubleshooting: The Server Environment

Kristin Marks

UNDERSTANDING AND ISOLATING NETWORK PROBLEMS

Monday morning is often a bad time for network administrators. It is not so much that the weekend is over, but that our servers have had days on their own to experience failures, glitches, sometimes altogether dying. There is a world of problems waiting to happen to our servers. In this chapter, we'll examine three independent areas where problems may arise. These areas are the physical environment, the electrical environment, and physical security.

PHYSICAL ENVIRONMENTAL ISSUES

Computers and networking equipment have operating limits. If you take the time to inspect the manuals which come with any piece of hardware, you'll find information about properly operating it. We can categorize the physical operating environment into three areas:

- Temperature
- Air quality
- Magnetism

Temperature

Nearly all pieces of electronic equipment have two sets of acceptable temperature ranges. One is the power off, or cold, temperature range. This range will vary from manufacturer to manufacturer. It states the minimum and maximum temperatures the equipment can reach while *not* powered up. When a component is shipped, it should never reach a temperature out of this range. Also, if the system is delivered to a loading dock, it should not

0-8493-9823-1/00/$0.00+$.50

sit out in the cold or under the sun so long that this range is exceeded at either end.

Exceeding the nonoperating temperature range can cause the system permanent damage. Just as certain foods will spoil when they are out in the open too long, server system components have the same limits. Damaged pieces may fail immediately upon use — if we are lucky — or they can develop some sort of intermittent failure. Only swapping out components will tell if a component has gone bad.

The second temperature range to respect is the operating temperature. This range is nearly always narrower than the nonoperating range. It describes the acceptable ambient temperatures at which the system may be safely used.

Although PC-based server systems are far more liberal than their mainframe counterparts, they do require care in terms of location. Hot stuffy rooms with no ventilation are bad choices. With only hot air present, the cooling fan will only be able to push hot air past a chassis. The result can be the early and unexpected failure of a component.

Many sites shut down their air conditioning systems over weekends and holidays. Make sure that the temperature in the server room does not skyrocket during these time periods. You can get a monitoring thermometer at most hardware stores. All it needs to do is report the high and low temperature marks seen during a time interval. Make a point of placing it in various network equipment locations to avoid some unpleasant surprises.

When bringing in a new server, be sure to give it ample time to adjust to room temperature before switching on the power. If the temperature differences between the cold machine and the warm room are just right (or wrong), condensation will form inside the machine. Powering it up then can cause a number of events, all of them unpleasant. Give it some time. Just because you are ready to play with the new toy does not mean it is ready for you.

Air Quality

Air quality may seem like a relatively trivial issue, but it affects the life span and reliability of network systems. Three basic components of air quality must be controlled for the long-term stability of equipment — humidity, dust, and pollutants.

While it is rare for poor air quality to swiftly degrade a system, humidity can come close. Low humidity promotes static. Excess levels of moisture in the air can cause several problems, too. Among them are short circuits, or electrical arcing. This happens when a system is cool and then moved into a high-humidity setting. The result could be continuing condensation within the system. Remember, your mother always told you water and electricity don't mix well.

Nearly all file servers have a pest living inside them — a "dust bunny." While dust seems to be a mere nuisance, it is a hazard to the equipment. Dust conducts electricity. It will enhance the probability of a static discharge problem or a short circuit. When these problems strike, they will most likely be silent killers which do not announce themselves.

Placing filters over the cooling fan vents does little to stop dust accumulation in the long run. Air can freely flow in though floppy disk openings, and other locations in a system unit. Only controlling the environment in which the equipment resides can fully do the job. Avoid situating equipment in areas with concrete floors or rooms that get dusty quickly. Also, make it a point to open up the system units of all servers periodically to blow out the dust bunny with a CO_2 spray. Be sure not to flash-freeze any components by spraying too close. Hold the spray at a good distance.

Another airborne pollutant is cigarette smoke. If smoking is permitted in your environment, then you should schedule more frequent preventative maintenance cleanings. Over time (about a year, for a heavy smoker) the nicotine and tar in the smoke will form an oily, conductive film on the computer's internal parts. When that happens, repair (cleaning) is difficult, sometimes impossible.

Pollutants and corrosives affect the longevity of network equipment. Airborne chemicals can cause all the same problems as humidity and dust, plus they can corrode surfaces. Scenarios particularly at risk are manufacturing and other environments which use chemicals. Printing shops, photographic developers, and publishers all use lots of chemicals in the course of a normal business day.

In these settings, either shelter the network equipment in a refined environment, or use industrial-grade systems. Although we don't usually see them in standard consumer advertisements, you can buy hardware designed to run in an impure environment. They cost more, but do offer a solution.

Magnetism

Magnetism is the final factor to examine in regard to the physical environment. Unlike air temperature and humidity, magnetism can directly and immediately affect network equipment. Magnetism can come from many sources, such as fluorescent lighting fixtures and large electric motors. But the primary source is electric power cabling. Many networking wiring centers do double-duty as wiring centers for electrical distribution. This is, unfortunately, a big mistake.

Placing network wiring close to electrical power can cause high-quality network Category 5 cabling to be essentially useless. Keeping a file server next to a power transformer can cause the CPU, disks, and LAN cards to stop working or become erratic. Keep these systems in separate locations.

ELECTRICAL ENVIRONMENTAL ISSUES

Electrical problems can be classified into two main sources: transients (includes spikes) and service disruption.

Isolating transients and power disruptions as the cause of a network failure involves a lot of research. While power disruptions are easy to see when they occur — the lights go out — determining that it was an upstream factor in a subsequent problem is very hard to establish. The same is true of transients — abnormalities in power conditions. Unless a component dies immediately after a known surge, we are left to swap parts and make our best guess.

To verify power problems, we need the correct equipment. Every network administrator should have a power monitor and a volt/ohm meter. A power meter is a device that plugs into an outlet. Over time it collects data on the volts and amperes running through the circuit and usually graphs this information on a paper tape, like an EKG chart. Power meters are good at catching power problems which are transient in nature, happen at night, or take a while to develop. By reviewing the tape, a technician can see the continuity and quality of power to a system.

Another tool useful when power problems are suspected is a volt/ohm meter. If a problem is ongoing, a volt/ohm meter can be used to quickly discern whether the voltage of a circuit is too high or too low.

When all else fails, you can call your power company and ask them if they can verify that there were or were not any problems. Most power companies are public utilities and are required by law to track the service they provide and log all problems.

Transients

Power transients are short, abnormal power conditions. The sources of power transients may trace back to the power company, devices sharing the same outlet as your network equipment, or even static electricity. These are among the various type of anomalies:

- Static
- Lightning
- Electromagnetic interference (EMI)
- Radio frequency interference (RFI)
- Low voltage conditions
- Power spikes

Static. Static, or electrostatic discharge (ESD), is perhaps the most common power problem you will encounter. Static can strike anywhere, any time. ESD can send several thousand volts of electricity through you and your equipment. While it is true that ESD does not typically pose a personal safety threat, servers can be greatly harmed. The circuitry and chips inside are designed to work at relatively low voltage and amperage levels. Electrostatic discharge easily exceeds acceptable limits of system operation. Even though the zap may last only a few nanoseconds, it is long enough to burn a hole in small, sensitive circuits.

The damage done may be felt immediately, or at any time after the shock. The damage done may, if you are lucky, be seen immediately by a malfunction. Or, the damage could be much more subtle and intermittent in nature. For example, what if ESD attacks a file server in such a way that just one small part of memory gets a glitch. The damage from the static may not surface until the server tries to access that part of memory and locks up.

The basic cause of ESD is movement. ESD comes from any number of sources: altitude, humidity, carpeting, air vents, clothing, and office furniture. Anything that moves can generate an electrical field. You do not need to be wearing a big woolly sweater and 100 percent polyester clothes to create a static charge. ESD can be generated by walking past some equipment, or even air movement. Geography and climate also affect static. At sea level in a warm climate, with modest humidity, you may not see much ESD. However, if you are in a tall building on a windy day, with strong humidity controls governing the air quality, static will flourish.

Now that we've just told you that simply walking up to your server can kill it, what do we do? Static prevention and management are the keys. Static prevention consists of suppressing the factors which favor static build-up. Make sure that antistatic measures are used in the physical environment near the servers. Antistatic strategies include use of antistatic carpeting, antistatic sprays, and providing local humidity. Total office climate control is seldom a viable option. Many buildings are required by law to provide certain levels of temperature and humidity. The reason for strict climate controls is to prevent airborne diseases. Unfortunately, the conditions that best discourage germs also encourage ESD.

Next, make sure the equipment is well grounded. Get rid of any 3-pin to 2-pin power plugs. Grounding a device depends on its getting to earth ground. These adapters are bad items to use in a network environment.

Last, manage ESD. This means you accept that it is there and can damage your equipment, but you're going to fight it. Here are the measures you can use to combat static damage:

- Whenever you handle electrical equipment, wear an antistatic wrist strap. These straps connect at the wearer's wrist and have a copper lead which must be connected to an electrical ground. This provides a discharge path for the electricity which is safe.
- Use antistatic keyboard mats and ensure they are properly grounded.
- Place static mats under the equipment, especially at the benches where PCs are repaired or assembled. These mats perform the same function as antistatic wrist straps, except they hook up to the equipment. This mat will have a lead which must be grounded to be effective.
- Before handling equipment, touch a grounded metal chassis. This will have the effect of discharging any accumulated voltage potential into a safe avenue.
- Refrain from handling chips and other fragile electronic items by their metal leads.
- Use shielded static bags as much as possible. These are the silver-tinted bags that most components are shipped in.

Lightning. Lightning is an electrical problem which can, in most cases, be managed satisfactorily. Granted, there is little you can do to prevent lightning, short of moving underground. But there are specific steps you can take to mitigate its effects when it does strike too close.

Lightning damage has a lot in common with ESD damage. It is possible that you may return to work after a stormy night and find a hole in the ceiling and a pile of melted steel and plastic where the fileserver used to be. The damage done by lightning is typically very subtle — a tiny scar deep within the electronics of a system. Very rarely is the damage actually visible to the unaided eye. As with ESD, lightning sends excessive voltage and amperage through available conductors in search of ground. If the path of least resistance is through your network cable or power circuits, then that is the path it runs. If any of that discharge gets into the guts of electronic equipment, the game is over.

As with ESD, the damage may be immediate and permanent, or it could be intermittent. It is nearly impossible to prove that a part was killed by lightning. We can only swap components and make judgment calls based on the time proximity of the failure to known lightning strikes.

Here are some specific steps you can take to effectively prevent lightning from ruining your day:

- Ensure that the building has lightning rods in place. These provide an alternative path for lightning, besides the building, its electrical wiring systems, and your network media.
- Place all network equipment under surge suppression. Don't go cheap on this item.

- Properly ground all network equipment. This includes earth ground for electrical feeds and network wiring. Many vendors consider the warranty void if the equipment power feeds and data cabling are not grounded as they recommend.

Electromagnetic Interference. EMI, as it is called, can be a tough gremlin to deal with in networking. Electromagnetic interference is caused when the electromagnetic field of one device reacts unfavorably with other devices. The result can be anything from a dysfunctional piece of equipment to slow network performance. The damage done by EMI is usually transitory. That does not mean that EMI cannot cause a big headache from an extended network outage.

Two common problems resulting from EMI are

- Slow network performance. This could come from cabling systems running too close to a source of EMI and breaking up the transmission. It could also come from a system malfunctioning and going through recoveries.
- Electronic equipment may mysteriously stop operation.

Essentially, anything which generates an electrical field radiates EMI. Since so many things in an office setting use electricity, we need to pare the list of suspects down a bit. Here is a list you can work with as EMI sources:

- Fluorescent light fixtures
- Other computer equipment
- Power conduits
- Photocopiers
- Electric motors
- Communications equipment
- Navigation systems
- Anything that uses high voltage (220+ volt systems)

Among these EMI sources, fluorescent lights are the most notorious. They make little or no noise and are always around. We think little of them because they are a natural part of nearly all offices. But they are great for radiating EMI.

Keeping EMI away from your network involves some up-front prevention and a little ongoing work. First, use good sense when situating your equipment in the first place. It is not a good idea to place your mission-critical file server next to the building's electrical transformers. Keep away from fluorescent lights and power conduits. As mentioned before, fluorescent light can disrupt network traffic. This can be seen with a network analyzer: incomplete frames and bad CRC checks.

Radio Frequency Interference. It is more likely that network equipment will be a source of RFI, radio frequency interference, than a victim. To discern if you are a victim of RFI, you'll need to get an electronic specialist, and will probably have to get an oscilloscope hooked to various pieces of network equipment. The odds are pretty good, though, that you're giving off more RFI than you're taking.

If radios and televisions in your workplace don't seem to be working as they should, a potential cause is workstations and servers. The only thing you can do about RFI is to buy low emission equipment in the first place and locate that equipment as far away from TVs, VCRs, and radios as you can.

Low Power. Low power, or brownout, conditions are another silent problem that lurks in the power systems that feed our servers. When the voltage drops on the feeds to systems, it can be just as bad as if a spike (discussed next) had occurred.

When power drops to low levels, the results are hard to predict. Some systems will shut down, others reboot, and a few just die. Most electronic systems rely on having a stream of voltage within certain levels. The power supply inside the system simply steps down the voltage and distributes it to various components, such as the hard drives and system boards. Two common effects of low voltage are

- Components will not have enough power and lock up.
- Parts with motors (such as hard drives) may burn out.

The bottom line is that things will start to fall apart fairly quickly and, unlike EMI and RFI, the damage may be permanent.

We can control some low power causes; others we have to deal with as a matter of reality. Among those we can prevent are

- Improper power feeds to servers
- Use of server power feeds to nonnetwork equipment.

Properly set up, equipment for the network should be on separate circuits from other office systems. This helps to isolate the network from drops and surges which may result from HVAC equipment, floor polishers, photocopiers, and refrigerators as they cycle on and off. Beyond this, make sure that people do not innocently use the same electrical wall outlet for their various pieces of office equipment.

UPS systems and surge suppressors are the order of the day. These devices can condition the power and ensure its availability in spite of whatever surprises life may have. Detecting power problems almost always requires the use of a power monitor.

Power Spikes. Power spikes are a common phenomenon in computing. Lightning and ESD are forms of power spikes. Voltages rise dramatically and briefly, and then return to normal levels. But ESD and lightning are not the only sources of high-voltage conditions on a network.

Among the many causes of power spikes are

- The power company (they have bad days, too)
- Lightning strikes on power company equipment
- Lightning strikes on telephone company equipment
- Short-circuits in site wiring
- Sunspots

We can take it for granted that the electric company has good days and bad. In general, most power companies give reliable service. Yet, occasional spikes to the power output occur. It can come from their own generation systems, or be induced, such as through a lightning strike on their wires. That much is obvious. What about the telephone systems? If you have workstations or servers hooked to the telephone lines, they are at risk, too. Lightning can hit a telephone wire and be transmitted down to your server or modem. The results: Zap — that item is dead.

One other less obvious source of spikes is sunspots. On the surface of the sun eruptions occur and spray out radiation which can take days to reach earth. When they do get here, all kinds of problems can occur. The EMI is measurably higher and you may notice it from poor radio or TV reception. Power lines get hit in a big way. Transmission lines act as giant antennae, which receive these EMI waves and convert them into voltages on their wires. We at the end of the line can feel a sun spot from suddenly higher voltages.

Disruption

Power disruption can be a dramatic event in the business world. When the power fails for just a few seconds, everyone gets mad, because they have to restart all their computers again, login to the network, and reload their programs. If the failure is prolonged, everyone is happy, because they'll probably get to go home early. Well, almost everyone will be happy. You'll probably be having fits. File servers do not do well when the power fails — short- or long-term. Faxes, copiers, and office lights can be turned off and on with relative impunity. But file servers are a bit more complex, and they certainly need to be shut down right. Failure to shut down a server properly often results in incomplete data being written to disk, transaction rollbacks, and recovery utilities needing to be run.

According to the National Power Laboratories, the average computer site will experience 300 power disturbances each year. Granted, they are not all power outages. Some are spikes, brownouts, and other anomalies. Still, you can expect some of the problems to cause good quality power to be unavailable for a significant period of time.

UPS Systems

Uninterruptible power supplies are a necessity for all servers. The cost of downtime is extremely high, and the hassle of repairing or replacing a server during business hours is something worth avoiding. UPS systems usually do double duty. They not only protect the load equipment against outage, but many systems will also effectively screen out power transients, like surges and low power conditions.

You may think a UPS is just a UPS, and that there are small variations in quality and brand name which separate the cheapest from the most costly. Wrong answer. There are several completely different types of UPS systems available on the market today. While there is a great deal of competition, especially as compared to just a few years ago, features and capabilities vary greatly. Some UPS systems are relatively advanced and incorporate surge control and clean operating system. Others may not be at all useful in protecting a network.

Building UPS systems typically cannot protect servers. Most building protection systems are geared to resume power after the mains fail. This means that when the power goes out, the diesel generator kicks in, and the power returns. This will not help your file servers at all.

On-line Power UPS Systems

On-line power systems are about the best you can get. They are also called double-conversion systems. This kind of UPS works by using the power feed to constantly charge the battery, as shown in Exhibit 53.1. The battery in turn runs the equipment. When the power fails, there is no transition time for the protected equipment — it is already running off the battery. Another benefit of on-line systems is that they condition the power to the load equipment. On-line systems stop virtually all power anomalies at the feed, and provide a clean, high-quality feed to the protected equipment.

How On-Line UPS Works. On-line UPS systems are good in that they provide constant power, with no fluctuation, even with a power failure. But they are not as *efficient* as other types of UPS systems. Power is lost as it goes into the battery and out to the server. This loss of power has one more negative aspect. On-line UPSs run hotter. Higher operating temperatures cause the battery to have a shorter life and may necessitate additional cooling for the server room. Batteries typically will last two to three years when used with on-line UPS systems.

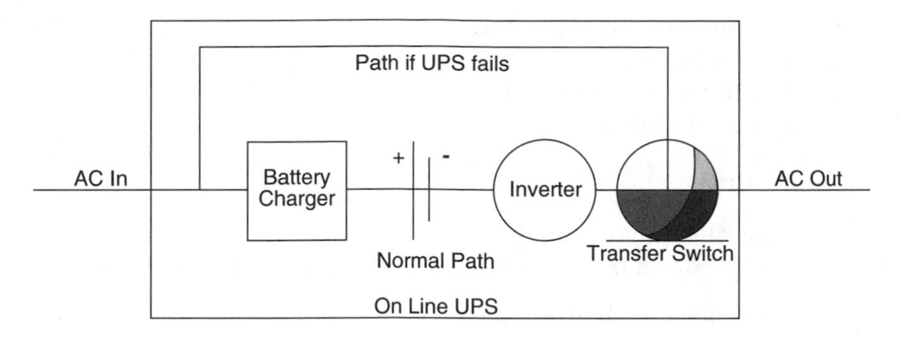

Exhibit 53.1 On-line power system.

Ferro UPS Systems. A ferro (or ferroresonant) UPS system uses a large, heavy ferroresonant power transformer as well as a battery. The power feed comes into the transformer and is then fed into the protected equipment. The incoming power feed also is used to keep the battery fully charged.

When a power failure occurs, the feed to the server is switched over to the battery system. The transformer acts as a buffer during the crucial moments when there is no incoming power and the battery has not yet engaged. This type of transformer can keep the feed into the server constant long enough for the battery to fully take over.

Standby UPS Systems. Standby UPS systems also employ batteries and transformers, but there is no on-line connection, and the transformer does not act as a buffer. Instead, during a power failure, a very fast switch changes the feed to a DC-to-AC converter which is attached to the battery. This now connects from the server to the battery. The switch, of course has to happen very quickly. A couple nanoseconds too long and servers will fail.

Nearly all standby systems are designed for use with computer systems, but you would do well to verify that this is the case before you buy. Standby systems have the advantage of allowing longer battery life than other U.S. types. The battery runs cooler, since it is not under any load. The disadvantage is that they offer no power conditioning.

Line Interactive UPS Systems. Line-interactive UPS systems are a bit like standby systems. There is no on-line feed from the battery, but there is a buffering function performed by the transformer. Line interactive UPS systems rely on fast cut-overs, as do standby systems, but instead of switching from direct power feed to an AC-to-DC converter, line interactive UPS systems reverse the function of the transformer to accomplish the same task.

They tend to cost less than standby systems; a cost savings is realized by making one component do double-duty. The transformer is already energized at the time of cut-over and is able to feed AC power to the protected equipment quickly.

Choosing a UPS

So, which one should you buy? Budget will be a definite limiting factor. The features you need will be determined largely by these factors:

- How long does the battery need to last?
- How many devices will be plugged into the UPS?
- How much power does the battery need?
- Does the server hooked to the UPS need special actions should the battery of the UPS run down?

A prime factor in the costing of a UPS system is the battery. The battery has two aspects to its capabilities.

Lifetime Amount of Power Stored. The lifetime of the battery means how many years it will last. This is determined by the type of UPS. On-line systems run hotter and degrade sooner. Standby systems are cooler and allow the battery to remain useful for more years. Having a long-lasting battery is important because it is one more thing that you'll have to maintain and eventually replace. Another factor is weight. These batteries are *very* heavy. Some are heavy enough to require two people to install and move.

Number of Devices Served. The power stored in the battery will determine how many devices are protected and the duration of the protection, should a failure occur. The measurements and calculations are relatively easy. Say, for example, you have two file servers. Each of them uses 250 watts of power. A 1500-watt power supply will protect one of them for 6 hours (6×250) or two of them for 3 hours. You have to judge the amount of protection time needed. By far, most outages last only a few seconds or minutes. Consult your local power company for outage data if you want to make a reasoned estimate.

Many UPS systems sold today interface with network operating systems by running special software on the server and making a data connection from the server to the UPS unit. One more critical function of power supply software is scheduling the server downtime prior to the expiration of battery backup. At a preassigned threshold of time or watts remaining, the software can issue the needed commands to shut down the server, instead of letting it crash.

The last feature to consider about a UPS system is insurance. Surprisingly, many UPS manufacturers will insure the equipment *properly* attached to any server. Most of them will insure the equipment up to $25,000. The damage insured against usually includes damage caused by: the UPS, failure of the UPS, and any other power anomalies the UPS allowed to get through. As of this writing, American Power Conversion (APC) and Sutton Designs, Inc. (SDI) both offer these policies with their equipment.

Surge Suppressor

A surge suppressor, or power conditioner, is a device that attaches to the power feed on one side and to the load equipment on the other. In-between, it acts as a filter against incoming power anomalies. By itself, it does not ensure the continuity of power. Although many UPS systems manufacturers incorporate power conditioning into their products, power spikes are the primary reason for the existence of surge suppressors. When these spikes get through, often some piece of equipment will get toasted.

Surge suppressors work much like a high-quality stereo radio. Stereo radios take in waves from the air and pass them through banks of filters to produce a high-quality sound. Power conditioners do the same thing, except they receive power from an outlet, which may have RFI, spikes, etc., and they feed to the protected equipment purified AC current. Their primary job is to suppress spikes. But they also may condition against RFI and even protect against low power (brownout) conditions.

Choosing a Surge Suppressor. When choosing a surge suppressor, it is important to resist the urge to go cheap. You can go to the local office supply store and find several devices labeled as surge suppressors costing $3 to $20. The odds are very good that these devices are of little or no value for protecting computers. The fact is that if your surge suppressor did not cost $50 or more, it is doing little more than acting as a multiple outlet strip.

A good indicator as to whether the suppressor you intend to buy is any good for sensitive electronic equipment may be found in the warranty. Many surge suppressor manufacturers offer a warranty to repair or replace any equipment which was damaged as a result of failure in their product.

Remember to use a grounded, three-pin power outlet when you plug in a surge suppressor. When a spike comes in, it must be absorbed and dissipated *somewhere*. If the third pin is not connected to ground, then the spike may still find its way into your computer equipment, or, more likely, damage the suppressor. Some suppressors have a small LED that goes off if the suppressor is failing. A failed suppressor might still work as a power strip, and thus hide the fact that it is no longer performing its primary function, suppression.

Some additional surge suppressor purchase criteria are

- How high a voltage can the suppressor take before it fails?

- How many hits can it sustain?
- What kind of diagnostics does it have?
- Does it automatically shut down if a component fails?
- Does it detect low power and shut down?

Many systems today will absorb upward of 6000 volts and keep on ticking. Models made several years ago could not safely sustain a hit this high. If they did, they would then be rendered useless ... a one shot deal. Modern suppression units should be able to take numerous hits and function perfectly.

This brings up another subject: how do you know it is working correctly? Testing them by sending 6000 volts through them might be a bit risky. Diagnostics are the key. Every good suppressor must have systems to confirm to you that they are alive and well. If they don't, you have no idea if the next hit will be stopped. Choose a unit which has a diagnostic switch or display which can confirm positively that things are working right.

Other Considerations. One of the sneaky ways that surges get into our network systems is through the telephone line. If you think about it, the telephone line is just as much a potential source of lightning spikes as is electrical wiring. If there is expensive communications equipment involved, or the telephone lines hook directly into a serial port on a server for remote monitoring or access, then you need to place surge suppressors there, also. Failing to do so leaves an obvious door open for damage to the network.

When choosing a surge suppressor, be sure to get one that protects all the wires in the jack. Depending on the type of telephone jack you have, there might be 2, 4, or 8 wires inside.

UNDERSTANDING PHYSICAL SECURITY

Servers convey and store information. The programs and data which comprise the daily work of network users allow the business to function. They are business assets. As assets, they need to be guarded against peril. More than likely, the security systems at your place of employment focus on the safe storage of cash, inventory, and paper records. The night watchman most likely does not concern him- or herself with network integrity. The job of guarding the network most likely falls to you.

Types of Threats

In terms of network risks, we face four different kinds of potential threats: disruption, disclosure, destruction, and corruption. Let's review them at a high level and review plans we can make to minimize the potential of being caught unaware by one of these threats.

Disruption means stopping the flow of data. Disruption can come from many sources. Power outages represent a potential disruption. A down server with a bad disk is also a disruption. In both cases the business function of the network is impeded or stopped.

Disclosure occurs when an unauthorized party gets access to network data. Disclosure can come from an outsider gaining access to trade secrets stored on a server, or it could be someone trying to find out the salary of a co-worker.

Destruction is the absolute loss of the data or systems. Of the two, the destruction of data is the most costly.

Corruption can be intentional or unintentional. With corruption, the data are still largely there, but they have been polluted and made unusable. Unless the corruption can be removed, the data have effectively been destroyed.

Evaluating Threats

There is no comprehensive formula for making your network safe for computing. Rather, implementing protection against the various threats you face consists of analyzing opportunities for problems on a point-by-point basis. Let's look at disclosure.

Losing data to a competitor, or allowing an insider to improperly access forbidden areas can cause loss of revenues, decreased market shares, and internal turmoil within a business. Network operating system security does much to protect you against this as far as user access to the server is concerned.

We are all familiar with the saying "Possession is 9/10ths of the law." The same applies to servers. Even if you do not have administrative or supervisory access, data on a server can still be accessed. Many utilities are sold today which allow someone who has physical access to circumvent security and gain access to files. These software tools may allow a technician to take down a server and copy data to a floppy disk. So, how do we protect network data against unwanted intrusion? The server must be physically secure. Placing the server in a locked room is a good start.

Threat Countermeasures

For every device and for every point of access to the network, we need to look at how disruption, disclosure, destruction, and corruption can cause problems. A realistic lengthy list of potential threats should be assembled. Next, make a list of actions to stop or inhibit each threat that matches up against the first list. As a sanity check, weigh the cost and effort associated with the countermeasure against the possible damage from and likelihood of the threat occurring.

Threats and their consequences vary from site to site. In a small commercial business, disclosure may not be a big deal. Making sure that the server is safely placed and that passwords are required may be good enough. On the other hand, at a military contractor's installation, it would be a much better idea to place the server in a data center with strict access controls and a guard at the entrance. The threat and the needed countermeasures are all relative to the setting.

The final step is to perform periodic reviews of the efficacy and necessity of all the threats and countermeasures. Times change and the way we approach network threats must continually evolve.

SUMMARY

This chapter has focused on troubleshooting the server environment. Good troubleshooting is more than having lots of knowledge and a willing attitude. As always, prevention planning is well worth the effort. As with the cases of power problems and general security, the problems are fewer if we put in place the appropriate countermeasures ahead of time.

To prevent problems, place servers in environmentally proper and secure environments. Look at your network and assess the threats which presently exist. Review the possible actions and make plans to stop a problem before it gets a foot in the door.

We have looked at many possible problems. There is no magic formula that can be applied to all servers. Documentation is one tool which shows up on both the prevention and resolution side of troubleshooting. For your network, this would make a good starting place for increasing up-time and decreasing complaints. Perhaps Monday morning won't be such a bad time of the week after all.

About the Editor

Gilbert Held is the Director of 4-Degree Consulting, a Macon, GA-based firm specializing in the areas of data communications, personal computing, and the application of technology. Gil is an award-winning author and lecturer, with over 40 books and 300 technical articles to his credit. Some of Gil's recent titles include *Frame Relay Networking, LAN Performance 2ed., Ethernet Networks 3ed., and Data Communications Networking Devices 4ed.*, all published by John Wiley & Sons of New York City and Chichester, England.

Gil is a frequent contributor to several industry publications and was selected to represent the United States at technical conferences in Moscow and Jerusalem. Gil received his BSEE from Pennsylvania Military College, MSEE from New York University, and the MSTM and MBA from the American University. Gil is currently a member of the adjunct faculty of Georgia College and State University, teaching several graduate courses covering various aspects of data communications.

Index

A

B

C

J

K

L

M

N

O

P

Q

R

T

U

V

W

Y

Z